Classics in Mathematics

Martin Aigner    Combinatorial Theory

Springer
*Berlin*
*Heidelberg*
*New York*
*Barcelona*
*Budapest*
*Hong Kong*
*London*
*Milan*
*Paris*
*Santa Clara*
*Singapore*
*Tokyo*

Martin Aigner

# Combinatorial Theory

Reprint of the 1979 Edition

Martin Aigner
Freie Universität Berlin
Institut für Mathematik II
Arnimallee 3
14195 Berlin
Germany

Originally published as Vol. 234 of the
*Grundlehren der mathematischen Wissenschaften*

Mathematics Subject Classification (1991): 05xx, 06xx

CIP data applied for

Die Deutsche Bibliothek – CIP-Einheitsaufnahme

**Aigner, Martin:**
Combinatorial theory / Martin Aigner - Reprint of the 1979 ed. - Berlin; Heidelberg; New York; Barcelona; Budapest; Hong Kong; London; Milan; Paris; Santa Clara; Singapore; Tokyo: Springer, 1997
(Classics in Mathematics)
ISBN 3-540-61787-6

ISBN 3-540-61787-6 Springer-Verlag Berlin Heidelberg New York


Printed in Germany



SPIN 10553966 41/3143-5 4 3 2 1 0 – Printed on acid-free paper

Martin Aigner

# Combinatorial Theory

Springer-Verlag
Berlin Heidelberg New York

Martin Aigner
II. Institut für Mathematik
Freie Universität Berlin
Königin-Luise-Strasse 24/26
1000 Berlin 33
Federal Republic of Germany

---

AMS Subject Classification (1980): 05xx, 06xx

---

With 123 Figures

**Library of Congress Cataloging in Publication Data**

Aigner, Martin, 1942–
Combinatorial theory.

(Grundlehren der mathematischen Wissenschaften; 234)
Bibliography: p.
Includes index.
1. Combinatorial analysis. I. Title. II. Series:
Die Grundlehren der mathematischen Wissenschaften
in Einzeldarstellungen; 234)
QA164.A36 511'.6 79-1011



Printed in the United States of America.

9 8 7 6 5 4 3 2 1

ISBN 0-387-90376-3 New York Heidelberg Berlin
ISBN 3-540-90376-3 Berlin Heidelberg New York

# Preface

It is now generally recognized that the field of combinatorics has, over the past years, evolved into a fully-fledged branch of discrete mathematics whose potential with respect to computers and the natural sciences is only beginning to be realized. Still, two points seem to bother most authors: The apparent difficulty in defining the scope of combinatorics and the fact that combinatorics seems to consist of a vast variety of more or less unrelated methods and results. As to the scope of the field, there appears to be a growing consensus that combinatorics should be divided into three large parts:

(a) *Enumeration*, including generating functions, inversion, and calculus of finite differences;
(b) *Order Theory*, including finite posets and lattices, matroids, and existence results such as Hall's and Ramsey's;
(c) *Configurations*, including designs, permutation groups, and coding theory.

The present book covers most aspects of parts (a) and (b), but none of (c). The reasons for excluding (c) were twofold. First, there exist several older books on the subject, such as Ryser [1] (which I still think is the most seductive introduction to combinatorics), Hall [2], and more recent ones such as Cameron–Van Lint [1] on groups and designs, and Blake–Mullin [1] on coding theory, whereas no comprehensive book exists on (a) and (b). Second, the vast diversity of types of designs, the very complicated methods usually still needed to prove existence or non-existence, and, in general, the rapid change this subject is presently undergoing do not favor a thorough treatment at this moment. I have also omitted reference to algorithms of any kind because I feel that presently nothing more can be said in book form about this subject beyond Knuth [1], Lawler [1], and Nijenhuis-Wilf [1].

As to the second point, that of systematizing the definitions, methods, and results into something resembling a theory, the present book tries to accomplish just this, admittedly at the expense of some of the spontaneity and ingenuity that makes combinatorics so appealing to mathematicians and non-mathematicians alike. To start with, mappings are grouped together into classes by placing various restrictions on them. To stick to the division outlined above, these classes of mappings are then counted, ordered, and arranged. The emphasis on ordering is well justified by the everyday experience of a combinatorist that most discrete structures, while perhaps lacking a simple algebraic structure, invariably admit a natural ordering. Following this program, the book is divided into three parts, the first part presenting the basic material on mappings and posets, in Chapters I and II, respectively, the second part dealing with enumeration in Chapters III to V, and the third part on the order-theoretical aspects in Chapters VI–VIII.

The arrangement of the material allows the reader to use the three parts almost independently and to combine several subsections into a course on special topics. For instance, Chapter II has been used as an introduction to finite lattices, Chapters VI and VII as a course on matroids, and parts of Chapter VII and Chapter VIII as a course on transversal theory and the major existence results. The exercises have been graded. Unmarked exercises can be solved without a great deal of effort; more difficult ones are marked with an asterisk (*). The symbol → indicates that the exercise is particularly helpful or interesting, but in no instance is the statement or the solution of an exercise necessary to the development of the subject. The references given at the end are, of course, by no means exhaustive; usually they have been included because they were used in one way or another in the preparation of the text. Books are indicated by an asterisk.

The German version of the present book appeared in two volumes—Kombinatorik I. Grundlagen und Zähltheorie; and II. Matroide und Transversaltheorie—as Springer Hochschultexts. Combining these two parts has been a more formidable task than I originally thought. Most of the material has been reorganized, with the major changes appearing in Chapter VIII due to many new results obtained in the last few years.

I had the opportunity of working as a research associate at the Department of Statistics of the University of North Carolina in the Combinatorial Year program 1968–1970. It was during this time that I first planned to write this book. Of the many people who have encouraged me since and furthered this work, I owe special thanks to G.-C. Rota, R. C. Bose, and T. A. Dowling for many hours of discussion; to H. Wielandt, H. Salzmann, and R. Baer for their constant support; to R. Weiss, G. Prins, R. H. Schulz, J. Schoene, and W. Mader, who read all or part of the manuscript; and finally to M. Barrett for her impeccable typing.

It is my hope that I have been able to record some of the many important changes that combinatorics has undergone in recent years while retaining its origins as an intuitively appealing mathematical pleasure.

*Berlin* M. Aigner
*September 1979*

# Contents

# Preliminaries

It seems convenient to list at the outset a few items that will be used throughout the book.

## 1. Sets

We use the symbols $\mathbb{N}$, $\mathbb{Z}$, $\mathbb{Q}$, $\mathbb{R}$, and $\mathbb{C}$ for the basic number systems, and set $\mathbb{N}_0 = \{0, 1, 2, \ldots\}$, $\mathbb{N}_n = \{1, 2, \ldots, n\}$; in chapter III the notation $\underline{n}$ for $\mathbb{N}_n$ is also used. $\delta_{ij}$ is the Kronecker symbol; $id_M$ stands for the identity mapping of a set $M$ onto itself and $2^M$ for the power set of $M$. The cardinality of a set $M$ is denoted by $|M|$ and we set $|M| = \infty$ whenever $M$ is infinite. For any set $M$ we use the symbol $M^k$ for the cartesian product, $M^k = \{(a_1, \ldots, a_k): a_i \in M\}$, and $M^{(k)}$ for the family of $k$-subsets of $M$, $M^{(k)} = \{A \subseteq M: |A| = k\}$. A finite set $M$ with $|M| = n$ is called an *n-set*. To define a set or a term we use $:=$ or $:\Leftrightarrow$.

The following rules are the basic tools for enumeration:

(i) *Rule of Equality*: If $N$ and $R$ are finite sets and if there exists a bijection between them, then $|N| = |R|$;
(ii) *Rule of Sums*: If $\{A_i: i \in I\}$ is a finite family of finite pairwise disjoint sets, then $|\bigcup_{i \in I} A_i| = \sum_{i \in I} |A_i|$;
(iii) *Rule of Products*: If $\{A_i: i \in I\}$ is a finite family of finite sets, then for the cartesian product $\prod_{i \in I} A_i$, $|\prod_{i \in I} A_i| = \prod_{i \in I} |A_i|$.

We use the symbols $A \dot{\cup} B$ or $\dot{\bigcup}_{i \in I} A_i$ to indicate that the sets involved are disjoint.

A *multiset* on $S$ is a set $S$ together with a function $r: S \to \mathbb{N}_0$ (giving the multiplicity of the elements of $S$). A convenient notation for a multiset $k$ on $S$ is $k = \{a^{k_a}: a \in S\}$ with $k_a := r(a)$, $a \in S$. The usual notions for sets can be carried over to multisets. For instance, if $k = \{a^{k_a}: a \in S\}$ and $l = \{a^{l_a}: a \in S\}$ then

$$k \subseteq l :\Leftrightarrow k_a \leq l_a \quad \text{for all} \quad a \in S,$$

$$k \cap l := \{a^{\min(k_a, l_a)}: a \in S\},$$

$$k \cup l := \{a^{\max(k_a, l_a)}: a \in S\}.$$

Clearly, the family of multisets on a set $S$ forms a lattice under inclusion; furthermore, this lattice is complete.

## 2. Graphs

An *undirected graph* $G(V, E)$ consists of a non-empty set $V$, called the *vertex-set* and a multiset $E$ of unordered pairs $\{a, b\}$ from $V$, called the *edge-set*. A *simple graph* is a graph that contains no *loops* $\{a, a\}$ and no *parallel edges* $\{a, b\}$, $\{a, b\}$, i.e., in which $E \subseteq V^{(2)}$ is an ordinary set. A *directed graph* or *digraph* $\vec{G}(V, E)$ is a non-empty set $V$ of vertices and a multiset $E$ of ordered pairs $(a, b)$ from $V$. The elements of $E$ are now called *arrows* or *directed edges*. An *orientation* of an undirected graph $G(V, E)$ is a rule which designates for each edge $k = \{a, b\}$ a direction $(a, b)$; we then write $a = k^-$, $b = k^+$. A graph is *finite* if both $V$ and $E$ are finite.

Except for the definition of a graph itself the terminology follows closely that of Harary [1]. (There, a graph means what we call a simple graph.) The reader is advised to consult chapter 2 in Harary's book for any term not previously defined. We shall, however, redefine most of the notions when they first appear, except for the most basic ones such as connected graph, path, circuit, etc. Whenever we simply use the term "graph" we always mean "undirected graph."

Two graphs $G(V, E)$ and $G'(V', E')$ are *isomorphic* if these exists a bijection $\phi: V \to V'$ such that $\{a, b\} \in E$ and $\{\phi(a), \phi(b)\} \in E'$ appear in $E$ and $E'$ with equal multiplicity. The *degree* $\gamma(v)$ of a vertex $v$ is the number of edges incident with $v$ where we count loops $\{v, v\}$ twice. Hence for a finite graph $G(V, E)$ we always have $\sum_{v \in V} \gamma(v) = 2|E|$.

Two important types of graphs are the *complete graphs* $K_n$ and the *complete bipartite graphs* $K_{m,n}$. $K_n$ is a simple graph with $n$ vertices with any two vertices joined by an edge. $K_{m,n}$ is a simple graph whose vertex-set is the union of two disjoint sets of cardinality $m$ and $n$ respectively, with two vertices being joined if and only if they are in different sets. A *bipartite graph* is any subgraph of a complete bipartite graph. We shall often denote a bipartite graph by $G(V_1 \cup V_2, E)$ to indicate the defining vertex-sets $V_1$, $V_2$, where every edge joins a vertex in $V_1$ with a vertex in $V_2$. The following rule is the single most useful tool in enumeration.

(iv) *Rule of "counting in two ways"*: Let $G(V_1 \cup V_2, E)$ be a finite bipartite graph with defining vertex-sets $V_1$ and $V_2$. Then

$$\sum_{v \in V_1} \gamma(v) = \sum_{v \in V_2} \gamma(v) \qquad (= |E|).$$

A bipartite graph $G(V_1 \cup V_2, E)$ can also be regarded as a directed graph with all edges directed from $V_1$ to $V_2$. In other words, bipartite graphs with defining vertex-sets $V_1$ and $V_2$ can be identified with *binary relations* between $V_1$ and $V_2$. For this reason, we often use the letter $R$ for the edge-set and in $G(V_1 \cup V_2, R)$ set $R(A) := \bigcup_{a \in A} \{y \in V_2 : (a, y) \in R\}$ for $A \subseteq V_1$, and similarly $R(B) := \bigcup_{b \in B} \{x \in V_1 : (x, b) \in R\}$ for $B \subseteq V_2$. For a singleton subset $\{a\}$, we simply write $R(a)$.

Bipartite graphs have two other important equivalent interpretations. A *set system* $(S, \mathfrak{A})$ is a set $S$ together with a family $\mathfrak{A}$ of not necessarily distinct subsets of $S$. Any set system $(S, \mathfrak{A})$ gives rise to its incidence graph $G(S \cup \mathfrak{A}, R)$ where

$(p, A) \in R :\Leftrightarrow p \in A$. Conversely, any bipartite graph $G(S \cup \mathfrak{A}, R)$ yields a set system $(S, \mathfrak{A})$ by identifying $A \in \mathfrak{A}$ with the set $R(A) \subseteq S$.

A set system $(S, \mathfrak{A})$ can also be described by a 0, 1-matrix $M = [m_{ij}]$ whose rows and columns are indexed by $S$ and $\mathfrak{A}$ respectively, with $m_{ij} = 1$ or 0 depending on whether $p_i \in A_j$ or $p_i \notin A_j$. $M$ is called the *incidence matrix* of $(S, \mathfrak{A})$. Conversely, any 0, 1-matrix gives rise to a set system by the reverse procedure.

**Example.**

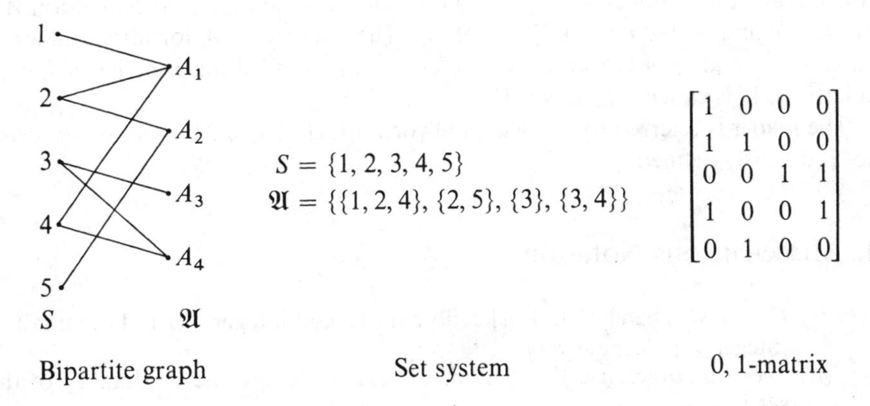


Bipartite graph Set system 0, 1-matrix

A graph which has no non-trivial circuits is called a *forest*. A connected forest is called a *tree*.

## 3. Posets

We employ the usual terminology as, for instance, in Birkhoff [1]. If $P$ is a poset then $P^*$ denotes the *dual poset* obtained by inverting the order relation of $P$. If $P$ contains a unique minimal element, then this element is called the *0-element*, denoted by 0; similarly, a unique maximal element is called the *1-element*, denoted by 1. We say, *b covers a* or *a is covered by b*, denoted by $a <\cdot b$, if $a < b$, and $a < x \leq b$ implies $x = b$. The *atoms* are the elements covering 0 (if 0 exists); the *co-atoms* are the elements covered by 1. We most often represent a poset $P$ by its diagram, which is the directed graph on $P$ with an arrow from $a$ to $b$ if and only if $b$ covers $a$. Whenever possible, we draw a diagram from the bottom up and omit the arrows. A *chain* is a poset in which any two elements are comparable. For the chain $\{a_1 < a_2 < \cdots < a_n\}$ we often use the short-hand notation $\{a_1, \ldots, a_n\}_<$. The *length* of a chain is one less than its cardinality. The length $l(a)$ of $a \in P$ is the length of the longest chain in $P$ with $a$ as last element. An *antichain* is a poset in which any two elements are incomparable. A chain in a poset $P$ is called *unrefinable* if any element of the chain is covered by its successor. If $L$ is a lattice then a non-empty subset $M$ is called a *sublattice* if $x, y \in M$ imply $x \wedge y \in M$, $x \vee y \in M$. A subset $M \subseteq L$ may be a lattice in its own right under the induced order relation, but we reserve the term sublattice for the former situation. An *interval* of a poset $P$ is any set $[a, b] := \{x \in P: a \leq x \leq b\}$ for $a, b \in P$. The *product* $\prod_{i \in I} P_i$ of posets $P_i$ is the poset on the cartesian product with the coordinatewise order relation. The product of lattices is again a lattice. The *sum* $\sum_{i \in I} L_i$ of lattices, each $L_i$ containing a 0-element, is the sublattice of the product $\prod_{i \in I} L_i$ consisting of those vectors with only a finite number of coordinates different from 0. $\mathbb{N}_0$, $\mathbb{N}$ and $\mathbb{N}_n$ are assumed to be endowed with the natural ordering unless otherwise stated. A *complete lattice* is one in which any non-empty subset has an infimum and a supremum. If $W$ is a word which contains only elements of a lattice $L$ and the symbols $\wedge$, $\vee$, $\leq$ and (,), then we obtain the dual expression $W^*$ by exchanging $\wedge$ with $\vee$ and $\leq$ with $\geq$. The validity of $W$ for all variables $x_i$ implies the validity of $W^*$ for all variables $x_i$. This is called the *principle of duality* in lattices. $W$ is called *self-dual* if $W^* = W$.

The reader is referred to Crawley–Dilworth [1, ch. 1 and 2] for all other terms not previously defined.

## 4. Miscellaneous Notation

(i) Let $a \in \mathbb{R}$. Then $\lfloor a \rfloor$ and $\lceil a \rceil$ denote the largest integer $\leq a$ and the smallest integer $\geq a$, respectively.

(ii) We sometimes use the symbol $\#\{\cdots\}$ to denote the cardinality of the set $\{\cdots\}$.

(iii) By a *partition* of a set $S$ we mean a disjoint union $S = \dot{\bigcup}_{i \in I} A_i$. We also use the notation $S = A_1 | A_2 | \cdots$ to indicate a partition. The sets $A_i$ are assumed to be non-empty unless otherwise stated.

(iv) To facilitate the summation notation, we shall often indicate the summation index by a dot underneath. For instance, let $M = [m_{ij}]$ be an $n \times n$-matrix. Then $\sum_{1 \leq \underset{\cdot}{i} < j \leq n} m_{ij} = m_{1j} + m_{2j} + \cdots + m_{j-1j}$.

Chapter I

# Mappings

The starting point for all our considerations is the following: We are given two sets, usually denoted by $N$ and $R$, and a mapping $f: N \to R$ satisfying certain conditions. The triple $(N, R, f)$ is called a *morphism*. Our program is to arrange mappings into *classes*, and then to *count* and *order* the resulting classes of mappings.

Accordingly, our first task will be to collect conditions of combinatorial significance which we want to impose on the mappings—and this is the content of the present chapter.

## 1. Classes of Mappings

Let $(N, R, f)$ be a morphism. $N$ and $R$ will, in most instances, be finite sets, and we shall use the letters $n = |N|$ and $r = |R|$, respectively, for their cardinalities. A common way to describe $f$ is by the expression

$$f = \begin{pmatrix} \cdots & a & \cdots \\ \cdots & f(a) & \cdots \end{pmatrix} \qquad (a \in N).$$

We call this the *standard representation* of $f: N \to R$. Most of the time the domain $N$ will be totally ordered in some natural way, but, of course, any ordering is possible. For example, the three expressions

$$\begin{pmatrix} 1 & 2 & 3 & 4 \\ a & a & b & b \end{pmatrix}, \begin{pmatrix} 1 & 4 & 3 & 2 \\ a & b & b & a \end{pmatrix}, \begin{pmatrix} 4 & 3 & 1 & 2 \\ b & b & a & a \end{pmatrix}$$

all represent the same mapping.

### A. Classification

With $f: N \to R$ we associate the *image* $\operatorname{im}(f)$ and the *kernel* $\ker(f)$:

$$\operatorname{im}(f) := \bigcup_{a \in N} f(a)$$

$$\ker(f) := \dot{\bigcup_{b \in \operatorname{im}(f)}} f^{-1}(b).^{(1)}$$

(1) For simplicity we write $f^{-1}(b)$ instead of the more precise $f^{-1}(\{b\})$. Similar abbreviations will be used later, e.g., $A \cup q$ for $A \cup \{q\}$.

Thus, the kernel of $f$ is the partition of $N$ induced by the equivalence relation

$$a \approx a' :\Leftrightarrow f(a) = f(a') \qquad (a, a' \in N).$$

It is convenient to postulate the *empty mapping* $f_\varnothing$ with $\mathrm{im}(f_\varnothing) = \varnothing$ and undefined kernel.

The mapping $f: N \to R$ is called *surjective* if $\mathrm{im}(f) = R$, and *injective* if $\ker(f) = 0$ (in the lattice of partitions of $N$—see section 2.B), i.e., if $a \neq a' \Rightarrow f(a) \neq f(a')$ for all $a, a' \in N$. Mappings that are both surjective and injective are called *bijective*.

With these definitions we obtain a first set of classes of mappings:

$$\begin{aligned} \mathrm{Map}(N, R) &:= \{f: N \to R, f \text{ arbitrary}\}, \\ \mathrm{Sur}(N, R) &:= \{f: N \to R, f \text{ surjective}\}, \\ \mathrm{Inj}(N, R) &:= \{f: N \to R, f \text{ injective}\}, \\ \mathrm{Bij}(N, R) &:= \{f: N \to R, f \text{ bijective}\}. \end{aligned}$$

When $N$ and $R$ are both finite sets, we have the following obvious but important rules concerning their cardinalities:

$$\begin{aligned} \mathrm{Sur}(N, R) \neq \varnothing &\Rightarrow |N| \geq |R|, \\ \mathrm{Inj}(N, R) \neq \varnothing &\Rightarrow |N| \leq |R|, \\ \mathrm{Bij}(N, R) \neq \varnothing &\Rightarrow |N| = |R|. \end{aligned}$$

For a mapping $f: N \to N$ of a finite set $N$ into itself the notions surjective, injective, and bijective coincide. For infinite sets this is no longer true. For example, $f: \mathbb{N} \to \mathbb{N}$, $f(k) = 2k$, is injective, but not surjective.

Suppose that both sets $N$ and $R$ are endowed with a partial order. $f \in \mathrm{Map}(N, R)$ is called *monotone* if it preserves the order relation, i.e., if $a \leq_N b \Rightarrow f(a) \leq_R f(b)$ for all $a, b \in N$, and it is called *antitone* if $a \leq_N b \Rightarrow f(a) \geq_R f(b)$ for all $a, b \in N$. The family of monotone mappings constitutes another important class:

$$\mathrm{Mon}(N, R) := \{f: N \to R, f \text{ monotone}\}.$$

Observe that if $N$ is totally unordered we simply have $\mathrm{Mon}(N, R) = \mathrm{Map}(N, R)$, regardless of the order on $R$. Any monotone or antitone mapping maps chains onto chains.

**Example.** Let $N = \{1 < 2 < 3 < 4\}$ and

$$R = \begin{array}{ccc} & d & \\ b & & c \\ & a & \end{array}.$$

$\begin{pmatrix} 1 & 2 & 3 & 4 \\ a & b & c & d \end{pmatrix}$ is monotone, whereas $\begin{pmatrix} 1 & 2 & 3 & 4 \\ d & b & c & a \end{pmatrix}$ is antitone.

Another well-known class of mappings arises in the context of algebra. Suppose there are algebraic systems of the same type defined on $N$ and $R$, e.g., groups, rings, or vector spaces over the same scalar domain. A mapping $f\colon N \to R$ which preserves all operations is called a *homomorphism*, and we denote the class of all homomorphisms from $N$ into $R$ by

$$\mathrm{Hom}(N, R) := \{f\colon N \to R, f \text{ homomorphism}\}.$$

By combining the classes we have encountered so far it is clear what we mean by a surjective monotone mapping or an injective homomorphism, etc. We shall see that most of the combinatorial counting problems can be phrased in terms of one or more of these classes.

## B. Representation

There are two particularly useful and suggestive interpretations of a morphism $(N, R, f)$. Let $N$ be totally ordered by some fixed order. We regard the elements of $N$ as places of a word and say the *place* $i \in N$ is occupied by the *letter* $l \in R$ if $f(i) = l$. The mapping is thus regarded as a *word* of length $n$ with letters from the *alphabet* $R$, indexed by $N$. Now order the elements of $R$ by some total order and regard them as *boxes*. If $f(a) = b$ we say that the object $a \in N$ has been *sorted* into the box $b$, or that the box $b$ contains $a$. In this way we interpret $f\colon N \to R$ as an *occupancy pattern* of the boxes $R$ by the objects $N$.
In summary:

$$f\colon N \to R = \begin{cases} \text{mapping from } N \text{ into } R; \\ \text{word in } R \text{ indexed by } N; \\ \text{occupancy of } R \text{ by } N. \end{cases}$$

Suppose $N = \{a_1, a_2, \ldots, a_n\}_<$ is totally ordered. Then the mapping

$$f = \begin{pmatrix} a_1 & \cdots & a_n \\ f(a_1) & \cdots & f(a_n) \end{pmatrix}$$

can be unambiguously represented by the word $f(a_1)f(a_2)\ldots f(a_n)$. We call this the *word representation* of $f$ (relative to the given total order on $N$). Similarly, if $R = \{b_1, \ldots, b_r\}_<$ is totally ordered then $f^{-1}(b_1) \,\dot\cup\, f^{-1}(b_2) \,\dot\cup \cdots \dot\cup\, f^{-1}(b_r)$ is called the *occupancy representation* of $f$. In most cases $N$ or $R$ will be the set $\{1, \ldots, n\}$ or $\{1, \ldots, r\}$ endowed with the natural order.

**Example.** Let $N = \{1 < 2 < 3\}$, $R = \{a < b < c\}$. We list the set $\mathrm{Map}(N, R)$ by giving the word representation of its members:

| | | | | | |
|---|---|---|---|---|---|
| *aaa* | *acc* | *aba* | *caa* | *cbc* | *acb* |
| *aab* | *ccc* | *baa* | *cac* | *ccb* | *bac* |
| *abb* | *bbc* | *bab* | *cca* | | *cab* |
| *bbb* | *bcc* | *bba* | *bcb* | | *bca* |
| *aac* | *abc* | *aca* | *cbb* | | *cba*. |

The monotone mappings are those in the first two columns, the last column together with *abc* gives the bijective mappings. Hence we have

$$|\text{Map}(N, R)| = 27, \qquad |\text{Bij}(N, R)| = 6, \qquad |\text{Mon}(N, R)| = 10.$$

The reader may set up a similar list using the occupancy representation.

The terms introduced in the beginning can now be interpreted as special words or occupancies, e.g., a mapping $f \in \text{Inj}(N, R)$ is called a *strict word* and $f \in \text{Sur}(N, R)$ a *full occupancy*. Let $N = \{1 < 2 < \cdots < n\}$ and $R$ be an arbitrary poset. $\text{Mon}(N, R)$ consists of all words $b_1 b_2 \ldots b_n$ in $R$ with $b_1 \leq b_2 \leq \cdots \leq b_n$. Hence we speak of $\text{Mon}(N, R)$ as the class of *monotone words*. If, in particular, $R$ is totally ordered, then the monotone words of length $n$ are precisely the *multisets* in $R$ of cardinality $n$, and it follows that there are just as many monotone words of length $n$ in $R$ as $n$-multisets in $R$. See the example above where the 3-multisets of $R$ are listed in the first two columns. If we restrict ourselves to *strict monotone* words we obtain the family of all *subsets* of $R$ of cardinality $n$.

## EXERCISES I.1

1. Let $f: N \to R$. Prove:
   (i) $\text{im}(f)$ minimal $\Leftrightarrow \ker(f)$ maximal.
   (ii) $\text{im}(f)$ maximal and $|N| < |R| \Rightarrow f$ injective.
   (iii) $\text{im}(f)$ maximal and $|N| > |R| \Rightarrow f$ surjective.
   (iv) $\text{im}(f)$ maximal and $|N| = |R| \Rightarrow f$ bijective.
2. Show that for $f: N \to N$ with $|N| < \infty$ the concepts injective, surjective, and bijective are equivalent, but that this is not true if $N$ is infinite.

→ 3. Let $N_<$ and $R_<$ be posets. Describe $\text{Mon}(N, R)$ when $N_<$ is an antichain and when $R_<$ is an antichain.

→ 4. Show that there are precisely 5 non-isomorphic posets with 3 elements and 16 with 4 elements. How many are isomorphic to their dual? How many are lattices?

→ 5. Show that a directed graph $\vec{G}(V, E)$ is the diagram of some poset on $V$ if and only if for any directed path $a_0 \to a_1 \cdots \to a_t$ of length $t \geq 2$ we always have $(a_0, a_t) \notin E$ and $(a_t, a_0) \notin E$.

6. Show that a poset is a chain if and only if all subposets are lattices.
7. Find a bijection from $[\mathscr{C}(1)]^2$ to $\mathscr{C}(3)$ which is monotone but preserves neither infima or suprema, where $\mathscr{C}(n)$ is the chain of length $n$.

→ 8. Let $N = \{1, 2, 3\}_<$ and $R = \{1, 2, 3, 4\}_<$. Compute $|\text{Map}(N, R)|$, $|\text{Inj}(N, R)|$, $|\text{Sur}(N, R)|$, and $|\text{Mon}(N, R)|$.

9. Let $G$ and $H$ be groups. Prove that a partition $\pi$ of $G$ is the kernel of a homomorphism $f: G \to H$ if and only if the following holds: If $a$ and $a'$

lie in a common block of $\pi$ and similarly for $b$ and $b'$ then $a \cdot b$ and $a' \cdot b'$ also lie in a common block of $\pi$. Deduce from this the homomorphism theorem for groups.

10. Let $N$ and $R$ both be cyclic groups of order 9. What is $|\mathrm{Hom}(N, R)|$?

→11. Suppose $V$ and $W$ are vector spaces over $GF(2)$ of dimension $m$ and $n$, respectively. Compute $|\mathrm{Hom}(V, W)|$.

→12. In how many ways can we sort the elements 1, 2, 3 into the boxes $A$, $B$ if, in addition, we require that the boxes are linearly ordered? For example,

$$\underset{A}{1, 2, 3} \,|\, \underset{B}{\varnothing}$$

is different from

$$\underset{A}{1, 3, 2} \,|\, \underset{B}{\varnothing};$$

similarly $2|1, 3$ is different from $2|3, 1$. (Answer: 24)

→13. Suppose $v = \sum_{i=1}^{b} l_i - b + 1$ objects are sorted into $b$ boxes $B_1, \ldots, B_b$. Show that for some $i$, box $B_i$ contains at least $l_i$ objects. This is called the "pigeon hole principle."

14. By using ex. 13 prove that a sequence of $mn + 1$ distinct integers contains either an increasing subsequence of length greater than $m$ or a decreasing subsequence of length greater than $n$.

15. Prove that there are two people in New York City who have precisely the same number of hairs on their head. (Hint: Use the pigeon hole principle.)

## 2. Fundamental Orders

We mentioned at the outset that our main object was to count and order classes of mappings. Later on we shall see that a good many of the counting problems consist in evaluating certain coefficients of the underlying order structure. So, let us first find out what order relations arise in connection with our general set-up.

### A. Inclusion

First we may compare mappings by looking at their images. Let $f, g$ be mappings with $\mathrm{im}(f) \subseteq R$, $\mathrm{im}(g) \subseteq R$. Regarding $\mathrm{im}(f)$, $\mathrm{im}(g)$ as multisets $\{b^{|f^{-1}(b)|}: b \in R\}$, $\{b^{|g^{-1}(b)|}: b \in R\}$, we obtain a natural relation

$$f \subseteq g :\Leftrightarrow |f^{-1}(b)| \leq |g^{-1}(b)| \quad \text{for all } b \in R.$$

This inclusion relation $\subseteq$ is obviously reflexive and transitive but, in general, not antisymmetric since it disregards the nature of the elements which are mapped into

$R$. In other words, $\subseteq$ only takes into account *how many* elements are mapped onto a particular $b \in R$ but not which elements.

The simplest class of posets arising in this way are the *lattices of multisets of a set* $R$ ordered by inclusion. All these lattices are complete and it is clear how meet and join are computed. For multisets $k = \{b^{k_b} : b \in R\}$, $l = \{b^{l_b} : b \in R\}$ we have

$$k \wedge l = \{b^{\min(k_b, l_b)} : b \in R\}$$

$$k \vee l = \{b^{\max(k_b, l_b)} : b \in R\}.$$

In most instances we shall only be interested in the sublattice of all *finite multisets* of $R$, denoted $\mathscr{M}(R)$.

**1.1 Proposition.** *We have $\mathscr{M}(R) \cong \sum_r \mathbb{N}_0$, where $\sum_r \mathbb{N}_0$ is the r-fold lattice sum of $\mathbb{N}_0, r = |R|$.*

*Proof.* Denote multisets by $k = \{b^{k_b} : b \in R\}$. Then $\phi : \mathscr{M}(R) \to \sum_r \mathbb{N}_0$ given by

$$\phi(k) = (\ldots, k_b, \ldots)$$

is an isomorphism. $\square$

**Examples.** If $R$ consists of just a single element $a$, as in Figure 1.1, then $\mathscr{M}(R) \cong \mathbb{N}_0$.

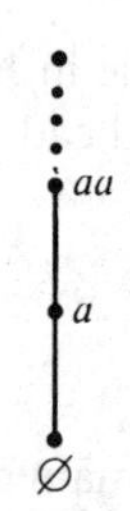


Figure 1.1

For $R = \{a, b\}$, as in Figure 1.2, we obtain $\mathscr{M}(R) \cong \mathbb{N}_0^2$:

We denote the chain of length $n$ (with cardinality $n + 1$) by $\mathscr{C}(n)$, i.e., $\mathscr{C}(n) \cong \{0 < 1 < 2 < \cdots < n\}$.

**1.2 Proposition.** *Let $R$ be countable. Then $\mathscr{M}(R) \cong \mathscr{T}$, the lattice of positive integers ordered by divisibility, i.e.,*

$$m \underset{\mathscr{T}}{\leq} n :\Leftrightarrow m | n \qquad (m, n \in \mathbb{N}).$$

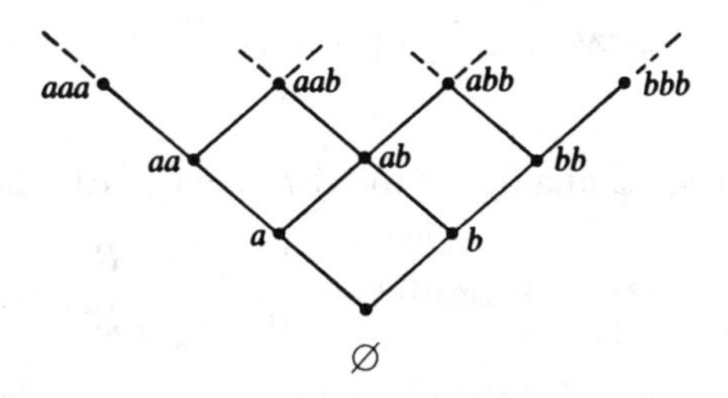


Figure 1.2

*Proof.* Let $R = \{b_1, b_2, \ldots\}$ and write multisets of $R$ in the form $k = b_1^{k_1} b_2^{k_2} \ldots$. Numbering the primes in ascending order $p_1 = 2$, $p_2 = 3, \ldots$, we obtain the isomorphism $\phi: \mathcal{M}(R) \to \mathcal{T}$,

$$\phi(k) = p_1^{k_1} p_2^{k_2} p_3^{k_3} \ldots . \quad \square$$

Figure 1.3 depicts the interval $[1, 360]$ of $\mathcal{T}$. From the decomposition $360 = 2^3 3^2 5$ we deduce $[1, 360] \cong \mathscr{C}(3) \times \mathscr{C}(2) \times \mathscr{C}(1)$. In section II.4 we shall take a closer look at the divisor lattice $\mathcal{T}$.

If in a class of mappings we restrict ourselves to injective mappings, the images $\operatorname{im}(f) \subseteq R$ become merely subsets of $R$. Hence we obtain the *lattice* $2^R$ *of all subsets of a set* $R$ ordered by inclusion. Again, our interest will focus on the lattice $\mathscr{B}(R)$ of all *finite subsets* of $R$.

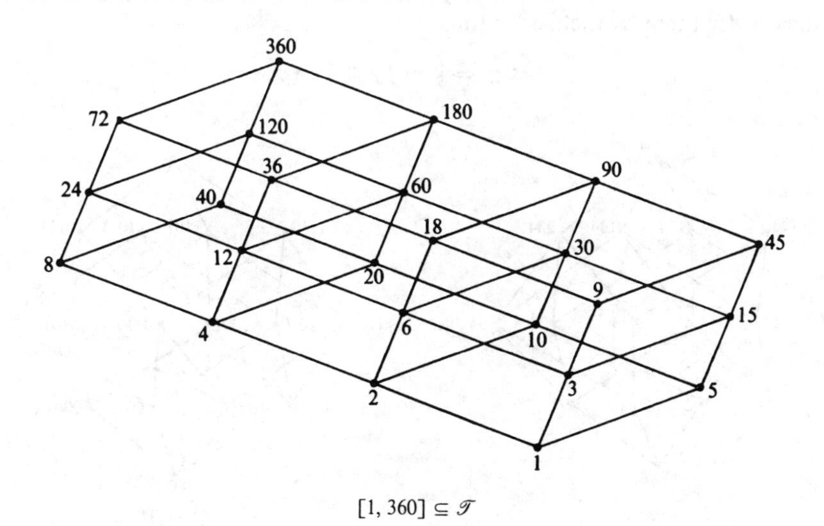


$[1, 360] \subseteq \mathcal{T}$

Figure 1.3

**1.3 Proposition.** *We have* $\mathscr{B}(R) \cong \sum_r \mathscr{C}(1)$, *where* $\sum_r \mathscr{C}(1)$ *denotes the r-fold lattice sum of* $\mathscr{C}(1)$, $r = |R|$.

*Proof.* The isomorphism $\phi$ analogous to **1.1** has in this case the simple form

$$\phi(A) = (\ldots, \varepsilon_b, \ldots) \quad \text{with } \varepsilon_b = \begin{cases} 1 & \text{if } b \in R \\ 0 & \text{if } b \notin R \end{cases} \qquad (A \subseteq R). \quad \square$$

The vector $(\ldots, \varepsilon_b, \ldots)$ is called the *characteristic vector* of the subset $A$.

Clearly, all lattices introduced so far depend up to isomorphism only on the cardinality of $R$. Hence it makes sense to introduce the symbols $\mathscr{M}(r)$, $2^r$ and $\mathscr{B}(r)$. A set $R$ of cardinality $r < \infty$ is called an *r-set*, and $\mathscr{B}(r)$ the *Boolean algebra of rank r*.

**Example.** Figure 1.4 shows the Boolean algebra $\mathscr{B}(4)$ and its isomorphic lattice $[\mathscr{C}(1)]^4$, where $R = \{1, 2, 3, 4\}$.

For brevity of notation, parentheses in the subsets of $R$ were omitted, similarly in the vectors of $[\mathscr{C}(1)]^4$.

Summarizing our results so far, we see that by comparing images of mappings we obtain the class of products of chains as a first class, fundamental to the whole of combinatorial order theory. In the next chapter we shall describe chain products and their sublattices intrinsically by the distributive property. The various characterizations arising there will then lead to the main questions of combinatorial order theory to be discussed in chapters VI to VIII.

## B. Refinement

Let us now compare mappings by their kernels. Let $f$ and $g$ be mappings both defined on a set $N$. Viewing the binary relations $\ker(f)$ and $\ker(g)$ as subsets of $N^2$ we may order them by inclusion; thus

$$f \preccurlyeq g :\Leftrightarrow \ker(f) \subseteq \ker(g).$$

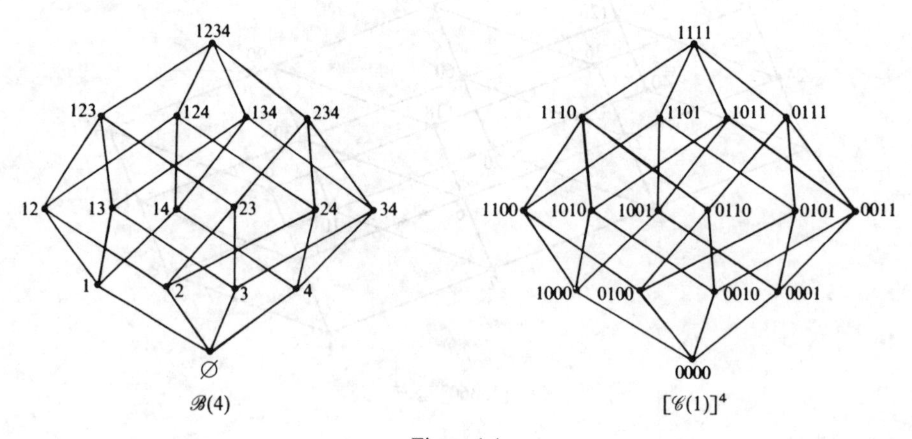


Figure 1.4

Again, $\leqslant$ is reflexive and transitive, but in general not antisymmetric since nothing is said about the images of $f$ and $g$, respectively. In other words, only the *blocks* of the kernel partition matter, not the elements onto which the blocks are mapped.

Thus we obtain as our second fundamental class the *lattices of partitions of a set* $N$. We shall usually denote partitions by lowercase Greek letters $\pi, \sigma, \rho, \tau, \ldots$. By definition $\pi \leqslant \sigma$ means that any two elements $a$, $b$ that are in a block of $\pi$ (or equivalently stand in the equivalence relation $(\pi)$—see below) are also in a block of $\sigma$. Thus $\pi \leqslant \sigma$ is equivalent to the statement: every block of $\pi$ is wholly contained in a block of $\sigma$, or, conversely, every block of $\sigma$ fully decomposes into blocks of $\pi$. For this reason, we say $\pi$ is a refinement of $\sigma$, and call $\leqslant$ the *refinement relation*. Other expressions used are $\pi$ is finer than $\sigma$, or $\sigma$ is coarser than $\pi$. Obviously, the partition consisting of just the single block $N$ is the unique coarsest partition, whereas the finest partition is the one in which all blocks are singletons. Further $\pi \mathrel{\lessdot} \sigma$ if and only if $\sigma$ consists of the same blocks as $\pi$ save one pair of blocks of $\pi$ which are joined into a single block in $\sigma$. It follows that $\pi \mathrel{\lessdot} \sigma$ implies that the number of blocks in $\sigma$ is precisely one less than that of $\pi$, assuming that both these numbers are finite.

With these remarks we can easily describe meet and join in a partition lattice. Denote by $(\pi)$ the equivalence relation associated with the partition $\pi$ of $N$. Then for any partitions $\pi, \sigma$ of $N$ and $a, b \in N$

$$a(\pi \wedge \sigma)b \Leftrightarrow a(\pi)b \text{ and } a(\sigma)b$$

$$a(\pi \vee \sigma)b \Leftrightarrow \exists a = u_0, u_1, \ldots, u_t = b \text{ such that } u_i(\pi)u_{i+1} \text{ or } u_i(\sigma)u_{i+1} \text{ for all } i.$$

If $A_1, \ldots, A_n$ are the blocks of the partition $\pi$, we shall write either

$$\pi = \dot{\bigcup_{i=1}^{n}} A_i \quad \text{or} \quad \pi = A_1|A_2|\cdots|A_n.$$

**Example.** Let $N = \{1, 2, \ldots, 11\}$, $\pi = 1, 2|3, 4, 5|6, 7, 8|9, 10, 11$, $\sigma = 1, 3, 4|2, 5|6, 7, 11|8|9, 10$. Then $\pi \wedge \sigma = 1|2|3, 4|5|6, 7|8|9, 10|11$, $\pi \vee \sigma = 1, 2, 3, 4, 5|6, 7, 8, 9, 10, 11$.

Again we shall be mainly interested in the sublattice consisting of all *finite partitions* of a set $N$, i.e., of partitions into a finite number of blocks, denoted by by $\mathscr{P}(N)$. Apparently $\mathscr{P}(N)$ depends up to isomorphism only on the cardinality of $N$, hence we may unambiguously introduce the symbol $\mathscr{P}(n)$. A partition into $k$ blocks is called a *k-partition*, and $\mathscr{P}(n)$ the *partition lattice of order n* if $n < \infty$. The lattice $\mathscr{P}(4)$ is illustrated in Figure 1.5.

As we have seen, after chain products, partition lattices comprise the second fundamental class in combinatorial lattice theory. The analogous question of describing all sublattices of partition lattices constituted for a long time one of the major unsolved problems in finite lattice theory. It was conjectured by Whitman

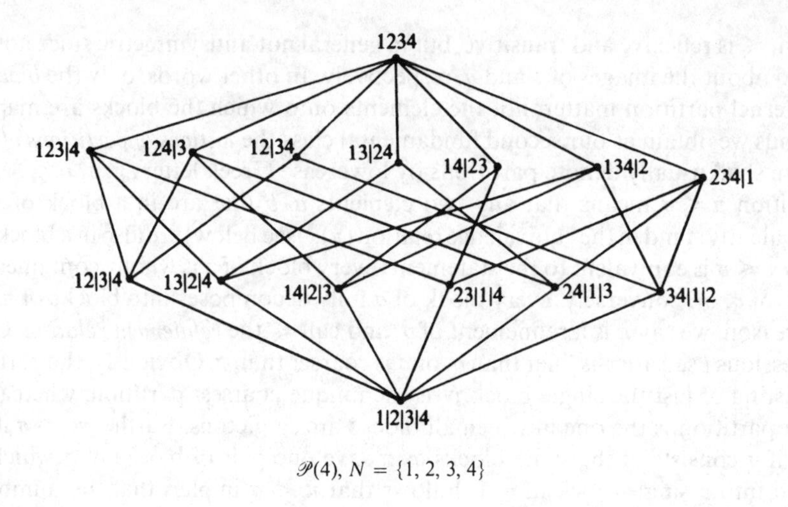


$\mathscr{P}(4)$, $N = \{1, 2, 3, 4\}$

Figure 1.5

and recently proved by Pudlák and Tůma [1] that any finite lattice can be embedded as a sublattice in a finite partition lattice. See Crawley–Dilworth [1, pp. 96–104] for a discussion of related questions.

Algebra provides a ready source of examples of sets endowed with the refinement relation. Let $A$ be any algebraic system. The equivalence relation corresponding to the kernel of a homomorphism from $A$ to another algebraic system of the same type is called a *congruence relation*. The set $\mathscr{CR}(A)$ of all congruence relations of $A$ ordered by refinement gives a complete lattice. Since the infimum of arbitrary congruence relations is again a congruence relation it follows that the intersection of arbitrary congruence classes (which may belong to different congruence relations) is again a congruence class. Hence the set of all *congruence classes* of $A$ ordered by inclusion constitutes another complete lattice denoted by $\mathscr{CC}(A)$. For the basic questions on how to characterize lattices $\mathscr{CR}$ and $\mathscr{CC}$ the interested reader is referred to Jónsson [3] or Wille [1].

For the familiar algebraic structures such as groups, rings and vector spaces a congruence relation $\Theta$ is completely determined by the block $0_\Theta$ containing the neutral element. From $\pi \preccurlyeq \sigma \Leftrightarrow 0_\pi \subseteq 0_\sigma$ it follows that $\mathscr{CR}(A)$ is isomorphic to the lattice of all $0_\Theta$'s ordered by inclusion. In the case of a group $A$, for instance, we obtain

$$\mathscr{CR}(A) \cong \mathscr{N}(A),$$

where $\mathscr{N}(A)$ is the lattice of all *normal subgroups* of $A$ ordered by inclusion.

**Example.** Let $C(n)$ be the cyclic group of order $n < \infty$ and $g$ a generator of $C(n)$. $C(n)$ is abelian and possesses a unique subgroup of order $k$ for every $k|n$, this subgroup being generated by $g^{n/k}$. For $i|n, j|n$ we have

$$\langle g^i \rangle \subseteq \langle g^j \rangle \Leftrightarrow j|i.$$

Hence if we let $\langle g^i \rangle$ correspond to $i \in \mathbb{N}$ for all $i|n$, we obtain $\mathcal{N}(C(n)) \cong [1, n]^*$, where $[1, n]^*$ is the dual lattice of the interval $[1, n] \cong \mathcal{T}$. But obviously $[1, n]^* \cong [1, n]$ by $i \to n/i$, and thus

$$\mathcal{N}(C(n)) \cong [1, n] \subseteq \mathcal{T}.$$

For our purposes vector spaces over division rings furnish the most interesting examples. In a vector space $V$ (say, left vector space) over $K$ congruence relations $\Theta$ correspond bijectively to *subspaces* $0_\Theta$, whence we obtain

$$\mathcal{CR}(V) \cong \mathcal{L}(V),$$

where $\mathcal{L}(V)$ is the lattice of all subspaces of $V$ ordered by inclusion;

$$\mathcal{CC}(V) \cong \mathcal{A}(V),$$

where $\mathcal{A}(V)$ is the lattice of all cosets of subspaces of $V$ ordered by inclusion. Later we shall discuss the intimate relationship between projective geometries and the lattices $\mathcal{L}(V)$, and affine geometries and the lattices $\mathcal{A}(V)$.

As vector spaces $V$ are determined up to isomorphism by the division ring $K$ and the dimension $n$, we may unambiguously define $V(n, K)$ and $\mathcal{L}(n, K)$, respectively, as the *left vector space* and *left vector space lattice of rank n over* $K$; similarly for $\mathcal{A}(n, K)$. Unless otherwise stated, vector space will mean left vector space. Of particular interest are the vector spaces and their lattices of finite rank over a finite field $GF(q)$. In this case we abbreviate the notation to $V(n, q)$, $\mathcal{L}(n, q)$, and $\mathcal{A}(n, q)$.

Figure 1.6 depicts the lattice $\mathcal{L}(3, 2)$. For simplicity the 0-vector was omitted from all subspaces. The 2-dimensional subspaces are the set unions of the 1-dimensional subspaces contained in them. Let us look at the lattice $\mathcal{A}(2, 2)$, $V = \{(0, 0), (1, 0), (0, 1), (1, 1)\}$, in Figure 1.7. The 0-space induces 4 cosets, each

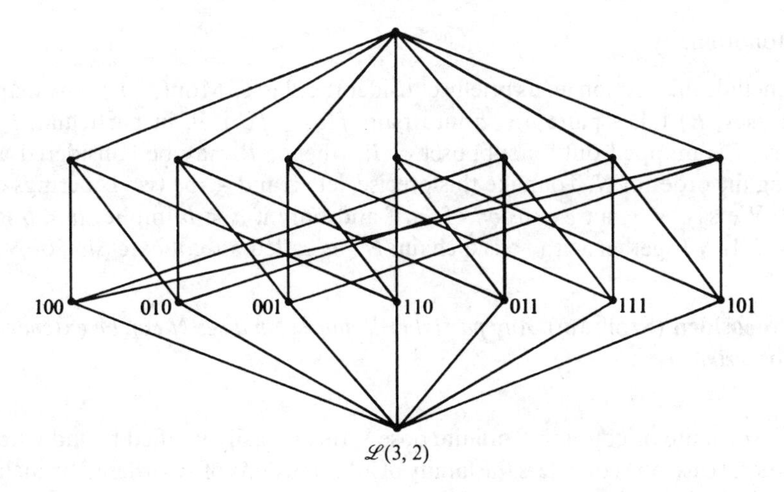


Figure 1.6

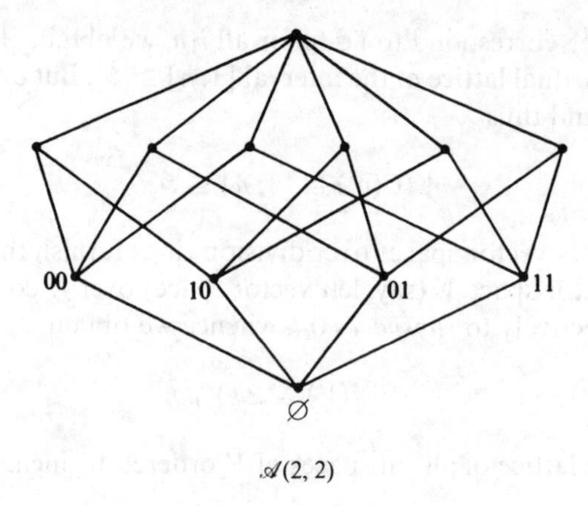


$\mathscr{A}(2, 2)$

Figure 1.7

containing a single vector. Each of the three 1-dimensional subspaces induces 2 cosets, hence we obtain 6 more lattice elements beside $\varnothing$ and $V$.

Let us summarize our findings up to now: Inclusion and refinement of mappings give rise to the following five classes of lattices: *chains* $\mathscr{C}(n)$, the *divisor lattice* $\mathscr{T}$, *Boolean algebras* $\mathscr{B}(n)$, *partition lattices* $\mathscr{P}(n)$ and *vector space lattices* $\mathscr{L}(n, K)$. These five classes will motivate many of the basic questions in combinatorial order theory and suggest approaches and methods to be tried in more general situations. On the other hand, they will also furnish the prime examples for which we shall deduce concrete results from the general theory. We shall return for a closer look at these lattices in section II.4.

## C. Monotonicity

To conclude this section let us briefly consider the classes $\mathrm{Mon}(N, R)$. Any mapping $f \in \mathrm{Mon}(N, R)$ takes pairs $a \leq b$ into pairs $f(a) \leq f(b)$. If, in particular, $f$ is injective, $N$ is mapped onto a subposet of $R$, whence $R$ may be considered as extending the order of $N$. To make this precise, let $<$ and $\prec$ be two orderings of the set $N$. We say $\prec$ is an *extension* of $<$ if and only if $a < b$ implies $a \prec b$ for all $a, b \in N$. If $N$ together with $\prec$ is a chain, $N_{\prec}$ is called a *total extension* of $N_{<}$.

**1.4 Proposition** (Szpilrajn). *Any partial ordering $<$ on a set $N$ can be extended to a total ordering.*

*Proof.* For finite or countably infinite posets, this is easily verified by induction. In the general case one considers the family of all extensions of $<$ ordered by inclusion and deduces from Zorn's lemma the existence of maximal elements. $\square$

Let $N$ and $R$ be finite posets of the same cardinality and $R$ a chain. In view of what we just said it is clear that the bijective monotone mappings from $N$ to $R$ correspond biuniquely to the total extensions of $N$. The number of these extensions is of considerable combinatorial significance and will be discussed in section III.4.

Suppose now $N$ is a chain and $R$ an arbitrary poset. In this case $\mathrm{im}(f)$ will be a chain in $R$ for any $f \in \mathrm{Mon}(N, R)$, or rather a *multichain* if we regard $\mathrm{im}(f)$ as a multiset. As before we shall be mainly interested in finite multichains $a = a_1 a_2 \ldots a_k$; it is convenient to list $a$ in decreasing order $a_1 \geq a_2 \geq \cdots \geq a_k$. In order to compare multichains it is useful to define a new 0-element $\hat{0}$ in $R$ with $\hat{0} < b$ for all $b \in R$, denoting the new poset by $\hat{R}$, and to write the multichain $a = a_1 a_2 \ldots a_k$ as vector $(a_1, a_2, \ldots, a_k, a_{k+1}, \ldots)$ where $a_i = \hat{0}$ for all $i > k$.

**1.5 Proposition.** *The set of all finite multichains of a poset R together with the relation*

$$a \leq b :\Leftrightarrow a_i \leq b_i \quad \textit{for all } i \in \mathbb{N}$$

*is a partially ordered set. If, in particular, R is a lattice, then this poset is a sublattice of the product* $\prod_{i=1}^{\infty} R_i$, *where* $R_i \cong \hat{R}$ *for all i.*

*Proof.* One only has to observe that the meet and join of two monotonically decreasing words are both monotonically decreasing. □

**Example.** Let $R = \mathbb{N}$ with the natural ordering, hence $\hat{R} \cong \mathbb{N}_0$. The elements of the lattice in **1.5** are sequences of non-negative integers $a = (a_1, a_2, \ldots)$, $a_1 \geq a_2 \geq \cdots$ with $a_j = 0$ for $j > n_0$, and the order relation is given by

$$(a_1, a_2, \ldots) \leq (b_1, b_2, \ldots) :\Leftrightarrow a_i \leq b_i \text{ for all } i \in \mathbb{N}.$$

The resulting lattice is called the *Young lattice* $\mathcal{Y}$ because of its close connection to a class of mappings called Young tableaux. We shall touch upon this subject in III.4. Notice that $\mathcal{Y}$, being a sublattice of a chain-product (**1.5**), is distributive according to a remark made earlier.

There is another very useful interpretation of the lattice $\mathcal{Y}$. Let $n \in \mathbb{N}$ and $n = n_1 + n_2 + \cdots + n_k$ be a decomposition of $n$ into positive integers $n_i$. We call $n_1 + \cdots + n_k$ a *k-partition* of $n$ into the *parts* or *summands* $n_1, n_2, \ldots, n_k$. Sometimes we shall use the term *number-partition* to distinguish these partitions from the set-partitions introduced earlier. Since the order of the parts $n_i$ does not matter, we adopt the convention of writing them in decreasing magnitude $n_1 \geq n_2 \geq \cdots \geq n_k$. It is clear now that the elements $(n_1, n_2, \ldots, n_k, 0, 0, \ldots) \in \mathcal{Y}$ can be identified with the number-partitions:

$$(n_1, n_2, \ldots, n_k, 0, \ldots) \leftrightarrow n_1 + n_2 + \cdots + n_k.$$

Thus $\mathcal{Y}$ can be viewed as the *lattice of number-partitions* ordered by magnitude of parts. The empty partition $\varnothing$ is the 0-element in $\mathcal{Y}$ and we have

$$n_1 + \cdots + n_k \leq m_1 + \cdots + m_l \Leftrightarrow k \leq l \text{ and } n_i \leq m_i \ (i = 1, \ \ldots, k).$$

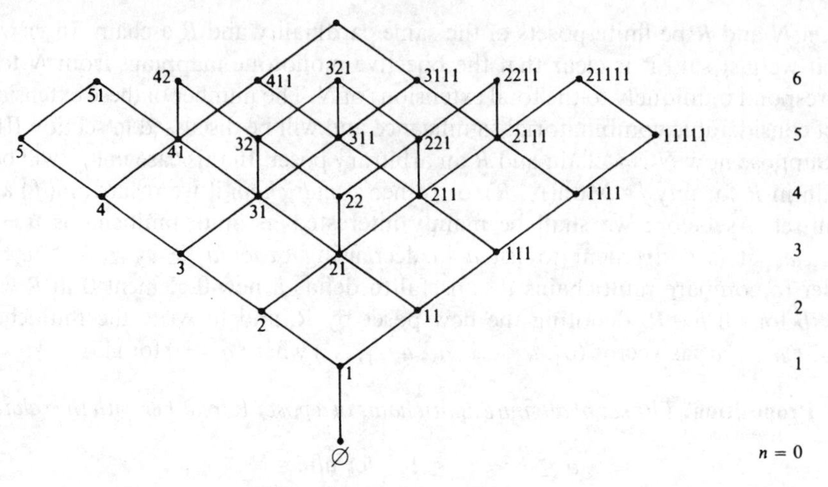


Figure 1.8. The Young lattice $\mathscr{Y}$.

Figure 1.8 gives the bottom part of $\mathscr{Y}$ where for simplicity we use the short notation $n_1 n_2 \ldots n_k$ for the partition $n_1 + \cdots + n_k$. Notice that with every set-partition $\pi = A_1 | A_2 | \cdots | A_k$ of an $n$-set we can associate a number-partition $z(\pi) := n_1 + n_2 + \cdots + n_k$ with $n_i = |A_i|$, $i = 1, \ldots, k$. Hence the refinement relation in $\mathscr{P}(n)$ induces an ordering on the partitions of $n$:

$$\sum n_i \leq \sum m_i :\Leftrightarrow \exists \pi, \sigma \in \mathscr{P}(n) \text{ with } \pi \preccurlyeq \sigma \text{ and } z(\pi) = \sum n_i, z(\sigma) = \sum m_i.$$

The reader may verify that this in fact gives a poset and that the definition of $\leq$ is equivalent to

$$\sum_{i=1}^{k} n_i <\cdot \sum_{i=1}^{l} m_i :\Leftrightarrow l = k - 1 \text{ and the } m_i\text{'s coincide with the } n_j\text{'s except for one pair } n_a, n_b \text{ and one } m_c \text{ for which } m_c = n_a + n_b.$$

The resulting poset is called the *dominance order* $\mathscr{D}(n)$. The poset $\mathscr{D}(7)$ is illustrated in Figure 1.9.

EXERCISES I.2

1. Show that every interval of a Boolean algebra $\mathscr{B}(S)$ is again a Boolean algebra. (Hint: For $A \subseteq B$ show $[A, B] \cong \mathscr{B}(B - A)$.)
2. Prove that $\mathscr{B}(n)$ is a complemented lattice for $n < \infty$. That is, for every $a \in \mathscr{B}(n)$ there exists $a' \in \mathscr{B}(n)$ with $a \wedge a' = 0$ and $a \vee a' = 1$.
3. Can ex. 2 be generalized to $\mathscr{M}(r)$, to arbitrary chain products?
4. Show that every interval $[m, n] \subseteq \mathscr{T}$ is isomorphic to a finite chain product.

→ 5. Let $L$ and $M$ be lattices. A mapping $\phi: L \to M$ is called an inf-*homomorphism* if $\phi(x \wedge_L y) = \phi(x) \wedge_M \phi(y)$. A sup-*homomorphism* is similarly defined.

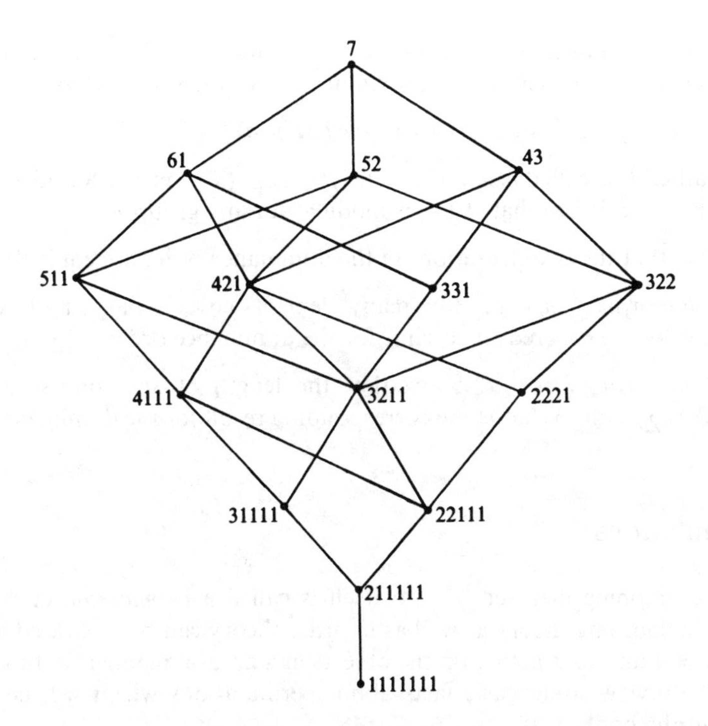


Figure 1.9. The dominance order $\mathscr{D}(7)$.

Show:

(i) Any inf-homomorphism or sup-homomorphism is monotone.
(ii) The converse to (i) is false.
(iii) For bijections the concepts inf-homomorphism, sup-homomorphism, lattice isomorphism, and order-isomorphism are equivalent.

→ 6. Show that a poset $P$ is a complete lattice if and only if $P$ has a 1-element and $\inf A$ exists for all $A \subseteq P$.

→ 7. A lattice $L$ is called *distributive* if $x \wedge (y \vee z) = (x \wedge y) \vee (x \wedge z)$ for all $x, y, z \in L$. Show that any sublattice of a distributive lattice is distributive; similarly, for the dual lattice and the product of distributive lattices.

8. Let $\pi, \sigma \in \mathscr{P}(n)$ and $b(\pi)$ and $b(\sigma)$ be the number of blocks of $\pi$ and $\sigma$, respectively. Show that $b(\pi \wedge \sigma) + b(\pi \vee \sigma) \geq b(\pi) + b(\sigma)$.

→ 9. Let $\pi, \sigma \in \mathscr{P}(n)$ and $G(V, E)$ be the undirected bipartite graph with the defining vertex-sets corresponding to the blocks $A_i$ of $\pi$ and $B_j$ of $\sigma$ respectively, and $\{A_i, B_j\} \in E$ if and only if $A_i \cap B_j \neq \varnothing$. Show:

(i) $|V| = b(\pi) + b(\sigma)$
(ii) $|E| = b(\pi \wedge \sigma)$
(iii) $b(\pi \vee \sigma) = \#$ {connected components of $G$}.
(iv) Deduce from (i), (ii), (iii) a new proof of ex. 8.

10. Verify in detail that the intersection of congruence classes is again a congruence class. Describe the supremum of two congruence classes.

11. How many atoms does $\mathscr{L}(n, q)$ have? $\mathscr{A}(n, q)$?

12. A lattice $L$ is called *modular* if $x \wedge (y \vee z) = (x \wedge y) \vee z$ for all $x, y, z \in L$ with $z \leq x$. Prove that $\mathscr{N}(A)$ is modular for any group $A$.

→13. Verify that the two definitions of the dominance order are equivalent.

→14. Let $n = n_1 n_2 \ldots n_t \in \mathscr{Y}$. How many elements cover $n$ in $\mathscr{Y}$; by how many elements is $n$ covered? The same for the dominance order.

15. Let $n = n_1 n_2 \ldots n_t \in \mathscr{Y}$. Show that the length of any longest $0,n$-chain in $\mathscr{Y}$ is $\sum_{i=1}^{t} n_i$. What is the corresponding result for the dominance order?

## 3. Permutations

A bijective mapping of a set $N$ onto itself is called a *permutation* of $N$. Many problems in counting theory as well as in order theory can be rephrased in terms of a set of permutations acting on the objects in a natural manner. In this section we want to review some basic facts about permutations which will be needed throughout the book.

### A. Algebraic Properties

The set of permutations of $N$ forms, under composition of mappings, a group called the *symmetric group* $S(N)$, where the composition $g \circ f$ of two permutations $g$ and $f$ is defined by $(g \circ f)(a) := g(f(a))$. The identity in $S(N)$ will be denoted by *id*. As always we shall mainly be concerned with finite sets $N$. Since $S(N)$ is determined up to isomorphism by the cardinality of $N$, we can unambiguously use the shorthand notation $S_n$. For $n < \infty$, $S_n$ is called the *symmetric group of degree n* and each of its subgroups a *permutation group of degree n*; most of the time we choose $N = \{1, 2, \ldots, n\}$. The degree is thus the number of elements being permuted. Fixing $\{1, \ldots, n\}$ in the natural order we usually write a permutation

$$f = \begin{pmatrix} 1 & 2 & \cdots & n \\ f(1) & f(2) & \cdots & f(n) \end{pmatrix}$$

as the word $f(1)f(2)\cdots f(n)$.

**Example.** $S_3$ consists of $id = 123, 132, 213, 231, 312$, and $321$, and we have, e.g., $312.213 = 132 \neq 321 = 213.312$; thus $S_3$ is not commutative.

Another common way to represent $f \in S_n$ is by decomposing $f$ into *cycles*

$$f = (1, f(1), f^2(1), \ldots)(i, f(i), f^2(i), \ldots) \ldots .$$

The order in which the cycles are written down is irrelevant. The *length* of a cycle is the number of elements contained in it. Cycles of length 1 are called *trivial* and are often left out in the cycle decomposition of $f$. The trivial cycles thus correspond precisely to the *fixed points* of the permutation. The *type* of $f \in S_n$ is the expression

$$\text{type}(f) := 1^{b_1(f)} 2^{b_2(f)} \ldots n^{b_n(f)}$$

where $b_i(f)$ is the number of cycles of $f$ of length $i$.

It should be remembered that type($f$) is just a notational device and not a product. Notice that $b_1(f)$ counts the number of fixed points of $f$. Clearly

$$\sum_{i=1}^{n} b_i(f) = \# \{\text{cycles of } f\}, \qquad \sum_{i=1}^{n} i b_i(f) = n.$$

**Example.** Let

$$f = \begin{pmatrix} 1 & 2 & 3 & 4 & 5 & 6 & 7 & 8 & 9 \\ 2 & 5 & 6 & 4 & 8 & 7 & 3 & 1 & 9 \end{pmatrix}.$$

Then $f = (1, 2, 5, 8)(3, 6, 7)(4)(9)$, and thus type($f$) $= 1^2 3^1 4^1$.

A permutation $f \in S_n$ consisting of one cycle of length $k$ and otherwise only trivial cycles is called *k-cyclic*. An *n*-cyclic permutation is called a *circular permutation* and a 2-cyclic permutation a *transposition*. Hence the cycle decomposition of $f$ is a factorization of $f$ into cyclic permutations. A transposition exchanges a pair $(i, j)$ of letters and leaves the rest unchanged. If $N = \{1, 2, \ldots, n\}$, then a transposition which exchanges two adjacent elements $(i, i + 1)$ is called a *standard transposition*.

Let $G$ be a permutation group on $N$. The relation

$$i \approx j :\Leftrightarrow \exists g \in G \text{ with } j = g(i)$$

is an equivalence relation on $N$ whose equivalence classes are called *G-orbits*. Thus, $i$ and $j$ lie in the same orbit if and only if there is a permutation in $G$ which takes $i$ into $j$. Notice that for a cyclic group $G = \langle f \rangle$, the $G$-orbits are precisely the cycles of $f$. $G$ is said to be *transitive* if there is only one $G$-orbit, namely the whole set $N$. We denote by $0(i)$ the $G$-orbit containing $i \in N$ and by $G(i)$ the subgroup of $G$ fixing $i$. $G(i)$ is called the *stabilizer* of $i$.

**1.6 Lemma.** *Let $G$ be a permutation group of a finite set $N$. Then*

$$|0(i)||G(i)| = |G| \quad \textit{for all } i \in N.$$

*Proof.* $0(i)$ consists of all distinct elements $g(i)$, $g \in G$. Since

$$g(i) = h(i) \Leftrightarrow h^{-1}g(i) = i \Leftrightarrow h^{-1}g \in G(i) \Leftrightarrow gG(i) = hG(i)$$

we infer that distinct images $g(i)$ give rise to distinct cosets $gG(i)$ and conversely. Hence the cardinalities of these two sets are the same, giving $|0(i)| = |G|/|G(i)|$. □

As a corollary we deduce the following fact, called *Burnside's lemma*, which will prove important in many counting problems.

**1.7 Proposition.** *Let G be a permutation group on a finite set N. Then*

$$k = \frac{1}{|G|} \sum_{g \in G} b_1(g)$$

*where $b_1(g)$ is the number of fixed points of g and k is the number of G-orbits.*

*Proof.* Counting the number of pairs $(g, i)$, $g \in G$, $i \in N$ with $g(i) = i$ in two ways, we obtain

$$\sum_{g \in G} b_1(g) = \sum_{i \in N} |G(i)| = |G| \sum_{i \in N} \frac{1}{|0(i)|} = |G|k. \quad \square$$

Let $G$ and $H$ be permutation groups on sets $N$ and $R$, respectively. $G$ and $H$ give rise to another permutation group $H^G = \{h^g : g \in G,\ h \in H\}$ on the set $\operatorname{Map}(N, R)$ where

$$h^g(f) := hfg \quad \text{for all } f \in \operatorname{Map}(N, R).$$

We call $H^G$ the *power group* on $\operatorname{Map}(N, R)$ induced by $G$ and $H$. The orbits of $H^G$ are thus the equivalence classes under the relation on $\operatorname{Map}(N, R)$

$$f \approx f' :\Leftrightarrow \exists g \in G,\ h \in H \text{ with } f' = hfg.$$

We observe that for finite sets $N$, $R$ the order of $H^G$ is $|H^G| = |G||H|$ (if $|R| > 1$) whereas the degree equals $|\operatorname{Map}(N, R)| = |R|^{|N|}$.

## B. Combinatorial Properties

So far we have reviewed some algebraic facts and definitions which will be needed in subsequent chapters. Let us now discuss some combinatorial properties of individual permutations of the set $\{1, \ldots, n\}$. This study of permutations marked the very beginning of systematic research in combinatorial analysis and constitutes, for instance, a major portion of MacMahon's classic treatise [1]. For a detailed account the reader may also consult Foata [3].

We shall concentrate for the rest of this section on permutations $f$ of $\{1, \ldots, n\}$ represented as words. An *inversion* of $f = a_1 a_2 \ldots a_n \in S_n$ is a pair $(a_i, a_j)$ with $i < j$, $a_i > a_j$, and we set

$$I(f) := \{(a_i, a_j) : a_i, a_j \text{ inversion}\},$$

$$i(f) := |I(f)|.$$

$f$ is said to be *odd* or *even* depending on whether $i(f)$ is an odd or even integer. A *descent* of $f$ is an inversion $(a_i, a_{i+1})$ and we denote by $D(f)$ the set

$$\{j: a_j > a_{j+1}\}$$

of positions where the descents occur.

**Example.** Let $f = 5\,6\,2\,1\,7\,4\,3$, then

$$I(f) = \{(2, 1), (4, 3), (5, 1), (5, 2), (5, 3), (5, 4), (6, 1), (6, 2), (6, 3), (6, 4), (7, 3), (7, 4)\},$$

$$i(f) = 12,$$

$$D(f) = \{2, 3, 5, 6\}.$$

Let $t = (i, i + 1)$ be a standard transposition and $f = a_1 \dots a_n$. Since $f \cdot t = a_1 \dots a_{i-1}a_{i+1}a_i \dots a_n$ we have $i(f \cdot t) = i(f) \pm 1$ depending on whether $i \in D(f)$ or not. From this observation we can easily deduce the following result.

**1.8 Proposition.** *Every permutation $f \in S_n$ can be written as a product $f = \prod_{j=1}^{i(f)} t_j$ where the $t_j$'s are standard transpositions. If $f = \prod_{j=1}^{k} u_j$ is any other factorization into standard transpositions then $k - i(f)$ is an even integer $\geq 0$. In particular, $i(f)$ is the minimum number of standard transpositions required to factorize $f$.*

*Proof.* If $f = id$ then $i(f) = 0$ and there is nothing to prove. Suppose now $f \neq id$ with $j \in D(f)$, then $i(f \cdot (j, j + 1)) = i(f) - 1$. Hence if we multiply by appropriate standard transpositions $t_l$ on the right, the inversion index is reduced to 0, and we obtain $f . t_{i(f)} \dots t_1 = id$. Since $t^2 = id$ for any transposition, it follows that $f = t_1 \dots t_{i(f)}$.

Let $f = \prod_{j=1}^{k} u_j$ be any other factorization of $f$ into standard transpositions. Then $fu_k \dots u_1 = id$ and $i(fu_k \dots u_j) = i(fu_k \dots u_{j+1}) \pm 1$ for all $j$. Suppose there are exactly $p$ $u_i$'s which increase the number of inversions. Since $i(id) = 0$ this gives

$$i(f) + p - (k - p) = 0 \quad \text{or} \quad k - i(f) = 2p \geq 0. \quad \square$$

**1.9 Corollary.** *The standard transpositions generate the whole group $S_n$.* $\square$

A slight generalization of **1.9** is given in the following statement whose easy proof is left to the reader.

**1.10.** *Let $T$ be a set of $n - 1$ transpositions in $S_n$ and let $G_T$ be the graph with vertex set $\{1, \dots, n\}$ such that $i$ and $j$ are joined if and only if $(i, j) \in T$. Then $T$ generates $S_n$ if and only if $G_T$ is a tree.* $\square$

Another well-known consequence of **1.8** is the following proposition.

**1.11 Proposition.** *For any $f, g \in S_n$*

$$(-1)^{i(fg)} = (-1)^{i(f)}(-1)^{i(g)}.$$

*In particular, the set $A_n$ of even permutations forms a subgroup of $S_n$ of index 2, called the* alternating group $A_n$.

*Proof.* Let $f = \prod_{i=1}^k t_i, g = \prod_{j=1}^l u_j$ be factorizations into standard transpositions. Then $fg = t_1 \ldots t_k u_1 \ldots u_l$ and hence by **1.8**

$$i(f) = k - 2p, \; i(g) = l - 2q$$

$$i(fg) = k + l - 2r = i(f) + i(g) + 2(p + q - r).$$

Thus $\phi: S_n \to \{1, -1\}$, given by $\phi(f) = (-1)^{i(f)}$, is an epimorphism with kernel $A_n$. □

To conclude, let us derive a characterization of even permutations in terms of their cycle decompositions. Let $f$ be of type $1^{n-k}k$, i.e., $f$ is $k$-cyclic. Applying **1.8** it is readily verified that $f$ is even if and only if $k$ is odd. In particular, any transposition is odd.

**1.12 Proposition.** *Let $f \in S_n$ be of type $1^{b_1}2^{b_2} \ldots n^{b_n}$. Then*

$$f \in A_n \Leftrightarrow b_2 + b_4 + \cdots \text{ is even}$$

$$\Leftrightarrow n - \sum_{i=1}^n b_i \text{ is even.}$$

*Proof.* As $f$ can be written as a product of $b_i$ cyclic permutations of length $i$, $i = 1, \ldots, n$, by the remark just made we have

$$i(f) \equiv \sum_{i=1}^n (i-1)b_i \equiv b_2 + b_4 + \cdots \pmod 2$$

and

$$\sum_{i=1}^n (i-1)b_i = n - \sum_{i=1}^n b_i. \quad \square$$

EXERCISES I.3

1. Show that the order of $g \in S_n$ equals the smallest common multiple of the lengths of the cycles of $g$.

2. Show that a transitive permutation group of degree $n > 1$ contains an element of order $n$.

→ 3. Show that the alternating group $A_n$ is generated by the cyclic permutations $(1, 2, i)$, $i = 3, \ldots, n$.

4.* Prove that $A_n$ contains no non-trivial normal subgroup for $n > 4$. (Hint: A normal subgroup of $A_n$ which contains some $(1, 2, i)$ must be $A_n$.)

5. Show that in any permutation group $G$, either all permutations are even or there are equally many even and odd permutations.

→ 6. Let $|N| = |R| = 3$ and $G = S_3, H = C_3$ permutation groups on $N$ and $R$ respectively. Describe the power group $H^G$. What are the subgroups, cycle-types, etc.?

→ 7. Prove that every $g \in S_n, g \neq id$, is the product of disjoint cyclic permutations and that this factorization is unique up to order.

→ 8. Let $g \in S_n$ be of type $1^{n-k}k$. Show that $g$ is even if and only if $k$ is odd.

9. Prove **1.10**.

10.* Let $T = \{t_1, \ldots, t_{n-1}\}$ be a set of transpositions in $S_n$. Show that $g = \prod_{i=1}^{n-1} t_i$ is a circular permutation if and only if the graph $G_T$ in proposition **1.10** is a tree. (Denes)

→11. Let $N = \{1, 2, \ldots, n\}$ and $f \in S_n$ be given in word form $f = a_1 a_2 \ldots a_n$. We define $E(f) := \{(a_i, a_j): i < j, a_i < a_j\}$. Show that $f \leq g :\Leftrightarrow E(f) \subseteq E(g)$ is a partial order on $S_n$ and that this poset is in fact a lattice $\text{Per}(S_n)$ on $S_n$. (Guilbaud–Rosenstiehl)

→12. Show that in $\text{Per}(S_n)$ the following hold:
 (i) Let $f \in \text{Per}(S_n)$. The length of any maximal $0, f$-chain is equal to $\binom{n}{2} - i(f)$; the length of any maximal $f, 1$-chain is equal to $i(f)$.
 (ii) $f <\cdot\, g \Leftrightarrow g = f \cdot (i, i+1)$ for some $i \in D(f)$.
 (iii) $\text{Per}(S_n)$ is a complemented lattice. (See ex. I.2.2.)

13. Deduce proposition **1.8** from ex. 12.

14. Let $N = \{1, \ldots, n\}$. Show that the expression

$$\delta(f, g) := \max_{1 \leq i \leq n} |f(i) - g(i)|, \qquad f, g \in S(N),$$

is a metric on the set $S(N)$. Show further that the numbers $a_f(n, r) := \#\{g \in S(N): \delta(g, f) \leq r\}$, $n, r \in \mathbb{N}$, are independent of the choice of $f \in S(N)$ and satisfy the recursion $a(n, 2) = 2a(n-1, 2) + 2a(n-3, 2) - a(n-5, 2)$, $n \geq 6$.

15. For $f \in S(N)$ let $N(f) := \{i \in N: f(i) \neq i\}$. Show that $d(f, g) := |N(fg^{-1})|$ is a metric on $S(N)$ and describe the balls $\{g \in S(N): d(g, f) \leq r\}$.

## 4. Patterns

Let us return to the starting point of our discussion. We were studying maps $f: N \to R$ satisfying certain conditions. In the order relation induced by inclusion of images (regarded as multisets) we disregarded the nature of the elements of $N$ and counted only how many were mapped onto a particular element of $R$. In other words: The elements of $N$ are *indistinguishable* whereas the elements of $R$ are *fully distinguishable*. To phrase it differently: We are not interested in single mappings $f: N \to R$ but in classes of mappings induced by the equivalence relation

$$f \approx f' :\Leftrightarrow \text{im}(f) = \text{im}(f') \text{ as multisets.}$$

Comparing this equivalence relation with the notion of orbits introduced in section 3 we see that $\approx$ is precisely the orbit relation with respect to the power group $E(R)^{S(N)}$ where $S(N)$ is the symmetric group on $N$ and $E(R)$ the identity group on $R$. Hence we have

$$f \approx f' \Leftrightarrow \exists g \in S(N) \text{ with } f' = fg,$$

and the classes of mappings we want to study are the $E(R)^{S(N)}$-*orbits* of $\text{Map}(N, R)$, corresponding to the *n-multisets of R*.

Consider now the refinement relation. Here we are only interested in the blocks of $\ker(f)$ disregarding the nature of the elements of $R$ onto which these blocks are mapped. Hence in this case the elements of $N$ are *fully distinguishable* whereas those of $R$ are *indistinguishable*, with the equivalence relation on $\text{Map}(N, R)$ given by

$$f \approx f' :\Leftrightarrow \exists h \in S(R) \text{ with } f' = hf.$$

The classes of $\approx$ are just the $S(R)^{E(N)}$-*orbits* of $\text{Map}(N, R)$, corresponding to the *set-partitions of N into r blocks.*

If we let the full symmetric groups $S(N)$, $S(R)$ act on both $N$ and $R$, respectively, we obtain as $S(R)^{S(N)}$-orbits partitions of $N$ where only the cardinality of the blocks matters and not which elements are contained in the various blocks. Both the elements of $N$ and those of $R$ are *indistinguishable* and the equivalence classes, i.e., the $S(R)^{S(N)}$-orbits of $\text{Map}(N, R)$, correspond therefore to the *number-partitions of n into r parts.*

In the other extremal case, when $G = E(N)$, $H = E(R)$, all elements are *fully distinguishable*, hence every $E(R)^{E(N)}$-orbit of $\text{Map}(N, R)$ consists of a single mapping, and we obtain the set $\text{Map}(N, R)$ *of all mappings.*

We are thus led to the following general problem: Let $N$ and $R$ be finite sets and $G$, $H$ permutation groups acting on $N$ and $R$, respectively. The $H^G$-orbits are called *G, H-patterns* or simply *G-patterns* if $H = E(R)$.

**Problem.** *Count the number of G, H-patterns.*

**1.13** *Table of patterns with the most important ones set in italic.*

| $\lvert N\rvert = n < \infty, G \le S(N)$<br>$\lvert R\rvert = r < \infty, H \le S(R)$ | | | | |
|---|---|---|---|---|
| $f: N \to R$ | arbitrary | injective | surjective | bijective |
| $G = E(N), H = E(R)$ | ordered generalized $r$-partitions of $N$ | strict words in $R$ of length $n$ | ordered $r$-partitions of $N$ | *permutations* |
| $G = S(N)$, $H = E(R)$ — words | monotone words in $R$ of length $n$ = *$n$-multisets* of $R$ | monotone strict words in $R$ of length $n$ = *$n$-subsets* of $R$ | $n$-multisets comprising all of $R$ | trivial |
| $G = S(N)$, $H = E(R)$ — sortings | ordered generalized $r$-partitions of $n$ | ordered generalized $r$-partitions $n = \sum n_i$ with $n_i \le 1$ for all $i$ | ordered $r$-partitions of $n$ | trivial |
| $G = E(N), H = S(R)$ | generalized $r$-partitions of $N$ | trivial | *$r$-partitions* of $N$ | trivial |
| $G = S(N), H = S(R)$ | generalized $r$-partitions of $n$ | trivial | *$r$-partitions* of $n$ | trivial |
| $G = G(\pi), \pi \in \mathscr{P}(N)$<br>$H = H(\sigma), \sigma \in \mathscr{P}(R)$ | distributions | | | |

As we have seen, we can give the expression "a set $M$ of indistinguishable elements" a precise meaning by letting the symmetric group act on $M$. Let us generalize this principle: Suppose $M$ consists of pairwise disjoint blocks of elements such that two elements are distinguishable precisely when they are in different blocks. This means we consider the group $G$ of all those permutations of $M$ which permute the elements in each block but fix each block as a whole. Hence $G$ is the maximal subgroup of $S(M)$ whose orbits are the given blocks of $M$. For $\pi \in \mathscr{P}(M)$ let us denote this group by $G(\pi)$.

**Example.** $M = \{1, 2, 3, 4, 5, 6\}$, $\pi = 1|2, 3|4, 5, 6$, then

$$G(\pi) = \{id, (4, 5), (4, 6), (5, 6), (4, 5, 6), (4, 6, 5), (2, 3), (2, 3)(4, 5), (2, 3)(4, 6), (2, 3)(5, 6), (2, 3)(4, 5, 6), (2, 3)(4, 6, 5)\}.$$

In general, we have the following obvious fact: Let $\pi = A_1|A_2|\cdots|A_k \in \mathscr{P}(M)$ with $|A_i| = n_i$, then $G(\pi) \cong \prod_{i=1}^{k} S_{n_i}$.

**Definition.** Let $N$ and $R$ be finite sets. $G, H$-patterns with $G = G(\pi)$, $H = H(\sigma)$ for some $\pi \in \mathscr{P}(N)$, $\sigma \in \mathscr{P}(R)$ are called *distributions of $N$ into $R$*.

The name distribution is suggested by the interpretation of a mapping as a sorting of objects into boxes. Assume, for example, that $N$ consists of $a$ apples, $b$ eggs, and $c$ oranges, and that $R$ is made up of $k$ big boxes, $l$ boxes of medium size, and $m$ small boxes. If our task consists in sorting these objects into the boxes where we do not distinguish between objects or boxes of the same type, then the possible sortings correspond precisely to distributions in the sense just defined.

To conclude this introductory chapter let us classify various types of mappings according to the relevant permutation groups, using the dual interpretations of mappings as words and sortings. We have already introduced the notion of a *word*, *monotone word* (which may be identified with a *multiset*), *strict word* (= *subset*), *set-partition* and *number-partition*. Observe that a partition of a set into $\leq k$ blocks can always be regarded as a $k$-partition by adding the appropriate number of empty blocks. Similarly, a number-partition into $\leq k$ parts can be made into a $k$-partition by adding 0's. Partitions of this type are called *generalized set-* or *number-partitions*. Partitions where the order of the blocks (summands) matters are called *ordered set-* or *number-partitions*. For instance, $4 + 2 + 2 + 1$ and $4 + 2 + 1 + 2$ are different ordered partitions of 9 but represent the same unordered partition.

## Exercises I.4

1. Let $N = \{1, 2, 3, 4, 5\}$, $R = \{a, b, c\}$ and $\pi = 1, 2|3, 4|5 \in \mathscr{P}(N)$, $\sigma = a, b|c \in \mathscr{P}(R)$. How many $\pi, \sigma$-distributions of $N$ into $R$ are there? (Answer: 57.)

→ 2. Verify in table **1.13** (p. 27) that the number of monotone words in $R$ of length $n$ equals the number of $n$-subsets of an $(r + n - 1)$-set. (Hint: To each monotone word $f = a_1 a_2 \ldots a_n$ associate the word $f' = a_1 a_2 + 1$ $a_3 + 2 \ldots a_n + (n - 1)$.)

3. Show that the number of monotone surjective words of length $n$ in an $r$-set equals the number of $(r - 1)$-subsets of an $(n - 1)$-set.
4. Compute the total number of partitions of $n = 4, 5, 6, 7$.
5. Compute the total number of partitions of a set $N$, $|N| = 4, 5, 6, 7$.
6. → Show that the number of partitions of an $n$-set into 2 blocks is $2^{n-1} - 1$. Construct a mapping of these partitions into the Boolean algebra $\mathscr{B}(n - 1)$.
7. → Give a formula for the number of partitions of an $n$-set into 3 blocks.
8. How many ordered set-partitions does a set with 10 elements have?
9. → Let $|N| = 6$, $R = \{0, 1\}$ and $G = A_6 \leq S(N)$, $H = E(R)$. Discuss the group $H^G$ and describe the $G, H$-patterns. Notice that $f: N \to R$ can be identified with the characteristic set $\{a \in N: f(a) = 1\} \subseteq N$, whence $H^G$ may be thought of as a group acting on the Boolean algebra $\mathscr{B}(N)$.
10. Verify the trivial entries in table **1.13** (p. 27).

## Notes

This chapter was intended to convince the reader that many of the basic notions and problems in enumerative combinatorics can be presented in the context of mappings and orders. As a consequence, the present book differs from previous texts such as Ryser [1], Riordan [1, 2] and Hall [2] on two essential points. First, all these books and most others (with the notable exception of Berge [2]) split the material into existence and enumeration results whereas here a unified approach is attempted. Secondly, the idea of interpreting the fundamental combinatorial concepts as certain mappings, which has come to be accepted in recent years, prompted me in several instances to depart from the classical terminology in favor of terms related to mappings. As examples, the reader may compare in the following list the terms used in this text with those employed in, say, Riordan [1].

| | |
|---|---|
| $\text{Inj}(N, R)$ | $n$-permutations of $r$ objects (in the older literature called $n$-variations; the name $n$-permutation was reserved for $n = r$); |
| $\text{Map}(N, R)$ | $n$-permutations of $r$ objects with repetitions or $n$-samples; |
| $\text{Mon}(N, R) \cap \text{Inj}(N, R)$ | $n$-combinations of $r$ objects; |
| $\text{Mon}(N, R)$ | $n$-combinations of $r$ objects with repetitions or $n$-selections. |

Other books on combinatorics that should be consulted are the classic treatise by MacMahon [1] and the recent one by Comtet [1]. As general references on posets and lattices, the reader is referred to Birkhoff [1] and Crawley–Dilworth [1], and on permutation groups, to Wielandt [1].

Chapter II

# Lattices

After having introduced in chapter I the main classes of mappings, let us now study in some depth the various types of lattices encountered there. It seems appropriate to proceed from the most special class of distributive lattices to the more general types, modular, semimodular, and geometric lattices, particularly in view of the fact that each of the three characterizations we shall derive for distributive lattices will lead directly to an important branch of combinatorial theory: *matroids* and *combinatorial geometries* to be discussed in chapters VI and VII, the *inversion calculus* (chapters III and IV), and *transversal theory* (chapter VIII).

As always in combinatorics we have to make some finiteness condition on the structure under consideration. For our purposes the following general assumption seems best suited:

(F) *Any chain between any two elements of a poset is finite.*

Two important types of posets where (F) holds are the *locally finite* posets (all intervals are finite), abbreviated to l.f., and the posets in which *all* chains are finite, abbreviated to c.f. Notice that (F) implies that all intervals are complete lattices if the poset is a lattice. Condition (F) will be assumed throughout the book and it will, in general, not be mentioned explicitly in the statement of a theorem. One other word concerning terminology: We shall often compare lattices by mapping one of them into the other. $N$ is called a *sublattice* of $M$ if $N \subseteq M$ is a lattice and meets and joins in $N$ coincide with those in $M$. Let us say, $L$ can be *embedded* in $M$ if there exists a lattice isomorphism $\phi$ (i.e., $\phi$ is injective and preserves meets and joins) from $L$ onto a sublattice of $M$.

Most of the order relations encountered in chapter I, for instance, inclusion and refinement, were induced by comparing certain sets. This suggests the introduction of the concept of *rank* (generalizing the cardinality of a set or dimension of a subspace) and of *atomicity* (an abstraction of the fact that every set is the union of its one-element subsets or that every space is the union of its one-dimensional subspaces).

**Definition.** A poset $P$ is said to satisfy the *Jordan–Dedekind chain condition* (JD-condition for short) if any two maximal chains between the same elements have the same finite length. If, in particular, all maximal chains with endpoint $a$ have the same finite length, then this common length is called the *rank* $r(a)$ of $a$. Dually we have the concept of *corank* of $a$.

The following proposition is obvious.

**2.1 Proposition.** *Let $P$ be a poset with 0. If $P$ satisfies the* JD-*condition then we have for the rank function $r: P \to \mathbb{N}_0$:*
(i) $r(0) = 0$,
(ii) $a <\cdot\, b \Rightarrow r(b) = r(a) + 1$.

*Conversely, if $P$ admits a function $r: P \to \mathbb{N}_0$ obeying (i) and (ii) then $P$ satisfies the* JD-*condition with $r$ as its rank function.* □

**Definition.** A lattice $L$ is called *atomic* or a *point lattice* if every element $a \in L$ is a supremum of atoms, i.e., $a = \sup\{p \in L: 0 <\cdot\, p \leq a\}$.

To emphasize the geometric origin of many problems, atoms will also be called *points* of the lattice. Thus in a lattice with rank function $r$, points $p$ are characterized by $r(p) = 1$.

## 1. Distributive Lattices

### A. Representation as a Lattice of Ideals

**Definition.** A lattice is called *distributive* if the following identities hold:

$$x \wedge (y \vee z) = (x \wedge y) \vee (x \wedge z),$$

$$x \vee (y \wedge z) = (x \vee y) \wedge (x \vee z).$$

Either identity implies the other and they can be readily generalized to finitely many variables:

$$\bigvee_{i=1}^{m} x_i \wedge \bigvee_{j=1}^{n} y_j = \bigvee_{i,j=1}^{m,n} (x_i \wedge y_j),$$

$$\bigwedge_{i=1}^{m} x_i \vee \bigwedge_{j=1}^{n} y_j = \bigwedge_{i,j=1}^{m,n} (x_i \vee y_j).$$

Notice that since the defining identities are duals of each other the dual lattice of a distributive lattice is again distributive.

**Example.** Every lattice with at most 4 elements is distributive. Of the five lattices containing 5 elements, three are distributive, whereas two are not, shown in Figure 2.1.

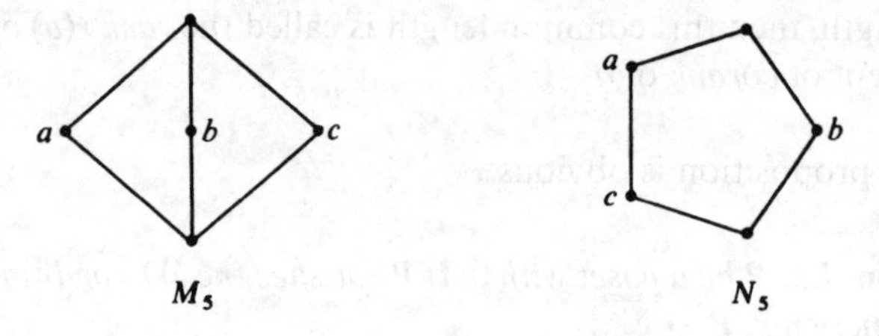


Figure 2.1

It follows immediately from the definition that every sublattice of a distributive lattice is distributive as is the product of distributive lattices. As a chain is trivially distributive it follows that every sublattice of a chain product is distributive. This includes in particular the lattices $2^S$ of all subsets of a set $S$ and the Boolean algebras $\mathscr{B}(n)$, the divisor lattice $\mathscr{T}$, and Young's lattice $\mathscr{Y}$. The main theorem proved in section B concerns the converse: Every distributive lattice is isomorphic to some sublattice of a chain product. To derive this result we need to go into the finite decomposition theory of lattices.

**Definition.** An element $p$ of a lattice $L$ is called *join-irreducible* (or just *irreducible* since with one exception only join-irreducibility will be considered) if for all $x, y \in L$

$$p = x \vee y \quad \text{implies} \quad p = x \quad \text{or} \quad p = y.$$

An expression $a = p_1 \vee \cdots \vee p_k$ where the $p_i$'s are irreducible elements is called a (finite) *decomposition* of $a$. The decomposition is *irredundant* if $a \neq p_1 \vee \cdots \vee p_{i-1} \vee p_{i+1} \vee \cdots \vee p_k$ for all $i$.

Obviously, in any irredundant decomposition the irreducible elements are pairwise incomparable, i.e., they form an antichain. Observe further that in any lattice $L$ with 0, decompositions exist for all $a \in L$ (and hence irredundant decompositions by deleting superfluous elements) because of our general assumption (F). Our main lemma states that in distributive lattices with 0 these irredundant decompositions are unique (up to order).

**2.2 Lemma.** *Let $p$ be an irreducible element of a distributive lattice. Then*

$$p \leq a_1 \vee \cdots \vee a_k \quad \textit{implies} \quad p \leq a_i \quad \textit{for some } i.$$

*Proof.* By distributivity, $p = p \wedge (\bigvee_{i=1}^{k} a_i) = \bigvee_{i=1}^{k} (p \wedge a_i)$ and thus $p = p \wedge a_i$, i.e., $p \leq a_i$ for some $i$, since $p$ is irreducible. □

Every atom is trivially irreducible. Hence if $\{p_1, p_2, \ldots\}$ is a set of atoms then $p_1 < p_1 \vee p_2 < p_1 \vee p_2 \vee p_3 < \cdots$ forms a strictly increasing chain by **2.2.**

Applying this remark to the atoms of an interval we conclude from (F) that any interval of a distributive lattice contains only finitely many atoms and thus by induction is itself finite.

**2.3 Corollary.** *A distributive lattice satisfying* (F) *is locally finite.* □

**2.4 Lemma.** *Let $L$ be a distributive lattice with* 0 *and $P \subseteq L$ the set of irreducible elements other than* 0. *Then any element $a \in L$ possesses a unique irredundant decomposition $a = p_1 \vee \cdots \vee p_k$ where we write $P(a) = \{p_1, \ldots, p_k\}$ with $P(0) = \emptyset$.*

*Proof.* By our earlier remarks only the uniqueness needs to be shown. Suppose

$$a = p_1 \vee \cdots \vee p_k = q_1 \vee \cdots \vee q_l$$

are two irredundant decompositions. For every $i$ we have $p_i \leq q_1 \vee \cdots \vee q_l$ and thus by **2.2** $p_i \leq q_j$ for some $j$. Again by **2.2** $q_j \leq p_h$ for some $h$, hence $p_i = q_j$ since the $p_i$'s are pairwise incomparable. Repeating this argument we conclude that the sets $\{p_1, \ldots, p_k\}$, $\{q_1, \ldots, q_l\}$ are in fact identical. □

Every element $a$ of a distributive lattice $L$ corresponds therefore to a unique subset (in fact an antichain) $P(a)$ of the poset $P$. The following theorem shows how to recover $L$ from the subposet $P$.

**Definition.** A subset $I$ of a poset $P$ is called an *ideal* of $P$ if for all $x, y \in P$

$$x \in I \quad \text{and} \quad y \leq x \quad \text{imply} \quad y \in I.$$

For $A \subseteq P$ the set $\{x \in L: x \leq a \text{ for some } a \in A\}$ is an ideal, the *ideal generated by $A$*. Ideals that are generated by single elements are called *principal ideals.*

The dual concept of ideal is called a *filter* of $P$.

Actually, what we just defined is usually called an order ideal. Since this is the only type of ideal (or filter) considered in this book we shall use the shorter term. The set-theoretic union of ideals is again an ideal, as is the intersection. Hence the family $\mathfrak{I}(P)$ of ideals of $P$ forms, under inclusion, a sublattice of $2^P$ and is therefore distributive. Clearly, the irreducible elements of $\mathfrak{I}(P)$ are precisely the principal ideals of $P$ (and $\emptyset$). Another sublattice is the lattice $\mathfrak{I}_f(P)$ of *finite* ideals. Again $\mathfrak{I}_f(P)$ is distributive, and the following first characterization of distributive lattices asserts that every distributive lattice obeying (F) is of this form.

**2.5 Theorem** (Birkhoff). *Let $L$ be a distributive lattice with* 0 *and $P \subseteq L$ the subposet of its irreducible elements $\neq 0$. Then $L \cong \mathfrak{I}_f(P)$ by means of the lattice isomorphism*

$$\phi: a \to I(a) = \{p \in P: p \leq a\} \qquad (a \in L).$$

*Conversely, every lattice* $\mathfrak{I}_f(P)$ *is distributive, with its poset of irreducible elements* $\neq 0$ *isomorphic to P. If, in particular, L is finite then*

$$L \cong \mathfrak{I}(P).$$

*Proof.* As every element of $L$ is determined by the irreducible elements it dominates, $\phi$ is injective. Let $I \in \mathfrak{I}_f(P)$ with $b = \sup I$. Then $I \subseteq I(b)$. But for any $p \in I(b)$ we have $p \leq \sup I$, hence $p \in I$ by **2.2**. Consequently, every finite ideal of $P$ is of the form $I(a)$, i.e., $\phi$ is a bijection. Suppose $a \leq b$ in $L$, then clearly $I(a) \subseteq I(b)$. Conversely, $I(a) \subseteq I(b)$ implies $a = \sup I(a) \leq \sup I(b) = b$, thus $\phi$ is a lattice isomorphism. In the distributive lattice $\mathfrak{I}_f(P)$ an element, i.e., an ideal, is irreducible if and only if it is empty or a principal ideal $I(a)$, $a \in P$. □

**2.6 Corollary.** *Two distributive lattices are isomorphic if and only if their posets of irreducible elements are isomorphic.* □

**2.6** has an interesting corollary. The dual concept of join-irreducible is *meet-irreducible*; i.e., $p$ is meet-irreducible if $p = x \wedge y$ implies $p = x$ or $p = y$ for all $x$, $y$.

**2.7 Corollary.** *In a finite distributive lattice L the subposets P of join-irreducible elements* $\neq 0$ *and Q of meet-irreducible elements* $\neq 1$ *are isomorphic.*

*Proof.* Let $L^*$ be the dual lattice of $L$. By the dual statement of **2.5**, $L^* \cong \mathfrak{F}(Q)$ where $\mathfrak{F}(Q)$ is the lattice of filters of $Q$ ordered by inclusion. Now clearly, $\phi: \mathfrak{I}(Q) \to \mathfrak{F}(Q)$ with $\phi(A) = Q - A$ is an anti-isomorphism between $\mathfrak{I}(Q)$ and $\mathfrak{F}(Q)$. Hence it follows that $\mathfrak{I}(Q) \cong L \cong \mathfrak{I}(P)$, and thus $Q \cong P$ by **2.6**. □

**Examples.** The lattice $L$ of Figure 2.2 is distributive with the join-irreducible elements $\neq 0$ circled and the meet-irreducible elements $\neq 1$ squared.

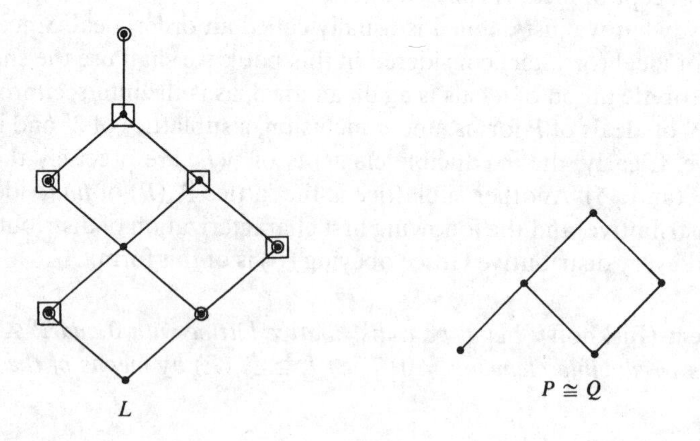


Figure 2.2

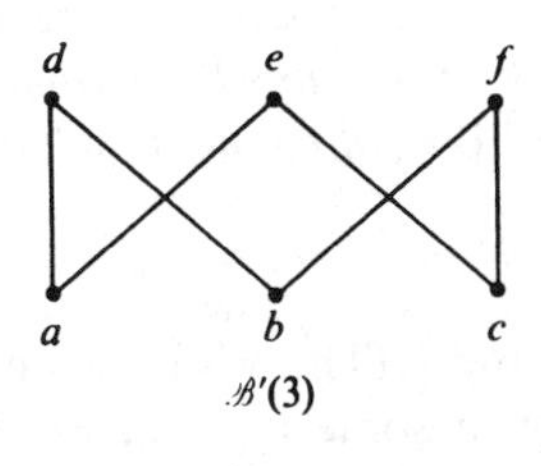


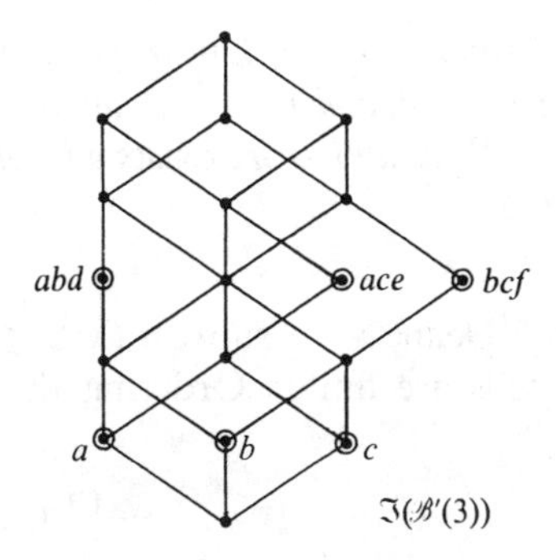


Figure 2.3

Ideals of the lattice $2^S$ are called *simplicial complexes*. By definition, they are families $\mathfrak{K} \subseteq 2^S$ such that $A \in \mathfrak{K}, B \subseteq A$ imply $B \in \mathfrak{K}$. A principal simplicial complex is called a *simplex*. Let $\mathscr{B}'(n) = \mathscr{B}(n) - \{0, 1\}$. Then it can be shown that $\mathfrak{J}(\mathscr{B}'(n))$ is isomorphic to the free distributive lattice with $n$ generators (cf. Birkhoff [1, p. 61]). The lattice $\mathfrak{J}(\mathscr{B}'(3))$ is illustrated in Figure 2.3.

## B. Chain Products and Coding

Let us return to theorem **2.5**. As the order relation in $\mathfrak{J}_f(P)$ is ordinary set-inclusion it follows that any distributive lattice is isomorphic to a sublattice of the lattice $\mathscr{B}(P)$ of all finite subsets of a set $P$. Since $\mathscr{B}(P)$ is itself a chain product we have proved our main result.

**2.8 Theorem.** *Any distributive lattice with* 0 *is isomorphic to a sublattice of a Boolean algebra and hence, in particular, to a sublattice of a chain product. Conversely, every sublattice of a chain product is distributive.* □

The two characterizations **2.5** and **2.8** suggest interesting further developments. How much can the uniqueness of decompositions in **2.4** be retained in more general classes of lattices? In attempting to answer this question we are led in a natural way to the concepts of dimension, basis and independence in modular and semimodular lattices to be discussed in chapter VI. Theorem **2.8**, on the other hand, gives rise to a very interesting extremal problem. Suppose $L$ is a finite distributive lattice. How many chains $C_i$ are necessary to assure an embedding of $L$ in the product $\prod_i C_i$? Without loss of generality we may take $C_i = \{0 < 1 < \cdots < c_i\}$, $c_i \in \mathbb{N}$, for all $i$. An embedding $\phi: L \to \prod_{i=1}^d C_i$ is called a *coding* of $L$ and $d$ the *dimension* of the coding. Thus we inquire about the *minimal dimension* among all possible codings of $L$. As an example, consider a finite Boolean algebra $\mathscr{B}(S)$. Clearly, $\phi: \mathscr{B}(S) \to \{0, 1\}^n$ mapping each subset onto its characteristic vector, is a coding of $\mathscr{B}(S)$ of dimension $n$. But it is plausible at first sight that we may be able to lower the dimension by allowing longer chains. That this cannot be done will follow from the next proposition.

**2.9 Proposition.** *Let $L$ be a distributive lattice with 0, $P$ the subposet of its irreducible elements $\neq 0$, and $P = \dot{\bigcup} P_i$ an arbitrary partition of $P$ into chains $P_i$. Setting $C_i = P_i \cup \{0\}$ for all $i$, there exists an isomorphism $\phi: L \to \prod_i C_i$ of $L$ onto a sublattice of $\prod_i C_i$.*

*Proof.* Define $x_i := \sup\{z \in C_i : z \leq x\}$ for all $x \in L$ and $i$. (The $x_i$'s exist since all intervals are finite.) Ordering the index set totally in some fashion, we define

$$\phi: L \to \prod_i C_i, \qquad \phi(x) = (\ldots, x_i, \ldots).$$

Notice that $x_i = 0$ for all but finitely many $i$ as $L$ is locally finite. Suppose $\phi(x) = \phi(y)$, i.e., $x_i = y_i$ for all $i$. Then $I(x) = I(y)$ and thus $x = y$. By **2.2**

$$\begin{aligned}(x \vee y)_i &= \sup\{z \in C_i : z \leq x \vee y\} = \sup\{z \in C_i : z \leq x \text{ or } z \leq y\} \\ &= x_i \vee y_i,\end{aligned}$$

hence $\phi(x \vee y) = \phi(x) \vee \phi(y)$, and similarly $\phi(x \wedge y) = \phi(x) \wedge \phi(y)$. $\square$

**Example.** The irreducible elements $\neq 0$ in the lattice of Figure 2.4 are $P = \{a, b, d, e, h\}$. Partitioning $P$ into the chains

$$\begin{array}{c} e \\ | \\ a \end{array} \qquad \begin{array}{c} d \\ | \\ b \end{array} \qquad \cdot\, h$$

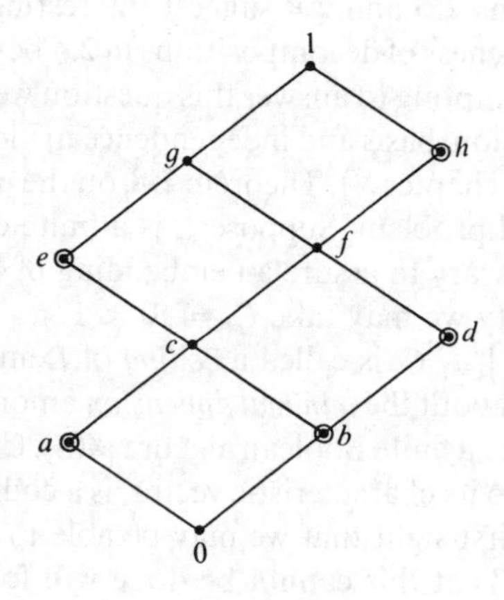


Figure 2.4

we obtain the coding:

$$\begin{array}{ll} 0 \to (0,0,0) & e \to (2,1,0) \\ a \to (1,0,0) & f \to (1,2,0) \\ b \to (0,1,0) & g \to (2,2,0) \\ c \to (1,1,0) & h \to (1,2,1) \\ d \to (0,2,0) & 1 \to (2,2,1). \end{array}$$

If instead we choose the partition

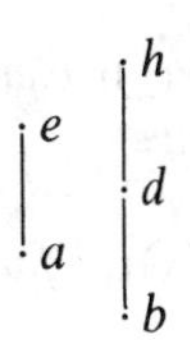

$$\begin{array}{cc} & h \\ e & \\ & d \\ a & \\ & b \end{array}$$

an optimal coding results:

$$\begin{array}{ll} 0 \to (0,0) & e \to (2,1) \\ a \to (1,0) & f \to (1,2) \\ b \to (0,1) & g \to (2,2) \\ c \to (1,1) & h \to (1,3) \\ d \to (0,2) & 1 \to (2,3). \end{array}$$

**2.9** tells us that *any* partition of $P$ into chains yields a coding of the lattice $L$. On the other hand, it is plain that any coding $\phi: L \to \prod_{i=1}^{d} C_i$ induces a partition of $P$ into chains (by running through each coordinate). Hence we have reduced the question as to the minimal dimension of a coding to the following problem in arbitrary finite posets: What is the minimum number $d(P)$ of disjoint chains into which a poset $P$ can be decomposed? $d(P)$ is called the *Dilworth number* of $P$. Quite obviously,

$$d(P) \geq \max_{A \text{ antichain}} |A|,$$

since any two elements of an antichain must appear in different chains. That, in fact, equality holds is one of the fundamental results in all of combinatorics and one of the starting points for the large and still growing field of *transversal theory* to be discussed in chapter VIII.

To return to the example $\mathscr{B}(S)$ notice that since $P = S$ is itself an antichain, we must have $d(\mathscr{B}(S)) \geq |S| = n$ for any coding; this means that $n$ is in fact the minimal possible dimension.

## C. The Rank Function

Our third characterization of distributive lattices involves the rank function. We will first prove the existence of a rank function in any distributive lattice with 0 using the representation theorem **2.5** and then characterize distributivity by means of certain regularity conditions on the rank. We use the notation $P(a)$, $I(a)$ and $\mathfrak{J}_f(P)$ as in section A.

**2.10 Proposition.** *Let $L$ be a distributive lattice with 0 and $x \to I(x)$, $x \in L$, the isomorphism in* **2.5**. *Then*

$$x <\cdot\, y \Leftrightarrow I(y) = I(x) \cup \{p\} \text{ for some } p \in P(y) - P(x).$$

*It follows that $y$ covers precisely $|P(y)|$ elements in $L$, for all $y \in L$. Further, $L$ possesses a rank function $r$, and we have*

$$r(x) = |I(x)| \quad \text{for all } x \in L.$$

*Proof.* If we delete from $I(y)$ an arbitrary maximal element $p$, i.e., $p \in P(y)$, then $I(y) - \{p\}$ is again in $\mathfrak{J}_f(P)$. Suppose conversely $x <\cdot\, y$ and $p \in I(y) - I(x)$. Then $p \leq q$ for some $q \in P(y)$, hence $I(x) \subseteq I(y) - \{q\}$ and thus $p = q$, $I(x) = I(y) - \{p\}$. To verify the last claim notice first $r(0) = 0 = |\varnothing|$. Now, for an arbitrary element $x \neq 0$ we have just shown that every maximal chain

$$0 <\cdot\, x_1 <\cdot\, x_2 <\cdot \cdots <\cdot\, x$$

corresponds to a finite chain of ideals

$$\varnothing <\cdot\, I(x_1) <\cdot\, I(x_2) <\cdot \cdots <\cdot\, I(x)$$

where the cardinality of the sets $I(x_i)$ increases by 1 at each step. □

**2.11 Corollary.** *For any finite distributive lattice $L$ we have $r(L) = |P|$, where $P$ is the set of irreducible elements $\neq 0$. $L$ is isomorphic to a sublattice of the Boolean algebra $\mathscr{B}(n)$, $n = |P|$, and $n$ is the smallest such integer.* □

The simple formula for the rank in **2.10** suggests the use of the cardinality function in deriving further properties of the rank.

**2.12** *Let $A_1, \ldots, A_t$ be finite subsets of a set $S$. Then*

$$\left|\bigcup_{i=1}^{t} A_i\right| = \sum_{i=1}^{t} |A_i| - \sum_{i<j}^{t} |A_i \cap A_j|$$
$$+ \sum_{i<j<k}^{t} |A_i \cap A_j \cap A_k| \mp \cdots + (-1)^{t-1} |A_1 \cap \cdots \cap A_t|.$$

*Proof.* Induction on $t$. □

Formula **2.12** leads to the following classification of lattices with rank function.

**Definition.** Let $L$ be a lattice with a rank function $r$. $r$ is said to possess the *regularity property* $(R_t)$, $t \geq 2$, if for all $x_1, \ldots, x_t \in L$:

$$(R_t) \qquad r(x_1 \vee \cdots \vee x_t) = \sum_{i=1}^{t} r(x_i) - \sum_{i<j}^{t} r(x_i \wedge x_j) \pm \cdots + (-1)^{t-1} r(x_1 \wedge \cdots \wedge x_t).$$

Observe that $(R_t)$ implies $(R_{t-1})$ and thus $(R_i)$ for all $i \leq t$. One just has to set $x_t = x_{t-1}$ and cancel superfluous summands. The next theorem will show conversely that $(R_3)$ implies $(R_t)$ for all $t \geq 3$. Hence we are left with the possibilities $(R_3)$ or $(R_2)$; these conditions characterize, in turn, *distributive* and *modular* lattices.

**2.13 Theorem.** *Let $L$ be a lattice with 0. $L$ is distributive if and only if it possesses a rank function $r$ satisfying $(R_3)$. In particular, $(R_3)$ implies $(R_t)$ for all $t$.*

*Proof.* If $L$ is distributive then the representation $L \cong \mathfrak{J}_f(P)$, **2.10**, and **2.12** imply that $r$ satisfies $(R_t)$ for all $t$. Suppose, conversely, that $L$ possesses a rank function $r$ obeying $(R_3)$. Since the inequality $x \wedge (y \vee z) \geq (x \wedge y) \vee (x \wedge z)$ holds in any lattice it suffices to show that for all $x, y, z \in L$

$$r(x \wedge (y \vee z)) = r((x \wedge y) \vee (x \wedge z)).$$

Now

$$\begin{aligned} r(x \wedge (y \vee z)) &= r(x) + r(y \vee z) - r(x \vee y \vee z) && (R_2)\\ &= r(x) + r(y) + r(z) - r(y \wedge z) - r(x \vee y \vee z) && (R_2)\\ &= r(x \wedge y) + r(x \wedge z) - r(x \wedge y \wedge z) && (R_3)\\ &= r((x \wedge y) \vee (x \wedge z)). \quad \square && (R_2) \end{aligned}$$

**2.14.** Let us illustrate the theorems of this section by looking at the lattices introduced in chapter I.

(i) In a chain all elements are irreducible, a property which clearly characterizes chains.
(ii) The irreducible elements $\neq 0$ of the lattices $2^S$ or $\mathscr{B}(S)$ are the one-element subsets, hence $r(A) = |A|$ for all $A \subseteq S$.
(iii) In $\mathscr{T}$ the irreducible elements $\neq 0$ are the prime powers $p^k$. For $n \in \mathbb{N}$ with prime factorization $n = p_1^{k_1} \ldots p_t^{k_t}$ we have $P(n) = \{p_1^{k_1}, \ldots, p_t^{k_t}\}$, $I(n) = \{p_i^{h_i} : 1 \leq h_i \leq k_i, i = 1, \ldots, t\}$ and thus $r(n) = \sum_{i=1}^{t} k_i$.

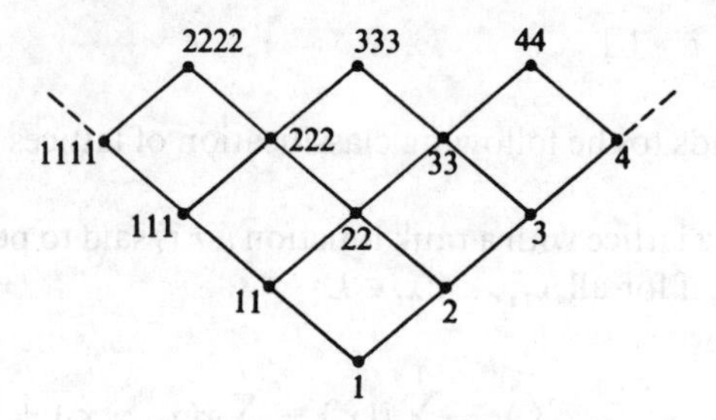


Figure 2.5

(iv) As a less trivial example consider the Young lattice $\mathcal{Y}$ introduced in section I.2. Here it is readily seen that precisely the sequences $n_1 n_2 \cdots n_t$ with all parts equal $n_1 = n_2 = \cdots = n_t$ are irreducible in $\mathcal{Y}$. (Proof?) For instance, the unique irredundant decomposition of $76554111 = 7 \vee 66 \vee 5555 \vee 44444 \vee 1111111$. For a sequence $n_1 n_2 \ldots n_t$ we therefore obtain

$$I(n_1 n_2 \ldots n_t) = \{\underbrace{ll \ldots l}_{i}: 1 \leq l \leq n_i, i = 1, \ldots, t\}$$

and hence $r(n_1 n_2 \ldots n_t) = \sum_{i=1}^{t} n_i$.

The poset $P$ of irreducible elements $\neq 0$ in $\mathcal{Y}$ is clearly isomorphic to $\mathbb{N}^2$ by means of $\phi: P \to \mathbb{N}^2$, $\phi(\underbrace{jj \ldots j}_{i}) = (i, j)$ (see Figure 2.5), from which we infer the important formula $\mathcal{Y} \cong \mathfrak{J}_f(\mathbb{N}^2)$. $\phi$ can be visualized by assigning to each partition $n_1 n_2 \ldots n_t$ a rectangular array containing $n_1$ unit boxes in the first row, $n_2$ boxes in the second row, ..., $n_t$ boxes in the $t$-th row. Denoting these boxes by $(1, 1), \ldots, (1, n_1); (2, 1), \ldots, (2, n_2); \ldots; (t, 1), \ldots, (t, n_t)$ we obtain precisely the ideal of $\mathbb{N}^2$ corresponding to $n_1 n_2 \ldots n_t$.

For the example 76554111 from above the corresponding array is given by Figure 2.6. The resulting array is called the *Ferrers diagram* of the partition. These diagrams will play an important role in the enumeration of partitions (see chapter III).

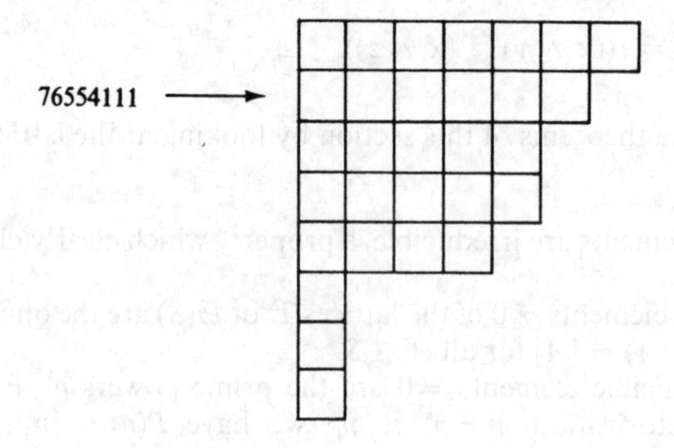


Figure 2.6

EXERCISES II.1

1. Show that a lattice $L$ is distributive if and only if $(x \wedge y) \vee (y \wedge z) \vee (z \wedge x) = (x \vee y) \wedge (y \vee z) \wedge (z \vee x)$ for all $x, y, z \in L$.

→ 2. Let $L$ be a finite distributive lattice and $P, P(x)$ be defined as in **2.4**. Let $\mathfrak{A}(P)$ be the family of antichains in $P$. Show that
(i) $\mathfrak{A}(P)$ consists of precisely the sets $P(x)$;
(ii) $A \to \sup A, x \to P(x)$ is a bijection between $\mathfrak{A}(P)$ and $L$.

→ 3. Prove that in a finite distributive lattice, for every $n$, the number of elements which cover $n$ elements equals the number of elements which are covered by $n$ elements. (Hint: Use **2.7** and ex. 2.)

4. Show that $\text{Aut } P \cong \text{Aut } \mathfrak{J}(P)$ for every finite poset $P$. Is this also true for infinite posets?

→ 5. Prove: A finite lattice $L$ is distributive if and only if $L$ has a rank function and $r(L) = |P| = |Q|$ where $P$ and $Q$ are defined as in **2.7**.

6. Let $L$ be a lattice. A *lattice polynomial* is an expression which contains only lattice elements, the operation symbols $\wedge$ and $\vee$ and parentheses. Let $a_1, \ldots, a_n \in L$. Show that the sublattice $L[a_1, \ldots, a_n]$ generated by the $a_i$'s (i.e., the smallest sublattice of $L$ containing $a_1, \ldots, a_n$) consists of all polynomials in $a_1, \ldots, a_n$. Show further that $L[a_1, a_2]$ contains at most 4 elements.

7.* Let $L$ be distributive.
(i) Prove that $L[a_1, a_2, a_3]$ contains at most 18 elements.
(ii) Prove that $\mathfrak{J}(\mathscr{B}'(3))$ is generated by 3 elements and has precisely 18 elements. (Hint: The generators are the maximal irreducible elements.)
(iii) Let $x_1, x_2, x_3$ be the generators of $\mathfrak{J}(\mathscr{B}'(3))$ and $a_1, a_2, a_3$ arbitrary elements of some distributive lattice $L$. Show that the mapping $\phi$: $x_i \to a_i$ can always be extended to a lattice homomorphism from $\mathfrak{J}(\mathscr{B}'(3))$ to $L$, hence that $\mathfrak{J}(\mathscr{B}'(3))$ is the *free distributive lattice* $FD_3$ with 3 generators.

8.* Generalize ex. 7: $FD_n \cong \mathfrak{J}(\mathscr{B}'(n))$.

9. Show:
(i) $n$ even $\Rightarrow |FD_n|$ even;
(ii)

$$\sum_{k=0}^{n} 2^{\binom{n}{k}} - (n+2) \leq |FD_n|.$$

→10. Let $n = n_1 n_2 \ldots n_t \in \mathscr{Y}$, where $\mathscr{Y}$ is the Young lattice. Show that $n$ is irreducible if and only if $n_1 = n_2 = \cdots = n_t$.

→11. Show that $\mathscr{Y}$ is the only locally finite distributive lattice with 0 for which, for all $n$, an element covers $n$ elements if and only if it is covered by $n + 1$ elements.

→12. Generalize the coding theorem to arbitrary locally finite lattices with 0 as follows. An element $a \in L$ is called *prime* if $a \leq x \vee y$ implies $a \leq x$ or $a \leq y$ for all $x, y \in L$. Prove:
(i) Every prime element is (supremum-) irreducible.
(ii) Let $Q := \{z \in L : z \text{ prime}, \; z \neq 0\}$, $Q(x) := \{z \in Q : z \leq x\}$ and let $Q = \dot{\bigcup}_i Q_i$ be a decomposition into disjoint chains, $D_i := Q_i \cup \{0\}$. Then there is a lattice homomorphism $\psi : L \to \prod_{i \in I} D_i$ such that $\psi(x) = \psi(y)$ if and only if $Q(x) = Q(y)$.
(iii) The following conditions are equivalent: (a) $L$ is distributive; (b) Every irreducible element is prime; (c) $Q(x) \neq Q(y)$ for all $x \neq y \in L$.
(Ky Fan)

13. Let $L$ be a lattice with 0, and $R$ a commutative ring with unity. A function $w : L \to R$ is called a *valuation* of $L$ in $R$ if $w(x \wedge y) + w(x \vee y) = w(x) + w(y)$ for all $x, y \in L$. Let $L$ be distributive and $w$ a valuation. Then $w(x_1 \vee \cdots \vee x_t) = \sum_{i=1}^{t} w(x_i) - \sum_{i<j}^{t} w(x_i \wedge x_j) \pm \cdots + (-1)^{t-1} w(x_1 \wedge \cdots \wedge x_t)$.

→14. Show that a valuation $w$ on a distributive lattice $L$ with 0 is uniquely determined by its values on the set of irreducible elements and that, conversely, any function $w : P \cup \{0\} \to R$ can be extended to a valuation. (Hint: Use induction on the rank.)

→15. The *characteristic* $\chi(L)$ of a distributive lattice $L$ is the valuation in $\mathbb{Z}$ given by $\chi(0) = 0$ and $\chi(p) = 1$ for all $p \in P$. Verify:
(i) $L = \mathscr{C}(n) : \chi(x) = 1$ for all $x \neq 0$;
(ii) $L = \mathscr{T} : \chi(p_1^{k_1} \ldots p_t^{k_t}) = t$;
(iii) $L = \mathscr{B}(n) : \chi(A) = ?$

## 2. Modular and Semimodular Lattices

### A. Modular Lattices

**Definition.** A lattice is called *modular* if for all elements $a, b, c$:

$$c \leq a \quad \text{implies} \quad a \wedge (b \vee c) = (a \wedge b) \vee c.$$

Every distributive lattice is modular since the first distributive identity reduces to the definition of modularity in the presence of $z \leq x$. The dual lattice of a modular lattice is modular, as are sublattices and direct products of modular lattices. Not every modular lattice is distributive, however, the smallest example being the lattice $N_5$ in Figure 2.1. Hence the question arises of how to weaken the characterizations of distributive lattices given in the last section so as to describe modular lattices. With the rank, it turns out that $(R_2)$ instead of $(R_3)$ provides the correct answer. The many similarities between Boolean algebras and vector space lattices suggest that embeddings of modular lattices into the subgroup lattice of a suitable abelian group may be the right generalization of **2.8**. This is indeed so for a restricted class of modular lattices, namely complemented Arguesian lattices (see Crawley–Dilworth [1, p. 124]), but unfortunately not every modular lattice can be embedded in a lattice of this type. We shall return to these questions in section 3.

The classical source of examples for modular lattices comes from algebraic systems like groups, rings, or vector spaces. The reader may verify that the lattice of normal subgroups of a group is modular as are the lattices of ideals of a ring and the vector space lattices $\mathscr{L}(V)$.

If in the definition we choose $c \in [a \wedge b, a]$ we obtain $(c \vee b) \wedge a = c$. The following lemma shows that we can, in fact, restrict ourselves to such triples $\{a, b, c\}$.

**2.15 Lemma.** *A lattice $L$ is modular if and only if for all $a, b \in L$ and all $z \in [a \wedge b, a]$ we have $(z \vee b) \wedge a = z$, or, equivalently, $(w \wedge a) \vee b = w$ for all $w \in [b, a \vee b]$.*

*Proof.* Since in the definition of modularity $\geq$ holds in any lattice, it suffices to verify $z \vee (a \wedge b) \geq (z \vee b) \wedge a$ for all $a, b, z \in L$, $z \leq a$. Suppose $a \wedge b \nleq z$, $z \leq a$. Then $z \vee (a \wedge b) \in [a \wedge b, a]$ and thus by hypothesis $z \vee (a \wedge b) = [(z \vee (a \wedge b)) \vee b] \wedge a \geq (z \vee b) \wedge a$. □

**2.15** suggests the use of the following monotone functions $\phi_b$, $\psi_a$ on $L$:

$$\phi_b(z) = z \vee b, \qquad \psi_a(w) = w \wedge a.$$

Notice that for any lattice $L$ and any $a, b \in L$:

**2.16.** *$\phi_b \psi_a \phi_b = \phi_b$ on $[a \wedge b, a]$ and $\psi_a \phi_b \psi_a = \psi_a$ on $[b, a \vee b]$, and thus*

*$\phi_b$ maps $[a \wedge b, a]$ injectively into $[b, a \vee b] \Leftrightarrow (z \vee b) \wedge a = z$, for all $z \in [a \wedge b, a]$,*

*$\psi_a$ maps $[b, a \vee b]$ injectively into $[a \wedge b, a] \Leftrightarrow (w \wedge a) \vee b = w$, for all $w \in [b, a \vee b]$.*

Recast in terms of the functions $\phi_b$ and $\psi_a$, **2.15** reads as follows:

**2.17.** *A lattice $L$ is modular if and only if $\phi_b$, $\psi_a$ are inverse lattice isomorphisms when restricted to $[a \wedge b, a]$ and $[b, a \vee b]$ respectively, for all $a, b \in L$.* □

The function $\phi_b$ transposes so to speak the interval $[a \wedge b, a]$ up to $[b, a \vee b]$, and similarly $\psi_a$ transposes $[b, a \vee b]$ down to $[a \wedge b, a]$ (see Figure 2.7). Hence we call two intervals $I$, $J$ *transposes* of each other if there exist $a, b \in L$ such that $I = [a \wedge b, a]$ and $J = [b, a \vee b]$. $I$, $J$ are called *projective* if there is a finite sequence $I = I_0, I_1, \ldots, I_t = J$ with $I_{i-1}$, $I_i$ transposes of each other for all $i$. Hence we have the corollary.

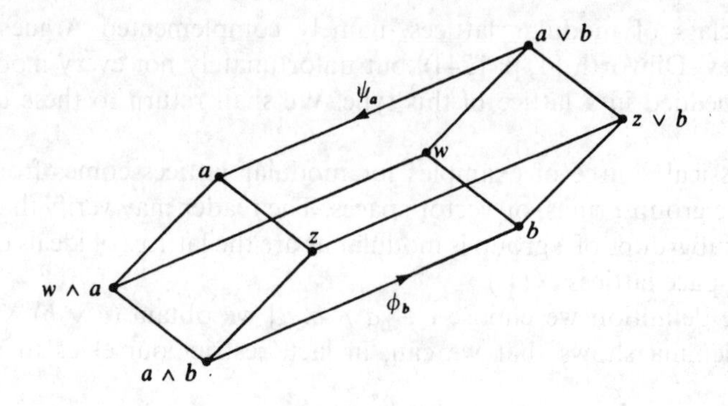


Figure 2.7

**2.18 Corollary.** *Projective intervals of a modular lattice are isomorphic.* □

Since by hypothesis (F) all chains between two elements are finite we can state **2.17** in the following concise form.

**2.19 Proposition.** *A lattice $L$ is modular if and only if for all $a, b \in L$:*

(i) $a \wedge b \lessdot a \Rightarrow b \lessdot a \vee b$,
(ii) $b \lessdot a \vee b \Rightarrow a \wedge b \lessdot a$.

Equivalently, $L$ is modular if and only if for all $a, b \in L$:

(i′) $a \wedge b \lessdot a, b \Rightarrow a, b \lessdot a \vee b$,
(ii′) $a, b \lessdot a \vee b \Rightarrow a \wedge b \lessdot a, b$. □

Let us return once more to the smallest non-distributive lattices $M_5$ and $N_5$ in Figure 2.1. $M_5$ is modular whereas $N_5$ is not. Surprisingly, the absence of these sublattices already characterizes, in turn, modularity and distributivity.

**2.20 Proposition.** *A lattice is modular if and only if it does not contain a sublattice isomorphic to $N_5$.*

*Proof.* Only the sufficiency needs to be shown. Let $L$ be non-modular. Then by **2.15** there are elements $a, b, c$ with $c \in [a \wedge b, a]$ and $(c \vee b) \wedge a > c$. We conclude that $a \wedge b, b, c, (c \vee b) \wedge a$ and $c \vee b$ are five distinct elements generating a sublattice isomorphic to $N_5$ (Figure 2.8). □

**2.21 Proposition.** *A lattice is distributive if and only if it does not contain a sublattice isomorphic to either $M_5$ or $N_5$.*

*Proof.* In view of **2.20** all we need to show is that every modular non-distributive lattice contains a sublattice isomorphic to $M_5$. Choose $x, y, z$ with $x \wedge (y \vee z) > (x \wedge y) \vee (z \wedge x)$. It readily follows that also $y \wedge (z \vee x) > (y \wedge z) \vee (x \wedge y)$

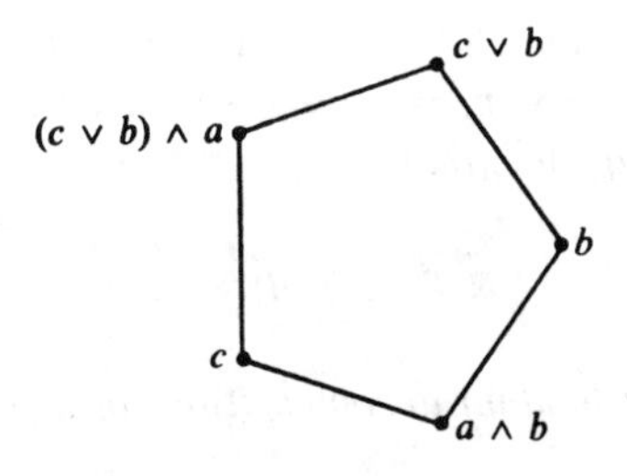


Figure 2.8

and $z \wedge (x \vee y) > (z \wedge x) \vee (y \wedge z)$. Applying the maps $\phi_{y \wedge z}$, $\phi_{z \wedge x}$ and $\phi_{x \wedge y}$ we obtain by modularity $d < a, b, c < e$ where

$$a = [x \wedge (y \vee z)] \vee (y \wedge z) \qquad d = (x \wedge y) \vee (y \wedge z) \vee (z \wedge x)$$
$$b = [y \wedge (z \vee x)] \vee (z \wedge x) \qquad e = (x \vee y) \wedge (y \vee z) \wedge (z \vee x).$$
$$c = [z \wedge (x \vee y)] \vee (x \wedge y)$$

Using the modular law it is easily shown that $\{a, b, c, d, e\}$ generates a sublattice isomorphic to $M_5$. □

As a consequence of the last two propositions we can derive the following useful characterizations of distributivity and modularity.

**2.22 Corollary.**

(i) *A lattice is distributive if and only if for all a, b, c:*

$$a \wedge c = b \wedge c, a \vee c = b \vee c \Rightarrow a = b.$$

(ii) *A lattice is modular if and only if for all a, b, c:*

$$b \leq a, a \wedge c = b \wedge c, a \vee c = b \vee c \Rightarrow a = b. \quad \square$$

Let us look for an analogue of the decomposition theorem **2.4** for modular lattices. Let $L$ be a modular lattice with 0 and $P$ the set of irreducible elements $\neq 0$. Hypothesis (F) implies the existence of finite decompositions $a = p_1 \vee \cdots \vee p_t$ for any $a \in L$, and **2.4** asserts the uniqueness of irredundant decompositions if $L$ is distributive. For arbitrary modular lattices this uniqueness property is no longer valid, but we can prove by means of the following exchange principle that the *number of elements* used in any irredundant decomposition is the same.

**2.23 Proposition.** *Let $L$ be a modular lattice with 0 and $P \subseteq L$ the set of irreducible elements $\neq 0$. If $a = p_1 \vee \cdots \vee p_s = q_1 \vee \cdots \vee q_t$ are two decompositions of $a \in L$, then for each $p_i$ there is a $q_j$ such that*

$$a = p_1 \vee \cdots \vee p_{i-1} \vee q_j \vee p_{i+1} \vee \cdots \vee p_s.$$

*In particular, any two irredundant decompositions of a contain the same number of elements.*

*Proof.* Set $\overline{p_i} = p_1 \vee \cdots \vee p_{i-1} \vee p_{i+1} \vee \cdots \vee p_s$ and $r_j = \overline{p_i} \vee q_j$ for $j = 1, \ldots, t$. As $[p_i \wedge \overline{p_i}, p_i] \cong [\overline{p_i}, p_i \vee \overline{p_i} = a]$ and $p_i$ is irreducible in $[p_i \wedge \overline{p_i}, p_i]$ we conclude that $a$ is irreducible in $[\overline{p_i}, a]$. Since $\overline{p_i} \leq r_j \leq a$, $q_j \leq r_j$, we have $a = q_1 \vee \cdots \vee q_t \leq r_1 \vee \cdots \vee r_t \leq a$, thus $a = r_1 \vee \cdots \vee r_t$. But $r_1, \ldots, r_t \in [\overline{p_i}, a]$, hence $a = r_j = p_1 \vee \cdots \vee p_{i-1} \vee q_j \vee p_{i+1} \vee \cdots \vee p_s$ for some $j$. □

**Example.** In the lattice $M_5$, $P$ is just the set of atoms and any of the three pairs of atoms gives an irredundant decomposition of 1. In the lattice of Figure 2.9, $P = \{a, b, c, d\}$ and $1 = a \vee b \vee c = a \vee d$ are two irredundant decompositions of 1. Hence the lattice is not modular.

## B. Semimodular Lattices

By weakening the conditions in **2.19** we are led to the concept of semimodularity.

**Definition.** A lattice is called *semimodular* if for all $a, b$

$$a \wedge b <\cdot\, a \Rightarrow b <\cdot\, a \vee b.$$

A lattice is called *lower semimodular* if dually for all $a, b$

$$b <\cdot\, a \vee b \Rightarrow a \wedge b <\cdot\, a.$$

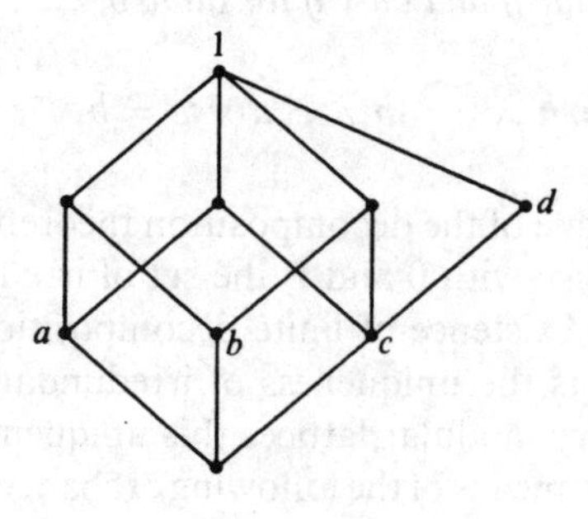


Figure 2.9

Hence by **2.19** a lattice is modular if and only if it is both semimodular and lower semimodular. Using the equivalent statement in **2.19** we have:

**2.24 Proposition.** *A lattice is semimodular if and only if for all a, b*

$$a \wedge b \lessdot a, b \Rightarrow a, b \lessdot a \vee b.$$

*The dual proposition holds for lower semimodular lattices.* □

**2.25 Corollary.** *Let L be a semimodular lattice. Then*

$$x \lessdot y \Rightarrow x \vee z = y \vee z \text{ or } x \vee z \lessdot y \vee z \text{ for all } x, y, z \in L.$$

*In particular, for any atom p and* $a \in L$, $p \not\leq a$, *we have* $a \lessdot a \vee p$. □

The class of semimodular lattices provides the right framework for a large number of algebraic-combinatorial problems in a lattice-theoretic setting. In particular, we shall prove an analogue of the exchange property **2.23** for arbitrary semimodular lattices, thus furnishing the fundamentals for a general theory of basis, dimension and independence.

Another source of examples apart from linear algebra is group theory. Let $G$ be a finite $p$-group (i.e., every element $\neq 1$ has order equal to a power of the prime $p$), and $A$, $B$ be two subgroups of $G$ with $A, B \lessdot A \vee B$ in the subgroup lattice $\mathscr{U}(G)$. Then index $[A \vee B : A] = [A \vee B : B] = p$, and thus $[A : A \wedge B] \leq [A \vee B : B] = p$, i.e., $[A : A \wedge B] = p$. Similarly $[B : A \wedge B] = p$, from which $A \wedge B \lessdot A, B$ follows. $\mathscr{U}(G)$ is therefore lower semimodular. The reader may extend this result to arbitrary finite nilpotent groups and construct an example of a nilpotent group $G$ whose lattice $\mathscr{U}(G)$ is non-modular.

**2.26 Lemma.** *A semimodular lattice satisfies the* JD-*condition.*

*Proof.* We use induction on the length of maximal $a, b$-chains. If $a \lessdot b$ then there is only one such chain. Assume the following proposition for all $t \leq m - 1$ and all $a, b$: If *one* maximal $a, b$-chain has length $t$ then *all* of them have length $t$. Suppose $a = c_0 \lessdot c_1 \lessdot \cdots \lessdot c_m = b$, $a \lessdot d_1 \lessdot \cdots \lessdot d_n = b$ are two maximal $a, b$-chains. If $c_1 = d_1$, then $m = n$ by the induction hypothesis applied to $c_1, b$-chains. But if $c_1 \neq d_1$, then $c_1, d_1 \lessdot c_1 \vee d_1$ by semimodularity. By induction, every maximal $c_1, b$-chain has length $m - 1$ (see Figure 2.10).

But this implies that the length of every maximal $d_1, b$-chain is $m - 1$, and thus $m = n$. □

**2.27 Theorem.** *Let L be a lattice with* 0. *L is semimodular if and only if L possesses a rank function r such that for all* $x, y \in L$:

$$r(x \wedge y) + r(x \vee y) \leq r(x) + r(y).$$

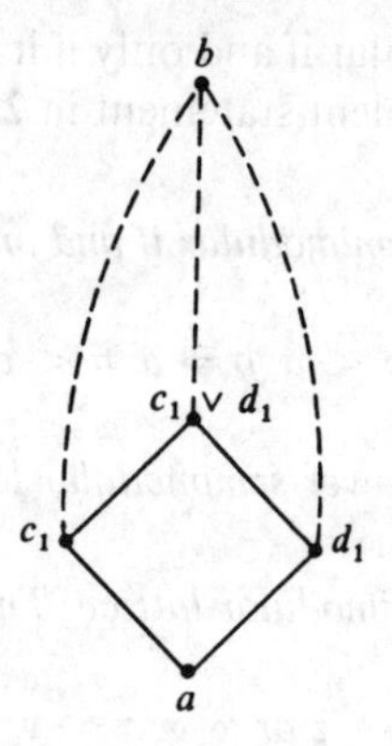


Figure 2.10

*L is modular if and only if for all x, y ∈ L:*

$$r(x \wedge y) + r(x \vee y) = r(x) + r(y).$$

*Proof.* A semimodular lattice with 0 possesses a rank function by **2.26**. Let $x \wedge y = c_0 <\cdot\, c_1 <\cdot \cdots <\cdot\, c_t = x$ be a maximal chain. Transposing by $y$ we infer from **2.25** that the *distinct elements* in $y \le c_1 \vee y \le c_2 \vee y \le \cdots \le c_t \vee y = x \vee y$ form a maximal $y, x \vee y$-chain. But this means $r(x) - r(x \wedge y) \ge r(x \vee y) - r(y)$. This inequality, on the other hand, trivially implies the semimodular property. □

The conditions **2.27** on the rank function are called the *semimodular inequality* and *modular equality* respectively.

**Examples.** Figure 2.11 shows the smallest semimodular lattice which is not modular.

The partition lattices $\mathscr{P}(n)$ are semimodular but not modular for $n > 3$. To see this take two partitions $\pi$, $\sigma$ of an $n$-set $S$ such that $\pi \wedge \sigma <\cdot\, \pi, \sigma$. Assuming $\pi \wedge \sigma = A_1|A_2|\cdots|A_t$ we can write without loss of generality

$$\pi = A_1 \cup A_2|A_3|\cdots|A_t$$

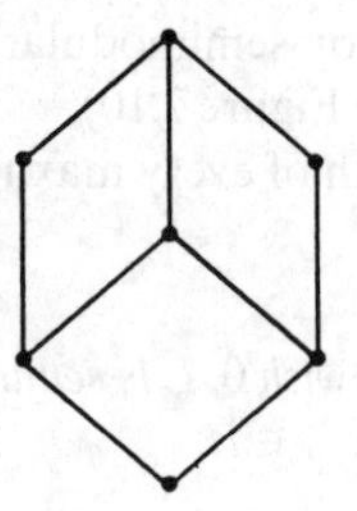

Figure 2.11

and

$$\sigma = A_1|A_2|A_3 \cup A_4|\cdots|A_t$$

or

$$\sigma = A_1 \cup A_3|A_2|\cdots|A_t.$$

In the first case $\pi \vee \sigma = A_1 \cup A_2|A_3 \cup A_4|\cdots|A_t$, and in the second $\pi \vee \sigma = A_1 \cup A_2 \cup A_3|A_4|\cdots|A_t$. Either way we have $\pi, \sigma <\cdot \pi \vee \sigma$ which means that $\mathscr{P}(n)$ is semimodular by **2.24**. Let $|S| \geq 4$, $a$, $b$, $c \in S$ and $\pi = \{a, b\}|S - \{a, b\}$, $\sigma = \{a, c\}|S - \{a, c\}$. Then $\pi \vee \sigma = 1$, $\pi \wedge \sigma = \{a\}|\{b\}|\{c\}|S - \{a, b, c\}$. Hence $\pi, \sigma <\cdot \pi \vee \sigma$ but $\pi \wedge \sigma$ is not covered by either $\pi$ or $\sigma$. We remarked in section I.2 that $\pi <\cdot \sigma$ implies that the number $b(\sigma)$ of blocks in $\sigma$ is precisely one less than the number $b(\pi)$ of blocks in $\pi$. Hence it follows by induction that $r(\pi) = n - b(\pi)$ for all $\pi \in \mathscr{P}(n)$.

The classical examples of semimodular lattices come from sets endowed with a closure operator satisfying the Steinitz exchange axiom. Recall that a mapping $A \to \bar{A}$ of the power set $2^S$ into itself is called a *closure operator* if for all $A, B \subseteq S$:

(i) $A \subseteq \bar{A}$,
(ii) $A \subseteq B \Rightarrow \bar{A} \subseteq \bar{B}$,
(iii) $\bar{\bar{A}} = \bar{A}$.

A subset $A \subseteq S$ with $A = \bar{A}$ is called *closed*. The family of closed subsets forms a complete lattice under inclusion with meet and join given by

$$A \wedge B = A \cap B \quad \text{and} \quad A \vee B = \overline{A \cup B}.$$

That is, the intersection of closed sets is always closed and $\overline{A \cup B}$ is the smallest closed set containing both $A$ and $B$.

If, in addition, $A \to \bar{A}$ satisfies the Steinitz exchange axiom:

(iv) For all $A \subseteq S$, $p, q \in S$:

$$p \notin \bar{A},\, p \in \overline{A \cup q} \Rightarrow q \in \overline{A \cup p},$$

then the complete lattice of closed subsets of $S$ is semimodular. (The reader can easily provide a proof of this statement; it will also follow from the theorems in the next section.)

**Examples.** Exchange systems as defined by (i)–(iv) abound in mathematics with roots in both algebra and topology. The first and most important example which immediately comes to mind (and from which the term Steinitz exchange axiom derives) is that of a vector space over a division ring. Let $V$ be the set of vectors and $A \to \bar{A}$ the linear closure, i.e., $v \in \bar{A}$ if and only if $v$ is *linearly dependent* on $A$. $A \to \bar{A}$ satisfies the exchange axiom (iv) and a subset $A$ is closed if and only if $A$ is a subspace of $V$. Hence we obtain as the lattice of closed subsets the vector space lattice $\mathscr{L}(V)$, introduced in chapter I, which by the modular equality **2.27** is even modular. Another classical example arises in the theory of field extensions. Let $F$ be an extension of $K$, and define the closure $A \to \bar{A}$, $A \subseteq F$, by putting into $\bar{A}$ all elements of $F$ which are *algebraically dependent* on $A$ over $K$. $A \to \bar{A}$ possesses the exchange property, hence the lattice of algebraically closed subfields of $F$ over $K$ is semimodular (but, in general, not modular). Both these examples, and in general, any lattice induced by a closure operator, are *point lattices*, i.e., every element is a join of points. Semimodular point lattices are the central object of study in combinatorial lattice theory; we shall turn our attention to this class of lattices in the next section.

EXERCISES II.2

1. Show that a lattice $L$ is modular if and only if $[(x \wedge z) \vee y] \wedge z = [(y \wedge z) \vee x] \wedge z$ for all $x, y, z \in L$.

2.* A lattice $L$ is called *Arguesian* if for all $a_0, a_1, a_2, b_0, b_1, b_2 \in L$: $\bigwedge_{i=0}^{2} (a_i \vee b_i) \le (a_0 \wedge (c \vee a_1)) \vee (b_0 \wedge (c \vee b_1))$ where $c = (a_0 \vee a_1) \wedge (b_0 \vee b_1) \wedge [((a_0 \vee a_2) \wedge (b_0 \vee b_2)) \vee ((a_1 \vee a_2) \wedge (b_1 \vee b_2))]$. Show that the normal subgroup lattices $\mathscr{N}(G)$ are Arguesian (and hence the vector space lattices). (Jónsson)

3. Verify: Arguesian $\Rightarrow$ modular.

→ 4. Prove in detail **2.16** and **2.17**.

5.* Let $L$ be a modular lattice with 0 without infinite chains and $a = c_0 \le c_1 \le \cdots \le c_s = b$, $a = d_0 \le d_1 \le \cdots \le d_t = b$ two $a$, $b$-chains. Prove:
   (i) The two chains can be refined by adding new elements $c_i = c_{i,0} \le c_{i,1} \le \cdots \le c_{i,k} = c_{i+1}$, $d_j = d_{j,0} \le d_{j,1} \le \cdots \le d_{j,l} = d_{j+1}$ such that the intervals $[c_{i,j-1}, c_{i,j}]$, $[d_{j,i-1}, d_{j,i}]$ are projective and hence isomorphic. (Ore)
   (ii) Application: Let $G$ be a group, $H$ a subgroup and $H \trianglelefteq H_1 \trianglelefteq \cdots \trianglelefteq H_s = G$, $H \trianglelefteq K_1 \trianglelefteq \cdots \trianglelefteq K_t = G$ two normal series, i.e., $H_i$ is a normal subgroup of $H_{i+1}$, for all $i$. Now formulate the group-theoretic result suggested by (i).
   (iii) Corollary: Let $H \lhd H_1 \lhd \cdots \lhd H_s = G$, $H \lhd K_1 \lhd \cdots \lhd K_t = G$ be two composition series, i.e., $H_i$ is a maximal normal subgroup of $H_{i+1}$, for all $i$. Then $s = t$ and the factor groups $H_{i+1}/H_i$ are isomorphic to the factor groups $K_{j+1}/K_j$ up to order. (Jordan–Hölder)

6.* Strengthen **2.23** as follows. Let $L$ be a modular lattice with 0 and $a = p_1 \vee \cdots \vee p_n = q_1 \vee \cdots q_n$ two irredundant decompositions of $a$.

Then:

(i) For each $p_i$ there is a $q_j$ such that

$$a = p_1 \vee \cdots \vee p_{i-1} \vee q_j \vee p_{i+1} \vee \cdots \vee p_n$$
$$= q_1 \vee \cdots \vee q_{j-1} \vee p_i \vee q_{j+1} \vee \cdots \vee q_n;$$

(ii) There is a permutation $\sigma$ of $\{1, \ldots, n\}$ such that, for each $i = 1, \ldots, n$,

$$a = p_1 \vee \cdots \vee p_{i-1} \vee q_{\sigma(i)} \vee p_{i+1} \vee \cdots \vee p_n.$$

(Dilworth)

→ 7. Strengthen **2.20** as follows: Let $L$ be semimodular, but not modular. Then $L$ contains a sublattice isomorphic to $N_5$ such that for $a$, $b$, $c$: $b \wedge c = b \wedge a \lessdot c \lessdot a, b \lessdot b \vee a = b \vee c$ (in the notation of Figure 2.1).

→ 8. Complete the proof of **2.21** and show that a modular non-distributive lattice contains a sublattice $M_5$ of rank 2.

9.* Is there a characterization of semimodular lattices by means of "forbidden" sublattices similar to **2.20** and **2.21**?

→10. Let $L$ be a c.f. semimodular lattice with point set $S$ and let $p_1, \ldots, p_k \in S$. Show that the following conditions are equivalent:
 (i) $p_i \not\leq \sup\{p_1, \ldots, p_{i-1}, p_{i+1}, \ldots, p_k\}$ for all $i$;
 (ii) $(p_1 \vee p_2 \vee \cdots \vee p_{i-1}) \wedge p_i = 0$ for all $i = 2, \ldots, k$;
 (iii) $r(p_1 \vee \cdots \vee p_k) = k$.

→11. Let $L$ be as in ex. 10, $a \in L$. Show that for any two irredundant decompositions $a = p_1 \vee \cdots \vee p_s = q_1 \vee \cdots \vee q_t$ where the $p_i$'s and $q_j$'s are *points* of $L$, the statement of **2.23** is valid.

12.* Strengthen ex. 11: Let $L$ be as in ex. 10. Prove that the statement of **2.23** holds for all $a \in L$ and all irredundant decompositions by means of irreducible elements if and only if for all $a \in L$, the sublattice $L_a$ generated by $\{p \in L: p \lessdot a\}$ is modular. (Dilworth)

→13. Generalize the concept of a closure from $2^S$ to arbitrary posets $P$ as follows: A mapping $x \to \bar{x}$ is a closure on $P$ if for all $x, y \in P$:
 (i) $x \leq \bar{x}$,
 (ii) $x \leq y \Rightarrow \bar{x} \leq \bar{y}$,
 (iii) $\bar{\bar{x}} = \bar{x}$.
 $Q = \{x \in P: x = \bar{x}\}$ is called the *quotient* of $P$ relative to the closure. Prove:
 (i) $x \to \bar{x}$ is a closure if and only if $x \leq \bar{x}$ and $x \leq \bar{y} \Rightarrow \bar{x} \leq \bar{y}$.
 (ii) If $L$ is a lattice, then the quotient $Q$ is a complete lattice and infima in $Q$ coincide with infima in $L$.
 (iii) Every complete lattice is the quotient of some lattice.

14. Consider the lattice of submodules of an $R$-module. Under what algebraic conditions is the exchange axiom **2.23** satisfied?

15. Construct a field extension $F$ over $K$ whose lattice of algebraically closed subfields is not modular.

## 3. Geometric Lattices

### A. Geometric Lattices and Matroids

**Definition.** A lattice is called *geometric* if it is

(i) a point lattice,
(ii) semimodular,
(iii) without infinite chains.

Any geometric lattice is complete and possesses a rank function. The purpose of the present section is twofold. First, to make clear the precise relationship between exchange systems defined at the end of the last section and geometric lattices, and secondly, to probe more deeply into the structure of geometric lattices concluding with an important characterization of indecomposable geometric lattices. In the course of our investigation we shall see that Boolean algebras and vector space lattices (of finite rank) are essentially all distributive and modular geometric lattices respectively, thereby underlining once again the central importance of these lattices.

**Examples.** Boolean algebras $\mathscr{B}(n)$ and vector space lattices $\mathscr{L}(n, K)$ are geometric for $n < \infty$. Let us consider the finite partition lattices $\mathscr{P}(n)$. We already know that $\mathscr{P}(n)$ is semimodular. The atoms of $\mathscr{P}(S)$, $|S| = n$, are all partitions $\pi_{a,b} = a, b|c|d|\ldots$ consisting of one block containing a pair $\{a, b\}$ and otherwise only of singletons. Hence if $\pi$ is an arbitrary partition of $S$, then $\pi = \bigvee_{a,b} \pi_{a,b}$ where the supremum is taken over all pairs $\{a, b\}$ lying in a block of $\pi$.

A simple but often useful criterion is the following proposition whose proof is left to the reader.

**2.28 Proposition.** *A c.f. lattice is geometric if and only if for all a, b*

$$a <\cdot b \Leftrightarrow \exists \text{ a point } p \text{ with } p \nleq a, b = a \vee p. \quad \square$$

The following definition is at the heart of combinatorial structure theory; it generalizes various notions such as independence, basis, rank and others abstracted from examples in geometry, graph theory and transversal theory. It will be our main object of study in chapters VI and VII.

**Definition.** A set $S$ together with a closure operator $A \to \bar{A}$ is called a *matroid* (or a *combinatorial pregeometry*) on $S$ if for all $A \subseteq S$, $p, q \in S$ the following holds:

(i) $p \notin \bar{A}, p \in \overline{A \cup q} \Rightarrow q \in \overline{A \cup p}$ (exchange axiom),
(ii) $\exists B \subseteq A$, $B$ finite with $\bar{B} = \bar{A}$ (finite basis axiom).

The matroid is called *simple* (or a *combinatorial geometry*) if, in addition,

(iii) $\overline{\varnothing} = \varnothing$ and $\bar{p} = p$ for all $p \in S$.

We customarily use the notation $\mathbf{M}$ for a matroid or $\mathbf{M}(S)$ if it is not otherwise clear on which set $\mathbf{M}$ is defined. The closed subsets are called *flats* or *subspaces* of $\mathbf{M}$ with $L$ or $L(S)$ denoting the lattice of flats. Both the terms matroid and pregeometry as well as flats and subspaces occur in the literature. Most of the time we shall prefer the shorter terms matroid and flat but will on occasion use the terms combinatorial geometry (or just geometry) and subspace to emphasize the geometric origin of some problem or method.

A few more definitions: Flats or rank 1 are called *points*, those of rank 2 *lines* and of rank 3 *planes*. Similarly, we call flats of corank 1, 2, 3, *copoints*, *colines*, and *coplanes*, respectively. In geometry, a copoint is more often called a *hyperplane*, a term which we shall also use occasionally. An element $p \in \overline{\varnothing}$ is called a *loop*, elements $p, q$ with $\bar{p} = \bar{q}$ *parallel*. Hence simple matroids possess neither loops nor parallel elements. By abuse of language we not only call the flats $\bar{p}$ points, but use this term for the elements of $S$ also.

To every matroid $\mathbf{M}$ on $S$ we associate a canonical simple matroid $\mathbf{M}_0$ in the following manner: Take as elements of $\mathbf{M}_0$ the flats of $\mathbf{M}$ of rank 1 with the closure on this new set induced by the closure of $\mathbf{M}$, i.e.,

$$\bar{A}^0 := \left\{ \bar{p} : p \in \overline{\bigcup_{\bar{q} \in A} q} - \overline{\varnothing} \right\}.$$

$\mathbf{M}_0$ is a simple matroid with its lattice of flats isomorphic to $L(S)$. $\mathbf{M}_0$ is called the *underlying simple matroid* (*geometry*) of $\mathbf{M}$. Hence, from the lattice-theoretic point of view, there is no loss of generality in restricting ourselves to simple matroids.

**Example.** An $n$-dimensional vector space $V$ over a division ring $K$ together with the linear closure (as mentioned in the last section) forms a matroid denoted by $\mathbf{M}(V(n, K))$. The 0-vector is the only loop, parallel vectors are scalar multiples of each other. The underlying geometry has as points the subspaces of $V$ of rank 1 and is called the *projective geometry* $\mathbf{PG}(n-1, K)$ of $g$-dimension $n-1$ ($=$rank $n$) over $K$ (see **2.32** below).

Notice that by the exchange property $\overline{A \cup p} \cdot\!> A$ for any $A \in L(S)$, $p \notin A$, since if $q \notin A$, $q \in \overline{A \cup p}$, then $p \in \overline{A \cup q}$, i.e., $\overline{A \cup p} = \overline{A \cup q}$. The following theorem provides the link between matroids and geometric lattices alluded to above.

**2.29 Theorem** (Birkhoff–Whitney).

(i) *Let* $\mathbf{M}(S)$ *be a matroid. Then the lattice of flats of* $L(S)$ *is geometric.*

(ii) *Let, conversely,* $L$ *be a geometric lattice with point set* $S$. *Then* $S$ *together with* $A \to \bar{A} := \{p \in S : p \leq \sup A\}$ *is a simple matroid* $\mathbf{M}(S)$, *and* $\phi$: $L \to L(S)$, $\phi(x) = \{p \in S : p \leq x\}$ *is a lattice isomorphism.*

*Hence geometric lattices correspond bijectively to simple matroids.*

*Proof.* Let **M** be a matroid on $S$ and $L(S)$ its lattice of flats. Since $A = \sup\{\bar{p}: p \in A\}$ for any $A \in L(S)$, $L(S)$ is a point lattice. To verify semimodularity, let $A, B \in L(S)$ with $A \cap B <\cdot A$. There exists $p \in A - B$ with $A = \overline{(A \cap B) \cup p}$ which implies that $\overline{A \cup B} = \overline{B \cup p} \cdot> B$. If infinite chains exist in $L(S)$, they must contain a countably infinite ascending chain or a countably infinite descending chain. Let $A_1 < A_2 < \cdots$ be an ascending chain in $L(S)$. Then by the finite basis property we can find a finite set $B \subseteq \bigcup A_i$ with $\bar{B} = \overline{\cup A_i}$. But then $B \subseteq A_m$ for some $m$, and thus $A_j \subseteq \overline{\cup A_i} = \bar{B} \subseteq A_m$ for all $j$, i.e., the chain terminates, in fact, with $A_m$. If $A_1 > A_2 > \cdots$ is a descending chain in $L(S)$, choose elements $a_i \in A_i - A_{i+1}$ and set $A = \{a_1, a_2, \ldots\}$, $S_i = \{a_i, a_{i+1}, \ldots\}$. As $\overline{S_{i+1}} \subseteq A_{i+1}$, we infer $a_i \notin \overline{S_{i+1}}$ for all $i$. Set $B_i = A - a_i$ and suppose $a_i \in \overline{B_i}$. Choosing $j$ to be maximal with respect to $a_i \in \overline{S_j - a_i}$ $(j \leq i - 1$ because $a_i \notin \overline{S_{i+1}})$ we conclude $a_i \notin \overline{S_{j+1} - a_i} = \overline{(S_j - a_i) - a_j}$ and hence by the exchange axiom $a_j \in \overline{S_{j+1}}$, a contradiction. Thus there is no finite subset of $A$ having the same closure as $A$, in violation of the finite basis axiom.

Conversely, let $L$ be a geometric lattice with point set $S$. It is readily verified that $A \to \bar{A} := \{p \in S: p \leq \sup A\}$ is a closure operator on $S$ with a subset being closed if and only if it is the set of *all* points lying beneath a certain lattice element of $L$. Since $L$ is a point lattice the mapping $\phi: L \to L(S)$ given by $\phi(x) = \{p \in S: p \leq x\}$ is bijective and, as it is monotone in both directions, it is a lattice isomorphism. It remains to prove that $A \to \bar{A}$ defines a matroid on $S$. But this follows immediately from **2.28** and the finite chain condition on $L$. □

**2.29** gives rise to a wealth of examples of matroids starting from geometric lattices, and vice versa. A fairly comprehensive account will follow later; for the moment let us just discuss some examples of matroids arising from our known geometric lattices.

**2.30.** Which matroids correspond to Boolean algebras $\mathscr{B}(n)$? Obviously, we must have $|S| = n$ with every subset of $S$ being closed. Such a matroid is called a *free matroid* or a *free geometry* of rank $n$, denoted by $\mathbf{FM}(n)$.

**2.31.** Consider the matroid **M** induced by the partition lattice on an $n$-set $V$. The point set of **M** is the set $V^{(2)}$ of unordered pairs of elements of $V$ with $\bar{A}$ being the *transitive closure* of a subset $A \subseteq V^{(2)}$. A very useful interpretation of this closure relation is as follows. Think of $V$ as the vertices and the point set $V^{(2)}$ as the edges of a complete graph. If $A \subseteq V^{(2)}$ and $A_1, \ldots, A_t$ are the connected components of the subgraph with edge-set $A$, then $\bar{A} = \bar{A}_1 \cup \cdots \cup \bar{A}_t$ where $\bar{A}_i$ is derived from $A_i$ by putting in all edges *within* the component $A_i$. This completion process of components can easily be extended to arbitrary graphs and forms the basis of an algebraic-geometric approach to many important problems in graph theory, to be discussed in section VII.3.

**Example.** Consider the complete graph with 10 vertices. Then the closure of $A$ is given in Figure 2.12.

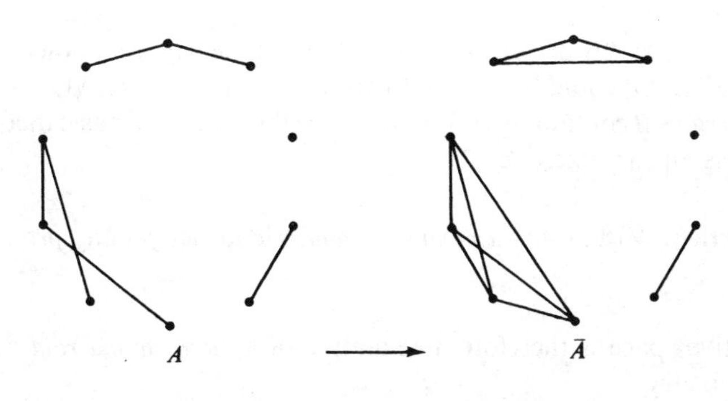


Figure 2.12

**2.32.** Let us recall the connection between abstract projective spaces and linear algebra. A *projective space* or *projective geometry* is a system consisting of a set $\mathfrak{P}$ of *points* together with certain subsets $\mathfrak{G}$ of $\mathfrak{P}$ called *lines* such that $(\mathfrak{P}, \mathfrak{G})$ satisfies the following axioms:

(i) Two distinct points lie on exactly one line.
(ii) If $P, Q, R$ form a triangle (i.e., they do not lie on a common line) and if the line $g$ intersects two sides of the triangle (but not in $P$, $Q$, or $R$), then $g$ also intersects the third side (see Figure 2.13).
(iii) Each line contains at least three distinct points.

A subset $A \subseteq \mathfrak{P}$ is called a *subspace* if for any two distinct points of $A$, it contains the whole line determined by them. It follows from (ii) that subspaces can also be introduced inductively using the concept of dimension: A point is a subspace of $g$-dimension 0, a line is a subspace of $g$-dimension 1. If $U$ is a subspace of $g$-dimension $k$ and if the point $P \notin U$, then $U$ together with all lines connecting $P$ with points in $U$ is a subspace of $g$-dimension $k + 1$.

(iv) $\mathfrak{P}$ has finite $g$-dimension.

If $\mathfrak{P}$ has $g$-dimension $n$, $(\mathfrak{P}, \mathfrak{G})$ is called a *projective space of $g$-dimension $n$*. Projective spaces of $g$-dimension 2 are commonly called *projective planes.*

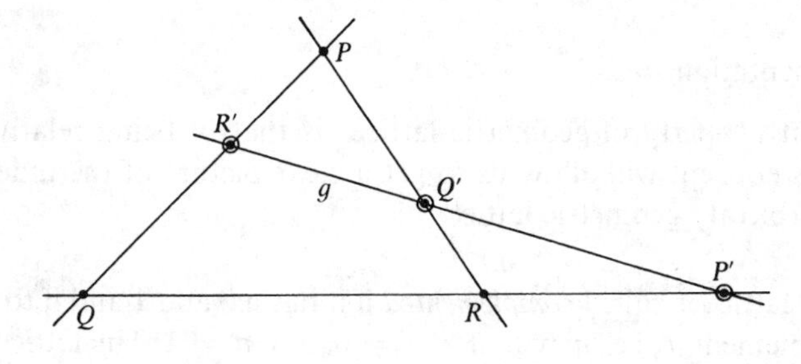


Figure 2.13

The lattice $\mathscr{L}(\mathfrak{P}, \mathfrak{G})$ of subspaces in $(\mathfrak{P}, \mathfrak{G})$ is clearly a complete lattice. Alternatively, $\mathscr{L}(\mathfrak{P}, \mathfrak{G})$ could be defined by the closure relation on $\mathfrak{P}$: $A \to \bar{A} = \bigcap B$ over all subspaces $B$ containing $A$. We now have the following basic theorem whose proof is left as an easy exercise.

**2.33 Proposition.** *$\mathscr{L}(\mathfrak{P}, \mathfrak{G})$ is a modular geometric lattice for any projective space $(\mathfrak{P}, \mathfrak{G})$.* □

A projective space is therefore an example of a *modular matroid*, in fact, of a modular geometry.

The converse of this theorem (apart from taking direct products) is one of the main results in this section. With this result, we have a complete description of modular geometric lattices (Theorem **2.55**). Hence the study of modular geometric lattices is reduced to the study of projective spaces. Let $V(n, K)$ be a vector space over the division ring $K$. Taking as $\mathfrak{P}$ and $\mathfrak{G}$ the families of rank 1 and rank 2 subspaces of $V$, the axioms for a projective space can be easily verified. The subspaces of this projective space are precisely the (vector-) subspaces of $V$, and thus the resulting projective space is just $\mathbf{PG}(n-1, K)$ introduced above **2.29**. A classical result of projective geometry asserts a partial converse (see, e.g., Baer [1, ch. 7]):

**2.34 Theorem.** *Any projective space of g-dimension $n \geq 3$ is isomorphic to $\mathbf{PG}(n, K)$ for some division ring $K$.* □

Projective spaces of $g$-dimension 0 or 1 are trivial since any point or line (with at least three points) will do. As for $g$-dimension 2 we have:

**2.35 Theorem.** *A projective plane is isomorphic to $\mathbf{PG}(2, K)$ for some division ring $K$ if and only if Desargues' law is satisfied.* (cf. **7.14**) □

The lattices of these "Desarguesian" planes are called *Arguesian* and can be described by translating Desargues' law into a lattice identity (see ex. II.2.2). The remaining non-Desarguesian planes have been the object of much study, but their classification is far from complete. For a detailed treatment consult M. Hall [1, ch. 20]. Other types of geometric structures (affine spaces, Möbius geometries, incidence geometries) giving rise to matroids will be discussed in chapters VI and VII.

## B. Complementation

An important property of geometric lattices is that of being relatively complemented. This concept will allow us to get a clear picture of the indecomposable parts of an arbitrary geometric lattice.

**Definition.** A lattice is called *complemented* if it has a 0 and 1 and if to each $a$ there exists a complement $a'$, i.e., $a'$ with $a \wedge a' = 0$, $a \vee a' = 1$. The lattice is *relatively complemented* if every interval is complemented.

**2.36 Proposition.** *A semimodular c.f. lattice is geometric if and only if it is relatively complemented.*

*Proof.* Let $L$ be geometric, $[a, b]$ an interval, $x \in [a, b]$ and $x' \in [a, b]$ arbitrary with $x \wedge x' = a$. If $x \vee x' = b$ we are finished. In case $x \vee x' < b$ choose a point $q \le b$, $q \nleq x \vee x'$ and set $x'' = x' \vee q$. Clearly $x'' \in [a, b]$, $x \vee x'' > x \vee x'$. We claim $x \wedge x'' = a$. Assume otherwise; then there is a point $p$ such that $x \wedge x' < (x \wedge x') \vee p \le x \wedge (x' \vee q)$. It follows that $p \nleq x'$, $p \le x' \vee q$ and thus $x' \vee p = x' \vee q$ by the exchange law, from which the contradiction $x \vee x' \vee q = (x \vee x') \vee p = x \vee x'$ results. Repeating this process we eventually arrive at an $x$-complement in $[a, b]$. Conversely, let $L$ be relatively complemented, $a \in L$ and suppose $b = \sup\{p \in L : 0 \lessdot p \le a\} < a$. Choose a $b$-complement $b'$ in $[0, a]$. Then clearly $b' > 0$, hence there is a point $q$ with $q \le b' \le a$, $q \nleq b$, contradicting the definition of $b$. □

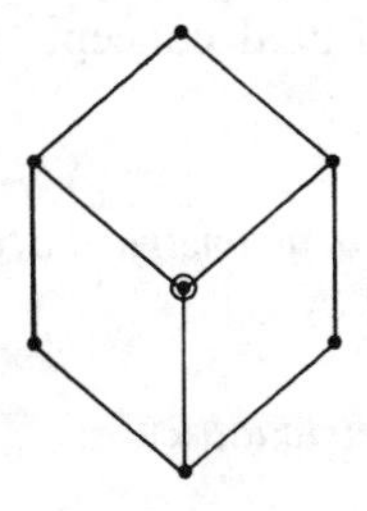

Figure 2.14

An arbitrary point lattice without infinite chains need not be complemented and *a fortiori* not relatively complemented. The lattice of Figure 2.14 is a lower semimodular point lattice, but the circled element possesses no complement. Conversely, a semimodular complemented lattice need not be a point lattice, let alone relatively complemented (see Figure 2.15). But we have:

**2.37 Proposition.** *Every complemented modular lattice is relatively complemented.*

*Proof.* Take $x \in [a, b]$ and suppose $y$ is a complement of $x$. Then $b \wedge (y \vee a) = (b \wedge y) \vee a$ is a relative $x$-complement in $[a, b]$. □

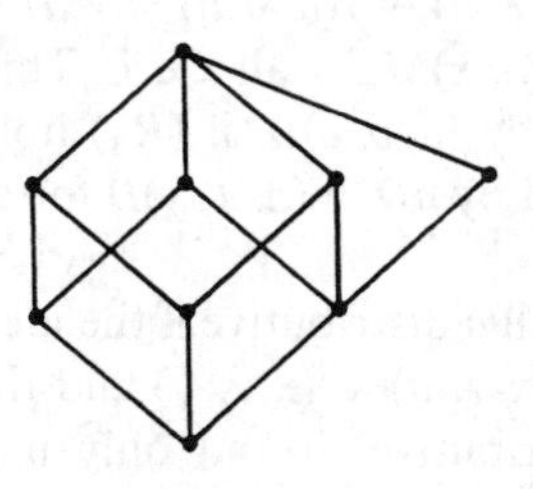

Figure 2.15

Notice that in **2.36** we have actually proved the following stronger statement which, in fact, holds in arbitrary relatively complemented c.f. lattices.

**2.38 Proposition.** *Let $L$ be a relatively complemented c.f. lattice, $a,b \in L$ and $x, y \in [a, b]$ with $x \wedge y = a$. Then we can find a relative $x$-complement $x'$ in $[a, b]$ such that $x' \geq y$.*

*Proof.* Suppose $x \vee y < b$ and choose $x'$ to be a relative $x \vee y$-complement in $[y, b]$. Claim: $x'$ is a relative $x$-complement with $x' \geq y$. We have $x \wedge x' = (x \wedge (x \vee y)) \wedge x' = x \wedge ((x \vee y) \wedge x') = x \wedge y = a$ and $x \vee x' = x \vee (y \vee x') = (x \vee y) \vee x' = b$. □

Relative complementation is clearly inherited by direct products and intervals. Further, since relative complementation is a self-dual property it follows that we obtain a dual proposition for every proposition in whose proof only relative complementation and other self-dual concepts (like the finite chain property) occur. For instance, we have:

**2.39 Corollary.** *Any interval of a geometric lattice is geometric as is the direct product of geometric lattices.* □

**2.40 Corollary.** *Let $L$ be a geometric lattice.*

(i) *Every element is the join of points; dually every element is the meet of copoints.*
(ii) *Let $[a, b] \subseteq L$ and $x, y \in [a, b]$ with $x \wedge y = a$. There exists a relative $x$-complement $x'$ in $[a, b]$ with $x' \geq y$. Dually, suppose $x, y \in [a, b]$ with $x \vee y = b$. There exists a relative $x$-complement $x'$ with $x' \leq y$.*

*In particular:*

(iii) *Let $a \in L$. Every point $p \nleq a$ can be extended to a complement $a'$ of $a$, i.e., $p \leq a'$. Dually, every copoint $h \ngeq a$ contains a complement $a'$ of $a$, i.e., $a' \leq h$.* □

Complementation provides a useful description of modular and distributive elements.

**Definition.** Let $L$ be a geometric lattice. The pair $a, b \in L$ is called a *modular pair*, denoted by $(a, b)M$, if $r(a \wedge b) + r(a \vee b) = r(a) + r(b)$. $a$ is called a *modular element*, denoted by $aM$, if $(a, x)M$ for all $x \in L$. Three elements $a, b, c \in L$ form a *distributive triple*, denoted by $(a, b, c)D$, if $(R_3)$ holds for $\{a, b, c\}$. $a$ is called a *distributive element*, denoted by $aD$, if $(a, x, y)D$ for all $x, y \in L$.

Equivalently we could call $a$ distributive if the identities $a \wedge (x \vee y) = (a \wedge x) \vee (a \wedge y)$, $x \wedge (a \vee y) = (x \wedge a) \vee (x \wedge y)$ and their duals hold for all $x, y$. Of course, $L$ is modular (distributive) if and only if all its elements are modular (distributive).

**2.41 Proposition.** *In a geometric lattice the following statements are equivalent:*

(i) $(a, b)M$.
(ii) *$b$ is a minimal relative complement of $a$ in $[a \wedge b, a \vee b]$ (or equivalently, $a$ is a minimal relative complement of $b$ in $[a \wedge b, a \vee b]$).*
(iii) *$\phi_b$ maps $[a \wedge b, a]$ injectively into $[b, a \vee b]$ (or equivalently, $\phi_a$ maps $[a \wedge b, b]$ injectively into $[a, a \vee b]$).*

*Proof.* (i) $\Rightarrow$ (ii). If $t < b$ were another complement of $a$ in $[a \wedge b, a \vee b]$, then $r(b) > r(t) \geq r(a \vee b) + r(a \wedge b) - r(a) = r(b)$, a contradiction. (ii) $\Rightarrow$ (iii). If $\phi_b$ is not injective on $[a \wedge b, a]$, there exists $z \in [a \wedge b, a]$ with $(z \vee b) \wedge a > z$ (by **2.16**). Suppose $t$ is a $((z \vee b) \wedge a)$-complement in $[z, a]$. Then $t \wedge b = a \wedge b, t \vee b = t \vee (z \vee b) = t \vee ((z \vee b) \wedge a) \vee b = a \vee b$, but $t < a$. (iii) $\Rightarrow$ (i). This follows directly from **2.25**. □

**2.42 Proposition.** *In a geometric lattice the following statements are equivalent:*

(i) $aM$.
(ii) *The complements of $a$ form an antichain.*
(iii) *$\phi_b, \psi_a$ are inverse isomorphisms between $[a \wedge b, a], [b, a \vee b]$ for all $b$, i.e., $(z \vee b) \wedge a = z$ for $z \in [a \wedge b, a]$ and $(w \wedge a) \vee b = w$ for $w \in [b, a \vee b]$.*

*Proof.* Exercise. □

Let us emphasize one point: $aM$ implies $[a \wedge b, a] \cong [b, a \vee b]$ for all $b$, but in general *not* $[a \wedge b, b] \cong [a, a \vee b]$. In the lattice of Figure 2.16, $a$ is modular, but $[a \wedge b, b]$ is not isomorphic to $[a, a \vee b]$. A very useful corollary of **2.41** is the following characterization of modular geometric lattices.

**2.43 Corollary.** *A geometric lattice is modular if and only if $h \wedge l > 0$ for every copoint $h$ and every line $l$.*

*Proof.* The necessity follows at once from the modular rank equality. Now suppose $L$ is not modular. Since $L$ is semimodular **2.19**(ii) must be violated, hence there

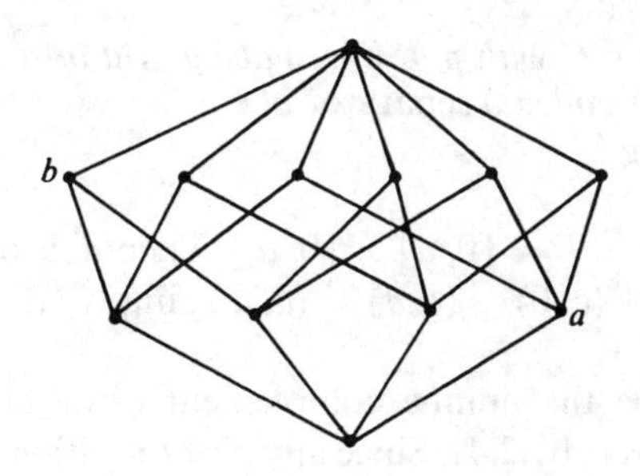


Figure 2.16

exist elements $a, b$ with $b \lessdot a \vee b$, $a \wedge b < a$ but not $a \wedge b \lessdot a$. We may choose $a$ in such a way that $r(a) = r(a \wedge b) + 2$. Let $h$ be a minimal relative complement of $a \vee b$ in $[b, 1]$ and $l$ a minimal relative complement of $a \wedge b$ in $[0, a]$. By **2.41**, $h$ is a copoint, $l$ is a line, and we have $h \wedge l = h \wedge a \wedge l = (h \wedge (a \vee b)) \wedge a \wedge l = (b \wedge a) \wedge l = 0$. □

Another consequence of **2.41** states that a geometric lattice is to some extent already determined by the structure of its upper intervals, i.e., of its intervals containing 1.

**2.44 Proposition.** *Any interval of a geometric lattice can be embedded in an upper interval.*

*Proof.* Consider $[a, b]$ and let $c$ be a minimal complement of $b$ in $[a, 1]$. $\phi_c: [a, b] \to [c, 1]$ is the required embedding. □

We turn to distributive elements and their characterization. Recall that the product $A \times B$ of lattices $A$ and $B$ is the set $\{(a, b): a \in A, b \in B\}$ of all ordered pairs together with the coordinate-wise order relation $(a, b) \leq_{A \times B} (c, d)$ if and only if $a \leq_A c$ and $b \leq_B d$. The notation $L = A \times B$ means $A$ and $B$ are sublattices of $L$, $A \cap B = \{0\}$ and $L$ is isomorphic to the direct product $A \times B$. $A$ and $B$ are, of course, lower intervals $[0, a]$, $[0, b]$ of $L$ which leads us to the following definition.

**Definition.** Let $L$ be a lattice with 0 and 1. The *center* $Z(L)$ consists of all elements $a$ for which a decomposition $L = [0, a] \times [0, a']$ exists. In other words, central elements are maximal elements of some direct factor of $L$. Of course, there is always the trivial decomposition $L = [0, 1] \times [0, 0]$, i.e., 0 and 1 are in $Z(L)$ for any lattice $L$. If $Z(L) = \{0, 1\}$, $L$ is called *indecomposable*.

**2.45 Theorem.** *Let $L$ be a geometric lattice with point set $S$ and copoint set $C$. The following propositions are equivalent:*

(i) *$a \in Z(L)$.*
(ii) *$a$ possesses a unique complement.*
(iii) *There exists $a'$ with $a \wedge a' = 0$, $r(a) + r(a') = r(1)$ such that $p \leq a$ or $p \leq a'$ for all $p \in S$.*
(iv) *For all $p \in S, h \in C$ with $p \nleq h$ we have $p \leq a$ or $a \leq h$. (An element $a$ with this property is called a* separator *of $L$.)*
(v) *$a$ is distributive.*

*Proof.* (i) ⇒ (ii). Suppose $L = [0, a] \times [0, a']$. Then $a'$ is a complement of $a$, and $(a, 0) \wedge (c, d) = (0, 0)$, $(a, 0) \vee (c, d) = (a, a')$ imply $a \wedge c = c = 0$, $0 \vee d = d = a'$.

(ii) ⇒ (iii). Let $a'$ be the unique complement of $a$. Trivially $a \wedge a' = 0$ and further $r(a) + r(a') = r(1)$ by **2.41**. Since any point $p$ with $p \nleq a$ can be extended to a complement of $a$ (**2.40**(iii)) we must have $p \leq a'$.

(iii) $\Rightarrow$ (iv). Let $p \in S$, $h \in C$ with $p \not\leq h$. If $p \not\leq a$ then $p \leq a'$ and hence $a' \not\leq h$. Suppose $a \not\leq h$. Then, by the hypothesis on the points, $h = (h \wedge a) \vee (h \wedge a')$, and thus $r(1) - 1 = r(h) = r((h \wedge a) \vee (h \wedge a')) \leq r(h \wedge a) + r(h \wedge a') \leq r(a) + r(a) - 2 = r(1) - 2$, a contradiction.

(iv) $\Rightarrow$ (v). We shall just verify that $a \wedge (x \vee y) = (a \wedge x) \vee (a \wedge y)$. As in this formula $\geq$ always holds, it remains, in light of **2.40**(i), to show that $h \geq (a \wedge x) \vee (a \wedge y)$ implies $h \geq a \wedge (x \vee y)$ for all $h \in C$. We claim $h \geq a \wedge x$ implies $h \geq a$ or $h \geq x$ for any $x \in L$. Suppose on the contrary $h \not\geq a$, $h \not\geq x$. Then there exists $q \in S$ with $q \leq x$, $q \not\leq h$. But by the hypothesis $q \leq a$, and thus $q \leq a \wedge x \leq h$, contrary to the assumption. We therefore obtain the implications: $h \geq (a \wedge x) \vee (a \wedge y) \Rightarrow h \geq a \wedge x$ and $h \geq a \wedge y \Rightarrow h \geq a$ or $h \geq x \vee y \Rightarrow h \geq a \wedge (x \vee y)$.

(v) $\Rightarrow$ (i). Clearly, $a$ has a unique complement, say $a'$, by **2.22**(i). Since proposition (iii) is symmetric in $a$ and $a'$, we infer from (iii) $\Rightarrow$ (v) $a'$ is distributive, too. Let us define $\phi: [0, a] \times [0, a'] \to L$ by setting $\phi(x, y) = x \vee y$. If $\phi(x, y) = \phi(w, z)$, i.e., $x \vee y = w \vee z$, then $x = x \vee 0 = (a \wedge x) \vee (a \wedge z) = a \wedge (x \vee z) = a \wedge (w \vee z) = (a \wedge w) \vee (a \wedge z) = w$, and similarly $y = z$. Hence $\phi$ is injective and since $\phi$ trivially preserves joins, it remains to show that $\phi$ is surjective. But for $z \in L$ we have $\phi(a \wedge z, a' \wedge z) = (a \wedge z) \vee (a' \wedge z) = z \wedge (a \vee a') = z$. $\square$

Observe that conditions (i), (ii), (iv), and (v) are equivalent in arbitrary relatively complemented lattices without infinite chains. We call $a \in Z(L)$ and its unique complement $a'$ *complementary separators*. The following statements are straightforward corollaries of **2.45** whose proofs are left to the reader.

**2.46 Proposition.** *Let $L$ be a geometric lattice and $a \in Z(L)$.*

(i) *If $b \leq a$, then $b \in Z(L)$ if and only if $b \in Z([0, a])$.*
(ii) *If $(c, d)M$ and $a$ is a separator of both the intervals $[c, 1]$ and $[d, 1]$, then $a$ is also a separator of $[c \wedge d, 1]$.*
(iii) *If $a$, $a'$ are complementary separators of $L$, and $c \leq d$, then $(a \vee c) \wedge d$, $(a' \vee c) \wedge d$ are complementary separators of the interval $[c, d]$.* $\square$

**2.46**(i) tells us that any decomposition $L = L_1 \times L_2$ can be refined to a *unique* finest decomposition where all factors are indecomposable geometric lattices.

**2.47 Theorem.** *Let $L$ be a geometric lattice with center $Z(L)$, and $\{a_1, \ldots, a_t\}$ the minimal elements $>0$ in $Z(L)$. Then $L = \prod_{i=1}^{t} [0, a_i]$ is the unique decomposition of $L$ into indecomposable sublattices.* $\square$

## C. Indecomposable Geometric Lattices

In the remainder of this section we shall describe indecomposable geometric lattices in terms of their point sets. Generalizing from concepts in projective geometry, we call two points $p, q$ *perspective* if $p = q$ or if $p$ and $q$ possess a common complement. $S$ and $C$ will again denote the set of points and copoints respectively, and we write $C_a := \{h \in C : h \geq a\}$, $H_a := C - C_a$ for $a \in L$. Since complements of points must be copoints, $p$ and $q$ are perspective if and only if $H_p \cap H_q \neq \emptyset$.

Perspectivity is clearly a reflexive and symmetric relation on $S$. The main theorem of this section asserts that perspectivity is also transitive, i.e., an equivalence relation, with the minimal separators being the joins of the perspectivity classes. First we need a few preliminary lemmas.

**2.48 Lemma.** *Let $L$ be an indecomposable geometric lattice, $0 \neq a \in L$ such that $[a, 1]$ is indecomposable. Then $[b, 1]$ is indecomposable for at least one $b \in L$ with $b \lessdot a$.*

*Proof.* The assertion is trivial for $a = 1$, hence we may assume $0 < a < 1$. Let $0 = c_k \lessdot c_{k-1} \lessdot \cdots \lessdot c_0 = a$ be an arbitrary maximal $0, a$-chain and set $D(c_i) := \{x \in L: c_i \leq x \lessdot a\}$. By **2.40**(i)

$$D(c_k) \supsetneqq D(c_{k-1}) \supsetneqq \cdots \supsetneqq D(c_1) \supsetneqq D(a) = \varnothing.$$

Choose $z_i \in D(c_i) - D(c_{i-1})$ for $i = 1, \ldots, k$. Then $c_i = z_1 \wedge z_2 \wedge \cdots \wedge z_i$. Claim: At least one of the intervals $[z_i, 1]$ is indecomposable. Assume otherwise. Let $a_i, a_i'$ be non-trivial complementary separators of $[z_i, 1]$ for all $i$. Then $a_i \vee a$, $a_i' \vee a$ are complementary separators of $[a, 1]$ for all $i$, by **2.46**(iii). Now $[a, 1]$ is indecomposable whence we may assume without loss of generality that $a \vee a_i = a$, $a \vee a_i' = 1$, i.e., $a = a_i, a_i' \in C$ for all $i$. We assert that $a$ is a separator of all intervals $[c_i, 1]$. Since $c_1 = z_1$ this is true for $i = 1$. Suppose our assertion is true for $j = 1, \ldots, i-1$. Then $a$ is a separator of $[c_{i-1}, 1]$ and $[z_i, 1]$ .But clearly, $c_{i-1}$ and $z_i$ are a modular pair with $c_i = c_{i-1} \wedge z_i$, whence $a$ is also a separator of $[c_i, 1]$ by **2.46**(ii), and thus by induction of all $c_i$. But since $c_k = 0$ this implies the decomposability of $L$, contrary to the hypothesis. □

**2.49 Lemma.** *Let $L$ be indecomposable and $l$ a line of $L$ with $[l, 1]$ indecomposable. Then there exist at least* two *points $p, q \leq l$ such that $[p, 1]$ and $[q, 1]$ are indecomposable.*

*Proof.* The assertion is trivial for $l = 1$, hence we assume $l < 1$. By the previous lemma, $[p, 1]$ is indecomposable for at least one point $p \leq l$. First suppose $l$ contains at least two more points, say $q$ and $s$. Setting $c_1 = z_1 = q, z_2 = s$, we infer as in the proof of **2.48** that $[q, 1]$ or $[s, 1]$ must be indecomposable as well. Now suppose $l$ contains only one other point $q$ beside $p$, and that $[q, 1]$ is decomposable. Then $l$ and some $h \in C$ must be non-trivial separators of $[q, 1]$ by the same argument as in **2.48**. We shall derive a contradiction by showing that $p$ and $h$ are complementary separators of $L$. Let us verify condition (iii) in **2.45**. Obviously, $p \wedge h = 0$ and $r(p) + r(h) = r(1)$. Choose $s \in S$, different from $p$ and $q$. Then $q \lessdot q \vee s \not\leq l$. But this implies $q \vee s \leq h$ and thus $s \leq h$ since $l$ and $h$ are complementary separators in $[q, 1]$. □

Let us call two copoints $c$, $h$ *neighbours* if $c \wedge h$ is a coline and if there is at least one other copoint beside $c$ and $h$ covering $c \wedge h$.

**2.50 Lemma.** *Let $L$ be indecomposable. Then for each pair $c \in C$, $p \in S$ there exists a neighbour $h$ of $c$ with $h \in H_p$.*

*Proof.* We use induction on the rank of $L$. For $r(L) = 1$ or $2$ there is nothing to prove. Suppose the proposition holds for all indecomposable geometric lattices $L'$ with $r(L') \leq r(L) - 1$. By the last two lemmas, for any pair $c \in C$, $p \in S$ there exists a point $q$, $q \neq p$, $q \leq c$, such that $[q, 1]$ is indecomposable. Set $l = p \vee q$. By the induction assumption applied to $[q, 1]$ there is a copoint $h \in H_l \cap C_q$ such that $h$ is a neighbour of $c$. But now clearly $H_l = H_p \cup H_q$, and thus $h \in H_p$. □

**2.51 Theorem** (Maeda). *Let $L$ be a geometric lattice with center $Z(L)$.*

(i) *Perspectivity is an equivalence relation on the points.*
(ii) *The minimal elements $> 0$ in $Z(L)$ are precisely the joins of the perspectivity classes of* (i).
(iii) *$L$ is indecomposable if and only if any two points are perspective.*

*Proof.* We may assume $r(L) \geq 2$. By **2.47**, every point $p$ is contained in precisely one minimal central element $> 0$, say $a_p$. All three parts of the theorem will follow from the following proposition: Points $p$ and $q$ are perspective if and only if $a_p = a_q$. Suppose $a_p \neq a_q$. Then there is a decomposition $L = [0, a] \times [0, a']$ with $p \leq a$, $q \leq a'$. But this implies $H_p \cap H_q = \varnothing$ by **2.45**(iv), i.e., $p$ and $q$ are not perspective. Let conversely $p \neq q$ be points with $a_p = a_q$. We want to find an element $l$ with $p \nleq l$, $q \nleq l$ and $l \lessdot a_p$. Choose any $k \in L$ with $p \nleq k \lessdot a_p$. If $q \nleq k$ we set $l = k$, so assume $q \leq k$. Applying **2.50** to the indecomposable lattice $[0, a_p]$ we infer the existence of $m \in L$, with $q \nleq m \lessdot a_p$ such that $k$ and $m$ are neighbours in $[0, a_p]$. If $p \nleq m$ we may set $l = m$, hence assume $p \leq m$. But now any element $l \lessdot a_p$ distinct from $k$ and $m$ and covering $k \wedge m$ (at least one such element exists by the definition of neighbours) must satisfy $p \nleq l, q \nleq l$. Denote by $a_p'$ the separator complementary to $a_p$. Then $h = l \vee a_p'$ is a copoint, and we have $a_p \wedge h = l$ (**2.45**(v)), thus $p \nleq h$, $q \nleq h$, i.e., $p$ and $q$ are perspective. □

As a corollary to the main theorem we obtain the following alternative description of the center.

**2.52 Corollary.** *Let $L$ be a geometric lattice, $Z(L)$ its center and $a^\perp$ the set of complements of $a \in L$. Then* $\inf a^\perp \in Z(L)$ *for all $a$. Conversely, every center element $a$ is of the form* $\inf a'^\perp$ *by* **2.45**. *In particular,* $\inf a^\perp = 0$ *for all $a = 0$ if and only if $L$ is indecomposable.*

*Proof.* Since operations in $L = \Pi\,[0, a_i]$ are performed coordinatewise it suffices to prove the proposition for indecomposable lattices. Let $a \neq 0, 1$, and $p, q$ points with $p \leq a, q \nleq a$. By **2.51**, there is a copoint $h$ with $p \nleq h$ and $q \nleq h$. Since $a \vee h = 1$ we can find $b \in a^\perp$ such that $b \leq h$ (**2.40**(iii)). Hence for every point $q$ with $q \nleq a$ there exists a complement of $a$ not containing $q$. But this means precisely $\inf a^\perp = 0$. The converse is obvious. □

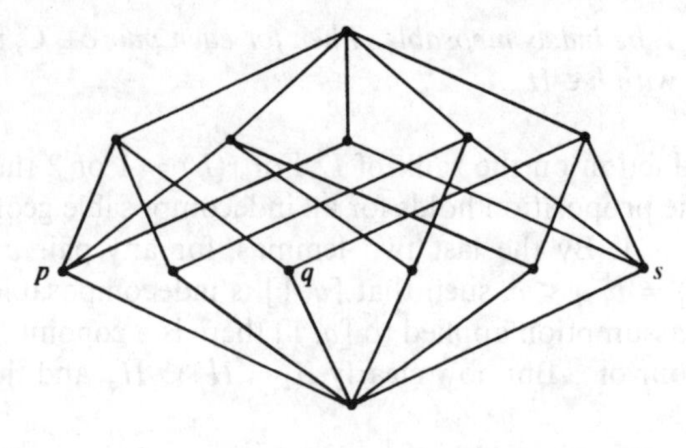


Figure 2.17

Notice that **2.51** and **2.52** are false in arbitrary relatively complemented lattices. The lattice in Figure 2.17 is relatively complemented, lower semimodular and indecomposable. The points $p$ and $q$ are perspective, as are $q$ and $s$, but $p$ and $s$ are not. Also, inf $p^{\perp} = s > 0$. As an application of **2.51** we are going to determine all distributive and modular geometric lattices.

**2.53 Theorem.** *The distributive geometric lattices are precisely the Boolean algebras of finite rank.*

*Proof.* Since distributive lattices are uniquely complemented (**2.22**(i)), no two points can be perspective, whence the lattice $\mathscr{B}(1)$ is the only indecomposable distributive lattice. Since distributivity is inherited by direct products, the theorem follows. □

Turning to modular lattices, let us first discuss a simple characterization of perspectivity in modular geometric lattices which is interesting in its own right.

**2.54 Lemma.** *In a modular geometric lattice, two points $p \neq q$ are perspective if and only if the line $p \vee q$ contains at least one other point beside $p$ and $q$.*

*Proof.* Let $p \neq q$ be perspective with $h$ as common complement. By the modular law $s = (p \vee q) \wedge h$ is a point, different from $p$, $q$ and contained in $p \vee q$. Let conversely $s$ be a point in $p \vee q$ different from $p$ and $q$ and let $l$ be a complement of $p \vee q$. Then the copoint $h = l \vee s$ is a common complement of $p$ and $q$ since for instance $p \leq h$ would imply $h = l \vee s \vee p = l \vee (p \vee q) = 1$. □

**2.55 Theorem** (Birkhoff). *The modular geometric lattices are precisely the direct products of projective space lattices $\mathscr{L}(\mathfrak{P}, \mathfrak{G})$.*

*Proof.* In view of **2.33** it remains to show that any indecomposable modular geometric lattice of rank at least 2 is isomorphic to the subspace lattice of some projective space $(\mathfrak{P}, \mathfrak{G})$. Let $L$ be such a lattice. We take $\mathfrak{P}$ as the set of points of $L$ and $\mathfrak{G}$ as the set of lines (viewed as the collection of points contained in it). Axiom

(i) in **2.32** is trivially satisfied, (iii) is true by **2.54**. To check axiom (ii) suppose $p, q, s \in \mathfrak{P}$ are non-collinear and $l$ is a line intersecting $p \vee q$ and $p \vee s$ in the points $s' \neq p, q$ and $q' \neq p, s$ respectively. Since $l \vee q \vee s = q' \vee s' \vee q \vee s = p \vee q \vee s$, $l \vee q \vee s$ has rank 3 and we have $r(l \wedge (q \vee s)) = r(l) + r(q \vee s) - r(l \vee q \vee s) = 2 + 2 - 3 = 1$. Hence $l$ intersects the third side $q \vee s$ as well. Now it is an easy matter to verify that the elements of $L$, viewed as sets of points, coincide with the subspaces of $(\mathfrak{P}, \mathfrak{G})$ defined inductively in **2.32**. □

In view of **2.34** and **2.35** we can supplement **2.55** by the following theorem.

**2.56 Corollary.** *The modular geometric lattices are precisely the direct products of lattices, each of which is a vector space lattice $\mathscr{L}(n, K)$ for some n and K or the subspace lattice of a non-Desarguesian projective plane. The Arguesian geometric lattices are precisely the direct products of lattices each of which is a vector space lattice $\mathscr{L}(n, K)$ for some n and K.* □

Of the five fundamental classes of lattices introduced in chapter I, chains, the divisor lattice, Boolean algebras, vector space lattices, and partition lattices, the last three have also played a prominent role in our discussion of geometric lattices. Further interesting results concern questions of embeddings.

(1) *Every finite distributive lattice can be embedded in a finite complemented distributive lattice, i.e., in a Boolean algebra $\mathscr{B}(n)$. (See Theorem* **2.8.**)
(2) *This result cannot be strengthened to finite modular lattices.*

Choose the subspace lattice of a finite non-Desarguesian projective plane **P** (such planes exist!) and the lattice $M_5$ of Figure 2.1. We designate a point $a$ in **P** and a copoint $b$ in $M_5$ and identify $a$ with the 1-element of $M_5$ and $b$ with the 0-element of **P** as in Figure 2.18. The resulting lattice $L$ is clearly modular. Suppose $L$ is embedded in a finite complemented modular lattice, i.e., in a modular geometric lattice $M$ by **2.38**. $M$ is a direct product as in **2.56** and from the construction of $L$ it is easily seen that $L$ is embedded in one of the factors of $M$. Since $r(L) = 4$ it follows that this factor is a vector space lattice and hence Arguesian, in contradiction to the choice of **P**.

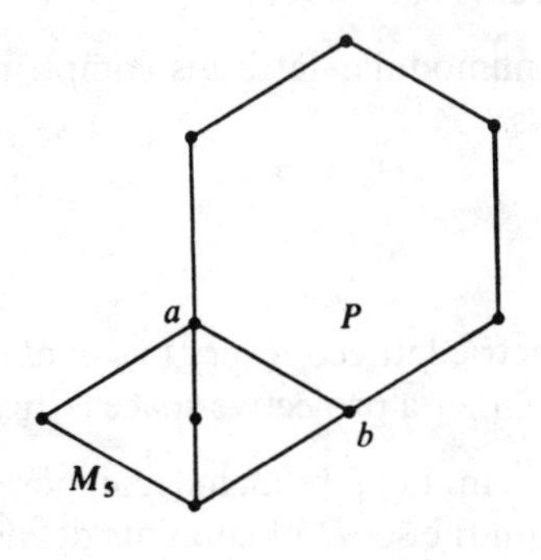


Figure 2.18

It is also not true that every finite Arguesian lattice can be embedded in a finite complemented Arguesian lattice, i.e., in a direct product of vector space lattices; see exercise II.3.15 for an example.

(3) *Any finite lattice can be embedded in a finite partition lattice* $\mathscr{P}(n)$.

EXERCISES II.3

→ 1. Prove **2.28**.

→ 2. Verify the details in the definition of the simple matroid underlying a given matroid.

3. If we weaken the finite basis axiom of a matroid to the statement:

$$\bar{A} = \bigcup_{B \subseteq A, |B| < \infty} \bar{B}$$

for all $A \subseteq S$, then the resulting structure is called a *finitary matroid*. Prove the result corresponding to **2.29** for finitary matroids.

4. Find finitary matroids which do not satisfy the finite basis axiom.

5. Verify the inductive definition of the subspaces of projective spaces in **2.32**.

→ 6. Prove **2.33**.

→ 7. Verify the axioms **2.32** for $\mathbf{PG}(n, K)$ where $K$ is a division ring.

8. Show that for a finite projective plane $(\mathfrak{P}, \mathfrak{G})$, there exists some $n \geq 2$ such that:

(i) $|\mathfrak{P}| = |\mathfrak{G}| = n^2 + n + 1$,
(ii) every line contains $n + 1$ points and through every point there pass $n + 1$ lines.

$n$ is called the *order* of the plane. What is the order of the projective plane $\mathbf{PG}(2, q)$?

→ 9. Prove: In a finite modular geometric lattice the number of $k$-elements subsets of points whose join is 1 equals the number of $k$-element subsets of copoints whose meet is 0, for every $k$.

10. Show that a c.f. semimodular lattice is complemented if and only if 1 is a supremum of points.

→ 11. Prove **2.42**.

12. Prove **2.46**.

13. Prove: If in a geometric lattice $L$ every line contains the same finite number of points, then $L$ is either a projective space lattice or a Boolean algebra.

→ 14. Prove the path theorem: Let $L$ be an indecomposable geometric lattice. Then the graph $G(H, E)$ on the set $H$ of copoints defined by $\{c, d\} \in E :\Leftrightarrow c, d$ are neighbors in the sense of **2.50** is connected. (Tutte)

→15. Choose two projective planes **PG**(2, $p$) and **PG**(2, $q$) over distinct prime fields $GF(p)$ and $GF(q)$ and glue them together as in Figure 2.18 by designating elements $a$ and $b$. Prove that the resulting lattice is Arguesian but that it cannot be embedded in an Arguesian complemented lattice.

## 4. The Fundamental Examples

To close these two introductory chapters let us collect some basic facts about the five fundamental classes of lattices introduced in chapter I. Most of these results are perfectly obvious but will nevertheless prove very useful in the chapters to come. We shall concentrate on two points. First, all these lattices possess a rank function and hence admit a classification of the elements according to their rank. For a poset $P$ with a rank function, the set $P^{(k)} := \{a \in P : r(a) = k\}$ is called the *k-level* of $P$. (See Footnote (1)). For instance, $P^{(0)}$ is the set of minimal elements of $P$ and $P^{(1)}$ the set of atoms. For finite posets $P$, the level-numbers $|P^{(k)}|$ will be of considerable combinatorial significance. Secondly, we shall investigate the interval structure of $P$. Let us denote by Int($P$) the set of all intervals of $P$. Recall that an interval is called *lower* (*upper*) if it contains a minimal (maximal) element. If $P$ has a rank function, we say $[a, b] \in \mathrm{Int}(P)$ has *rank k* or *length k* if $r(b) - r(a) = k$.

Let us quote another useful result whose proof is an easy exercise:

**2.57.** $(\prod_i P_i)^* \cong \prod_i P_i^*$ *for any posets* $P_i$. □

A. Chains

$\mathscr{C}(n)$ denotes the chain of length $n$, $n \in \mathbb{N}$, and $\mathscr{C}(\infty)$ the countable chain isomorphic to $\mathbb{N}_0$. Whenever advantageous we shall identify $\mathscr{C}(n)$ with $\{0, 1, \ldots, n\}_<$ and $\mathscr{C}(\infty)$ with $\mathbb{N}_0$. $\mathscr{C}(n)$ is a distributive lattice.

**2.58.** $|\mathscr{C}(n)| = n + 1$ for $n < \infty$, $|\mathscr{C}(\infty)| = \infty$, and $\mathscr{C}(n) \cong \mathscr{C}^*(n)$ for $n < \infty$. We have $r(i) = i$ for all $i \in \mathbb{N}_0$, hence $|\mathscr{C}(n)^{(k)}| = 1$ for all $k$ and $n$, and $r(\mathscr{C}(n)) = n$.

**2.59.** $[i, j] \cong \mathscr{C}(j - i) = \mathscr{C}(r[i, j])$ for all $i, j \in \mathbb{N}_0$, $i \leq j$. Hence two intervals $[i, j]$, $[k, l]$ are isomorphic if and only if they have the same rank. Partitioning Int($\mathscr{C}(n)$) into its isomorphism classes we see that each isomorphism class is uniquely determined by the rank of its members. Thus we may associate to each isomorphism class the symbol $(n)$ with

$$[i, j] \in (n) :\Leftrightarrow j - i = n \quad (n \in \mathbb{N}_0).$$

$(n)$ is called the *type* of the interval.

(1) The notation $S^{(k)}$ is also used for the family of $k$-subsets of a set $S$. Since it will always be clear from the text whether we are talking about a set or a ranked poset this should cause no confusion.

B. The Divisor Lattice

The divisor lattice $\mathcal{T}$ is distributive, being isomorphic to the sum of countably many copies of $\mathbb{N}_0$. (See Proposition **1.2**.)

**2.60.** If $n = p_1^{k_1} \dots p_t^{k_t} \in \mathbb{N}$, then $r(n) = \sum_{i=1}^{t} k_i$ in $\mathcal{T}$ (**2.14**(iii)). All levels except $\mathcal{T}^{(0)}$ contain countably many elements with $\mathcal{T}^{(k)}$ consisting of all natural numbers whose prime factorization contains $k$ (not necessarily distinct) primes.

**2.61.** $[l, m] \cong [1, m/l]$ for all $[l, m] \in \operatorname{Int}(\mathcal{T})$, by means of $\phi: i \to i/l$. If $n = p_1^{k_1} \dots p_t^{k_t}$, then $[1, n] \cong \prod_{i=1}^{t} \mathscr{C}(k_i)$ under the isomorphism $\phi: p_1^{i_1} \dots p_t^{i_t} \to (i_1, \dots, i_t)$. Thus, in particular, $[l, m]^* \cong [l, m]$ and $|[l, m]| = |[1, m/l]| = \sum_{i=1}^{t} (k_i + 1)$ if $m/l = p_1^{k_1} \dots p_t^{k_t}$. Let us again introduce the *type* $(n)$ of an interval $[l, m]$ by

$$[l, m] \in (n) :\Leftrightarrow \frac{m}{l} = n \quad (n \in \mathbb{N}).$$

Non-isomorphic intervals have different type, but not conversely. For integers $m = p_1^{k_1} \dots p_t^{k_t}$, $n = q_1^{k_1} \dots q_t^{k_t}$ with the same number of prime factors and equal exponents (apart from the order) we have $[1, m] \cong [1, n] \cong \prod_{i=1}^{t} \mathscr{C}(k_i)$ but, in general, $(m) \neq (n)$. Example: $12 = 2^2 3$, $45 = 3^2 5$, thus $[1, 12] \cong [1, 45] \cong \mathscr{C}(2) \times \mathscr{C}(1)$, but $(12) \neq (45)$. In other words, the types as just defined give a finer partition of $\operatorname{Int}(\mathcal{T})$ than isomorphism of intervals.

C. Boolean Algebras

$\mathscr{B}(n)$ denotes the lattice of subsets of a finite $n$-set, $n < \infty$, ordered by inclusion and $\mathscr{B}(\infty)$ the lattice of all finite subsets of a countable set. We shall usually use the letters $A, B, C, \dots$ for subsets and $S$ for the whole set. $\mathscr{B}(n)$ is a distributive complemented lattice if $n < \infty$.

**2.62.** For $n < \infty$, we have $\mathscr{B}(n) \cong \mathscr{B}^*(n)$ by mapping each subset onto its complement. Further, $\mathscr{B}(n) \cong [\mathscr{C}(1)]^n$, hence $|\mathscr{B}(n)| = 2^n$, and $r(A) = |A|$ for all $A \subseteq S$. $\mathscr{B}(S)^{(k)}$ consists therefore of all $k$-subsets of $S$, and we have $r(\mathscr{B}(n)) = n$. $|\mathscr{B}(n)^{(k)}| =: \binom{n}{k}$ are, of course, the *binomial coefficients*, and it follows that

$$\binom{n}{k} = \binom{n}{n-k} \quad \text{for all } k, n \in \mathbb{N}_0,$$

$$2^n = \sum_{k=0}^{n} \binom{n}{k}.$$

For $n = \infty$, all levels above 0 are infinite.

**2.63.** $[A, B] \cong \mathscr{B}(|B - A|) = \mathscr{B}(r[A, B])$ for all $A \subseteq B \subseteq S$. Hence the isomorphism classes are uniquely determined by their *type* $(n)$ where

$$[A, B] \in (n) :\Leftrightarrow |B - A| = n \quad (n \in \mathbb{N}_0).$$

## D. Vector Space Lattices

$\mathscr{L}(n, K)$ denotes the lattice of subspaces of an $n$-dimensional vector space over a division ring $K, n < \infty$. $\mathscr{L}(\infty, K)$ denotes the lattice of finite-dimensional subspaces of a countable-dimensional vector space over $K$. If $K$ is the finite field $GF(q)$ we use the short-hand notation $\mathscr{L}(n, q)$. We shall usually use the letters $U, W, Z, \ldots$ for the subspaces and $V$ for the whole space. Algebraic dimension and rank of a space are synonymous and will be denoted by $\dim V$ or $r(V)$. $\mathscr{L}(n, K)$ is an indecomposable modular complemented lattice.

**2.64.** The dual vector space of a left vector space $V$ over $K$ is the right vector space $\mathrm{Hom}(V, K)$ over $K$ of all linear transformations from $V$ into $K$. If $\dim V < \infty$, we know from linear algebra that $\dim V = \dim \mathrm{Hom}(V, K)$. Setting

$$U^0 := \{f \in \mathrm{Hom}(V, K) : f(U) = 0\},$$

we have that $\phi: U \to U^0$, $U \subseteq V$, is a bijection between $\mathscr{L}(V)$ and $\mathscr{L}(\mathrm{Hom}(V, K))$ which reverses the inclusion relation, i.e., $U \subseteq W \Leftrightarrow U^0 \supseteq W^0$. Hence $\mathscr{L}(V) \cong \mathscr{L}^*(\mathrm{Hom}(V, K))$. If, in particular, $K$ is commutative, we know $V \cong \mathrm{Hom}(V, K)$ thus $\mathscr{L}(V) \cong \mathscr{L}(\mathrm{Hom}(V, K))$, i.e., $\mathscr{L}(n, K) \cong \mathscr{L}^*(n, K)$. $\mathscr{L}(n, K)^{(k)}$ consists of all $k$-dimensional subspaces of an $n$-space, and we have $r(\mathscr{L}(n, K)) = n$. For finite fields, the numbers $|\mathscr{L}(n, q)^{(k)}| =: \binom{n}{k}_q$ are called the *Gaussian coefficients* and $|\mathscr{L}(n, q)| =: G_{n,q}$ the *Galois numbers*. We infer from the arguments above that

$$\binom{n}{k}_q = \binom{n}{n-k}_q \quad \text{for all } k, n \in \mathbb{N}_0,$$

$$G_{n,q} = \sum_{k=0}^{n} \binom{n}{k}_q .$$

For $n = \infty$, all levels above 0 are infinite.

Notice that because of **2.56** and $\mathscr{L}(n, q) \cong \mathscr{L}^*(n, q)$ (since every finite division ring is commutative), any finite Arguesian complemented lattice is isomorphic to its dual. This is, however, not true for arbitrary finite modular geometric lattices since there exist non-Desarguesian projective planes whose subspace lattices are not isomorphic to their duals (see Crawley–Dilworth [1, p.132]).

**2.65.** $[U, W] \cong \mathscr{L}(k, K)$ with $k = r(W) - r(U)$ for all subspaces $U \subseteq W \subseteq V$. Hence the isomorphism classes of $\mathrm{Int}(\mathscr{L}(n, K))$ are again uniquely determined by the *type* $(n)$ where

$$[U, W] \in (n) :\Leftrightarrow r(W) - r(U) = n \quad (n \in \mathbb{N}_0).$$

## E. Partition Lattices

$\mathscr{P}(n)$ denotes the lattice of partitions of an $n$-set, $n < \infty$, and $P(\infty)$ the lattice of all finite partitions of a countable set. We shall usually use the letters $\pi, \sigma, \rho, \tau, \ldots$ to denote partitions of an $n$-set $S$ and $b(\pi)$ for the number of blocks in $\pi$. $\mathscr{P}(n)$ is an indecomposable geometric lattice if $n < \infty$.

**2.66.** For $n < \infty$, we have $r(\pi) = n - b(\pi)$ for all $\pi \in \mathscr{P}(n)$. Hence the $(n-k)$-level of $\mathscr{P}(n)$ consists of all partitions with exactly $k$ blocks. The numbers $|\mathscr{P}(n)^{(n-k)}| =: S_{n,k}$ are called the *Stirling numbers of the second kind* (for reasons to be made clear in the next chapter) and the numbers $|\mathscr{P}(n)| =: B_n$ the *Bell numbers.* Thus we have

$$B_n = \sum_{k=1}^{n} S_{n,k} \quad \text{for all } k, n \in \mathbb{N}.$$

The lattice $\mathscr{P}(\infty)$ does not have a 0-element and hence no rank function. The dual lattice $\mathscr{P}^*(\infty)$ has a rank function with the $k$-level consisting of all $k$-partitions.

**2.67.** Let $n < \infty$ and $\pi = A_1|A_2|\cdots|A_{b(\pi)} \in \mathscr{P}(n)$ with $|A_i| = n_i$, $\sum_{i=1}^{b(\pi)} n_i = n$. Since any partition $\sigma \in [\pi, 1]$ is obtained by combining two or more blocks of $\pi$, we have $[\pi, 1] \cong \mathscr{P}(b(\pi))$. On the other hand, any partition $\tau \in [0, \pi]$ is obtained by partitioning the blocks $A_i$ into smaller parts so that we have $[0, \pi] \cong \prod_{i=1}^{b(\pi)} \mathscr{P}(n_i)$. In summary:

$$[\pi, 1] \cong \mathscr{P}(b(\pi)),$$

$$[0, \pi] \cong \prod_{i=1}^{b(\pi)} \mathscr{P}(n_i),$$

$$[\pi, \sigma] \cong \prod_{i=1}^{b(\sigma)} \mathscr{P}(m_i) \quad \text{for some } m_i\text{'s with } \sum_{i=1}^{b(\sigma)} m_i = b(\pi).$$

Defining the *type* of the interval $[\pi, \sigma]$ by the formal expression

$$\text{type}[\pi, \sigma] := 1^{b_1}2^{b_2}\ldots n^{b_n} \quad \text{if } [\pi, \sigma] \cong \prod_{i=1}^{n} [\mathscr{P}(i)]^{b_i}$$

we see that the isomorphism classes of $\text{Int}(\mathscr{P}(n))$ are uniquely determined by the types associated with it.

Similarly, we define the *type* of a single partition $\pi$ by

$$\text{type}(\pi) := 1^{b_1}2^{b_2}\ldots n^{b_n} \quad \text{if } [0, \pi] \cong \prod_{i=1}^{n} [\mathscr{P}(i)]^{b_i},$$

i.e., if $\pi$ contains exactly $b_i$ blocks of cardinality $i$ ($i = 1, \ldots, n$). Notice that the types of $\pi$ correspond bijectively to the possible number-partitions of $n$ (or the elements of the dominance order $\mathscr{D}(n)$):

$$1^{b_1}2^{b_2}\ldots n^{b_n} \leftrightarrow \underbrace{1 + \cdots + 1}_{b_1} + \underbrace{2 + \cdots + 2}_{b_2} + \cdots + \underbrace{n + \cdots + n}_{b_n}.$$

Corresponding results hold for the lattice $\mathscr{P}(\infty)$.

EXERCISES II.4

1. Prove **2.57**.

→ 2. Verify the details of **2.58** and **2.59**.

→ 3. Verify the details of **2.60** and **2.61** and determine Aut $\mathcal{T}$ and its orbits.

→ 4. Verify the details of **2.62** and **2.63** and determine Aut $\mathcal{B}(n)$ and its orbits.

5. Let $S$ be a finite set. Is there another lattice isomorphism $\phi: \mathcal{B}^*(S) \to \mathcal{B}(S)$ beside the complementation $A \to S - A$?

→ 6. Verify the details of **2.64** and **2.65** and determine Aut $\mathcal{L}(n, k)$ and its orbits.

7. Show that for any point $p \in \mathcal{L}(n, K)$, we have $[p, 1] \cong \mathcal{L}(n-1, K)$.

→ 8. Show that the affine lattices $\mathcal{A}(n, K)$ are geometric; determine Aut $\mathcal{A}(n, K)$ and its orbits. Describe the closure operator corresponding to $\mathcal{A}(n, K)$.

9.* Show that for any point $p \in \mathcal{A}(n, K)$, we have $[p, 1] \cong \mathcal{L}(n-1, K)$ and for any copoint $c$, $[0, c] \cong \mathcal{A}(n-1, K)$. Describe the levels of $\mathcal{A}(n, q)$ as in **2.64** and **2.65**.

→ 10. Verify the details in **2.66** and **2.67**.

→ 11. Determine Aut $\mathcal{P}(n), n < \infty$, and show that the orbits of Aut $\mathcal{P}(n)$ correspond bijectively to the types of the partitions.

12. Continuation of ex. I.2.9: Show that $\pi, \sigma \in \mathcal{P}(n)$, $n < \infty$, form a modular pair in $\mathcal{P}(n)$ if and only if the graph $G$ in ex. I.2.9 is a forest.

→ 13. Prove that the modular elements in $\mathcal{P}(n)$, $n < \infty$, are precisely the partitions of type $1^{n-k}k$ for $1 \leq k \leq n$.

14. Suppose the relations $(\pi)$ and $(\sigma)$ corresponding to $\pi, \sigma \in \mathcal{P}(n)$, $n < \infty$, commute, i.e., $(\pi)(\sigma) = (\sigma)(\pi)$. Prove that then $\pi, \sigma$ are a modular pair in $\mathcal{P}^*(n)$. Show, further, that $\pi, \sigma$ are a modular pair in $\mathcal{P}^*(n)$ if and only if $(\pi)(\sigma)(\pi) = (\sigma)(\pi)(\sigma)$.

15.* Show that $\mathcal{P}(n)$, $n < \infty$, is isomorphic to a sublattice of the subgroup lattice of the symmetric group $S_n$. (Hint: Associate to $\pi \in \mathcal{P}(n)$ the group $G(\pi)$ of section I.4).

## Notes

As mentioned in the introduction, this chapter was intended to prepare the ground for the subsequent theory by describing some of the combinatorial aspects of finite posets and lattices. Naturally, because of the representation theorem **2.5**, distributive lattices occupy the central position in any such discussion. The presentation in sections 1 and 2 is based on Birkhoff [1] and Crawley–Dilworth [1].

The formulation of the exchange principle in geometric lattices is contained explicitly or implicitly in the work of Birkhoff [5], Dilworth [1, 2], Whitney [7], Maeda [2, 3], and others. The development here is similar to that in Crapo–Rota [1] in sections 3.A and 3.B, and to that in Tutte [10] in section 3.C on indecomposable geometric lattices, although Tutte's terminology differs entirely from ours; see also Sasaki–Fujiwara [1] and Mac Lane [1]. For an excellent account of the geometric setting, the reader is referred to Jónsson [3].

Chapter III

# Counting Functions

Consider the family of $n$-subsets of a set $R$ or the family of $r$-partitions of a set $N$. In chapter I, these families were viewed as special patterns whereas in chapter II they appeared as levels of certain lattices. In this chapter we want to count patterns and lattice levels starting from table **1.13** on the one hand and the fundamental examples of section II.4 on the other. In accordance with the set-up of the book we shall concentrate more on deriving general counting principles rather than on supplying a large set of recursion and inversion formulae for well-known coefficients such as the binomial coefficients or various partition numbers. For a good collection of the latter the reader is referred to Riordan [1, 2] or to Knuth [1, vol. 1].

## 1. The Elementary Counting Coefficients

Let us determine the cardinalities of the fundamental patterns in **1.13**. Most of the proofs are straightforward; only a few will be given. $N$ and $R$ are finite sets of cardinality $n$ and $r$, respectively.

### A. The Number of all Mappings and of all Injective Mappings $f: N \to R$

**3.1.** *The number of mappings* $f: N \to R$ *is*

$$|\mathrm{Map}(N, R)| = r^n.$$

*Proof.* Every mapping $f: N \to R$ can be viewed as a word of length $n$ from the alphabet $R$. As there are $r$ possibilities for each of the $n$ letters, there are $\underbrace{r \cdot r \cdots r}_{n} = r^n$ such words. (Rule of Products) □

**Example.** Consider the set of $n \times n$-matrices over $\mathbb{Z}$ with all elements equal to 0, 1 or $-1$. Any such matrix $M$ can be thought of as a mapping $M: \{1, \ldots, n\}^2 \to \{0, 1, -1\}$, and conversely. Hence the total number of these matrices is $3^{n^2}$. (Rule of Equality)

**3.2.** *The number of injective mappings* $f: N \to R$ *is*

$$|\mathrm{Inj}(N, R)| = r(r-1)\cdots(r-n+1).$$

*Proof.* $\mathrm{Inj}(N, R)$ is by definition the set of strict words in $R$ of length $n$. There are $r$ possibilities for the first letter. Having chosen the first letter there are $r - 1$ possibilities for the second letter. In general, for every fixed $i$-tuple of letters in the first $i$ positions we can choose any of the remaining $r - i$ letters to fill the $i + 1$-st position. The result follows by induction. □

The expressions $r(r-1)\cdots(r-n+1)$ are called *falling factorials of length n*, denoted by

$$[r]_n := r(r-1)\cdots(r-n+1) \qquad (n \in \mathbb{N}_0)$$

with the convention $[r]_0 = 1$ for all $r$.

**Example.** Let $R = \{a, b, c, d\}$, $N = \{1, 2\}$. Then $[4]_2 = 12$, and the strict words are

$$\begin{array}{cccc} ab & ba & ca & da \\ ac & bc & cb & db \\ ad & bd & cd & dc. \end{array}$$

For $n = r$ we obtain the corollary:

**3.3.** *The number of permutations of an n-set is*

$$|S_n| = n(n-1)\cdots 3 \cdot 2 \cdot 1. \quad \square$$

The expression $n(n-1)\cdots 2 \cdot 1$ is called *n-factorial*, denoted by $n!$, with the convention $0! = 1$. Thus we have

$$[r]_n = \frac{r!}{(r-n)!} \quad \text{for all } r \geq n \geq 0.$$

B. The Number of $n$-subsets and of $n$-multisets of $R$

**3.4.** *The number* $\binom{r}{n}$ *of n-subsets of an r-set is*

$$\binom{r}{n} = \frac{[r]_n}{n!} = \frac{r!}{n!(r-n)!}.$$

*Proof.* The function $\phi: b_1 b_2 \ldots b_n \to \{b_1, b_2, \ldots, b_n\}$ maps the set of injective words in $R$ of length $n$ onto the set of all $n$-subsets of $R$. Each $n$-subset is the image under $\phi$ of precisely $n!$ injective mappings obtained by permuting the $b_i$'s. The result follows by **3.2.** □

The *binomial coefficients* $\binom{r}{n}$ have no combinatorial meaning for negative values of $r$. But we can formally extend **3.4** to negative arguments by using the definition of lower factorials:

$$\binom{-r}{n} = \frac{[-r]_n}{n!} = \frac{(-r)(-r-1)\cdots(-r-n+1)}{n!}$$
$$= (-1)^n \frac{r(r+1)\cdots(r+n-1)}{n!}.$$

The expressions $r(r+1)\cdots(r+n-1)$ are called *rising factorials of length n*, denoted by

$$[r]^n := r(r+1)\cdots(r+n-1) \qquad (n \in \mathbb{N}_0)$$

with the convention $[r]^0 = 1$.

Hence we have

**3.5.**
$$[-r]_n = (-1)^n[r]^n \qquad (n, r \in \mathbb{N}_0)$$
$$[-r]^n = (-1)^n[r]_n \qquad (n, r \in \mathbb{N}_0).$$

This extension of combinatorial identities to all integers and even to real or complex numbers will provide short proofs of many counting formulae. We shall return to this subject in the next section. **3.5**, relating the two counting sequences $\{[r]_n : n \in \mathbb{N}_0\}$ and $\{[r]^n : n \in \mathbb{N}_0\}$, is the first example of what is called a *combinatorial reciprocity law*. A study of general reciprocity and inversion theorems will be undertaken in section IV.2.

**3.6.** *The number of n-multisets in R (i.e., the number of monotone words of length n in R) is*

$$|\mathrm{Mon}(N, R)| = \frac{[r]^n}{n!}.$$

*Proof.* Let $N = \{1, \ldots, n\}$ and $R = \{1, \ldots, r\}$, both with the natural order. The mapping $\phi: b_1 b_2 \ldots b_n \to \{b_1, b_2 + 1, \ldots, b_n + (n-1)\}$ is a bijection between $\mathrm{Mon}(N, R)$ and the set of all $n$-subsets of $\{1, 2, \ldots, r+n-1\}$. □

**Example.** Let $N = \{1 < 2 < 3 < 4\}$, $R = \{a < b < c\}$. Then $|\mathrm{Mon}(N, R)| = [3]^4/4! = 15$, and the words are

| | | | | |
|---|---|---|---|---|
| *aaaa* | *aabb* | *abbb* | *accc* | *bbcc* |
| *aaab* | *aabc* | *abbc* | *bbbb* | *bccc* |
| *aaac* | *aacc* | *abcc* | *bbbc* | *cccc*. |

There is also a combinatorial meaning of the rising factorials alone. $[r]^n$ counts the number of sortings of $n$ objects into $r$ linearly ordered boxes. (See the exercises)

The binomial coefficients are probably the best known of all combinatorial counting numbers. Their applications abound in all branches of mathematics. Let us for the moment just list two basic properties.

**3.7 Proposition.** *Let $n \in \mathbb{N}$. The sequence $\{\binom{n}{k}: k = 0, \ldots, n\}$ of binomial coefficients satisfies:*

(i) $$\binom{n}{k} = \binom{n}{n-k} \qquad (k = 0, \ldots, n),$$

(ii) $$\binom{n}{0} < \binom{n}{1} < \cdots < \binom{n}{n/2} > \binom{n}{n/2+1} > \cdots > \binom{n}{n} \qquad (n \text{ even})$$

$$\binom{n}{0} < \binom{n}{1} < \cdots < \binom{n}{(n-1)/2} = \binom{n}{(n+1)/2} > \cdots > \binom{n}{n}$$

$(n$ *odd*$)$. □

Property (ii) states that the sequence $\{\binom{n}{k}: k = 0, \ldots, n\}$ rises to a maximum (which is possibly attained twice) and falls thereafter. Sequences that rise and fall in this fashion are called unimodal. To be precise:

**Definition.** A sequence $v_0, v_1, \ldots, v_n$ of real numbers is called *unimodal* if there exists an integer $M \geq 0$ such that

$$v_0 \leq v_1 \leq \cdots \leq v_{M-1} \leq v_M \geq v_{M+1} \geq \cdots \geq v_n.$$

We shall verify the unimodality property for other basic sequences in this and the next section. In chapter VIII we shall take the subject up again when we examine the level numbers of finite posets with a rank function.

**3.8 Binomial Theorem.** *Let $A$ be a commutative ring. Then for all $a, b \in A$*

$$(a+b)^n = \sum_{k=0}^{n} \binom{n}{k} a^k b^{n-k}.$$

*Proof.* Writing $(a+b)^n = (a+b)(a+b)\cdots(a+b)$ we see that the right-hand side is the sum of all products $a^k b^{n-k}$ where $a^k b^{n-k}$ appears as often as we can choose $k$ $a$'s from the $n$ factors, i.e., $\binom{n}{k}$ times. □

For $A = \dot{\mathbb{Z}}$, (i) $a = b = 1$ and (ii) $a = -1, b = 1$ we obtain:

**3.9.**

(i) $$\sum_{k=0}^{n} \binom{n}{k} = 2^n,$$

(ii) $$\sum_{k=0}^{n} (-1)^k \binom{n}{k} = 0 \qquad (n \geq 1).$$

The binomial coefficients admit several useful generalizations. Let us associate with each $A \subseteq N$ its characteristic function $f_A: N \to \{0, 1\}$, given by

$$f_A(a) = \begin{cases} 1 & \text{if } a \in A \\ 0 & \text{if } a \notin A. \end{cases}$$

The bijection $\phi: \mathscr{B}(N) \to \mathrm{Map}(N, \{0, 1\})$ with $\phi A = f_A$ maps the $k$-subsets of $N$ precisely onto the maps $f \in \mathrm{Map}(N, \{0, 1\})$ with $|f^{-1}(1)| = k$. This suggests the following generalization. Consider $\mathrm{Map}(N, R)$ and denote by

$$\binom{n}{k_1, \ldots, k_r}$$

the number of mappings $f: N \to R = \{b_1, \ldots, b_r\}$ with $|f^{-1}(b_i)| = k_i$ $(i = 1, \ldots, r)$. The numbers

$$\binom{n}{k_1, \ldots, k_r}$$

are called the *multinomial coefficients.* We note the following generalizations of **3.4**, **3.8**, and **3.9**(i).

**3.10 Proposition.** *We have*:

(i) $$\binom{n}{k_1, \ldots, k_r} = \begin{cases} \dfrac{n!}{k_1!k_2!, \ldots, k_r!} & \text{if } n = \sum_{i=1}^{r} k_i \\ 0 & \text{otherwise.} \end{cases}$$

*Let $A$ be a commutative ring, $a_1, \ldots, a_r \in A$. Then*

(ii) $$(a_1 + \cdots + a_r)^n = \sum_{(k_1, \ldots, k_r)} \binom{n}{k_1, \ldots, k_r} a_1^{k_1} a_2^{k_2} \ldots a_r^{k_r}.$$

(iii) $$\sum_{(k_1, \ldots, k_r)} \binom{n}{k_1, \ldots, k_r} = r^n.$$

*Proof.* We prove (i). Suppose $n = k_1 + k_2 + \cdots + k_r$. The $k_1$ elements which are mapped onto $b_1 \in R$ can be chosen in

$$\binom{n}{k_1}$$

ways. Assuming that we have picked these elements we can choose the $k_2$ elements which are mapped onto $b_2 \in R$ in

$$\binom{n-k_1}{k_2}$$

ways. Continuing in this way we find that

$$\begin{aligned}\binom{n}{k_1,\ldots,k_r} &= \binom{n}{k_1}\binom{n-k_1}{k_2}\cdots\binom{n-k_1-\cdots-k_{r-1}}{k_r} \\ &= \frac{n!}{k_1!(n-k_1)!}\frac{(n-k_1)!}{k_2!(n-k_1-k_2)!}\cdots\frac{(n-k_1-\cdots-k_{r-1})!}{k_r!(n-k_1-\cdots-k_r)!} \\ &= \frac{n!}{k_1!k_2!\cdots k_r!}. \quad \square\end{aligned}$$

C. The Number of $k$-subspaces of an $n$-dimensional Vector Space

Let us turn to vector space lattices $\mathscr{L}(n, q)$ over finite fields. In **2.64** the number of $k$-dimensional subspaces of an $n$-dimensional vector space $V$ over $GF(q)$ was called the *Gaussian coefficient* $\binom{n}{k}_q$. There are several close analogies between the numbers $\binom{n}{k}_q$ and the binomial coefficients $\binom{n}{k}$. Viewing, for instance, **3.11** below as a function of the real variable $q$, it is easily verified that $\binom{n}{k}_q$ tends to $\binom{n}{k}$ as $q$ tends to 1. The same relationship can be observed for most other formulae involving the numbers $\binom{n}{k}_q$.

**3.11.** *The number of k-dimensional subspaces of an n-dimensional vector space V(n, q) is*

$$\binom{n}{k}_q = \frac{(q^n-1)(q^{n-1}-1)\cdots(q^{n-k+1}-1)}{(q^k-1)(q^{k-1}-1)\cdots(q-1)} \qquad (k = 0, \ldots, n). \quad \square$$

*Proof.* Let us first determine the number $U_{n,k}$ of ordered $k$-tuples of linearly independent vectors in $V(n, q)$. As the first coordinate of such a $k$-tuple we can take any one of the $q^n - 1$ vectors different from 0. Any vector $v \neq 0$ spans a one-dimensional subspace containing $q$ vectors. Hence there are $q^n - q$ vectors linearly independent of $v$ and any one of them can be taken as the second coordinate. Let $w$ be one of these. The pair $\{v, w\}$ spans a two-dimensional subspace containing $q^2$ vectors. Hence there are $q^n - q^2$ vectors linearly independent of

$\{v, w\}$ and any one of them can be taken as the third coordinate. Continuing in this way we conclude that

$$U_{n,k} = (q^n - 1)(q^n - q^2) \cdots (q^n - q^{k-1}).$$

Each $k$-tuple of linearly independent vectors spans a $k$-dimensional subspace and, conversely, any $k$-dimensional subspace possesses $U_{k,k}$ ordered bases. Thus we obtain

$$\binom{n}{k}_q = \frac{U_{n,k}}{U_{k,k}},$$

and the theorem follows by cancelling powers of $q$. □

If we write **3.11** in the form

**3.12.**
$$\binom{n}{k}_q = \frac{\prod_{i=1}^{n} (q^i - 1)}{\prod_{i=1}^{k} (q^i - 1) \prod_{i=1}^{n-k} (q^i - 1)}$$

and compare this with **3.4** we see that $\prod_{i=1}^{n} (q^i - 1)$ is the "*q-analogue*" of the number $n!$. Later on we shall translate some other counting coefficients pertaining to sets into their $q$-analogues pertaining to vector spaces. For the moment, let us note the properties corresponding to **3.7** for the sequence $\binom{n}{k}_q$.

**3.13 Proposition.** *Let $n \in \mathbb{N}$. The sequence $\{\binom{n}{k}_q : k = 0, \ldots, n\}$ of Gaussian coefficients satisfies*:

(i) $$\binom{n}{k}_q = \binom{n}{n-k}_q \qquad (k = 0, \ldots, n),$$

(ii) $$\binom{n}{0}_q < \binom{n}{1}_q < \cdots < \binom{n}{n/2}_q > \binom{n}{n/2+1}_q > \cdots > \binom{n}{n}_q \qquad (n \text{ even})$$

$$\binom{n}{0}_q < \binom{n}{1}_q < \cdots < \binom{n}{(n-1)/2}_q = \binom{n}{(n+1)/2}_q > \cdots > \binom{n}{n}_q$$

*(n odd).*

*Thus the sequence $\{\binom{n}{k}_q : k = 0, \ldots, n\}$ is unimodal.*

## D. The Number of Surjective Mappings $f: N \to R$ and of $r$-partitions of $N$

In **2.66** we defined the Stirling numbers of the 2nd kind $S_{n,r}$ as the number of $r$-partitions of an $n$-set. There is no straightforward computation of the numbers $S_{n,r}$ as in the case of the binomial or Gaussian coefficients. We shall derive a formula in the next section where we also show that the sequence $\{S_{n,1}, \ldots, S_{n,n}\}$ is unimodal for any $n$.

**3.14.** *The number of surjective mappings $f: N \to R$ is*

$$|\mathrm{Sur}(N, R)| = r! S_{n,r}.$$

*Proof.* Any surjective mapping $f$ induces a unique partition of $N$, namely $\ker(f)$. Conversely, to any partition $\pi$ of $N$ there are precisely $r!$ surjective mappings $f$ with $\ker(f) = \pi$, obtained by permuting the elements of $R$. □

**3.15.** *Denote by* $\mathrm{Per}(b_1, \ldots, b_n)$ *the number of permutations of an n-set N of type* $1^{b_1} \ldots n^{b_n}$ *and by* $P(b_1, \ldots, b_n)$ *the number of partitions of N of type* $1^{b_1} \ldots n^{b_n}$. *Then*

(i) $$\mathrm{Per}(b_1, \ldots, b_n) = \begin{cases} \dfrac{n!}{b_1!b_2!\ldots b_n!1^{b_1}2^{b_2}\ldots n^{b_n}} & \text{if } n = \sum_{i=1}^{n} ib_i \\ 0 & \text{otherwise.} \end{cases}$$

(ii) $$P(b_1, \ldots, b_n) = \begin{cases} \dfrac{n!}{b_1!b_2!\ldots b_n!(2!)^{b_2}\ldots(n!)^{b_n}} & \text{if } n = \sum_{i=1}^{n} ib_i \\ 0 & \text{otherwise.} \end{cases}$$

*Proof.* Let $f$ be a permutation of type $1^{b_1} \ldots n^{b_n}$. Permuting the elements of $N$ yields all other permutations of this type, however with duplications. Duplications are produced in two ways: (a) by permuting whole cycles of the same length; this gives $b_1!b_2!\ldots b_n!$ duplications; (b) by cyclically permuting the elements within a cycle; this gives $2^{b_2} \ldots n^{b_n}$ duplications. As the duplications under (a) and (b) are independent, (i) results. Every permutation yields a unique partition of the same type, namely its cycle partition. Conversely, to every $\pi \in \mathscr{P}(N)$ of type $1^{b_1} \ldots n^{b_n}$ there are precisely $1!^{b_2}2!^{b_3} \ldots (n-1)!^{b_n}$ permutations with $\pi$ as their cycle decomposition. □

**Example.** $N = \{1, 2, 3, 4, 5\}$. There are $5!/2!3! = 10$ partitions of $N$ of type $1^2 3$:

123|4|5 124|3|5 125|3|4 134|2|5 135|2|4
145|2|3 234|1|5 235|1|4 245|1|3 345|1|2.

In general, we have $P(1^{n-k}k) = n!/k!(n-k)! = \binom{n}{k}$ which is also clear from the fact that we may pick all possible $k$-subsets of $N$ as the non-trivial block.

## E. The Number of $r$-partitions and Ordered $r$-partitions of $n$

**3.16.** *The number of ordered r-partitions of n is* $\binom{n-1}{r-1}$.

*Proof.* For $n = 1$ there is nothing to prove. If $n > 1$ then

$$\phi: n_1 + n_2 + \cdots + n_r \to (n_1, n_1 + n_2, \ldots, n_1 + \cdots + n_{r-1})$$

maps every ordered $r$-partition of $n$ onto a monotone strict word of length $r-1$ in $\{1, \ldots, n-1\}_<$. The inverse of $\phi$ is given by

$$\psi: (s_1, s_2, \ldots, s_{r-1}) \to s_1 + (s_2 - s_1) + \cdots + (s_{r-1} - s_{r-2}) + (n - s_{r-1}).$$

Hence the two sets have the same cardinality. □

**Example.** For $n = 7$, $r = 4$ there are $\binom{6}{3} = 20$ ordered 4-partitions of 7:

| | | | |
|---|---|---|---|
| 4 + 1 + 1 + 1 | 3 + 2 + 1 + 1 | 1 + 3 + 2 + 1 | 2 + 2 + 2 + 1 |
| 1 + 4 + 1 + 1 | 3 + 1 + 2 + 1 | 1 + 3 + 1 + 2 | 2 + 2 + 1 + 2 |
| 1 + 1 + 4 + 1 | 3 + 1 + 1 + 2 | 1 + 2 + 3 + 1 | 2 + 1 + 2 + 2 |
| 1 + 1 + 1 + 4 | 2 + 3 + 1 + 1 | 1 + 2 + 1 + 3 | 1 + 2 + 2 + 2. |
| | 2 + 1 + 3 + 1 | 1 + 1 + 3 + 2 | |
| | 2 + 1 + 1 + 3 | 1 + 1 + 2 + 3 | |

The number of unordered $r$-partitions of $n$ is usually denoted by $P_{n,r}$. We shall derive a simple recursion for the numbers $P_{n,r}$ in the next section. For the present, let us note some consequences that are easily seen from the concept of the Ferrers diagram introduced in **2.14**(iv). Recall that the Ferrers diagram of the partition $\alpha$: $n_1 + \cdots + n_r$ (where, as always, $n_1 \geq n_2 \geq \cdots \geq n_r$) is a rectangular array with $n_i$ boxes in row $i$.

**Example.** Consider the partition $\alpha$: $20 = 7 + 4 + 3 + 3 + 2 + 1$. The Ferrers diagram of $\alpha$ is shown in Figure 3.1. By counting the columns of the diagram we obtain a new partition $\alpha^*$ of 20, namely $\alpha^*: 20 = 6 + 5 + 4 + 2 + 1 + 1 + 1$. Clearly, this vertical counting can be done for any partition $\alpha$, and the partition thus obtained is called the *conjugate partition* $\alpha^*$. If $\alpha = \alpha^*$, the partition $\alpha$ is called *self-conjugate.*

The following facts about conjugate partitions are obvious.

**3.17 Proposition.** *Let* $\alpha: n_1 + \cdots + n_r$. *Then*

(i) $\alpha^*: n_1^* + \cdots + n_s^*$ *with* $n_j^* = |\{i: n_i \geq j\}|$.
(ii) $(\alpha^*)^* = \alpha$. *In particular,* $\phi: \alpha \to \alpha^*$ *is a bijection on the set of all partitions.* □

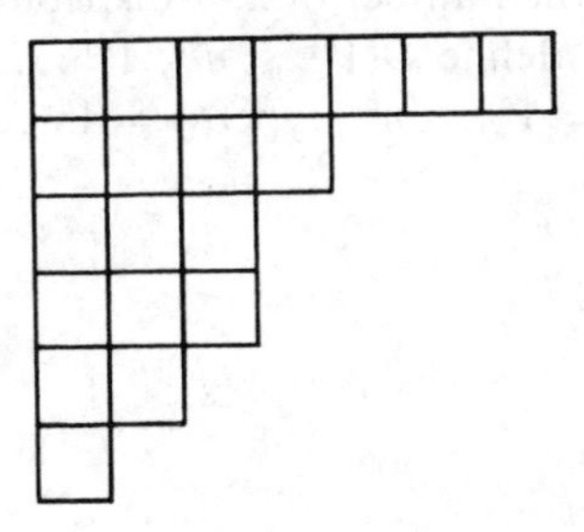

Figure 3.1

Let us note a very useful application of the notion of conjugacy.

**3.18.** *The number $P_{n,r}$ of r-partitions of n equals the number of partitions of n with greatest summand equal to r.* □

Many other interesting formulae can be obtained by looking at the Ferrers diagram of various types of partitions. The reader may consult Riordan [1, p. 114] for more information. Just to get an idea of the subject let us derive one particularly useful result concerning self-conjugate partitions.

**3.19.** *The number of self-conjugate partitions of n equals the number of partitions of n consisting of unequal odd summands.*

*Proof.* The assumption $\alpha = \alpha^*$ is equivalent to the fact that the Ferrers diagram is symmetric with respect to the main diagonal. Hence we may decompose the diagram into *hooks* as shown in Figure 3.2. Clearly, all these hooks have odd length, decreasing strictly as we proceed inward. The converse construction is equally clear. □

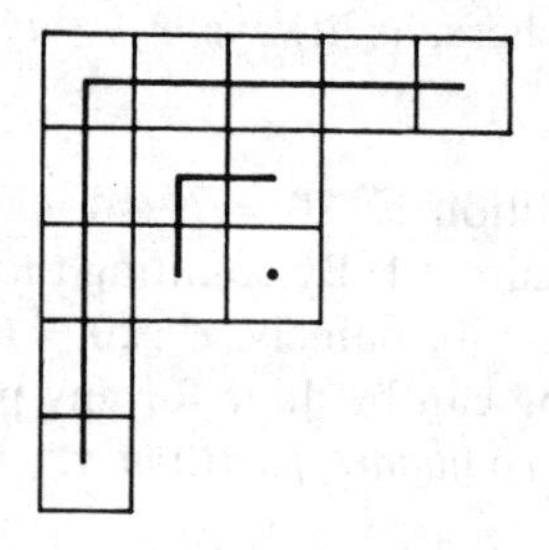

Figure 3.2

## F. The Number of Distributions

To conclude this section, let us discuss the distributions introduced in I.4. Recall that a distribution is a $G, H$-pattern with $G = G(\pi)$, $H = H(\sigma)$ for some partitions $\pi \in \mathscr{P}(N)$, $\sigma \in \mathscr{P}(R)$. Since the number of $\pi, \sigma$-distributions solely depends on the types of $\pi$ and $\sigma$ we may define $D(1^{b_1} \dots n^{b_n}, 1^{d_1} \dots r^{d_r})$ as the number of $\pi,\sigma$-distributions with $\text{type}(\pi) = 1^{b_1} \dots n^{b_n}$, $\text{type}(\sigma) = 1^{d_1} \dots r^{d_r}$.

**3.20.** *We have:*

(i) $D(1^n, 1^r) = r^n$
(ii) $D(n, 1^r) = \binom{r+n-1}{n}$
(iii) $D(1^n, r) = \sum_{k=1}^{r} S_{n,k}$
(iv) $D(n, r) = \sum_{k=1}^{r} P_{n,k}$
(v) $D(1^{b_1} \dots n^{b_n}, 1^r) = \prod_{i=1}^{n} \binom{r+i-1}{i}^{b_i}$.

*Proof.* (i) to (iv) have been shown before. As to (v), sort the blocks of $\pi$ into the boxes of $R$ and combine all these sortings. □

**3.21.** *Let $D_s(1^{b_1} \dots n^{b_n}, 1^{d_1} \dots r^{d_r})$ be the number of surjective $\pi$, $\sigma$-distributions. Then*

(i) $D_s(1^n, 1^r) = r!S_{n,r}$
(ii) $D_s(n, 1^r) = \binom{n-1}{r-1}$
(iii) $D_s(1^n, r) = S_{n,r}$
(iv) $D_s(n, r) = P_{n,r}$. □

## G. Summary

Corresponding to table **1.13** let us summarize the cardinalities of the fundamental patterns. The columns labeled trivial in **1.13** contain either 1 or 0 depending on whether $n \le r$ or $n > r$.

**3.22.** *Fundamental counting numbers*

| $f: N \to R$ | arbitrary | injective | surjective | bijective |
|---|---|---|---|---|
| $G = E(N)$, $H = E(R)$ | $r^n$ | $[r]_n$ | $r!S_{n,r}$ | $r!$ |
| $G = S(N)$, $G = E(R)$ | $\frac{[r]^n}{n!}$ | $\binom{r}{n}$ | $\binom{n-1}{r-1}$ | 1 |
| $G = E(N)$, $H = S(R)$ | $\sum_{k=1}^{r} S_{n,k}$ | 0 or 1 | $S_{n,r}$ | 1 |
| $G = S(N)$, $H = S(R)$ | $\sum_{k=1}^{r} P_{n,k}$ | 0 or 1 | $P_{n,r}$ | 1 |

### Exercises III.1

→ 1. Identities involving binomial coefficients. Verify:

(i) $\sum_{k\ge 0} \binom{n}{2k} = \sum_{k\ge 0} \binom{n}{2k+1} = 2^{n-1}$,
(ii) $\sum_{i=m}^{n} (-1)^{m+i} \binom{n}{i}\binom{i}{m} = \delta_{m,n}$,
(iii) $\sum_{i=0}^{n} \binom{2n+1}{2i} = 4^n$,
(iv) $\binom{r}{n}\binom{n}{k} = \binom{r}{k}\binom{r-k}{n-k}$.

→ 2. Let $\mathbb{N}_0^k$ be the product lattice, $k \in \mathbb{N}$, and $\alpha = (a_1, \dots, a_k)$, $\beta = (b_1, \dots, b_k) \in \mathbb{N}_0^k$ with $\alpha \le \beta$. Show that the number of maximal $\alpha$, $\beta$-chains in $\mathbb{N}_0^k$ equals the multinomial coefficient

$$\binom{\sum_{i=1}^{k} (b_i - a_i)}{b_1 - a_1, b_2 - a_2, \dots, b_k - a_k}.$$

3. Prove: (i)

$$\binom{n}{n_1, \ldots, n_k} = \sum_{i=0,\, n_i \neq 0}^{k} \binom{n-1}{n_1, \ldots, n_{i-1}, n_i - 1, n_{i+1}, \ldots, n_k},$$

(ii)
$$\sum_{(n_1, \ldots, n_k)} \binom{n}{n_1, \ldots, n_k} \cdot (-1)^{n_2 + n_4 + n_6 + \cdots} = \frac{1 - (-1)^k}{2}.$$

4. Verify **3.13**.

5. Prove:

$$S_{n,k} = \sum_{\substack{(b_1, \ldots, b_n) \\ \Sigma b_i = n-k}} 1^{b_1} 2^{b_2} \ldots k^{b_k}.$$

6. What is the number of sortings of $n$ objects into $r$ boxes such that $r - k$ of the boxes remain empty (ordered and unordered)?

→ 7. Show that the number of ordered $r$-partitions of $n$ of type $1^{k_1} \ldots n^{k_n}$ (i.e., there appear $k_i$ $i$'s in the partition for every $i$) is $r!/k_1! \ldots k_n!$ if $n = \sum_{i=1}^{n} ik_i$, $r = \sum_{i=1}^{n} k_i$. (Hint: Consider the proof of **3.15**.)

8. Is it clear that $\binom{n-1}{k-1}$ is the number of 0, 1-chains in $\mathscr{C}(n)$ of length $k$? What is the corresponding number in $\mathscr{B}(n)$? Show that the number of all 0, 1-chains in $\mathscr{C}(n)$ is $2^{n-1}$. What is this number in $\mathscr{B}(n)$?

→ 9. Extend **3.18**: The number of partitions of $n$ with at most $r$ summands equals the number of partitions of $n$ with largest summand $\leq r$ which, in turn, equals the number of $r$-partitions of $n + r$.

→10. Prove by means of the Ferrers diagram: $P_{n,k}$ equals the number of partitions of $n + \binom{k}{2}$ with exactly $k$ unequal summands.

11. Show that the number of partitions of $n$ into odd summands equals the number of partitions into unequal summands.

→12. Just as a chain $0 < a_1 < a_2 < \cdots < a_k = n$ in $\mathscr{C}(n)$ can be regarded as the ordered partition $a_1 + (a_2 - a_1) + \cdots + (n - a_{k-1})$ of $n$ (cf. ex. 8), a 1,$n$-chain in the lattice $\mathscr{T}$, say $1 < b_1 < b_2 < \cdots < b_k = n$, can be interpreted as an ordered factorization of $n$, $n = b_1 \cdot b_2/b_1 \cdot b_3/b_2 \cdots n/b_{k-1}$, into $k$ factors $\geq 2$. Let $n = p_1^3 p_2^2 p_3$ be the prime factorization of $n \in \mathbb{N}$. How many ordered factorizations of $n$ are there? How many unordered?

13. A number-partition $n = n_1 + \cdots + n_k$ is called *perfect* if every number $k \leq n$ can be represented uniquely as a sum of certain of the $n_i$'s, where we do not distinguish between equal summands. Example: $4 + 2 + 1$ is perfect since $1 = 1$, $2 = 2$, $3 = 2 + 1$, $4 = 4$, $5 = 4 + 1$, $6 = 4 + 2$, $7 = 4 + 2 + 1$. $4 + 2 + 2 + 1$ is not perfect since $5 = 4 + 1 = 2 + 2 + 1$. Show that the number of perfect partitions of $n$ equals the number of ordered factorizations of $n + 1$ into factors $\geq 2$.

14.* Consider a product $X_1X_2 \ldots X_n$. In how many ways can we insert parentheses such that there is no ambiguity as to the meaning of the product? This number is called the *Catalan number*, denoted by $c_n$. Example: $c_2 = 1$, $(X_1X_2)$; $c_3 = 2$, $(X_1X_2)X_3$, $X_1(X_2X_3)$; $c_4 = 5$, $(X_1X_2)(X_3X_4)$, $(X_1(X_2X_3))X_4$, $((X_1X_2)X_3)X_4$, $X_1((X_2X_3)X_4)$, $X_1(X_2(X_3X_4))$. (Answer: $c_n = \binom{2n-2}{n-1}/n$.)

15. How many of the expressions in the previous exercise are nested? Example: $X_1((X_2X_3)X_4)$ is nested, $(X_1X_2)(X_3X_4)$ is not.

## 2. Recursion and Inversion

For all except the most trivial combinatorial coefficients, explicit formulae are not easy to come by. Hence one is interested in recursions and in manipulating known coefficients (for instance, inverting them) in order to obtain new ones. The basic numbers in table **3.22** are all of the form $f(n, r)$, i.e., they depend on the parameters $n$ and $r$. We keep one of these parameters fixed, say $n$, and express $f(n, r)$ in terms of the numbers $f(n, i)$, $i < r$. From the initial values $f(n, 0)$, $f(n, 1)$ and the recursion one then obtains a complete table of the numbers $f(n, r)$.

**Example.** $\binom{n}{0} = 1$, $\binom{n}{k} = \binom{n-1}{k-1} + \binom{n-1}{k}$ for $k \geq 1$.

To prove this recursion, list the $k$-subsets of an $n$-set $N$ and choose some element $a \in N$. The $k$-subsets $A$ of $N$ split into two classes depending on whether $a \in A$ or $a \notin A$. Sets of the first type are of the form $B \cup \{a\}$ where $B$ runs through all $(k-1)$-subsets of $N - \{a\}$, hence there are $\binom{n-1}{k-1}$ of them. The sets of the second type are precisely the $k$-subsets of $N - \{a\}$, hence there are $\binom{n-1}{k}$ of them. The recursion now results by applying the rule of sums.

This method of splitting a set into smaller disjoint subsets is perhaps the most widely used elementary approach to obtain recursion formulae. Unfortunately, as simple and elegant as some of them are, proofs using this method are *ad hoc* in nature and soon become unpleasantly complicated as we consider sets with more structure. A direct proof of the recursion **3.34** for the Gaussian numbers, for instance, is already far from easy. We prefer a less direct but more systematic approach. The idea is to set up certain counting functions using table **3.22**, and then to derive recursion formulae for various coefficients by applying the trivial relations holding between these counting functions.

### A. The Elementary Counting Functions

Consider the expressions $r^n$, $[r]_n$ and $[r]^n$. If we replace $r$ by the real variable $x$ we obtain the real polynomials $x^n$, $[x]_n := x(x-1)\cdots(x-n+1)$, $[x]^n := x(x+1)\cdots(x+n-1)$, all of degree $n$. As before we set $x^0 = [x]_0 = [x]^0 = 1$.

The polynomials $x^n$, $[x]_n$ and $[x]^n$ are called the *standard polynomials*, *falling factorials*, and *rising factorials*, respectively. The sequences

$$\{x^n : n \in \mathbb{N}_0\}, \{[x]_n : n \in \mathbb{N}_0\} \quad \text{and} \quad \{[x]^n : n \in \mathbb{N}_0\}$$

are examples of what we call polynomial sequences.

**Definition.** A *polynomial sequence* is a family $\{p_n(x) : n \in \mathbb{N}_0\}$ of real polynomials such that $\deg(p_n(x)) = n$ for all $n$, with the convention $\deg(p(x)) = 0$ if and only if $p(x) \equiv a \neq 0$, and $\deg(p(x)) = -1$ if $p(x) \equiv 0$.

We denote by $\mathbb{R}[x]$ the set of all real polynomials in the variable $x$. $\mathbb{R}[x]$ with the usual addition and scalar multiplication is a vector space over $\mathbb{R}$. From the fact that the members of a polynomial sequence have strictly increasing degree we immediately infer the following result.

**3.23 Proposition.** *Any polynomial sequence is a basis of the vector space* $\mathbb{R}[x]$. □

According to **3.23** we shall sometimes speak of the *standard basis* $\{x^n : n \in \mathbb{N}_0\}$ or the *basis of falling factorials*, etc. The main idea is now the following. Since any two polynomial sequences $\{p_n(x)\}$ and $\{q_n(x)\}$ are bases of $\mathbb{R}[x]$, each $q_n(x)$ can be uniquely expressed as a linear combination of the polynomials $p_k(x)$, $0 \le k \le n$, and conversely each $p_n(x)$ can be uniquely expressed in terms of the $q_k(x)$, $0 \le k \le n$. The coefficients which appear in these linear combinations are called the *connecting coefficients* between the two sequences. The connecting coefficients provide all information needed to establish recursion and inversion formulae. The basic question of how to compute these connecting coefficients for arbitrary sequences $\{p_n(x)\}$ and $\{q_n(x)\}$ will be the subject of section 3. Let us now look at the elementary counting functions arising from table **3.22**.

**3.24 Proposition.** *For all* $n \in \mathbb{N}_0$:

(i) $x^n = \sum_{k=0}^{n} S_{n,k}[x]_k$
(ii) $[x]^n = \sum_{k=0}^{n} L'_{n,k}[x]_k$

*where* $S_{n,k}$ *are the Stirling numbers of the second kind, with the convention* $S_{0,0} = 1$, $S_{n,0} = 0$ *for* $n > 0$, *and* $L'_{n,k}$ *are the* (*signless*) Lah numbers, *given by*

$$L'_{0,0} = 1, \quad L'_{n,0} = 0 \quad \text{for } n > 0,$$

$$L'_{n,k} = \frac{n!}{k!}\binom{n-1}{k-1} \quad \text{for } k \ge 1.$$

*Proof.* Both (i) and (ii) are identities between polynomials of degree $n$. Since a polynomial of degree $n$ has at most $n$ roots it suffices to prove that the polynomials involved agree in at least $n + 1$ points. What we are going to show in fact is that they agree in every $x = r$, $r \in \mathbb{N}$. This method of translating a combinatorial identity of natural numbers into a polynomial identity will be the main tool in most of the proofs to follow.

To prove (i) let us classify the mappings $f: N \to R$ according to their images. Since any mapping has a unique image we obtain the partition

$$\operatorname{Map}(N, R) = \dot{\bigcup_{B \subseteq R}} \operatorname{Sur}(N, B),$$

and thus by **3.1** and **3.14**

$$r^n = \sum_{B \subseteq R} |B|!S_{n,|B|} = \sum_{k=0}^{n} \binom{r}{k} k!S_{n,k} = \sum_{k=0}^{n} S_{n,k}\,[r]_k.$$

But this means that the polynomials $x^n$ and $\sum_{k=0}^{n} S_{n,k}[x]_k$ agree in every $x = r \in \mathbb{N}$. Hence they must be identical. The analogous classification of $\operatorname{Mon}(N, R)$ where $N$ and $R$ are chains yields (ii). Here we have

$$\frac{[r]^n}{n!} = \sum_{k=0}^{n} \binom{r}{k}\binom{n-1}{k-1},$$

thus

$$[r]^n = \sum_{k=0}^{n} \frac{n!}{k!}\binom{n-1}{k-1}[r]_k. \quad \square$$

The (signed) *Lah numbers* $L_{n,k}$ are defined by the identities

$$[-x]_n = \sum_{k=0}^{n} L_{n,k}[x]_k.$$

Since by **3.5**

$$[-x]_n = (-1)^n[x]^n \quad \text{for all } n,$$

we immediately have from **3.24**(ii):

**3.25.**

$$L_{n,k} = (-1)^n \frac{n!}{k!}\binom{n-1}{k-1}.$$

Notice also that

$$[x]_n = \sum_{k=0}^{n} L_{n,k}[-x]_k,$$

i.e., the sequences $\{[x]_n\}$ and $\{[-x]_n\}$ have the same connecting coefficients in both directions.

Reversing the role of $x^n$ and $[x]_n$ in **3.24**(i) we now express the falling factorials $[x]_n$ in terms of the standard basis $\{x^n\}$.

**Definition.** The *Stirling numbers of the first kind* are the numbers $s_{n,k}$ given by

$$[x]_n = \sum_{k=0}^{n} s_{n,k} x^k \qquad (n, k \in \mathbb{N}_0),$$

with the convention

$$s_{0,0} = 1 \quad \text{and} \quad s_{n,0} = 0 \quad \text{for } n > 0.$$

**3.26 Proposition.** *We have:*

(i) $[x]^n = \sum_{k=0}^{n} |s_{n,k}| x^k$.
(ii) *The numbers* $s_{n,1}, s_{n,2}, \ldots, s_{n,n}$ *are all non-zero and have alternating sign, for every fixed* $n \in \mathbb{N}$.

*Proof.* Replace $x$ by $-x$ in the defining identity. Then

$$[-x]_n = (-1)^n [x]^n = \sum_{k=0}^{n} (-1)^k s_{n,k} x^k$$

thus

$$[x]^n = \sum_{k=0}^{n} (-1)^{n+k} s_{n,k} x^k.$$

Since $[x]^n = x(x+1)\cdots(x+n-1)$ it is clear that all coefficients $(-1)^{n+k} s_{n,k}$ must be $>0$ for $k \geq 1$ whence (i) and (ii) follow. □

Let us turn to the vector space lattices $\mathscr{L}(n, q)$ and the Gaussian coefficients $\binom{n}{k}_q$.

**Definition.** The polynomials

$$g_n(x) = (x-1)(x-q)\cdots(x-q^{n-1}) \qquad (n \in \mathbb{N})$$

with $g_0(x) \equiv 1$, are called the *Gaussian polynomials.*

Note that the polynomials $\{g_n(x)\}$ form a polynomial sequence.

**3.27 Proposition.** *We have*

(i) $x^n = \sum_{k=0}^{n} \binom{n}{k}(x-1)^k$,
(ii) $x^n = \sum_{k=0}^{n} \binom{n}{k}_q g_k(x)$.

*Proof.* (i) is just a special case of the binomial theorem. Notice that, in accordance with a previous remark, (ii) becomes (i) if we set $q = 1$. To prove (ii) consider $\mathrm{Hom}(V, W)$ where $V$ is an $n$-dimensional vector space over $GF(q)$ and $W$ is a vector space over $GF(q)$ of cardinality $r$. As any $f \in \mathrm{Hom}(V, W)$ is uniquely determined by the values it takes on a basis of $V$ we have

$$|\mathrm{Hom}(V, W)| = r^n.$$

Let us now classify $\mathrm{Hom}(V, W)$ according to the kernel subspace $f^{-1}(0) \subseteq V$. Given some subspace $U \subseteq V$ how many $f \in \mathrm{Hom}(V, W)$ exist with $f^{-1}(0) = U$? Let $u_1, \ldots, u_k$ be a basis of $U$ and extend it to a basis $u_1, \ldots, u_n$ of $V$. Then $f^{-1}(0) = U$ precisely when $f(u_i) = 0$ for $i = 1, \ldots, k$, and $f$ maps $u_{k+1}, \ldots, u_n$ onto linearly independent vectors in $W$. Hence we obtain

$$r^n = \sum_{U \subseteq V} (r-1)(r-q)\cdots(r-q^{n-r(U)-1}) = \sum_{k=0}^{n} \binom{n}{k}_q (r-1)\cdots(r-q^{n-k-1})$$

$$= \sum_{k=0}^{n} \binom{n}{k}_q (r-1)(r-q)\cdots(r-q^{k-1}) = \sum_{k=0}^{n} \binom{n}{k}_q g_k(r).$$

As $r$ can be any power of $q$, the polynomials $x^n$ and $\sum_{k=0}^{n} \binom{n}{k}_q g_k(x)$ agree in infinitely many values of $x$ and hence must be identical. □

B. Recursions

Using our knowledge of polynomial sequences and their connecting coefficients we can now establish some recursion formulae. Our method will always be the same: Any linear functional on $\mathbb{R}[x]$ is uniquely determined by the values it assumes on a basis. In particular, for a polynomial sequence $\{p_n(x)\}$ and a sequence $\{a_n\}$ of real numbers there is precisely one functional $L: \mathbb{R}[x] \to \mathbb{R}$ with $Lp_n(x) = a_n$ for all $n$. The idea is to translate relations between the polynomials $p_n(x)$ (which are trivial in most instances) into relations between the numbers $a_n$.

We note the following obvious polynomial recursions.

**3.28 Proposition.** *For all* $n \in \mathbb{N}_0$:

(i) $x^{n+1} = xx^n$
(ii) $[x]_{n+1} = x[x]_n - n[x]_n$
(iii) $[x]_{n+1} = x[x-1]_n$
(iv) $[x]^{n+1} = x[x]^n + n[x]^n$
(v) $[x]^{n+1} = x[x+1]^n$
(vi) $(x-1)^{n+1} = x(x-1)^n - (x-1)^n$
(vii) $g_{n+1}(x) = xg_n(x) - q^n g_n(x)$. □

**3.29** (Recursion for the Stirling Numbers of the Second Kind). *For all* $n, k \geq 0$:

(i) $S_{0,0} = 1$, $S_{n,0} = 0$ *for* $n > 0$,
(ii) $S_{n+1,k} = S_{n,k-1} + kS_{n,k}$,
(iii) $S_{n+1,k} = \sum_{j=1}^{n} \binom{n}{j} S_{j,k-1}$.

*Proof*. Consider the functionals $L_k$ on $\mathbb{R}[x]$, $k \in \mathbb{N}_0$, given by

$$L_k[x]_n = \delta_{n,k} \quad \text{(Kronecker symbol)}.$$

By **3.24**(i)

$$L_k x^n = L_k\left(\sum_{i=0}^{n} S_{n,i}[x]_i\right) = \sum_{i=0}^{n} S_{n,i} L_k[x]_i = S_{n,k}.$$

It follows from **3.28**(ii) that

$$L_k[x]_{n+1} = L_k x[x]_n - nL_k[x]_n$$

which according to the definition of $L_k$ is equivalent to

$$(+) \qquad L_{k-1}[x]_n = L_k x[x]_n - kL_k[x]_n \qquad (k \geq 1).$$

Since the falling factorials $[x]_n$ are a basis of $\mathbb{R}[x]$ we infer by linear extension that $(+)$ holds for *all* polynomials $p(x) \in \mathbb{R}[x]$, i.e.,

$$L_{k-1}p(x) = L_k xp(x) - kL_k p(x) \qquad (k \geq 1).$$

For $p(x) = x^n$ this yields

$$L_{k-1}x^n = L_k x^{n+1} - kL_k x^n$$

or

$$S_{n,k-1} = S_{n+1,k} - kS_{n,k}.$$

To obtain the recursion (iii) we use **3.28**(iii). In this case

$$L_k[x]_{n+1} = L_k x[x-1]_n,$$

i.e.,

$$L_{k-1}[x]_n = L_k x[x-1]_n \qquad (k \geq 1),$$

and thus for $k \geq 1$

$$L_{k-1}p(x) = L_k xp(x-1) \quad \text{for all } p(x) \in \mathbb{R}[x].$$

For $p(x) = (x+1)^n$ this gives

$$S_{n+1,k} = L_k x^{n+1} = L_{k-1}(x+1)^n = \sum_{j=0}^{n} \binom{n}{j} L_{k-1}x^j = \sum_{j=0}^{n} \binom{n}{j} S_{j,k-1}. \quad \square$$

From **3.29**(i)(ii) we obtain the complete table of the numbers $S_{n,k}$, and hence by **3.14** of the numbers $|\mathrm{Sur}(N, R)|$ as well. **3.29**(ii) is an example of a *triangular* recursion because the three numbers $S_{n+1,k}$, $S_{n,k-1}$ and $S_{n,k}$ form a triangle in the Stirling table. **3.29**(iii) is a *vertical* recursion; in this case the second parameter remains fixed. The table below shows the first rows of the table of Stirling numbers of the second kind. The empty cells are to be filled with 0's.

Stirling numbers of the second kind

| $S_{n,k}$ | $k = 0$ | 1 | 2 | 3 | 4 | 5 | 6 | 7 | 8 |
|---|---|---|---|---|---|---|---|---|---|
| $n = 0$ | 1 | | | | | | | | |
| 1 | 0 | 1 | | | | | | | |
| 2 | 0 | 1 | 1 | | | | | | |
| 3 | 0 | 1 | 3 | 1 | | | | | |
| 4 | 0 | 1 | 7 | 6 | 1 | | | | |
| 5 | 0 | 1 | 15 | 25 | 10 | 1 | | | |
| 6 | 0 | 1 | 31 | 90 | 65 | 15 | 1 | | |
| 7 | 0 | 1 | 63 | 301 | 350 | 140 | 21 | 1 | |
| 8 | 0 | 1 | 127 | 966 | 1701 | 1050 | 266 | 28 | 1 |

As a consequence of **3.29** we can say more about the sequence $\{S_{n,k}: k = 0, \ldots, n\}$.

**3.30 Proposition.** *The sequence* $\{S_{n,k}: k = 0, \ldots, n\}$ *is unimodal for all* $n \in \mathbb{N}_0$. *Let* $M(n) = \max\{k: S_{n,k} = \max\}$. *Then* $\{S_{n,k}\}$ *is of one of the following two types*:

(i) $S_{n,0} < S_{n,1} < \cdots < S_{n,M(n)} > S_{n,M(n)+1} > \cdots > S_{n,n}$,
(ii) $S_{n,0} < S_{n,1} < \cdots < S_{n,M(n)-1} = S_{n,M(n)} > \cdots > S_{n,n}$.

*Furthermore,* $M(n) = M(n-1) + \varepsilon(n)$ *with* $\varepsilon(n) = 0$ *or* 1.

*Proof.* We use induction on $n$. For $n = 0$ or 1 there is nothing to prove. Suppose our proposition holds for $i \leq n$. Then $M(i) \leq M(j)$ for $1 \leq i \leq j \leq n$. Let $2 \leq k \leq M(n)$. Then by **3.29**(ii)

$$S_{n+1,k} - S_{n+1,k-1} = (S_{n,k-1} - S_{n,k-2}) + k(S_{n,k} - S_{n,k-1}) + S_{n,k-1},$$

and the right hand side is positive by the induction hypothesis. Now suppose $M(n) + 2 \leq k \leq n + 1$. Then by **3.29**(iii)

$$S_{n+1,k} - S_{n+1,k-1} = \sum_{j=0}^{n} \binom{n}{j} (S_{j,k-1} - S_{j,k-2}).$$

Here the right hand side is negative because of the induction hypothesis and $M(j) \leq M(n)$ for all $j \leq n$. Hence $\{S_{n,k}\}$ is unimodal with $M(n+1) = M(n)$ or $M(n+1) = M(n) + 1$. □

It is not known whether $\{S_{n,k}\}$ always has a single maximum for $n \geq 3$. Some results concerning this problem and the value $M(n)$ have been established (see, e.g., Canfield [1]).
Recall the definition of the *Bell numbers* $B_n$:

$$B_n = |\mathscr{P}(n)| = \sum_{k=0}^{n} S_{n,k}.$$

**3.31** (Recursion for the Bell Numbers).

$$B_0 = 1$$

$$B_{n+1} = \sum_{k=0}^{n} \binom{n}{k} B_k.$$

*Proof.* Define the functional $L: \mathbb{R}[x] \to \mathbb{R}$ by

$$L[x]_k = 1 \quad \text{for all } k \in \mathbb{N}_0.$$

It follows from **3.24**(i) that

$$Lx^n = \sum_{k=0}^{n} S_{n,k} = B_n.$$

Applying the recursion $[x]_{n+1} = x[x-1]_n$ we deduce that

$$L[x]_{n+1} = L[x]_n = Lx[x-1]_n,$$

hence

$$Lp(x) = Lxp(x-1) \quad \text{for all } p(x) \in \mathbb{R}[x].$$

For $p(x) = (x+1)^n$ this yields

$$L(x+1)^n = \sum_{k=0}^{n} \binom{n}{k} Lx^k = \sum_{k=0}^{n} \binom{n}{k} B_k = Lx^{n+1} = B_{n+1}. \quad \square$$

Bell numbers

| | $n = 0$ | 1 | 2 | 3 | 4 | 5 | 6 | 7 | 8 |
|---|---|---|---|---|---|---|---|---|---|
| $B_n$ | 1 | 1 | 2 | 5 | 15 | 52 | 203 | 877 | 4140 |

The following recursion formulae are derived from **3.28** by using the same arguments as in the proofs of **3.29** and **3.31**. Accordingly, except for **3.35** whose proof requires a little more care, we will confine ourselves to listing the results.

**3.32** (Recursion for the Stirling Numbers of the First Kind). *For all* $n, k \geq 0$:

(i) $s_{0,0} = 1$, $s_{n,0} = 0$ *for* $n > 0$,
(ii) $s_{n+1,k} = s_{n,k-1} - ns_{n,k}$,
(iii) $s_{n+1,k} = \sum_{j=0}^{n} (-1)^j [n]_j s_{n-j,k-1}$. □

Stirling numbers of the first kind

| $s_{n,k}$ | $k = 0$ | 1 | 2 | 3 | 4 | 5 | 6 | 7 | 8 |
|---|---|---|---|---|---|---|---|---|---|
| $n = 0$ | 1 | | | | | | | | |
| 1 | 0 | 1 | | | | | | | |
| 2 | 0 | −1 | 1 | | | | | | |
| 3 | 0 | 2 | −3 | 1 | | | | | |
| 4 | 0 | −6 | 11 | −6 | 1 | | | | |
| 5 | 0 | 24 | −50 | 35 | −10 | 1 | | | |
| 6 | 0 | −120 | 274 | −225 | 85 | −15 | 1 | | |
| 7 | 0 | 720 | −1764 | 1624 | −735 | 175 | −21 | 1 | |
| 8 | 0 | −5040 | 13068 | −13132 | 6769 | −1960 | 322 | −28 | 1 |

It can be shown by an argument similar to that in **3.30**, that the sequence $\{|s_{n,k}| : k = 0, \ldots, n\}$ is unimodal for every $n$.

**3.33** (Recursion for the Binomial Coefficients). *For all* $n, k \geq 0$:

$$\binom{n}{0} = 1,$$

$$\binom{n+1}{k} = \binom{n}{k-1} + \binom{n}{k}. \quad \square$$

Binomial coefficients (Pascal's triangle)

| $\binom{n}{k}$ | $k = 0$ | 1 | 2 | 3 | 4 | 5 | 6 | 7 | 8 |
|---|---|---|---|---|---|---|---|---|---|
| $n = 0$ | 1 | | | | | | | | |
| 1 | 1 | 1 | | | | | | | |
| 2 | 1 | 2 | 1 | | | | | | |
| 3 | 1 | 3 | 3 | 1 | | | | | |
| 4 | 1 | 4 | 6 | 4 | 1 | | | | |
| 5 | 1 | 5 | 10 | 10 | 5 | 1 | | | |
| 6 | 1 | 6 | 15 | 20 | 15 | 6 | 1 | | |
| 7 | 1 | 7 | 21 | 35 | 35 | 21 | 7 | 1 | |
| 8 | 1 | 8 | 28 | 56 | 70 | 56 | 28 | 8 | 1 |

**3.34** (Recursion for the Gaussian Numbers). *For all n, $k \geq 0$, q a prime power*:

$$\binom{n}{0}_q = 1,$$

$$\binom{n+1}{k}_q = \binom{n}{k-1}_q + q^k \binom{n}{k}_q . \quad \square$$

**3.35** (Recursion for the Galois Numbers $G_{n,q}$). *For all $n \geq 0$, q a prime power*:

$$G_{0,q} = 1, \qquad G_{1,q} = 2,$$
$$G_{n+1,q} = 2G_{n,q} + (q^n - 1)G_{n-1,q}.$$

*Proof.* Since by definition $G_{n,q} = |\mathscr{L}(n, q)|$, $G_{0,q} = 1$ and $G_{1,q} = 2$ are clear. Now define the functional $L: \mathbb{R}[x] \to \mathbb{R}$ by

$$Lg_n(x) = 1 \quad \text{for all } n,$$

where $g_n(x)$ are the Gaussian polynomials. It follows from **3.27**(ii) that

$$Lx^n = G_{n,q}.$$

Hence the asserted recursion is equivalent to

$$(+) \qquad Lx^{n+1} = 2Lx^n + (q^n - 1)Lx^{n-1}.$$

Applying **3.28**(vii) we have

$$Lg_{n+1}(x) = Lxg_n(x) - q^n Lg_n(x)$$

and thus

$$(++) \qquad Lxg_n(x) = q^n + 1.$$

We introduce an operator $D_q$ on $\mathbb{R}[x]$, setting

$$D_q p(x) = \frac{p(qx) - p(x)}{x} \qquad (p(x) \in \mathbb{R}[x]).$$

$D_q$ is linear and it is immediately verified that

$$D_q x^n = (q^n - 1)x^{n-1}, \qquad D_q g_n(x) = (q^n - 1)g_{n-1}(x).$$

Using $D_q$ we can rewrite $(++)$ as

$$Lxg_n(x) = 2 + (q^n - 1) = 2Lg_n(x) + LD_q g_n(x).$$

Since $\{g_n(x)\}$ is a basis, this identity holds for all $p(x) \in \mathbb{R}[x]$, i.e.,

$$Lxp(x) = 2Lp(x) + LD_q p(x).$$

For $p(x) = x^n$ we obtain

$$Lx^{n+1} = 2Lx^n + LD_q x^n = 2Lx^n + (q^n - 1)Lx^{n-1}$$

which is precisely (+). □

For many combinatorial numbers one has to use longer recursions. As an example let us consider the partition numbers $P_{n,r}$.

**3.36** (Recursion for the Partition Numbers). *For all* $n \geq r \geq 1$:

$$P_{n,1} = P_{n,n} = 1,$$
$$P_{n,r} = P_{n-r,1} + P_{n-r,2} + \cdots + P_{n-r,r}.$$

*Proof.* The initial values $P_{n,1} = P_{n,n} = 1$ are clear. Let $\alpha$ be a partition of $n - r$ into $k \leq r$ parts, say $\alpha: n_1 + \cdots + n_k$. We define a new partition $\alpha': (n_1 + 1) + (n_2 + 1) + \cdots + (n_k + 1) + \underbrace{1 + \cdots + 1}_{r-k}$. $\alpha'$ is a partition of $n$ into $r$ parts, and the mapping $\phi: \alpha \to \alpha'$ is easily seen to be bijective. □

Partition numbers $P_{n,r}$

| $P_{n,r}$ | $r = 1$ | 2 | 3 | 4 | 5 | 6 | 7 | 8 |
|---|---|---|---|---|---|---|---|---|
| $n = 1$ | 1 | | | | | | | |
| 2 | 1 | 1 | | | | | | |
| 3 | 1 | 1 | 1 | | | | | |
| 4 | 1 | 2 | 1 | 1 | | | | |
| 5 | 1 | 2 | 2 | 1 | 1 | | | |
| 6 | 1 | 3 | 3 | 2 | 1 | 1 | | |
| 7 | 1 | 3 | 4 | 3 | 2 | 1 | 1 | |
| 8 | 1 | 4 | 5 | 5 | 3 | 2 | 1 | 1 |

## C. Inversion of Sequences

Let $\{p_n(x)\}$ and $\{q_n(x)\}$ be two polynomial sequences. What we called connecting coefficients can be considered as the elements of the two transformation matrices which transport one basis into the other. The fact that these two matrices are inverses of each other gives rise to inversion formulae for the corresponding connecting coefficients. Let us formulate this as a proposition.

**3.37 Proposition.** *Let* $\{p_n(x): n \in \mathbb{N}_0\}$ *and* $\{q_n(x): n \in \mathbb{N}_0\}$ *be polynomial sequences with connection coefficients* $a_{n,k}$, $b_{n,k}$, *i.e.*,

$$q_n(x) = \sum_{k=0}^{n} a_{n,k} p_k(x) \qquad (n \in \mathbb{N}_0)$$

$$p_n(x) = \sum_{k=0}^{n} b_{n,k} q_k(x) \qquad (n \in \mathbb{N}_0).$$

*If* $u_0, u_1, \ldots; v_0, v_1, \ldots$ *are real numbers, then*

$$v_n = \sum_{k=0}^{n} a_{n,k} u_k \ (n \in \mathbb{N}_0) \Leftrightarrow u_n = \sum_{k=0}^{n} b_{n,k} v_k \ (n \in \mathbb{N}_0).$$

*Proof.* The hypothesis implies that the matrices $A = [a_{n,k}]$, $B = [b_{n,k}]$ are inverses of each other, where we set $a_{n,k} = b_{n,k} = 0$ for all $n < k$. But this means

$$v = Au \Leftrightarrow u = Bv$$

for all vectors $u = (u_0, u_1, \ldots)$, $v = (v_0, v_1, \ldots)$. □

Using our knowledge of the connecting coefficients between our basic sequences (3.24, 3.25, 3.27) we obtain the following classical inversion formulae.

**3.38 Corollary.** *Let* $u_0, u_1, \ldots; v_0, v_1, \ldots$ *be real numbers.*

(i) *Binomial inversion*

$$v_n = \sum_{k=0}^{n} \binom{n}{k} u_k \ (n \in \mathbb{N}_0) \Leftrightarrow u_n = \sum_{k=0}^{n} (-1)^{n-k} \binom{n}{k} v_k \ (n \in \mathbb{N}_0).$$

(ii) *Stirling inversion*

$$v_n = \sum_{k=0}^{n} s_{n,k} u_k \ (n \in \mathbb{N}_0) \Leftrightarrow u_n = \sum_{k=0}^{n} S_{n,k} v_k \ (n \in \mathbb{N}_0).$$

(iii) *Lah inversion*

$$v_n = \sum_{k=0}^{n} L_{n,k} u_k \ (n \in \mathbb{N}_0) \Leftrightarrow u_n = \sum_{k=0}^{n} L_{n,k} v_k \ (n \in \mathbb{N}_0).$$

(iv) *Gauss inversion*

$$v_n = \sum_{k=0}^{n} \binom{n}{k}_q u_k \ (n \in \mathbb{N}_0) \Leftrightarrow u_n = \sum_{k=0}^{n} (-1)^{n-k} q^{\binom{n-k}{2}} \binom{n}{k}_q v_k \ (n \in \mathbb{N}_0).$$

*Proof.* Let us prove just the binomial inversion formula. From the binomial theorem we have

$$x^n = \sum_{k=0}^{n} \binom{n}{k}(x-1)^k, \qquad (x-1)^n = \sum_{k=0}^{n}(-1)^{n-k}\binom{n}{k}x^k.$$

Hence $\binom{n}{k}$, $(-1)^{n-k}\binom{n}{k}$ are the connecting coefficients between the sequences $\{x^n\}$, $\{(x-1)^n\}$, and the result follows from **3.37**. □

As an application of the binomial inversion formula we can derive an explicit expression for the Stirling numbers $S_{n,k}$.

**3.39.**

$$S_{n,k} = \frac{1}{k!}\sum_{i=0}^{n}(-1)^{k-i}\binom{k}{i}i^n.$$

*Proof.* By **3.24**(i)

$$k^n = \sum_{i=0}^{n}\binom{k}{i}(i!S_{n,i}) \quad \text{for a fixed } n \text{ and all } k \in \mathbb{N}_0.$$

**3.38**(i) now yields

$$k!S_{n,k} = \sum_{i=0}^{n}(-1)^{k-i}\binom{k}{i}i^n. \quad \square$$

Exercises III.2

1. Show:

   (i) $\sum_{m=0}^{n}\binom{m}{k} = \binom{n+1}{k+1}$.
   (ii) Use (i) and $m^2 = 2\binom{m}{2} + \binom{m}{1}$ to prove

   $$\sum_{m=0}^{n} m^2 = \tfrac{1}{6}n(n+1)(2n+1).$$

   (iii) $\sum_{m=0}^{n} m^3 = ?$

2.* Find a simple expression for $\sum_{k=0}^{n}\binom{m}{k}$, $n < m$.

3. Let $A_n = \sum_{k=0}^{n}[n]_k$ be the number of injective mappings into an $n$-set. Verify:

   (i) $A_n = nA_{n-1} + 1$,
   (ii) $A_n = n!\sum_{k=0}^{n}(1/k!)$

→ 4. Let $I_{n,k}$ be the number of permutations of $\{1,\ldots,n\}$ with exactly $k$ inversions (cf. section I.3.B). Prove:

   (i) $I_{n,0} = 1, I_{n,1} = n-1, I_{n,k} = 0$ for $k \geq n$;
   (ii) $I_{n,\binom{n}{2}-k} = I_{n,k}$;
   (iii) $I_{n,k} = I_{n,k-1} + I_{n-1,k}$ for $k < n$.

   Derive explicit expressions for $I_{n,k}$, $k = 2, 3, 4, 5$.

5. Verify the recursion for the Lah numbers:

   (i) $L'_{0,0} = L_{0,0} = 1$,
   (ii) $L'_{n+1,k} = L'_{n,k-1} + (n+k)L'_{n,k}$.
   (iii) $L_{n+1,k} = ?$
   (iv) $L_{n,k} = \sum_{j=0}^{n} (-1)^j s_{n,j} S_{j,k}$.

→ 6. A sequence $\{a_n\}$ of reals is called *log-concave* if $a_n^2 \geq a_{n-1}a_{n+1}$ for all $n \geq 2$. Show that $\{\binom{n}{k}\}$, $\{\binom{n}{k}_q\}$, $\{S_{n,k}\}$, $\{|s_{n,k}|\}$, and $\{L'_{n,k}\}$ are log-concave for fixed $n$.

→ 7. The *Fibonacci numbers* $F_n$ are defined by the recursion

$$F_0 = 0, \qquad F_1 = 1, \qquad F_n = F_{n-1} + F_{n-2} \qquad \text{for } n \geq 2.$$

   Hence: 0, 1, 1, 2, 3, 5, 8, 13, 21, ... Prove:

   (i) $F_{n-1}F_{n+1} - F_n^2 = (-1)^n$,
   (ii) $F_k | F_{nk}$,
   (iii) $\gcd(F_m, F_n) = F_{\gcd(m,n)}$,
   (iv) $(\frac{1}{2}(1+\sqrt{5}))^{n-2} \leq F_n \leq (\frac{1}{2}(1+\sqrt{5}))^{n-1}$.

8. Let $f_{n,k}$ be the number of $k$-subsets of $\{1, \ldots, n\}$ which do not contain a pair of consecutive integers. Prove:

   (i) $f_{n,k} = \binom{n-k+1}{k}$,
   (ii) $\sum_{k \geq 0} f_{n,k} = F(n+2)$.

→ 9. Let $f^*_{n,k}$ be as in the previous exercise except that, in addition, we do not allow the occurrence of 1 and $n$ together. Prove:

   (i) $f^*_{n,k} = n\binom{n-k}{k}/(n-k)$.

   Set $F^*_n := \sum_{k \geq 0} f^*_{n,k}$, $n \geq 1$.

   (ii) $F^*_1 = 1$, $F^*_2 = 3$, $F^*_n = F^*_{n-1} + F^*_{n-2}$.

   The numbers $F^*_n$ are called the *Lucas numbers*.

10. Show that $\sum_k \binom{n}{k} F_{m+k}$ is always a *Fibonacci number*.

11. Let the numbers $a(n, r)$ be defined as in ex. I.3.14. Show that $a(n, 1) = F_{n+1}$ (the $(n+1)$th Fibonacci number).

→ 12. We consider an example of a recursion of non-integral numbers. The numbers

$$H_n := 1 + \frac{1}{2} + \cdots + \frac{1}{n} \qquad (n \in \mathbb{N})$$

   are called the *harmonic numbers*. The series $\sum_{k \geq 1} (1/k)$ is called the harmonic series and is divergent, but only very weakly. Prove:

   (i) $H_1 = 1$, $H_2 = \frac{3}{2}$,
   (ii) $(n+1)H_n - n = \sum_{k=1}^{n} H_k$,

(iii) $\sum_{k=1}^{n} \binom{k}{m} H_k = \binom{n+1}{m+1}(H_{n+1} - 1/(m+1))$,
(iv) $1 + n/2 \le H_{2^n} \le 1 + n$,
(v) $H_n = |s_{n+1,2}|/n!$.

13.* The complex function $\zeta(s) := \sum_{n \ge 1} (1/n^s)$ is called the *Riemann $\zeta$-function.* If $r$ is an even integer, then

$$\zeta(r) = \tfrac{1}{2}|b_r| \frac{(2\pi)^r}{r!}$$

where the numbers $b_r$ are called the *Bernoulli numbers.* They also appear in the expansion

$$x(e^x - 1)^{-1} = \sum_{k \ge 0} \frac{b_k x^k}{k!}.$$

Show:

(i) $b_0 = 1, b_1 = -\frac{1}{2}, b_2 = \frac{1}{6}, b_3 = 0, b_4 = -\frac{1}{30}$,
(ii) $b_{2n+1} = 0$ for $n \ge 1$,
(iii) $\sum_k \binom{n}{k} b_k = b_n + \delta_{n,1}$,
(iv) $b_n = \sum_{k=0}^{n} ((-1)^k k! S_{n,k}/(k+1))$.

→14. Determine the numbers $a_n \in \mathbb{N}$ from the identity

$$n! = a_0 + a_1 n + a_2 n(n-1) + a_3 n(n-1)(n-2) + \cdots \qquad (n \in \mathbb{N}).$$

15. Prove the Gauss inversion formula using **3.27**(ii).

## 3. Binomial Sequences

As we mentioned before, to any two polynomial sequences $\{p_n(x)\}$ and $\{q_n(x)\}$ there exist uniquely determined connecting coefficients $c_{n,k}$ with

$$q_n(x) = \sum_{k=0}^{n} c_{n,k}\, p_k(x) \qquad (n \in \mathbb{N}_0).$$

We have computed these coefficients for the fundamental sequences $\{x^n\}$, $\{[x]_n\}$ and $\{[x]^n\}$ in the preceding section. For arbitrary sequences determining the connecting coefficients presents considerable difficulties, but there is an interesting class of sequences, called binomial sequences, where the problem is manageable. The theory of binomial sequences is the subject of this section.

**Definition.** A polynomial sequence $\{p_n(x): n \in \mathbb{N}_0\}$ is called *binomial* if

$$p_0(x) = 1$$

$$p_n(x+y) = \sum_{k=0}^{n} \binom{n}{k} p_k(x) p_{n-k}(y) \quad \text{for all } x, y \in \mathbb{R}.$$

**Examples.** The standard sequence $\{x^n\}$ is binomial since in this case the definition reduces precisely to the binomial theorem. Other examples are $\{[x]_n\}$, $\{[x]^n\}$ and the Lah polynomials $\sum_{k=0}^{n} L'_{n,k} x^k$; this can easily be verified and is also a consequence of the following theory.

## A. Normalized Sequences and Differential Operators

How can we tell whether a sequence is binomial? Our plan is to associate with each sequence a unique operator, called its basis operator, and to characterize binomial sequences in terms of their basis operators. This method is well-known in the theory of finite differences and is basically analogous to Taylor-type operators in calculus.

**Definition.** A polynomial sequence $\{p_n(x): n \in \mathbb{N}_0\}$ is called *normalized* if

$$p_0(x) = 1$$
$$p_n(0) = 0 \quad \text{for all } n \geq 1.$$

**3.40 Proposition.** *A binomial sequence is normalized; the converse is false.* □

The proof is easy and is left to the reader. $\{1, x, 2x^2, x^3, x^4, \ldots\}$ is an example of a normalized sequence which is not binomial, for

$$p_2(x + y) = 2(x + y)^2 = 2(x^2 + 2xy + y^2)$$

but

$$p_0(x)p_2(y) + 2p_1(x)p_1(y) + p_2(x)p_0(y) = 2y^2 + 2xy + 2x^2 = 2(x^2 + xy + y^2).$$

Let us call a linear transformation of the vector space $\mathbb{R}[x]$ into itself a *polynomial operator*. Throughout this section all polynomials and scalars are assumed to be real whenever not explicitly mentioned.

**Definition.** A polynomial operator $P$ is called a *differential operator* or *of differential type* if

$$\deg(Pp(x)) = \deg(p(x)) - 1 \qquad (p(x) \in \mathbb{R}[x]).$$

Recall the convention $\deg(p(x)) = 0$ if and only if $p(x) \equiv a \neq 0$.

The correspondence sequence-operator referred to above is spelled out in the following fundamental definition.

**Definition.** To every sequence $\{p_n(x): n \in \mathbb{N}_0\}$ there exists a unique polynomial operator $P$ satisfying

(i) $Pp_0(x) = 0$
(ii) $Pp_n(x) = np_{n-1}(x) \qquad (n \geq 1)$.

$P$ is called the *basis operator* of $\{p_n(x)\}$.

Conversely, if we are given an operator $P$, then any sequence $\{p_n(x)\}$ satisfying (i) and (ii) is called a *basis sequence* of $P$.

**3.41 Proposition.** *We have:*

(i) *Any polynomial sequence possesses a unique basis operator; this operator is of differential type.*
(ii) *Any differential operator possesses a unique normalized basis sequence.*
(iii) *The correspondence basis operator ↔ basis sequence is a bijection between the set of differential operators and the set of normalized sequences.*

*Proof.* (i) is clear from the definition of a basis operator. Conversely, let $P$ be a differential operator. We construct the corresponding normalized basis sequence $\{p_n(x)\}$ inductively. Set $p_0(x) \equiv 1$, $p_1(x) = (Px)^{-1}x$. Suppose the normalized polynomials $p_k(x)$ have been defined for $k = 0, \ldots, n-1$ with $p_{n-1}(x) = \sum_{i=1}^{n-1} b_i x^i$, $b_{n-1} \neq 0$, and $Px^k = c_{k-1}^{(k)} x^{k-1} + \cdots + c_0^{(k)}$, $c_{k-1}^{(k)} \neq 0$, for all $k$. Let us set $p_n(x) = a_n x^n + \cdots + a_1 x$. From

$$Pp_n(x) = \sum_{k=1}^{n} a_k Px^k = \sum_{k=1}^{n} a_k \left( \sum_{i=0}^{k-1} c_i^{(k)} x^i \right) = \sum_{i=0}^{n-1} \left( \sum_{k=i+1}^{n} a_k c_i^{(k)} \right) x^i$$

$$= np_{n-1}(x) = \sum_{i=1}^{n-1} (nb_i) x^i$$

we obtain the following linear equations by comparing coefficients:

$$\begin{aligned}
x^{n-1}&: a_n c_{n-1}^{(n)} &&= nb_{n-1} \\
x^{n-2}&: a_n c_{n-2}^{(n)} + a_{n-1} c_{n-2}^{(n-1)} &&= nb_{n-2} \\
&\ \ \vdots && \ \ \vdots \\
x\ &: a_n c_1^{(n)} + a_{n-1} c_1^{(n-1)} + \cdots + a_2 c_1^{(2)} &&= nb_1 \\
1\ &: a_n c_0^{(n)} + a_{n-1} c_0^{(n-1)} + \cdots + a_2 c_0^{(2)} + a_1 c_0^{(1)} &&= 0.
\end{aligned}$$

Since $c_{k-1}^{(k)} \neq 0$ for all $k$, the system has a unique solution in the $a_i$'s. Furthermore $a_n \neq 0$ because $b_{n-1} \neq 0$, i.e., $p_n(x)$ has degree $n$. Assertion (iii) is already contained in the definition of a basis operator. □

**Example.** Let us determine the basis operators for the sequences $\{x^n\}$, $\{[x]_n\}$ and $\{[x]^n\}$. Quite obviously, the basis operator of the standard sequence $\{x^n\}$ is the ordinary differential operator which shall be called the *standard operator*, denoted by $D$. Applying **3.28**(ii) and (iv) we deduce

$$n[x]_{n-1} = x[x]_{n-1} - [x]_n + [x]_{n-1} = [x+1]_n - [x]_n,$$
$$n[x]^{n-1} = [x]^n - x[x]^{n-1} + [x]^{n-1} = [x]^n - [x-1]^n.$$

**Definition.** Let $a \in \mathbb{R}$. The polynomial operator $E^a$, defined by $E^a p(x) = p(x+a)$ is called the *translation* by $a$. The operator $I$ mapping every polynomial onto itself is called the *identity operator*. Trivially, $E^0 = I$, and we write $E := E^1$ for short. Notice that

$$E^n = \underbrace{E \cdots E}_{n}$$

and $E^{-n} = (E^{-1})^n$ for all $n \in \mathbb{N}_0$, where the right-hand side is the ordinary operator product. With these definitions we have, by the above identities

$$\begin{array}{ll} D & \text{is the basis operator of } \{x^n\}, \\ \Delta := E - I & \text{is the basis operator of } \{[x]_n\}, \\ \nabla := I - E^{-1} & \text{is the basis operator of } \{[x]^n\}. \end{array}$$

$\Delta$ and $\nabla$ are called the *forward* and *backward difference operators*, respectively.

**3.42 Theorem.** *Let $\{p_n(x): n \in \mathbb{N}_0\}$ be a normalized sequence with basis operator $P$. Then*

$$q(x) = \sum_{k \geq 0} \frac{[P^k q(x)]_{x=0}}{k!} p_k(x) \qquad (q(x) \in \mathbb{R}[x])$$

*is the unique representation of $q(x)$ in terms of the basis $\{p_n(x): n \in \mathbb{N}_0\}$.*

*Proof.* By linear extension, it suffices to prove the theorem for some basis of $\mathbb{R}[x]$, for instance the basis $\{p_n(x)\}$. Iterating the defining relation $Pp_n(x) = np_{n-1}(x)$ we obtain

$$P^k p_n(x) = [n]_k p_{n-k}(x),$$

and thus

$$[P^n p_n(x)]_{x=0} = n!$$
$$[P^k p_n(x)]_{x=0} = 0 \quad \text{for } k < n.$$

This can be written as

$$p_n(x) = \sum_{k \geq 0} \frac{[P^k p_n(x)]_{x=0}}{k!} p_k(x)$$

and the result follows. □

**Examples.** For $\{x^n\}$ and $D$ we have

$$q(x) = \sum_{k \geq 0} \frac{q^{(k)}(0)}{k!} x^k \qquad (q(x) \in \mathbb{R}[x])$$

which is of course just the Taylor expansion of calculus. Consider now $[x]_n$ and $\Delta$. From

$$\Delta^k = (E - I)^k = \sum_{i=0}^{k} (-1)^{k-i} \binom{k}{i} E^i$$

we infer

$$q(x) = \sum_{k \geq 0} \frac{\sum_{i=0}^{k} (-1)^{k-i} \binom{k}{i} q(i)}{k!} [x]_k \qquad (q(x) \in \mathbb{R}[x]).$$

For $q(x) = x^n$ we know from **3.24**(i) that the coefficient of $[x]_k$ is precisely $S_{n,k}$. Hence we have

$$S_{n,k} = \frac{1}{k!} \sum_{i=0}^{k} (-1)^{k-i} \binom{k}{i} i^n,$$

in accordance with **3.39**.

Using $\{[x]^n\}$ and $\nabla$ we obtain similarly

$$q(x) = \sum_{k \geq 0} \frac{\sum_{i=0}^{k} (-1)^{i} \binom{k}{i} q(-i)}{k!} [x]^k \qquad (q(x) \in \mathbb{R}[x]).$$

Our derivation of formula **3.39** for the Stirling numbers $S_{n,k}$ suggests the following general inversion principle for normalized sequences.

**3.43 Proposition.** *Let $\{p_n(x): n \in \mathbb{N}_0\}$ and $\{q_n(x): n \in \mathbb{N}_0\}$ be two normalized sequences with basis operators $P$ and $Q$, respectively. The matrices $[[Q^k p_n(x)]_{x=0}/k!]$ and $[[P^k q_n(x)]_{x=0}/k!]$ are inverses of each other. Hence for any sequences $u_0, u_1, \ldots;$ $v_0, v_1, \ldots$ of real numbers:*

$$v_n = \sum_{k=0}^{n} \frac{[Q^k p_n(x)]_{x=0}}{k!} u_k \; (n \in \mathbb{N}_0) \Leftrightarrow u_n = \sum_{k=0}^{n} \frac{[P^k q_n(x)]_{x=0}}{k!} v_k \; (n \in \mathbb{N}_0).$$

*Proof.* Immediate from **3.42** and **3.37**. □

As we shall on many occasions evaluate polynomials $p(x)$ at some point $a$, it is useful to introduce the following linear functionals.

**Definition.** Let $a \in \mathbb{R}$. The linear functional $L_a\colon \mathbb{R}[x] \to \mathbb{R}$, defined by $L_a\colon p(x) \to p(a)$ is called the *evaluation at a*. For the evaluation at 0 we set $L := L_0$ for short.

Notice that

$$L_a = LE^a$$

where $E^a$ is the translation introduced above. Using $L$ we can rewrite the fundamental expansion formula **3.42** as

**3.44.**

$$q(x) = \sum_{k \geq 0} \frac{LP^k q(x)}{k!} p_k(x)$$

*or in operator form*

$$I = \sum_{k \geq 0} \frac{p_k(x)}{k!} LP^k.$$

Formally, the last expression is a sum involving infinitely many terms. But it follows from the definition of a differential operator that for any given polynomial only a *finite* number of summands are different from 0. Hence, no special convergence arguments are necessary.

Let us now characterize binomial sequences by means of their associated basis operators.

**Definition.** A polynomial operator $P$ is called *shift invariant* if $PE^a = E^aP$ for all $a \in \mathbb{R}$. A shift invariant operator satisfying in addition $Px = c \neq 0$ is called a *Delta operator*.

It is perhaps useful to pause for a moment and see what the condition $PE^a = E^aP$ means. It means that for any $q(x) \in \mathbb{R}[x]$

$$P(q(x + a)) = Pq(x + a),$$

i.e., $P$ applied to $q(x + a)$ (as a polynomial in $x$) equals $Pq$ at the point $x + a$. As an example, let $P$ be defined by $Px^n = x^n$ for all $n \neq 2$, $Px^2 = x$. Then

$$\begin{aligned} PE^a x^2 &= P(x + a)^2 = P(x^2 + 2ax + a^2) = (2a + 1)x + a^2 \\ E^a P x^2 &= E^a x = x + a, \end{aligned}$$

hence $PE^a x^2 \neq E^a P x^2$ for $a \neq 0$.

**3.45 Theorem.** *A polynomial sequence is binomial if and only if its basis operator is a Delta operator.*

*Proof.* Let $\{p_n(x)\}$ be a binomial sequence with basis operator $P$. Since $P$ is of differential type we have $Px = c \neq 0$. To verify the shift invariance of $P$ we need only consider the basis polynomials $p_n(x)$. For $n = 0$ there is nothing to prove, so assume $n \geq 1$. By the definition of a binomial sequence we have

$$E^y p_n(x) = p_n(x+y) = \sum_{k \geq 0} \binom{n}{k} p_{n-k}(y) p_k(x) = \sum_{k \geq 0} \frac{L_y P^k p_n(x)}{k!} p_k(x)$$
$$= \sum_{k \geq 0} \frac{LE^y P^k p_n(x)}{k!} p_k(x).$$

From this we infer

$$E^y P p_n(x) = E^y n p_{n-1}(x) = \sum_{k \geq 0} \frac{LE^y P^k n p_{n-1}(x)}{k!} p_k(x) = \sum_{k \geq 0} \frac{LE^y P^{k+1} p_n(x)}{k!} p_k(x)$$
$$= \sum_{k \geq 1} \frac{LE^y P^k p_n(x)}{k!} k p_{k-1}(x) = P\left(\sum_{k \geq 0} \frac{LE^y P^k p_n(x)}{k!} p_k(x)\right) = PE^y p_n(x).$$

Now let $P$ be a Delta operator. It is easily verified that $P$ is of differential type. Applying **3.44** to $q(x) = p_n(x+y)$ (as a polynomial in $x$) we have

$$p_n(x+y) = \sum_{k=0}^{n} \frac{LP^k p_n(x+y)}{k!} p_k(x) = \sum_{k=0}^{n} \frac{LP^k E^y p_n(x)}{k!} p_k(x)$$
$$= \sum_{k=0}^{n} \frac{LE^y P^k p_n(x)}{k!} p_k(x) \quad \text{(by hypothesis)}$$
$$= \sum_{k=0}^{n} \frac{L_y [n]_k p_{n-k}(x)}{k!} p_k(x) = \sum_{k=0}^{n} \binom{n}{k} p_{n-k}(y) p_k(x). \quad \square$$

**Examples.** The standard operator $D$ is quite obviously shift invariant as are $\Delta = E - I$ and $\nabla = I - E^{-1}$. Hence all three are Delta operators and we obtain the following classical identities:

$$(x+y)^n = \sum_{k=0}^{n} \binom{n}{k} x^k y^{n-k}$$

$$[x+y]_n = \sum_{k=0}^{n} \binom{n}{k} [x]_k [y]_{n-k}$$

$$[x+y]^n = \sum_{k=0}^{n} \binom{n}{k} [x]^k [y]^{n-k}.$$

As $D$ is a Delta operator, so are all operators $E^aD = DE^a$. Operators $E^aD$ are called *Abel operators*. Their associated sequences are $\{x(x - an)^{n-1}\}$ as can easily be verified (and will also follow from a later result). Hence we obtain another well-known identity

$$(x + y)(x + y - an)^{n-1} = \sum_{k=0}^{n} \binom{n}{k} x(x - ak)^{k-1} y(y - a(n - k))^{n-k-1}.$$

B. Shift Invariant Operators

Any operator of the form $\sum_{k \geq 0} c_k D^k$ is clearly shift invariant since $D$ is. That, conversely, every shift invariant operator can be so described is our next major result.

Let $\mathscr{S}$ be the set of all shift invariant operators. $\mathscr{S}$ is an algebra over $\mathbb{R}$ with the usual addition, scalar multiplication and operator product. We note the following simple result:

**3.46.** *Let $P_1, P_2 \in \mathscr{S}$. Then $P_1 = P_2$ if and only if $LP_1 = LP_2$.*

*Proof.* Notice first the trivial fact that two polynomial operators $Q_1$ and $Q_2$ are identical if and only if $L_y Q_1 = L_y Q_2$ for all $y \in \mathbb{R}$. Thus we have for $P_1, P_2 \in \mathscr{S}$ and all $y \in \mathbb{R}$

$$\begin{aligned} LP_1 = LP_2 &\Rightarrow LP_1 E^y = LP_2 E^y \Rightarrow LE^y P_1 = LE^y P_2 \Rightarrow L_y P_1 \\ &= L_y P_2 \Rightarrow P_1 = P_2. \quad \square \end{aligned}$$

**3.47 Expansion Theorem.** *Let $P \in \mathscr{S}$. Then*

$$P = \sum_{k \geq 0} \frac{a_k}{k!} D^k,$$

*where $a_k = [Px^k]_{x=0}$ for all $k$. Conversely, any expansion $\sum_{k \geq 0} (a_k/k!)D_k$ is in $\mathscr{S}$.*

*Proof.* It suffices to consider the sequence $\{x^n\}$. Write $Q = \sum_{k \geq 0} (a_k/k!)D^k$ with $a_k$ defined as in the theorem. Then

$$LPx^n = a_n$$

$$LQx^n = L\left(a_0 I + a_1 D + \cdots + \frac{a_n}{n!} D^n\right) x^n = \frac{a_n}{n!} n! = a_n.$$

We infer $LP = LQ$, and thus $P = Q$ by **3.46**. $\square$

**3.47** suggests the following relationship between $\mathscr{S}$ and the algebra $\mathscr{F}$ of formal power series in one variable endowed with the ordinary product of two series.

**3.48 Proposition.** *Let $\mathscr{F}$ be the algebra of formal power series over $\mathbb{R}$ in the variable $t$. Then*

$$\Phi: \sum_{k \geq 0} \frac{a_k}{k!} t^k \to \sum_{k \geq 0} \frac{a_k}{k!} D^k$$

*is an algebra isomorphism between $\mathscr{F}$ and $\mathscr{S}$. It follows, in particular, that any two operators in $\mathscr{S}$ commute.*

*Proof.* $\Phi$ is clearly injective and also surjective by **3.47**. Let $f(t) = \sum_{k \geq 0} (a_k/k!)t^k$, $g(t) = \sum_{k \geq 0} (b_k/k!)t^k \in \mathscr{F}$ and set $\Phi(f) = F$, $\Phi(g) = G$. Then

$$(f \cdot g)(t) = \sum_{n \geq 0} \left( \sum_{k=0}^{n} \frac{a_k b_{n-k}}{k!(n-k)!} \right) t^n = \sum_{n \geq 0} \left( \sum_{k=0}^{n} \binom{n}{k} \frac{a_k b_{n-k}}{n!} \right) t^n.$$

It remains to be shown that

$$[(FG)x^n]_{x=0} = \sum_{k=0}^{n} \binom{n}{k} a_k b_{n-k}.$$

We have

$$\begin{aligned} [(FG)x^n]_{x=0} &= \left[ \left( \sum_{k \geq 0} \frac{a_k}{k!} D^k \cdot \sum_{l \geq 0} \frac{b_l}{l!} D^l \right) x^n \right]_{x=0} \\ &= \left[ \left( \sum_{k \geq 0} \sum_{l \geq 0} \frac{a_k b_l}{k! l!} D^{k+l} \right) x^n \right]_{x=0} \\ &= \sum_{k=0}^{n} \frac{a_k b_{n-k}}{k!(n-k)!} n! = \sum_{k=0}^{n} \binom{n}{k} a_k b_{n-k}. \quad \square \end{aligned}$$

The expansion theorem is the single most powerful result in this theory, connecting operator methods with the concept of generating functions. Notice that in **3.47** we may replace $D$ and $\{x^n\}$ by any Delta operator $Q$ and basis sequence $\{q_n(x)\}$, using precisely the same argument. This generalization is useful for some applications, so let us state it as a proposition.

**3.49 Proposition.** *Let $P$ be a shift invariant operator and $Q$ a Delta operator with basis sequence $\{q_n(x): n \in \mathbb{N}_0\}$. Then*

$$P = \sum_{k \geq 0} \frac{a_k}{k!} Q^k,$$

*where $a_k = [Pq_k(x)]_{x=0}$ for all $k$.* $\square$

Let $P$ be a shift invariant operator. The power series $p(t)$ with $P = p(D)$ under the isomorphism **3.48** is called the *indicator* of $P$. The following facts are immediate consequences of **3.47** and **3.48**.

**3.50 Corollary.** *Let* $P = \sum_{k \geq 0} (a_k/k!)D^k \in \mathscr{S}$. *Then*

(i) $\deg(Pq(x)) \leq \deg(q(x))$ *for all* $q(x) \in \mathbb{R}[x]$.
(ii) $P$ *is invertible* $\Leftrightarrow a_0 \neq 0 \Leftrightarrow \deg(Pq(x)) = \deg(q(x))$ *for all* $q(x) \in \mathbb{R}[x]$.
(iii) $P$ *is a Delta operator* $\Leftrightarrow a_0 = 0,\ a_1 \neq 0 \Leftrightarrow P = DT$ *where* $T \in \mathscr{S}$ *and* $T$ *is invertible.* □

**Definition.** Let $P \in \mathscr{S}$ with indicator $p(t)$. The operator $P' \in \mathscr{S}$ whose indicator is the derivative $p'(t)$ of $p(t)$ is called the *derivative* of $P$.

From the isomorphism **3.48** we infer the following formulae.

**3.51 Proposition.** *Let* $P, Q \in \mathscr{S}$. *Then*
(i) $(P + Q)' = P' + Q'$.
(ii) $(cP)' = cP'$.
(iii) $(PQ)' = PQ' + P'Q$.
(iv) $(P^n)' = nP^{n-1}P'$ *for all* $n \in \mathbb{Z}$, *where* $P$ *is assumed to be invertible if* $n < 0$.
(v) *The derivative of a Delta operator is invertible.* □

We want to derive an expression for the derivative in terms of the operator only. Define the *multiplication operator* $\underline{x}$ by

$$\underline{x}\colon p(x) \to xp(x) \qquad (p(x) \in \mathbb{R}[x]).$$

**3.52 Proposition.** *Let* $P \in \mathscr{S}$. *Then*

$$P' = P\underline{x} - \underline{x}P.$$

*Proof.* It is easily seen that $P\underline{x} - \underline{x}P \in \mathscr{S}$. By **3.47**, $P = \sum_{k \geq 0} (a_k/k!)D^k$ with $a_k = LPx^k$, hence by the definition of $P'$

$$P' = \sum_{k \geq 1} \frac{LPx^k}{(k-1)!} D^{k-1}.$$

On the other hand, we trivially have $L\underline{x}P = 0$ for any polynomial operator $P$ and hence by **3.47** again

$$P\underline{x} - \underline{x}P = \sum_{k \geq 0} \frac{L(P\underline{x} - \underline{x}P)x^k}{k!} D^k = \sum_{k \geq 0} \frac{LP\underline{x}x^k}{k!} D^k = \sum_{k \geq 1} \frac{LPx^k}{(k-1)!} D^{k-1},$$

and thus $P' = P\underline{x} - \underline{x}P$. □

**Examples.** For the translations $E^a$ we have $E^a = \sum_{k\geq 0} (LE^a x^k/k!)D^k = \sum_{k\geq 0} (a^k/k!)D^k = e^{aD}$. (This is, of course, nothing but the Taylor expansion of $p(x + a)$.) This implies

$$E^a = e^{aD}, \qquad \Delta = e^D - I, \qquad \nabla = I - e^{-D},$$
$$(E^a)' = aE^a, \qquad \Delta' = E, \qquad \nabla' = E^{-1}.$$

If we are given a binomial sequence $\{p_n(x)\}$, then we can think of the basis operator $P$ as determined by the action $Pp_n = np_{n-1}(x)$. As an important application of the concept of derivative we are now going to establish two closed formulae for the basis sequence of a given Delta operator.

**3.53 Theorem.** *Let $P = DT$ be a Delta operator. The basis sequence $\{p_n(x): n \in \mathbb{N}_0\}$ of $P$ is given by*

(i) $p_0(x) = 1$
$p_n(x) = x(P')^{-1}p_{n-1}(x) \qquad (n \geq 1),$ *(Rodrigues's formula)*

(ii) $p_0(x) = 1$
$p_n(x) = xT^{-n}x^{n-1} \qquad (n \geq 1),$ *(Steffensen's formula)*

*Proof.* Since by **3.51**(v) $P'$ is an invertible shift invariant operator we see that the polynomials $\{p_n(x)\}$ as defined in (i) form a normalized sequence. Hence to prove (i) it remains to be shown that $Pp_n(x) = np_{n-1}(x)$ for all $n > 0$. Notice first that

$$P\underline{x}(P')^{-1} = (\underline{x}P + P')(P')^{-1} = \underline{x}P(P')^{-1} + I.$$

Using induction on $n$ we have (recalling that shift invariant operators commute)

$$\begin{aligned} Pp_n(x) &= (P\underline{x}(P')^{-1})p_{n-1}(x) = \underline{x}P(P')^{-1}p_{n-1}(x) + p_{n-1}(x) \\ &= (n-1)x(P')^{-1}p_{n-2}(x) + p_{n-1}(x) = np_{n-1}(x). \end{aligned}$$

As to (ii), observe that by **3.51**

$$\begin{aligned} \underline{x}T^{-n+1} &= T^{-n+1}\underline{x} - (T^{-n+1})' = T^{-n+1}\underline{x} + (n-1)T^{-n}T' \\ &= T^{-n}(T\underline{x} + (n-1)T'). \end{aligned}$$

Using (i) and induction on $n$ this yields

$$\begin{aligned} p_n(x) &= \underline{x}(P')^{-1}p_{n-1}(x) = \underline{x}(P')^{-1}xT^{-n+1}x^{n-2} \\ &= \underline{x}(P')^{-1}T^{-n}(T\underline{x} + (n-1)T')x^{n-2} = \underline{x}(P')^{-1}T^{-n}(T + T'D)x^{n-1} \\ &= \underline{x}(P')^{-1}T^{-n}P'x^{n-1} = xT^{-n}x^{n-1}. \quad \square \end{aligned}$$

The following slight generalization is easily obtained by applying **3.53** twice.

**3.54 Corollary.** *Let $P = DT$, $Q = DU$ be two Delta operators with basis sequences $\{p_n(x): n \in \mathbb{N}_0\}$ and $\{q_n(x): n \in \mathbb{N}_0\}$ respectively. Then*

(i) $q_n(x) = x(TU^{-1})^n x^{-1} p_n(x) \qquad (n \geq 1)$
(ii) $q_n(x) = Q'(P')^{-1} U^{-n-1} T^{n+1} p_n(x) \qquad (n \geq 1)$. $\square$

**Examples.** Let $P = DE^a$ be the Abel operators mentioned above. Their basis sequences are by **3.53**(ii)

$$p_n(x) = xE^{-an}x^{n-1} = x(x - an)^{n-1},$$

i.e., the Abel polynomials. For $P = \Delta$ we know that $\Delta' = E$. Hence by **3.53**(i)

$$p_n(x) = xE^{-1}p_{n-1}(x) = xp_{n-1}(x-1) = x(x-1)p_{n-2}(x-2) = \cdots = [x]_n.$$

The *central difference operator* $\delta$ is defined by

$$\delta: p(x) \rightarrow p(x + \tfrac{1}{2}) - p(x - \tfrac{1}{2}).$$

$\delta$ is a Delta operator since $\delta = E^{-1/2}\Delta$. If we set $\Delta = DT$ as in **3.50**(iii), then $\delta = D(E^{-1/2}T)$. The basis sequence $q_n(x)$ of $\delta$ can now be calculated from **3.54**(i)

$$\begin{aligned} q_n(x) &= x(TT^{-1}E^{1/2})^n x^{-1}[x]_n = xE^{n/2}[x-1]_{n-1} \\ &= x\left[x + \frac{n}{2} - 1\right]_{n-1} \qquad \text{for } n \geq 1. \end{aligned}$$

The operator Lg defined by

$$\text{Lg}: p(x) \rightarrow -\int_0^\infty e^{-t}\frac{dp}{dx}(x+t)dt$$

is called the *Laguerre operator*. It is readily verified that Lg is shift invariant. To compute the indicator of Lg notice that

$$\text{Lg} = -\int_0^\infty e^{-t}e^{tD}D\,dt,$$

hence

$$\text{Lg} = -e^{-t(I-D)}\frac{D}{D-I}\Bigg|_0^\infty = \frac{D}{D-I} = D\left(\frac{I}{D-I}\right).$$

Applying **3.53**(ii) we obtain as basis sequence $\{l_n(x)\}$ of Lg

$$l_n(x) = x(D - I)^n x^{n-1} = x \sum_{k=0}^{n} (-1)^k \binom{n}{k} D^{n-k} x^{n-1}$$

$$= \sum_{k=0}^{n} (-1)^k \frac{n!}{k!} \binom{n-1}{k-1} x^k = \sum_{k=0}^{n} (-1)^k L'_{n,k} x^k.$$

On the other hand, $D - I = e^x D e^{-x}$ and thus $(D - I)^n = e^x D^n e^{-x}$ which yields the usual expression for $l_n(x)$

$$l_n(x) = xe^x \left(\frac{d}{dx}\right)^n (e^{-x} x^{n-1}).$$

The polynomials $l_n(x)$ are called the *Laguerre polynomials.*

C. Connecting Coefficients

Let us return to the starting point of our discussion. Given two binomial sequences $\{p_n(x)\}$, $\{q_n(x)\}$, we want to find the connecting coefficients $c_{n,k}$ where

$$q_n(x) = \sum_{k=0}^{n} c_{n,k} p_k(x) \qquad (n \in \mathbb{N}_0).$$

To phrase it differently: Given $\{p_n(x)\}$, $\{q_n(x)\}$, we wish to compute the polynomials $r_n(x)$ where

$$r_n(x) = \sum_{k=0}^{n} c_{n,k} x^k.$$

Let the operator $Z$ be defined by $Z(x^n) = p_n(x)$ for all $n$. $Z$ is obviously invertible and we have

$$Z(r_n(x)) = q_n(x) \qquad (n \in \mathbb{N}_0).$$

Hence our problem can be formulated as follows: Given the binomial sequences $\{p_n(x)\}$, $\{q_n(x)\}$ and the operator $Z$ defined by $Z(x^n) = p_n(x)$, compute the sequence $\{Z^{-1} q_n(x)\}$.

Our method will be to derive from the operators $P$ and $Q$ associated with $\{p_n(x)\}$, $\{q_n(x)\}$ the operator $R$ and then to compute $\{r_n(x)\}$ using the formulae in **3.53**.

Consider the set $\mathscr{S}$ of shift invariant operators and the isomorphic set $\mathscr{F}$ of formal power series in the variable $t$. Apart from the usual product, two power series can also be multiplied using composition of power series. Let $f(t) = \sum_{k \geq 1} f_k t^k$, $g(t) = \sum_{k \geq 1} g_k t^k$, then

$$(g \circ f)(t) := g(f(t)) = \sum_{k \geq 1} g_k \left( \sum_{i \geq 1} f_i t^i \right)^k.$$

This composition is associative and possesses the series $t$ as two-sided identity. The following simple result is of key importance to our theory.

**3.55 Proposition.** *Let $f(t) = \sum_{n\geq 0} f_n t^n \in \mathscr{F}$ with $f_0 = 0$, $f_1 \neq 0$. There exists a unique series $\hat{f}(t) \in \mathscr{F}$ with $f(\hat{f}(t)) = \hat{f}(f(t)) = t$.*

*Proof.* Notice first that the coefficient of $t^n$ in an arbitrary product $g \circ f$ is given by

$$\sum_{k=1}^{n} g_k \sum_{i_1+\cdots+i_k=n} f_{i_1} f_{i_2} \ldots f_{i_k}.$$

It follows from this expression that the coefficients of $\hat{f}$ such that $\hat{f}(f(t)) = t$ are uniquely given by

$$\hat{f}_1 = \frac{1}{f_1}$$

$$\hat{f}_n = -\frac{1}{f_1^n} \sum_{k=1}^{n-1} \hat{f}_k \sum_{i_1+\cdots+i_k=n} f_{i_1} f_{i_2} \ldots f_{i_k} \qquad (n \geq 2).$$

That $\hat{f}$ is a two-sided inverse of $f$ follows at once from associativity. □

Let $P = p(D) = \sum_{k\geq 0} (a_k/k!)D^k$ be a Delta operator. We know from **3.50**(iii) that $a_0 = 0$, $a_1 \neq 0$. Hence by **3.55** there is a unique Delta operator $\hat{P} = \hat{p}(D)$ with $\hat{p}(p(D)) = p(\hat{p}(D)) = D$.

**3.56 Theorem.** *Let $S = s(D)$, $P = p(D)$ be Delta operators with basis sequences $\{s_n(x): n \in \mathbb{N}_0\}$ and $\{p_n(x): n \in \mathbb{N}_0\}$ respectively, and let the operator $Z$ be defined by $Z(s_n(x)) = p_n(x)$. Then the following holds:*

(i) *$Z$ maps any binomial sequence onto a binomial sequence.*
(ii) *If $Z$ maps $r_n(x)$ onto $q_n(x)$ and $R = r(D)$, $Q = q(D)$ are the associated basis operators, then $Q = q(D) = r(\hat{s}(p(D)))$.*

*Proof.* $Z$ is obviously invertible. Suppose $\{r_n(x)\}$ is a binomial sequence and that $q_n(x) = Zr_n(x)$ for all $n$. Then

$$q_0(x) = Zr_0(x) = Z1 = Zs_0(x) = p_0(x) = 1$$

$$q_n(x) = Zr_n(x) = Z\left(\sum_{k=1}^{n} r_{n,k} s_k(x)\right) = \sum_{k=1}^{n} r_{n,k} p_k(x)$$

and thus

$$q_n(0) = \sum_{k=1}^{n} r_{n,k} p_k(0) = 0 \quad \text{for all } n \geq 1.$$

Hence $\{q_n(x)\}$ is a normalized sequence. Let $Q = q(D)$ be its basis operator. We want to show that $Q = ZRZ^{-1}$. We have

$$(ZR)1 = (QZ)1 \quad \text{and}$$
$$(ZR)r_n(x) = Z(nr_{n-1}(x)) = nq_{n-1}(x) = Qq_n(x) = QZr_n(x) \qquad (n \geq 1).$$

Hence $ZR = QZ$, i.e., $Q = ZRZ^{-1}$. Applying the same argument to $S$ and $P$ in place of $R$ and $Q$ we obtain similarly $P = ZSZ^{-1}$, and thus $P^k = ZS^kZ^{-1}$ for all $k$. Let $R = g(S) = \sum_{k \geq 1} (a_k/k!)S^k$ be the expansion **3.49** of $R$ in terms of $S$. (Note that $a_0 = 0$, $a_1 \neq 0$.) Then

$$Q = ZRZ^{-1} = Zg(S)Z^{-1} = Z\left(\sum_{k \geq 1} \frac{a_k}{k!} S^k\right)Z^{-1} = \sum_{k \geq 1} \frac{a_k}{k!} P^k = g(P).$$

$Q$ as an expansion of $P$ is therefore shift invariant and hence a Delta operator.

Finally, we infer from $R = g(S)$ that

$$r(D) = g(s(D))$$
$$r(\hat{s}(D)) = g(D),$$

and thus

$$Q = q(D) = g(p(D)) = r(\hat{s}(p(D))). \quad \square$$

**3.57 Corollary.** *Let $P$ be a Delta operator with basis sequence $\{p_n(x): n \in \mathbb{N}_0\}$. The basis sequence $\{\hat{p}_n(x): n \in \mathbb{N}_0\}$ of the operator $\hat{P}$ is given by*

$$\hat{p}_n(x) = \sum_{k=0}^{n} \frac{[P^k x^n]_{x=0}}{k!} x^k.$$

*Proof.* In **3.56**, set $S = D$, $R = \hat{P}$. Then $Q = q(D) = \hat{p}(p(D)) = D$. This means that the operator $Z$ defined by $Z(x^n) = p_n(x)$ satisfies $Z\hat{p}_n(x) = x^n$. Hence

$$\hat{p}_n(x) = Z^{-1}x^n \qquad (n \geq 0)$$
$$x^n = Z^{-1}p_n(x) \qquad (n \geq 0).$$

By **3.42**

$$x^n = \sum_{k=0}^{n} \frac{[P^k x^n]_{x=0}}{k!} p_k(x)$$

and the result follows by applying $Z^{-1}$ to both sides of this equation. $\square$

We come to our main theorem.

**3.58 Theorem** (Mullin–Rota). *Let $\{p_n(x): n \in \mathbb{N}_0\}$ and $\{q_n(x): n \in \mathbb{N}_0\}$ be binomial sequences with connecting coefficients $c_{n,k}$*

$$q_n(x) = \sum_{k=0}^{n} c_{n,k} p_k(x) \qquad (n \in \mathbb{N}_0)$$

*and basis operators $P = p(D)$ and $Q = q(D)$ respectively. The polynomials $r_n(x) = \sum_{k=0}^{n} c_{n,k} x^k$ are then the basis sequence corresponding to the Delta operator $R = q(\bar{p}(D))$.*

*Proof.* Set $S = D$ in **3.56**. The operator $Z$ defined by $Z(x^n) = p_n(x)$ satisfies $Zr_n(x) = q_n(x)$. Hence $Q = q(D) = r(p(D))$, and thus $R = r(D) = q(\bar{p}(D))$. $\square$

**Examples.** Let us compute the connecting coefficients $c_{n,k}$ in $x^n = \sum_{k=0}^{n} c_{n,k}[x]_k$. The polynomials $e_n(x) = \sum_{k=0}^{n} c_{n,k} x^k$ correspond to the operator $\hat{\Delta}$ whence by **3.57**

$$e_n(x) = \sum_{k=0}^{n} \frac{[\Delta^k x^n]_{x=0}}{k!} x^k = \sum_{k=0}^{n} S_{n,k} x^k \qquad \text{(cf. 3.42)}.$$

The polynomials $e_n(x)$ are called the *exponential polynomials*. Let $Z$ be the operator defined by $Z[x]_n = x^n$. Then $Zx^n = e_n(x)$ and by the recursion $[x]_{n+1} = x[x-1]_n$

$$ZxE^{-1}[x]_n = Zx[x-1]_n = xx^n = xZ[x]_n \qquad (n \in \mathbb{N}_0).$$

Through linear extension we infer

$$ZxE^{-1}p(x) = xZp(x) \qquad (p(x) \in \mathbb{R}[x]).$$

For $p(x) = (x+1)^n$ this yields a recursion for the exponential polynomials

$$e_{n+1}(x) = x \sum_{k=0}^{n} \binom{n}{k} e_k(x).$$

Since $e_n(1) = B_n$ (Bell number) we obtain at the point $x = 1$ the recursion **3.31** for the Bell numbers. Another recursion results from **3.53**(i). Since $\Delta = e^D - I$ we have $\hat{\Delta} = \log(I + D)$, hence $\hat{\Delta}' = I/(I + D)$. Applying **3.53**(i) we infer $e_n(x) = x(I + D)e_{n-1}(x)$, i.e.,

$$e_n(x) = x(e_{n-1}(x) + e'_{n-1}(x)) \qquad (n \geq 1).$$

Now let us look at $[x]^n = \sum_{k=0}^{n} c_{n,k}[x]_k$. $\hat{\Delta} = \log(I + D)$, $\nabla = I - e^{-D}$ imply

$$\begin{aligned} R = r(D) &= I - e^{-\log(I+D)} = I - (I + D)^{-1} = I - (I - D + D^2 \mp \cdots) \\ &= D(I - D + D^2 \mp \cdots) = D(I + D)^{-1}. \end{aligned}$$

**3.53**(ii) yields

$$r_n(x) = x(I + D)^n x^{n-1} = x \sum_{k=0}^{n} \binom{n}{k} D^k x^{n-1} = \sum_{k=0}^{n} \binom{n}{k} [n-1]_k x^{n-k}$$
$$= \sum_{k=0}^{n} \binom{n}{k} [n-1]_{n-k} x^k = \sum_{k=0}^{n} \frac{n!}{k!} \binom{n-1}{k-1} x^k,$$

in agreement with **3.24**(ii).

Our final theorem gives a compact characterization of binomial sequences, underlining the intimate relationship between binomial sequences and exponential series, a subject which we shall study in depth in chapter V.

**3.59 Theorem.** *A polynomial sequence $\{p_n(x): n \in \mathbb{N}_0\}$ is binomial if and only if*

$$\sum_{n \geq 0}^{n} \frac{p_n(x)}{n!} t^n = e^{xg(t)}$$

*for some formal power series $g(t) = \sum_{k \geq 0} g_k t^k$ with $g_0 = 0$, $g_1 \neq 0$. In fact, when the sequence is binomial $g(t) = \hat{p}(t)$ is the indicator of the operator $\hat{P}$.*

*Proof.* The identity $e^{(x+y)g(t)} = e^{xg(t)} e^{yg(t)}$ implies

$$\sum_{n \geq 0} \frac{p_n(x+y)}{n!} t^n = \sum_{k \geq 0} \frac{p_k(x)}{k!} t^k \sum_{l \geq 0} \frac{p_l(y)}{l!} t^l$$

and hence, by comparing coefficients, the binomiality of $\{p_n(x)\}$.

If, on the other hand, $\{p_n(x)\}$ is binomial we have by **3.57**

$$p_n(x) = \sum_{k=0}^{n} \frac{[\hat{P}^k x^n]_{x=0}}{k!} x^k$$

and thus by **3.47**

$$\sum_{n \geq 0} \frac{p_n(x)}{n!} D^n = \sum_{k \geq 0} \frac{x^k}{k!} \left( \sum_{n \geq 0} \frac{[\hat{P}^k x^n]_{x=0}}{n!} D^n \right) = \sum_{k=0} \frac{x^k \hat{p}(D)^k}{k!} = e^{x\hat{p}(D)}.$$

Now apply the isomorphism **3.48**. □

**Examples.** Let us look once more at the exponential polynomials $e_n(x) = \sum_{k=0}^{n} S_{n,k} x^k$. By the theorem just proved

$$\sum_{n \geq 0} \frac{e_n(x)}{n!} t^n = e^{x(e^t - 1)}.$$

Comparing coefficients of $t^n$ one easily obtains

$$e_n(x) = e^{-x} \sum_{k \geq 0} \frac{x^k k^n}{k!}.$$

For $x = 1$ this yields two formulae for the Bell numbers:

$$\sum_{n \geq 0} \frac{B_n}{n!} t^n = e^{e^t - 1} \qquad \text{(Bell)}$$

$$B_n = \frac{1}{e} \sum_{k \geq 0} \frac{k^n}{k!} \qquad \text{(Dobinski)}.$$

We shall re-derive the first of these identities in chapter V where we interpret the right hand side as the *exponential generating function* for the Bell numbers.

As a final example consider the Laguerre polynomials $l_n(x)$. It is immediate that $\mathrm{Lg} = \widehat{\mathrm{Lg}}$ for the Laguerre operator Lg. Hence we have

$$\sum_{n \geq 0} \frac{l_n(x)}{n!} t^n = e^{x(t/(t-1))}.$$

## Exercises III.3

1. Prove **3.40**.

2. Verify Abel's identity: For all $a \in \mathbb{R}$

$$(x + y)^n = \sum_{k=0}^{n} \binom{n}{k} x(x - ka)^{k-1}(y + ka)^{n-k}.$$

→ 3. Show that the following operators are shift invariant:

(i) $B: p(x) \to \int_x^{x+1} p(t)dt$ (Bernoulli operator)
(ii) $H: p(x) \to \sqrt{2/\pi} \int_{-\infty}^{+\infty} e^{-t^2/2} p(x + t)dt$ (Hermite operator)
(iii) $Eu: p(x) \to \frac{1}{2}(p(x) + p(x + 1))$ (Euler operator).

4. Let $B_r$ be defined by

$$B_r: p(x) \to \int_x^{x+r} p(t)dt.$$

In particular, $B_1 = B$ is the Bernoulli operator. Show:

(i) $B_r$ is shift invariant,
(ii) $B_r = (E^r - I)B/\Delta$.

Expand $B_r$ in terms of $\Delta$ (according to **3.49**).

→ 5. Verify the first terms in Simpson's formula:

$$\int_x^{x+2} p(t)dt = 2\left(I + \Delta + \frac{\Delta^2}{6} - \frac{\Delta^4}{180} + \frac{\Delta^6}{180} \pm \cdots\right)p(x).$$

6. Let $P$ be a shift invariant operator and $P^{(k)}$ its $k$-th derivative. Show:

$$\underline{P}x^n = \sum_{k\geq 0} \binom{n}{k} x^{n-k} \underline{P^{(k)}}.$$

7. Fill in the details in the proof of **3.54**.

→ 8. Let $\{p_n(x)\}$ and $\{q_n(x)\}$ be binomial sequences, and let the operator $T$ be defined by $Tp_n(x) = q_n(x)$. Show that $T^{-1}$ exists and that $U \to TUT^{-1}$ is an automorphism of the algebra $\mathscr{S}$. (Mullin–Rota)

→ 9. Prove **3.59** by expanding $E^a$ in terms of $P = p(D)$.

10. Let $l_n(x) = \sum_{k=0}^{n} l_{n,k} x^k$ for all $n \in \mathbb{N}$ where $l_n(x)$ is the Laguerre polynomial. Formulate the Laguerre inversion formula.

11.* The *Bernoulli polynomials* $b_n(x)$ are defined by

$$te^{tx}(e^t - 1)^{-1} = \sum_{n\geq 0} \frac{b_n(x)}{n!} t^n.$$

Prove:

(i) $b_n(0) = b_n$ (the Bernoulli number of ex. III.2.13),
(ii) $b_n(x) = \sum_k \binom{n}{k} b_{n-k} x^k$,
(iii) $x^n = \sum_k \binom{n}{k} (n + k - 1)^{-1} b_k(x)$.

Derive inversion formulae from (ii) and (iii).

12. We consider the analogous situation for vector spaces. Let $q$ be a prime power. The *Euler translation* $\mathscr{E}^a$ is defined by $\mathscr{E}^a p(x) = p(q^a x)$. An operator $P$ is called an *Eulerian operator* if $\mathscr{E}^a P = q^{-a} P \mathscr{E}^a$ for all $a \in \mathbb{R}$ and if $Px^n \neq 0$ for all $n > 0$. Prove that for an Eulerian operator $P$

$$\deg(Pp(x)) = \deg(p(x)) - 1 \qquad (p(x) \in \mathbb{R}[x]).$$

→13.* The sequence $\{p_n(x)\}$ is called an *Eulerian sequence* if $p_0(x) = 1$ and

$$p_n(xy) = \sum_{k=0}^{n} \binom{n}{k}_q p_k(x) y^k p_{n-k}(y) \qquad (n \in \mathbb{N}).$$

$P$ is the *basis operator* corresponding to $\{p_n(x)\}$ if $Pp_n = (q^n - 1)p_{n-1}$, and, conversely, $p_n(x)$ is then a *basis sequence* of $P$. Prove:

(i) if $\{p_n(x)\}$ is Eulerian, then the basis operator is Eulerian;
(ii) if $P$ is an Eulerian operator, then $P$ has a unique Eulerian basis sequence;
(iii) the correspondence in (i) and (ii) is a bijection.

(Andrews)

→14. Verify that the operator $D_q$ used in the proof of **3.35** is Eulerian. Expand $D_q$ in terms of $\mathscr{E}$. What is the basis sequence?

15. $Q$ is called *Euler shift invariant* if $Q\mathscr{E}^{q^a} = \mathscr{E}^{q^a}Q$ for all $a \in \mathbb{R}$. Prove the statement analogous to **3.49**: Let $P$ be Euler shift invariant and $Q$ an Eulerian operator with basis sequence $\{q_n(x)\}$. Then

$$P = \sum_{k \geq 0} \frac{a_k}{\prod_{i=1}^{k} (q^i - 1)} (xQ)^k, \quad \text{where } a_k = [Pq_k(x)]_{x=1}.$$

(Andrews)

## 4. Order Functions

According to the general frame work set out in chapter I let us now study counting functions which arise in connection with *monotone* mappings. The main problem is to determine $|\mathrm{Mon}(N, R)|$ where $N$ and $R$ are finite posets. It is, of course, too much to expect that a satisfactory solution can be found for arbitrary posets; but under the assumption that one or both of $N$ and $R$ is a chain the problem becomes manageable.

### A. The Order Polynomial

In the following, $P$ will always denote a finite poset with $|P| = n$. To shorten the notation we shall use $\underline{n}$ to denote the chain $\{1 < 2 < \cdots < n\}$.

**Definition.** Let $P$ be a finite poset.

$$\omega(P; x) := |\mathrm{Mon}(P, \underline{x})|$$

is called the *order polynomial* of $P$. That is, $\omega(P; r)$ counts the number of monotone mappings from $P$ into $\underline{r}$ for all $r \in \mathbb{N}$. If $P$ is empty we set $\omega(P; x) = 1$.

The definition anticipates that $\omega(P; x)$ is a polynomial in $x$, a fact which we are going to show first.

Define $e_k$ as the number of monotone surjective mappings from $P$ into $\underline{k}$, i.e.,

$$e_k := |\mathrm{Mon}(P, \underline{k}) \cap \mathrm{Sur}(P, \underline{k})| \qquad (k \in \mathbb{N}).$$

Note that $e_n =: e(P)$ is precisely the number of *total extensions* of $P$ (cf. section I.2).

**3.60 Proposition.** *For a non-empty poset P*

$$\omega(P; x) = \sum_{k=1}^{n} \frac{e_k}{k!} [x]_k.$$

*In particular, $\omega(P; x)$ is a polynomial of degree n with leading coefficient $e(P)/n!$ and constant term 0.*

*Proof.* There are $e_k\binom{x}{k}$ monotone mappings $f: P \to \underline{x}$ with $|\mathrm{im}(f)| = k$. $\square$

**3.61 Proposition.** *Let $\mathfrak{I}(P)$ denote the lattice of ideals of P. Then*

$$e_k = \#\{\varnothing, P\text{-chains in } \mathfrak{I}(P) \text{ of length } k\} \qquad (k \in \mathbb{N}).$$

*In particular:*

$$e_1 = 1, \qquad e_2 = |\mathfrak{I}(P)| - 2, \qquad e(P) = \#\{\text{maximal chains in } \mathfrak{I}(P)\}.$$

*Proof.* Let $f \in \mathrm{Mon}(P, \underline{k})$ be surjective. The sets $J_i := \{a \in P: f(a) \le i\}$ are ideals in $P$, whence we obtain the chain

$$\varnothing < J_1 < J_2 < \cdots < J_k = P.$$

Conversely, let such a chain be given.

$$J_1 | J_2 - J_1 | J_3 - J_2 | \cdots | P - J_{k-1}$$

is a $k$-partition of $P$, and $f: P \to \underline{k}$ defined by $f(a) = i$ if $a \in J_i - J_{i-1}$ is monotone and surjective. This correspondence between mappings $f: P \to \underline{k}$ and chains in $\mathfrak{I}(P)$ is clearly bijective, and the result follows. $\square$

If we consider $\mathrm{Mon}(\underline{x}, P)$ where $P$ is an arbitrary finite poset we obtain another counting polynomial, called the *chain polynomial* of $P$. (See the exercises for more details.)

**Examples.** Let us look at the two simplest examples where $P = n\mathscr{C}(0)$ is an antichain and where $P = \mathscr{C}(n-1)$ is a chain respectively. In the first case, any mapping is monotone whereas in the second case the monotone mappings correspond bijectively to the $n$-multisets in $\underline{x}$ (cf. **1.13**). Hence

$$\omega(n\mathscr{C}(0); x) = x^n = \sum_{k=0}^{n} S_{n,k}[x]_k, \qquad e_k = k!S_{n,k},$$

$$\omega(\mathscr{C}(n-1); x) = \frac{[x]^n}{n!} = \sum_{k=0}^{n} \frac{L'_{n,k}}{n!}[x]_k, \qquad e_k = \binom{n-1}{k-1}.$$

Obviously, $\mathfrak{I}(n\mathscr{C}(0)) \cong \mathscr{B}(n)$. A chain of length $k$ in $\mathscr{B}(n)$ corresponds to an ordered $k$-partition of an $n$-set. In the second case, $\mathfrak{I}(\mathscr{C}(n-1)) \cong \mathscr{C}(n)$ with chains of length $k$ corresponding to ordered $k$-partitions of $n$.

The following definition gives the appropriate generalization of injective mappings between ordered sets.

**Definition.** Let $N$ and $R$ be finite posets. A monotone mapping $f: N \to R$ is called *strict* if $a <_N b$ implies $f(a) <_R f(b)$ for all $a, b \in N$.

Analogous to $\omega(P; x)$ we define the *strict order polynomial* of $P$

$$\overline{\omega}(P; x) := |\overline{\text{Mon}}(P, \underline{x})|,$$

where $\overline{\text{Mon}}$ denotes the set of strict monotone mappings.

If we set

$$\bar{e}_k := |\overline{\text{Mon}}\,(P, \underline{k}) \cap \text{Sur}(P, \underline{k})|,$$

we obtain, as in **3.60**,

**3.62.**

$$\overline{\omega}(P, x) = \sum_{k=1}^{n} \frac{\bar{e}_k}{k!}[x]_k \qquad (P \text{ non-empty}).$$

The following proposition is the key result leading to an explicit expression of the coefficients $e_k$, $\bar{e}_k$ in terms of certain permutation indices associated with the mappings $f \in \text{Mon}(P, \underline{x})$. First we choose an indexing of the elements $a_1, a_2, \ldots, a_n$ of $P$ such that the chain $a_1 < a_2 < \cdots < a_n$ is an extension of the order on $P$, i.e., $a_i <_P a_j \Rightarrow i < j$. To each of the $e(P)$ total extensions $a_{i_1} < a_{i_2} < \cdots < a_{i_n}$ of $P$ we associate the permutation $i_1 i_2 \ldots i_n$. Let $E(P)$ be the set of all these permutations. By construction, the identity is always in $E(P)$, and we have $|E(P)| = e(P)$.

**3.63 Proposition** (Stanley). *Let $P$ be a finite poset, $|P| = n$.*

(i) *To every $f \in \text{Mon}(P, \underline{x})$ there is a unique permutation $\pi = i_1 i_2 \ldots i_n \in E(P)$ such that*

(a) $f(a_{i_1}) \le f(a_{i_2}) \le \cdots \le f(a_{i_n})$
(b) $i_s > i_{s+1} \Rightarrow f(a_{i_s}) < f(a_{i_{s+1}})$.

*Conversely, any mapping $f: P \to \mathbb{N}$ satisfying (a) and (b) for some $\pi \in E(P)$ is monotone.*

(ii) *To every $\bar{f} \in \overline{\text{Mon}}(P, \underline{x})$ there is a unique permutation $\bar{\pi} = j_1 j_2 \ldots j_n \in E(P)$ such that*

($\bar{a}$) $\bar{f}(a_{j_1}) \le \bar{f}(a_{j_2}) \le \cdots \le \bar{f}(a_{j_n})$
($\bar{b}$) $j_s < j_{s+1} \Rightarrow \bar{f}(a_{j_s}) < \bar{f}(a_{j_{s+1}})$.

*Conversely, any mapping $\bar{f}: P \to \mathbb{N}$ satisfying ($\bar{a}$) and ($\bar{b}$) for some $\bar{\pi} \in E(P)$ is strictly monotone.*

*Proof.* Let us first prove the converse statements. Any $f: P \to \mathbb{N}$ which satisfies ($a$) or ($\bar{a}$) is obviously monotone. Suppose $\bar{f}$ satisfies ($\bar{b}$) too. Whenever $a_{j_t} <_P a_{j_u}$ then $j_t < j_u$ and $t < u$. Hence in the sequence $j_t, j_{t+1}, \ldots, j_u$ there must be two consecutive indices $j_s, j_{s+1}$ with $j_s < j_{s+1}$ whence we have by $(\bar{a}), (\bar{b})$ $\bar{f}(a_{j_t}) \leq \bar{f}(a_{j_s}) < \bar{f}(a_{j_{s+1}}) \leq \bar{f}(a_{j_u})$, i.e., $\bar{f}(a_{j_t}) < \bar{f}(a_{j_u})$. Thus $\bar{f}$ is strictly monotone.

Now let $f: P \to \underline{x}$ be monotone and surjective, where without loss of generality $\text{im}(f) = \underline{k}$. To find a suitable $\pi \in E(P)$ let us list the elements of $P$ writing first those of $f^{-1}(1)$ then those of $f^{-1}(2)$ and so on until $f^{-1}(k)$. Condition (b) says that $f(a_{i_s}) = f(a_{i_{s+1}})$ implies $i_s < i_{s+1}$. Hence within a group $f^{-1}(i)$ the elements must be written with rising index. Altogether this gives the permutation

$$\pi = i_1 \dots i_r j_1 \dots j_s \dots l_1 \dots l_t$$

with $f^{-1}(1) = \{a_{i_1}, \dots, a_{i_r}\}$, etc., and $i_1 < i_2 < \cdots < i_r$, $j_1 < \cdots < j_s$, etc. The sequence $\{a_{i_1}, \dots, a_{l_t}\}$ is a total extension of $P$, hence $\pi \in E(P)$, and it is clear from the construction that $\pi$ is uniquely determined by (a) and (b). (ii) is proved by an analogous argument. □

**3.63** gives a method for computing the numbers $e_k$ and $\bar{e}_k$. Recall that a *descent* of the permutation $i_1 i_2 \dots i_n$ is a pair $i_s, i_{s+1}$ with $i_s > i_{s+1}$. For $k = 0, \dots, n-1$ set

$$w_k := \#\{\text{permutations in } E(P) \text{ with precisely } k \text{ descents}\}.$$

Since only the identity has no descents we have $w_0 = 1$.

**3.64 Proposition.** *Let $P$ be a finite poset, $e_k$, $\bar{e}_k$ and $w_k$ defined as before. Then*

(i) $e_k = \sum_{j=0}^{n-1} w_j \binom{n-1-j}{n-k}$
(ii) $\bar{e}_k = \sum_{j=0}^{n-1} w_j \binom{j}{n-k}$
(iii) $w_k = (-1)^k \sum_{i=1}^{n} (-1)^{i+1} e_i \binom{n-i}{n-1-k} = (-1)^k \sum_{i=1}^{n} (-1)^{i+1} \bar{e}_i \binom{i-1}{n-1-k}$.

*Proof.* Suppose $\pi \in E(P)$ possesses $j$ descents. How many mappings $f \in \text{Mon}(P, \underline{x})$ with $|\text{im}(f)| = k$ are associated with $\pi$ under the correspondence **3.63**? According to **3.63**(i), at every descent $i_s > i_{s+1}$ we must have $f(a_{i_s}) < f(a_{i_{s+1}})$. Therefore, if $j \geq k$, then no $f \in \text{Mon}(P, \underline{x})$ belongs to $\pi$. Suppose now $j < k$. From the set of $n - 1 - j$ pairs which do not form a descent we can choose, in all possible ways, $k - 1 - j$ pairs $i_l < i_{l+1}$ with $f(a_{i_l}) < f(a_{i_{l+1}})$, and precisely these mappings $f$ will belong to $\pi$. Since by **3.63**(i), all mappings $f \in \text{Mon}(P, \underline{x})$ with $|\text{im}(f)| = k$ arise exactly once if $\pi$ runs through $E(P)$, formula (i) follows. (ii) is established in analogous fashion. (iii) is then easily verified by inversion. □

**Example.** For the poset $P$ in Figure 3.3 we have $E(P) = \{12345, 12354, 13245, 13524, 13254\}$, with $w_0 = 1$, $w_1 = 3$, $w_2 = 1$. Consider the monotone surjective mappings $f: P \to \underline{3}$. We have

$$e_3 = 6w_0 + 3w_1 + w_2 = 16.$$

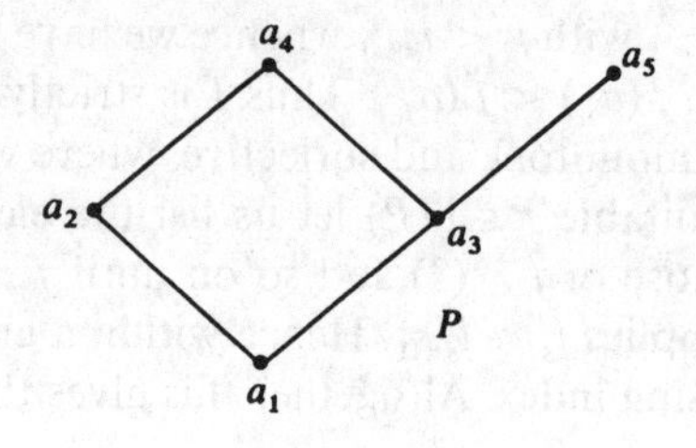


Figure 3.3

Written as words these mappings are:

| | | | | | |
|---|---|---|---|---|---|
| 11123 | 11223 | 11233 | 12223 | 12233 | 12333 |
| 11132 | 11232 | 12232 | | | |
| 12123 | 12133 | 13233 | | | |
| 12131 | 13132 | 13232 | | | |
| 12132. | | | | | |

The five rows correspond via **3.63** to the five permutations in $E(P)$ listed above.

Next, consider the strict monotone surjective mappings $\bar{f}: P \to \underline{4}$. We have $\bar{e}_4 = w_1 + 2w_2 = 5$, and the mappings are

| | | |
|---|---|---|
| 12344 | | 12354 |
| 12234 | corresponding to | 13245 |
| 13243 | the permutation | 13524 |
| 13244, 12243 | | 13254. |

By substituting **3.64** into $\omega(P; x)$ and $\bar{\omega}(P; x)$ and using the binomial identity $\binom{x+y}{n} = \sum_{k=0}^{n} \binom{x}{k}\binom{y}{n-k}$ proved in section 3 we obtain the following expressions for the order polynomials.

**3.65 Theorem** (Stanley). *Let $P$ be a finite poset with $|P| = n$. Then*

(i) $\omega(P; x) = \sum_{j=0}^{n-1} \binom{x+n-1-j}{n} w_j$
(ii) $\bar{\omega}(P; x) = \sum_{j=0}^{n-1} \binom{x+j}{n} w_j$
(iii) $\omega(P; -x) = (-1)^n \bar{\omega}(P; x)$
(iv) $e_k = (-1)^n \sum_{i=k}^{n} (-1)^i \binom{i-1}{k-1} \bar{e}_i$. $\square$

**3.65**(iii) is another example of a combinatorial reciprocity theorem generalizing formula **3.5**.

**Examples.** For the poset $P$ in Figure 3.3 we have

$$\omega(P;x) = \binom{x+4}{5} + 3\binom{x+3}{5} + \binom{x+2}{5} = \frac{(x+2)(x+1)^3x}{4!}.$$

In particular, $\omega(P;2) = |\mathfrak{J}(P)| = e_2 + 2 = 9$. **3.65**(iii) gives

$$\bar{\omega}(P;x) = \frac{(x-2)(x-1)^3x}{4!}.$$

Consider $P = \mathscr{C}(n-1)$. In this case $E(\mathscr{C}(n-1))$ consists of the identity only, whence the expressions **3.65**(i)(ii) reduce to our well-known formulae

$$|\mathrm{Mon}(\mathscr{C}(n-1), \underline{x})| = \frac{[x]^n}{n!}$$

$$|\overline{\mathrm{Mon}}(\mathscr{C}(n-1), \underline{x})| = \frac{[x]_n}{n!}.$$

**3.65**(iii) is the identity **3.5** and finally, **3.65**(iv), gives a new proof of **3.16**.

Consider now the antichain $P = n\mathscr{C}(0)$. $E(P)$ consists of all permutations of $\underline{n}$. Hence **3.65**(i) and (ii) can be written as

$$x^n = \sum_{k=0}^{n-1} W_{n,k}\binom{x+k}{n},$$

where $W_{n,k}$ is the number of permutations of $\underline{n}$ with precisely $k$ descents. Note that $W_{n,k} = W_{n,n-1-k}$. The *Eulerian numbers* $A_{n,k}$ are commonly defined by $A_{n,k} := W_{n,k-1}$ $(k = 1, \ldots, n)$.

**3.66** (Recursion for the Eulerian numbers $A_{n,k}$). *For all* $n, k \geq 1$:

(i) $x^n = \sum_{k=1}^n A_{n,k}\binom{x+k-1}{n}$.
(ii) $A_{n,k} = A_{n,n-k+1}$.
(iii) $A_{n,1} = 1$, $A_{n,k} = 0$ *for* $n < k$,
$A_{n+1,k} = (n-k+2)A_{n,k-1} + kA_{n,k} \qquad (k \geq 2)$.

*Proof.* We have already seen (i) and (ii). Trivially,

$$x = (n-k+1)\frac{x+k}{n+1} + k\frac{x+k-1-n}{n+1}$$

whence

$$
\begin{aligned}
x^{n+1} = xx^n &= \sum_{k=1}^{n} (n-k+1)A_{n,k}\frac{x+k}{n+1}\binom{x+k-1}{n} \\
&\quad + \sum_{k=1}^{n} kA_{n,k}\frac{x+k-1-n}{n+1}\binom{x+k-1}{n} \\
&= \sum_{k=2}^{n+1} (n-k+2)A_{n,k-1}\binom{x+k-1}{n+1} + \sum_{k=1}^{n} kA_{n,k}\binom{x+k-1}{n+1} \\
&= \sum_{k=1}^{n+1} [(n-k+2)A_{n,k-1} + kA_{n,k}]\binom{x+k-1}{n+1}.
\end{aligned}
$$

Since the polynomials $\binom{x+k-1}{n+1}$, $k = 1, \ldots, n+1$, are linearly independent in the vector space $\mathbb{R}[x]$, the result follows from (i). □

Eulerian numbers

| $A_{n,k}$ | $k = 1$ | 2 | 3 | 4 | 5 | 6 | 7 | 8 |
|---|---|---|---|---|---|---|---|---|
| $n = 1$ | 1 | | | | | | | |
| 2 | 1 | 1 | | | | | | |
| 3 | 1 | 4 | 1 | | | | | |
| 4 | 1 | 11 | 11 | 1 | | | | |
| 5 | 1 | 26 | 66 | 26 | 1 | | | |
| 6 | 1 | 57 | 302 | 302 | 57 | 1 | | |
| 7 | 1 | 120 | 1191 | 2416 | 1191 | 120 | 1 | |
| 8 | 1 | 247 | 4293 | 15619 | 15619 | 4293 | 247 | 1 |

Let us make a brief detour to describe an interesting application of the order polynomial to finite graphs. Let $G(V, E)$ be a finite undirected graph with vertex-set $V$ and edge-set $E$.

A *coloring* of $G$ is a mapping $c: V \to C$ from the set of vertices into a set of colors such that $c(a) \neq c(b)$ for all edges $\{a, b\} \in E$. The theory of colorings of graphs and surfaces is among the most thoroughly studied subjects in graph theory and will be presented in detail in section VII.3. Here we are going to discuss briefly the *chromatic polynomial* $c(G; x)$ of $G$. $c(G; x)$ is defined as the function which counts at the point $x = r$ the number of $r$-colorings, i.e., of colorings with $r$ or less colors. Of particular interest is the *chromatic number* $\mathrm{chrom}(G)$ of $G$ which is the minimum number of colors necessary such that a coloring of $G$ exists. Hence

$$\mathrm{chrom}(G) = \min_{k \in \mathbb{N}} k \quad \text{such that } c(G; k) > 0.$$

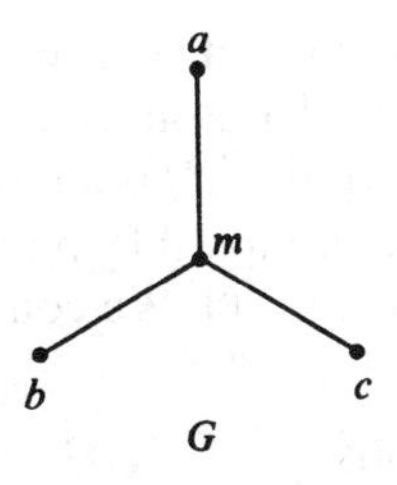


Figure 3.4

That $c(G, x)$ is indeed a polynomial is a consequence of the result we are about to prove.

**Example.** To compute the chromatic polynomial of the graph of Figure 3.4 we proceed as follows. First we color the vertex $m$ with any of the $x$ possible colors. Once a color is chosen, $a$, $b$ and $c$ can be colored with any of the remaining $x - 1$ colors, whence we obtain $c(G; x) = x(x - 1)^3$ and $\operatorname{chrom}(G) = 2$.

An *orientation* $\mathcal{O}$ of a graph $G$ is an assignment of an arrow to each edge of $G$. We write $u \underset{\mathcal{O}}{\to} v$ if the edge $\{u, v\}$ receives the direction from $u$ to $v$. The orientation $\mathcal{O}$ is called *acyclic* if the oriented graph possesses no directed cycles, i.e., no sequence $u_0 \to u_1 \to u_2 \to \cdots \to u_t = u_0$.

Claim. For all $r \in \mathbb{N}$, $c(G; r)$ is equal to the number of pairs $(f, \mathcal{O})$ where $f: V \to \{1, \ldots, r\}$ and $\mathcal{O}$ is an acyclic orientation of $G$ such that

$$(+) \qquad u \underset{\mathcal{O}}{\to} v \Rightarrow f(u) < f(v) \qquad (u, v \in V).$$

Clearly, in any pair $(f, \mathcal{O})$ satisfying (+), $f$ is an $r$-coloring. Conversely, if $f$ is an $r$-coloring, then it induces an acyclic orientation $\mathcal{O}$ by defining $u \underset{\mathcal{O}}{\to} v$ whenever $f(u) < f(v)$. This correspondence is bijective, which proves our claim.

Let us now similarly define $\bar{c}(G; x)$ as the number of pairs $(f, \mathcal{O})$ as above such that

$$(++) \qquad u \underset{\mathcal{O}}{\to} v \Rightarrow f(u) \leq f(v) \qquad (u, v \in V).$$

**3.67 Proposition** (Stanley). *Let $G(V, E)$ be a finite undirected graph. The functions $c(G; x)$ and $\bar{c}(G; x)$ are polynomials in $x$ of degree $|V|$, and we have*

$$c(G; -x) = (-1)^{|V|}\bar{c}(G; x).$$

*It follows, in particular, that $|c(G; -1)|$ is the number of acyclic orientations of $G$.*

*Proof.* Let $\mathcal{O}$ be an acyclic orientation of $G$. The reflexive and transitive closure $\hat{\mathcal{O}}$ of $\mathcal{O}$ is a partial ordering of $V$ and, what is more, a mapping $f: V \to \mathbb{N}$ is compatible with $\mathcal{O}$ in the sense of $(++)$ if and only if $f$ is a monotone mapping from $\hat{\mathcal{O}}$ into $\mathbb{N}$. Similarly, a mapping $f: V \to \mathbb{N}$ is compatible in the sense of $(+)$ if and only if $f$ is a strict monotone mapping from $\hat{\mathcal{O}}$ into $\mathbb{N}$. We conclude that

$$c(G; x) = \sum_{\mathcal{O}} \bar{\omega}(\hat{\mathcal{O}}; x), \qquad \bar{c}(G; x) = \sum_{\mathcal{O}} \omega(\hat{\mathcal{O}}; x)$$

and hence the proposition follows by applying **3.65**(iii) to each of the summands. □

## B. Standard Tableaux

A particularly interesting class of monotone mappings with applications to the representation theory of the symmetric group (see Young [1] or Robinson [1]) arises in connection with number-partitions. Let $\alpha: \alpha_1 \geq \cdots \geq \alpha_t$ be a partition of $n$. Recall that the Young lattice $\mathcal{Y}$ satisfies $\mathcal{Y} \cong \mathfrak{J}(\mathbb{N}^2)$ where a number-partition $\alpha \in \mathcal{Y}$ is mapped onto its Ferrers diagram $J(\alpha) \in \mathfrak{J}(\mathbb{N}^2)$, $J(\alpha) = \{(i, j): 1 \leq i \leq t, 1 \leq j \leq \alpha_i\}$. A monotone mapping $f: J(\alpha) \to \mathbb{N}$ thus assigns natural numbers to the boxes $(i, j)$ in $J(\alpha)$ in such a way that the numbers increase monotonically in each row and each column of $J(\alpha)$.

**Definition.** Let $\alpha: \alpha_1\alpha_2 \ldots \alpha_t$, $\alpha_1 \geq \cdots \geq \alpha_t$, be a partition of $n$. A *Young tableau with frame* $\alpha$ (and cardinality $n$) is an array $n_{ij}$ of positive integers, $1 \leq i \leq t$, $1 \leq j \leq \alpha_i$, such that

$$j \leq j' \Rightarrow n_{ij} \leq n_{ij'} \text{ for all } i,$$
$$i \leq i' \Rightarrow n_{ij} \leq n_{i'j} \text{ for all } j.$$

If the numbers $n_{ij}$ are all distinct and precisely the integers $1, \ldots, n$, then the Young tableau is called a *standard tableau*. Let $\mathrm{St}(\alpha)$ denote the set of all standard tableaux with frame $\alpha$.

**Example.** Let $\alpha: 432211$

$$\begin{array}{llll} 2 & 2 & 4 & 5 \\ 3 & 6 & 7 & \\ 3 & 7 & & \\ 8 & 8 & & \\ 8 & & & \\ 9 & & & \end{array} \qquad \begin{array}{llll} 1 & 2 & 3 & 6 \\ 4 & 7 & 9 & \\ 5 & 11 & & \\ 8 & 13 & & \\ 10 & & & \\ 12 & & & \end{array}$$

are two Young tableaux with frame $\alpha$ of which the second one is a standard tableau.

Let us list the standard tableaux of cardinality 2, 3 and 4.

| | | | | | | |
|---|---|---|---|---|---|---|
| $(n = 2)$ | 12 | 1 | | | | |
| | | 2 | | | | |
| $(n = 3)$ | 123 | 12 | 13 | 1 | | |
| | | 3 | 2 | 2 | | |
| | | | | 3 | | |
| $(n = 4)$ | 1234 | 123 | 124 | 134 | 12 | 13 |
| | | 4 | 3 | 2 | 34 | 24 |
| | 12 | 13 | 14 | 1 | | |
| | 3 | 2 | 2 | 2 | | |
| | 4 | 4 | 3 | 3 | | |
| | | | | 4 | | |

The main problem is to compute the number of standard tableaux with a given frame $\alpha$, i.e., the cardinality $|\mathrm{St}(\alpha)|$. Since standard tableaux correspond by definition to the total extensions of the poset $J(\alpha)$ we have $|\mathrm{St}(\alpha)| = e(J(\alpha))$ in the notation of **3.60**. We set $e(\alpha) := |\mathrm{St}(\alpha)|$. Applying **3.61**, we obtain the following result.

**3.68 Proposition.** *The number $e(\alpha)$ of standard tableaux with frame $\alpha$ equals the number of maximal $0, \alpha$-chains in the Young lattice $\mathscr{Y}$.* $\square$

The reader may verify **3.68** in figure 1.8 for small $\alpha$. To facilitate the induction arguments to follow let us agree on the following notation. For $\alpha_1\alpha_2 \ldots \alpha_t \in \mathscr{Y}$ let $e(\alpha, -i)$, $i = 1, \ldots, t$, be the number of standard tableaux with frame $\alpha_1\alpha_2 \ldots \alpha_i - 1 \ldots \alpha_t$. If the monotonicity condition is violated, i.e., if $\alpha_i - 1 < \alpha_{i+1}$, we set $e(\alpha, -i) = 0$. Similarly, $e(\alpha, +i)$, $i = 1, \ldots, t + 1$, is the number of standard tableaux with frame $\alpha_1\alpha_2 \ldots \alpha_i + 1 \ldots \alpha_t$ or $\alpha_1\alpha_2 \ldots \alpha_t 1$ if $i = t + 1$. Again $e(\alpha, +i) = 0$ if the monotonicity condition is violated, i.e., if $\alpha_i + 1 > \alpha_{i-1}$. If $\alpha$ is a partition of $n$ we write $|\alpha| = n$.

**Example.** Let $\alpha: 432211$. Then $e(\alpha, -3) = e(\alpha, -5) = 0$ and $e(\alpha, -i) > 0$ for $i = 1, \ldots, 6$, $i \neq 3, 5$. Also, $e(\alpha, +4) = e(\alpha, +6) = 0$ and $e(\alpha, +i) > 0$ for $i = 1, \ldots, 7$, $i \neq 4, 6$, $i \neq 4, 6$. (See Figure 3.5.) In general, we have the obvious fact:

$$e(\alpha, -i) = 0 \Leftrightarrow e(\alpha, +(i + 1)) = 0.$$

**3.69 Proposition.** *Let $\alpha: \alpha_1\alpha_2 \ldots \alpha_t$ be a partition of $n$. Then*

(i) $e(\alpha) = \sum_{i=1}^{t} e(\alpha, -i)$
(ii) $(n + 1)e(\alpha) = \sum_{i=1}^{t+1} e(\alpha, +i)$
(iii) $\sum_{|\alpha|=n} e^2(\alpha) = n!$.

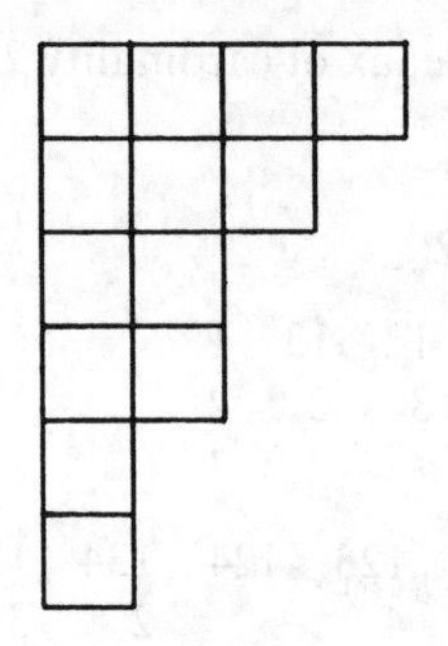

Figure 3.5

*Proof.* Let $A \in \mathrm{St}(\alpha)$. If the number $n$ appears in row $i$ of $A$, then obviously $e(\alpha, -i) \neq 0$. Thus after deletion of the box containing $n$ we obtain a standard tableau $\phi A$ of cardinality $n - 1$ and frame $(\alpha, -i)$. Conversely, let $B \in \mathrm{St}(\alpha, -i)$. By adding the box $(i, \alpha_i)$ and assigning to it the number $n$ we obtain a standard tableau $\psi B \in \mathrm{St}(\alpha)$. Since $\phi$ and $\psi$ are inverse mappings, (i) follows.

We prove (ii) by induction on $n$. For $n = 1$ the assertion is trivial. Consider $\alpha : \alpha_1 \dots \alpha_t$ with $|\alpha| = n$. From (i) we have

$$\sum_{i=1}^{t+1} e(\alpha, +i) = \sum_{i=1}^{t+1} \sum_{j=1}^{t+1} e((\alpha, +i), -j) = \sum_{j=1}^{t} \sum_{i=1}^{t+1} e((\alpha, -j), +i) + e(\alpha),$$

by observing that $e((\alpha, +i), -(t + 1)) = 0$ for $i = 1, \dots, t$, and

$$e((\alpha, +(t + 1)), -(t + 1)) = e(\alpha).$$

By the induction hypothesis and (i) this gives

$$\sum_{i=1}^{t+1} e(\alpha, +i) = \sum_{j=1}^{t} ne(\alpha, -j) + e(\alpha) = ne(\alpha) + e(\alpha) = (n + 1)e(\alpha).$$

(iii) is trivial for $n = 1$. Suppose it holds for all $k \leq n - 1$. Then

$$\sum_{|\alpha|=n} e^2(\alpha) = \sum_{|\alpha|=n} e(\alpha) \sum_j e(\alpha, -j) \quad \text{(by (i))}$$

$= \sum e(\gamma)e(\delta)$ where the sum runs over all pairs $(\gamma, \delta)$ with $|\gamma| = n$, $|\delta| = n - 1$ and $\gamma = (\delta, +i)$, $\delta = (\gamma, -j)$ for some pair $(i, j)$

$$= \sum_{|\beta|=n-1} e(\beta) \sum_i e(\beta, +i)$$

$$= \sum_{|\beta|=n-1} ne^2(\beta) \quad \text{(by (ii))}$$

$= n(n - 1)! = n!$ (induction hypothesis). $\square$

Formula **3.69**(iii) can also be interpreted in the following way. The number of *ordered pairs* of standard tableaux with equal frame and cardinality $n$ equals the number $n!$ of *permutations* of $\{1, \ldots, n\}$. Hence the question arises whether there exists a natural bijection $\Phi\colon S_n \to \bigcup_{|\alpha|=n} \mathrm{St}(\alpha)^2$ assigning to each permutation $g \in S_n$ a pair $(A, B)$, $A, B \in \mathrm{St}(\alpha)$ for some $\alpha$ with $|\alpha| = n$ such that

$$\Phi g = (A, B) \Leftrightarrow \Phi g^{-1} = (B, A).$$

We are going to give the construction of such a mapping $\Phi$ without proving all the details. (See Knuth [1, vol. 3, p. 52].)

Let $T$ be a Young tableau (not necessarily in standard form) filled with the distinct numbers $k_1, \ldots, k_r \in \mathbb{N}$. Let $a \in \mathbb{N}$ be different from the $k_i$'s. We form a new tableau $T \cup a$ according to the following rules:

(i) *Replace the smallest number $k_{i_1}$ in the first row of $T$ which is greater than $a$, by $a$. If all numbers in the first row are smaller than $a$, write $a$ at the end of the first row.*

(ii) *In the case where $k_{i_1}$ was replaced by $a$, put $k_{i_1}$ into the second row according to rule (i), and so forth.*

It is immediately clear that $T \cup a$ is again a Young tableau. Now take the permutation $g = a_1 a_2 \ldots a_n \in S_n$. We construct the tableau $A(g)$ inductively

$$A(g) = (\ldots(a_1 \cup a_2) \cup a_3) \cup \cdots \cup a_{n-1}) \cup a_n.$$

Parallel to $A(g)$ we form the standard tableau $B(g)$ by writing $i \in \mathbb{N}$ in the box which is added to $A(g)$ at the $i$-th step

$$(\ldots(a_1 \cup a_2) \cup \ldots) \cup a_{i-1} \to (\ldots(a_1 \cup a_2) \cup \ldots) \cup a_{i-1}) \cup a_i.$$

The mapping $\Phi\colon S_n \to \bigcup_{|\alpha|=n} \mathrm{St}(\alpha)^2$, given by

$$\Phi g = (A(g), B(g)) \qquad (g \in S_n)$$

has the required properties. $A(g)$ and $B(g)$ obviously have the same frame. $\Phi$ is bijective since the sequence $a_1 a_2 \ldots a_n$ can be uniquely recovered from the tableaux $A(g)$ and $B(g)$. The reader may provide a proof for

$$(+) \qquad \Phi g = (A, B) \Leftrightarrow \Phi g^{-1} = (B, A).$$

**Example.** Consider $g = 4263751$.

| | | | | | | | |
|---|---|---|---|---|---|---|---|
| | 4 | 2 | 26 | 23 | 237 | 235 | 135 |
| $A(g)$ | | 4 | 4 | 46 | 46 | 467 | 267 |
| | | | | | | | 4 |
| | 1 | 1 | 13 | 13 | 135 | 135 | 135 |
| $B(g)$ | | 2 | 2 | 24 | 24 | 246 | 246 |
| | | | | | | | 7 |

Let us construct $g$ from the pair $(A(g), B(g))$. We know that 4 was placed last. Hence 4 must have been replaced by 2 in the second row which, in turn, must have been replaced by 1 in the first row. Thus 1 must be the last element in the word $g$; and so on.

Property (+) implies, in particular, that

$$\Phi g = (A, A) \Leftrightarrow g = g^{-1} \Leftrightarrow g^2 = id.$$

Hence we obtain the following corollary.

**3.70 Proposition.** *We have*

$$\sum_{|\alpha|=n} e(\alpha) = \#\{g \in S_n : g^2 = id\} = \sum_{k=0}^{\lfloor n/2 \rfloor} \frac{n!}{k!(n-2k)!2^k}.$$

*Proof.* The permutations $g$ with $g^2 = id$ are precisely those of cycle type $1^{n-2k}2^k$ for some $k$, $0 \leq k \leq \lfloor n/2 \rfloor$. Now apply **3.15**(i). □

Several interesting corollaries can be deduced from the construction $g \to (A(g), B(g))$. Let us mention just one of them.

**3.71 Proposition** (Schensted). *The number of permutations in $S_n$ which contain a longest increasing subword of length s and a longest decreasing subword of length t, equals $\sum_\alpha e^2(\alpha)$ where the summation runs over all t-partitions of n with greatest summand s.*

*Proof.* Suppose $g \in S_n$ satisfies the condition of the theorem. If we can show that $A(g)$ has $s$ columns and $t$ rows the result will follow. Call the sequence of integers inserted into the $j$-th box of the first row of $A(g)$ the *$j$-th fundamental sequence of $g$*. The construction is set up so that every such sequence is decreasing and that every number appears in precisely one fundamental sequence. On the other hand, whenever a number $k$ is inserted into the $j$-th box of row 1 of $A(g)$, the adjacent number in box $j - 1$ must be smaller than $k$ and must precede $k$ in the word $g$. Hence the length of row 1 of $A(g)$, i.e., the number of columns, is precisely $s$. The second part of the theorem follows easily upon observing that an $i$-th row is added to $A(g)$ when an $i$-th element is inserted for the first time in some fundamental sequence. □

**Example.** In the permutation $g = 4263751$ the fundamental sequences are 421, 63, 75. A longest decreasing subword is 421 whereas the longest increasing subwords constructed as in the preceding proof are 237 and 235.

Let us return to the computation of $e(\alpha)$. Obviously, $e(\alpha) = e(\alpha^*)$ where $\alpha^*$ is the conjugate partition of $\alpha$. This suggests introducing some concept related to both sets of numbers $\alpha_i$ and $\alpha_i^*$. Let $\alpha : \alpha_1 \ldots \alpha_t \in \mathcal{Y}$. For every $(i, j)$, $1 \leq i \leq t$, $1 \leq j \leq \alpha_i$,

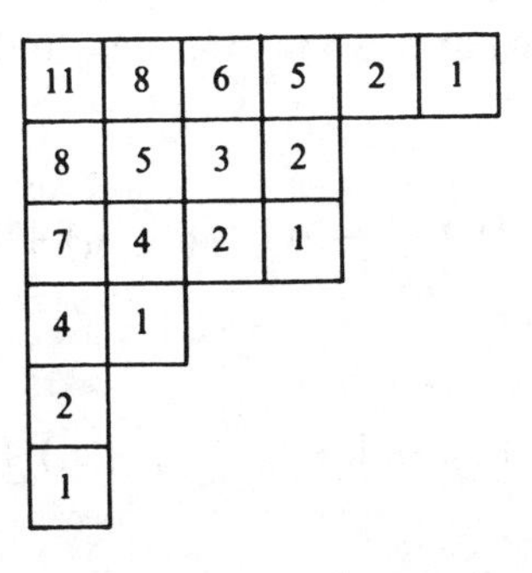

| 11 | 8 | 6 | 5 | 2 | 1 |
|---|---|---|---|---|---|
| 8 | 5 | 3 | 2 | | |
| 7 | 4 | 2 | 1 | | |
| 4 | 1 | | | | |
| 2 | | | | | |
| 1 | | | | | |

Figure 3.6

the *hook length* $h_{i,j}$ is the number of boxes to the right of and below $(i, j)$ including the box $(i, j)$ itself (cf. the proof of **3.19**). Hence

$$h_{i,j} = (\alpha_i - j) + (\alpha_j^* - i) + 1.$$

In particular, we set

$$x_i := h_{i,1} = \alpha_i + t - i \qquad (i = 1, \ldots, t).$$

**Example.** The hook lengths of $\alpha$: 644211 are shown in Figure 3.6 with the numbers $x_i$ listed in the first column.

**3.72 Lemma.** *Let $\alpha: \alpha_1\alpha_2 \ldots \alpha_t \in \mathcal{Y}$. For all i, the sequence of the $x_i = \alpha_i + t - i$ numbers*

$$x_i = h_{i,1}, h_{i,2}, \ldots, h_{i,\alpha_i}, x_i - x_{i+1}, x_i - x_{i+2}, \ldots, x_i - x_t$$

*is a permutation of* $\{1, 2, \ldots, x_i\}$.

*Proof.* We have to show that these numbers are distinct and lie between 1 and $x_i$. Notice first that

$$x_i = h_{i,1} > h_{i,2} > \cdots > h_{i,\alpha_i} \geq 1,$$

since the hook lengths decrease at each step by at least one. Similarly,

$$x_i > x_{i+1} > \cdots > x_t \geq 1,$$

hence

$$1 \leq x_i - x_{i+1} < \cdots < x_i - x_t < x_i.$$

It remains to be proved that any two numbers $h_{i,j}$ and $x_i - x_k$ are distinct.

(a) $j \le \alpha_k$. Then

$$h_{i,j} \ge (\alpha_i - \alpha_k + 1) + (k - i) > (\alpha_i - \alpha_k) + (k - i) = x_i - x_k.$$

(b) $j > \alpha_k$. Then

$$h_{i,j} \le (\alpha_i - \alpha_k) + (k - 1 - i) < (\alpha_i - \alpha_k) + (k - i) = x_i - x_k. \qquad \square$$

**3.73 Theorem** (Frame–Thrall–Robinson). *For an n-partition* $\alpha\colon \alpha_1\alpha_2\ldots\alpha_t$

$$e(\alpha) = \frac{n!}{\prod_{i,j} h_{i,j}}.$$

*Proof.* By **3.72**, we have for all $i = 1, \ldots, t$

$$\prod_{j=1}^{\alpha_i} h_{i,j} = \frac{x_i!}{\prod_{\underset{\cdot}{k} > i} (x_i - x_k)}.$$

Hence the theorem is equivalent to

$$e(\alpha) = n! \frac{\prod_{i<k} (x_i - x_k)}{\prod_{\underset{\cdot}{i}=1}^{t} x_i!},$$

and this is the formula we want to prove. We use induction on $n$. For $n = 1$ the assertion is obvious. Now let $\alpha\colon \alpha_1 \ldots \alpha_t$ be an $n$-partition. By **3.69**(i) and the induction hypothesis we have[1]

$$\begin{aligned} e(\alpha) &= \sum_{j=1}^{t} e(\alpha, -j) \\ &= \sum_{j=1}^{t} (n-1)! \prod_{\substack{\underset{\cdot}{i}<\underset{\cdot}{k} \\ i,k \ne j}} (x_i - x_k) \frac{\prod_{\underset{\cdot}{i}<j} (x_i - (x_j - 1)) \prod_{\underset{\cdot}{k}>j} (x_j - 1 - x_k)}{(\prod_{\underset{\cdot}{k} \ne j} x_k!)(x_j - 1)!} \\ &= (n-1)! \sum_{j=1}^{t} \frac{\prod_{\underset{\cdot}{i}<\underset{\cdot}{k}} (x_i - x_k)}{\prod_{\underset{\cdot}{k} \ne j} (x_j - x_k)} \frac{x_j}{\prod_{\underset{\cdot}{i}=1}^{t} x_i!} \prod_{\underset{\cdot}{k} \ne j} (x_j - 1 - x_k) \\ &= n! \frac{\prod_{\underset{\cdot}{i}<\underset{\cdot}{k}} (x_i - x_k)}{\prod_{\underset{\cdot}{i}=1}^{t} x_i!} \sum_{j=1}^{t} \frac{x_j}{n} \prod_{\underset{\cdot}{k} \ne j} \frac{(x_j - 1 - x_k)}{(x_j - x_k)}. \end{aligned}$$

[1] Recall that $\underset{\cdot}{k}$ indicates that $k$ is the running index.

Set $f(x) := \prod_{k=1}^{t}(x - x_k)$. Then $f'(x_j) = \prod_{k \neq j}(x_j - x_k)$, and we may write the last equation in the form

$$e(\alpha) = n! \frac{\prod_{i<k}(x_i - x_k)}{\prod_{i=1}^{t} x_i!} \sum_{j=1}^{t} \frac{-x_j}{n} \frac{f(x_j - 1)}{f'(x_j)}.$$

Thus it remains to be shown that

$$(+) \qquad \sum_{j=1}^{t} \frac{x_j f(x_j - 1)}{f'(x_j)} = -n.$$

Consider the polynomial $x^2 f(x - 1)$ and denote by $q(x)$, $r(x)$ the quotient and rest polynomials respectively, upon division by $f(x)$. Hence

$$(++) \qquad x^2 f(x - 1) = f(x)q(x) + r(x), \quad \text{with } \deg(r(x)) \leq t - 1.$$

We have $r(x_j) = x_j^2 f(x_j - 1)$ for $j = 1, \ldots, t$, thus

$$r(x) = \sum_{j=1}^{t} \left( r(x_j) \prod_{k \neq j} \frac{x - x_k}{x_j - x_k} \right) = \sum_{j=1}^{t} \left( x_j^2 f(x_j - 1) \frac{f(x)}{(x - x_j) f'(x_j)} \right)$$

and

$$q(0) = -\frac{r(0)}{f(0)} = \sum_{j=1}^{t} \frac{x_j f(x_j - 1)}{f'(x_j)}.$$

On the other hand, $q(x)$ is a polynomial of degree 2, say $q(x) = q_2 x^2 + q_1 x + q_0$. Comparing coefficients in $(++)$ we obtain

$$x^{t+2}: \quad 1 = q_2,$$

$$x^{t+1}: \quad -\sum_{j=1}^{t}(x_j + 1) = -\sum_{j=1}^{t} x_j - t = -\sum_{j=1}^{t} x_j + q_1 \Rightarrow q_1 = -t,$$

$$x^t: \quad \sum_{i<j}(x_i + 1)(x_j + 1) = \sum_{i<j} x_i x_j + (t - 1)\sum_{i=1}^{t} x_i + \binom{t}{2}$$

$$= \sum_{i<j} x_i x_j + \sum_{i=1}^{t} x_i + q_0 \Rightarrow q_0 = \binom{t}{2} - \sum_{i=1}^{t} x_i.$$

Finally, recall the definition of $x_i$, $x_i = \alpha_i + t - i$. We infer that

$$\sum_{i=1}^{t} x_i = \sum_{i=1}^{t} \alpha_i + \binom{t}{2} = n + \binom{t}{2},$$

and thus

$$q_0 = -n,$$

$$\sum_{j=1}^{t} \frac{x_j f(x_j - 1)}{f'(x_j)} = q(0) = q_0 = -n,$$

which is precisely $(+)$. □

**Example.** Consider $\alpha$: 2211. The hook lengths are given in Figure 3.7. Hence $e(\alpha) = 6!/5 \cdot 4 \cdot 2 \cdot 2 = 9$. The standard tableaux with frame 2211 are

| | | | | | | | | |
|---|---|---|---|---|---|---|---|---|
| 12 | 12 | 12 | 13 | 13 | 13 | 14 | 14 | 15 |
| 34 | 35 | 36 | 24 | 25 | 26 | 25 | 26 | 26 |
| 5 | 4 | 4 | 5 | 4 | 4 | 3 | 3 | 3 |
| 6 | 6 | 5 | 6 | 6 | 5 | 6 | 5 | 4 |

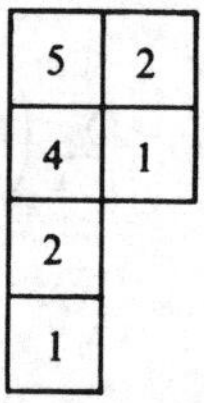

| | |
|---|---|
| 5 | 2 |
| 4 | 1 |
| 2 | |
| 1 | |

Figure 3.7

**Example.** As a final example let us determine the number of permutations in $S_9$ whose longest increasing subwords have length 4 and whose longest decreasing subwords have length 3. The only possible standard tableaux with 3 rows and 4 columns, together with their hook lengths, are shown in Figure 3.8. Hence, by **3.71** and **3.73**, the number of these permutations is

$$\left(\frac{9!}{6 \cdot 5 \cdot 4 \cdot 3 \cdot 3 \cdot 2 \cdot 2}\right)^2 + \left(\frac{9!}{6 \cdot 5 \cdot 4 \cdot 3 \cdot 3 \cdot 2}\right)^2 = 35280.$$

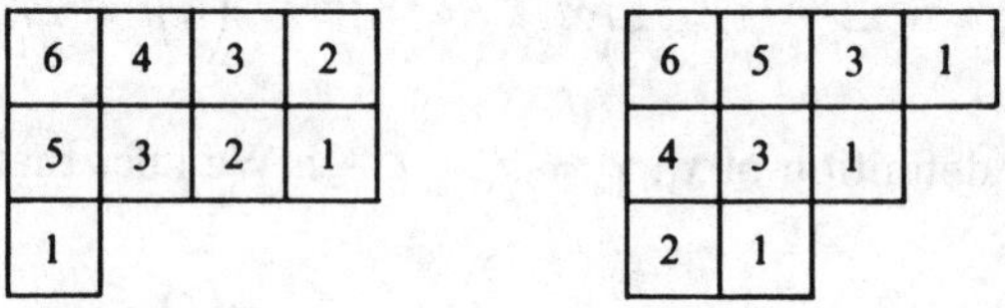

| | | | |
|---|---|---|---|
| 6 | 4 | 3 | 2 |
| 5 | 3 | 2 | 1 |
| 1 | | | |

| | | | |
|---|---|---|---|
| 6 | 5 | 3 | 1 |
| 4 | 3 | 1 | |
| 2 | 1 | | |

Figure 3.8

## Exercises III.4

→ 1. Let $P$ be a finite poset with $|P| = n$. We call $\kappa(P; x) := |\mathrm{Mon}(\underline{x}, P)|$ the *chain polynomial* of $P$ where $\underline{x} = \{1, \ldots, x\}$. Prove that

$$\kappa(P; x) = \sum_{k=0}^{l} \frac{u_k}{k!} [x-1]_k,$$

where

$$u_k = \# \{\text{chains in } P \text{ of length } k\},\ l = l(P).$$

2. Determine $\kappa(P; x)$ for $P = \mathscr{C}(n), n\mathscr{C}(0), \mathscr{C}(3) \times \mathscr{C}(2)$, and $\mathscr{B}(n)$.

3. Compute directly $\omega(P; x)$ and $\overline{\omega}(P; x)$ for $P = \mathscr{B}(4)$ and $\mathscr{B}(5)$.

→ 4. Establish in detail the inversion used in the proof of **3.64**(iii).

→ 5.* Let $P$ be a finite poset, $|P| = n > 0$ and $l(P) = l$. Show that

(i) $\omega(P; m+1) = |\mathfrak{J}(P \times \underline{m})|$ for all $m \in \mathbb{N}$,
(ii) $\omega(P; 1) = 1$,
(iii) $\omega(P; 0) = \omega(P; -1) = \cdots = \omega(P; -l) = 0$,
(iv) $(-1)^n \omega(P; -l-m) \geq \omega(P; m) > 0$ for all $m \in \mathbb{N}$,
(v) $\omega(P; -l-1) = (-1)^n$ if and only if every element of $P$ is contained in a maximal chain of length $l$,
(vi) $(-1)^n \omega(P; -l-m) = \omega(P; m)$ for all $m \in \mathbb{N}$ if and only if

$$(-1)^n \omega(P; -l-2) = \omega(P; 2)$$

if and only if every maximal chain in $P$ has length $l$. (Stanley)

(Hint: To prove (i) use an argument as in **3.61**. To prove (iv)–(vi) find appropriate mappings from $\mathrm{Mon}(N, \underline{m})$ to $\mathrm{Mon}(N, \underline{m+l})$ and use **3.65**(iii).)

6. Let $P$ be as in the previous exercise and assume that every maximal chain in $P$ has length $l$. Prove for the coefficients $e_k = |\mathrm{Mon}(P, \underline{k}) \cap \mathrm{Sur}(P, \underline{k})|$:

(i) $2e_{n-1} = (n + l - 1)e(P)$,
(ii) $2\bar{e}_{n-1} = (n - l - 1)e(P)$,
(iii) $\sum_{k=1}^{n} e_k = 2^l \sum_{k=1}^{n} \bar{e}_k$;
(iv) the coefficient of $x^{n-1}$ in $\omega(P; x)$ is $le(P)/2(n-1)!$.

(Stanley)

→ 7.* Let $P$ be a finite poset, $|P| = n$, and $\omega(P; x) = \sum_{k=0}^{n-1} w_k \binom{x+n-1-k}{n}$ its order polynomial. The *Euler polynomial* $\varepsilon(P; x)$ of $P$ is defined by

$$\varepsilon(P; x) := \sum_{k=0}^{n-1} w_k x^{k+1}.$$

Hence, in particular, $\varepsilon_n(x) := \varepsilon(n\mathscr{C}(0); x) = \sum_{k=1}^{n} A_{n,k}x^k$. Prove:

(i) $1 + \sum_{n\geq 1} (\varepsilon_n(x)/n!)t^n = (1 - x)(1 - xe^{t(1-x)})^{-1}$,
(ii) $\varepsilon_n(x) = (1 - x)^{n+1} \sum_{k\geq 0} k^n x^k$,
(iii) $\varepsilon_n(x) = nx\varepsilon_{n-1}(x) + x(1 - x)\varepsilon'_{n-1}(x)$.

8. Verify the values:

(i) $\varepsilon_1(x) = x$, $\varepsilon_2(x) = x^2 + x$, $\varepsilon_3(x) = ?$
(ii) $A_{n,2} = 2^n - (n + 1)$,
(iii) $A_{n,3} = 3^n - (n + 1)2^n + \binom{n+1}{2}$.

9. Find the chromatic polynomials $c(G; x)$ and $\bar{c}(G; x)$ for the graphs $C_n$ consisting of a single circuit of length $n$ and the graphs $W_n$ consisting of a circuit of length $n$ and a single vertex which is joined to the $n$ vertices of the circuit. $W_n$ is called the *wheel* of length $n$.

10. Generalize **3.68**. Let $\alpha \leq \beta \in \mathscr{Y}$. What is the number of maximal $\alpha, \beta$-chains in $\mathscr{Y}$?

11. Find all standard tableaux of shape $3 + 2 + 2$.

→12.* Fill in the details of the construction $g \to (A(g), B(g))$ and prove that $\Phi(g) = (A, B) \Leftrightarrow \Phi(g^{-1}) = (B, A)$. (Hint: Induction.)

13. Let $t_n$ be the number of $g \in S_n$ with $g^2 = id$. Show:

(i) $t_1 = 1$, $t_2 = 2$,
(ii) $t_{n+1} = t_n + nt_{n-1}$.

14. Verify the identity

$$\sum_{n\geq 0} t_n \frac{x^n}{n!} = \exp\left(x + \frac{x^2}{2}\right).$$

→15. Use **3.71** to prove that every permutation $a_1 a_2 \ldots a_{n^2+1}$ of $\mathbb{N}_{n^2+1}$ contains either a monotonically decreasing or a monotonically increasing subsequence of length $n + 1$.

## Notes

As in chapter I, the emphasis in the present chapter again lies on a unified presentation of various counting coefficients and functions rather than on deriving a large number of individual formulae and identities. For the latter, the reader is referred to Knuth [1, vols. 1 and 3], Riordan [1, 2], and Comtet [1]. The operator approach taken in sections 2 and 3 has been known for a long time and was studied in depth in recent years by Mullin–Rota [1], Garsia [1], and others; see the bibliography in Roman–Rota [1]. For the analogous discussion on vector spaces, see Andrews [1] and Goldman–Rota [1]. Section 4 is based on the work of Stanley [4, 7], Knuth [1, 2], Schensted [1], Foata–Schützenberger [1], and others; see also the excellent survey article of Stanley [3].

Chapter IV

# Incidence Functions

After having computed the level numbers and the total number of elements of some important lattices in the last chapter we now consider counting functions and inversion formulae in an arbitrary poset. Our method of study will be to associate with the poset $P$ an algebraic object called the incidence algebra $\mathbb{A}(P)$, and to investigate its structure and subobjects.

The two most important features of the notion of incidence algebra are that it explains the appearance of certain formal power series in connection with counting procedures—we shall study this in depth in the next chapter—and that it provides a general framework for the inversion calculus on posets. By this we mean the following: Let $f: P \to \mathbb{R}$ be a real function on the poset $P$. The *sum function* $g$ of $f$ is defined by

$$g(x) := \sum_{y \leq x} f(y) \qquad (x \in P).$$

Problem: Is it possible to determine a general inversion formula on $P$ which computes $f$ from the sum function $g$? This is indeed so, and the answer is the following: There exists an element $\mu$ of the incidence algebra, called the Möbius function of $P$, such that

$$g(x) = \sum_{y \leq x} f(y)\, (x \in P) \Leftrightarrow f(x) = \sum_{y \leq x} g(y)\mu(y, x)\, (x \in P).$$

Hence $\mu$ alone completely determines the inversion calculus on $P$ which is therefore aptly called *Möbius inversion* on $P$. Möbius inversion and methods for computing the Möbius function on an arbitrary poset will comprise the main part of this chapter. We shall see that all the inversion formulae found above are special cases of Möbius inversion on a suitable poset. Thinking of the sum function as the discrete analogue of the indefinite integral in calculus, we may view Möbius inversion as the discrete counterpart of the derivative. It will be interesting to note some close parallels with calculus, such as integration by parts and substitution of variables, in the methods for computing the Möbius function.

## 1. The Incidence Algebra

All posets $P$ in this section will be locally finite. We are going to consider functions from $P^2$ into a field of characteristic 0 (usually the real numbers) or, more generally, an integral domain containing the rationals.

### A. Definition and Structure

Let $P$ be a locally finite poset and $K$ a field of characteristic 0. We define

$$\mathbb{A}_K(P) := \{f: P^2 \to K : x \not\leq y \Rightarrow f(x, y) = 0\}.$$

When there is no danger of confusion we set $\mathbb{A}(P) = \mathbb{A}_K(P)$ for short. $\mathbb{A}_K(P)$ is a vector space over $K$ in the usual way, where, for $f, g \in \mathbb{A}_K(P)$, $r \in K$

$$(f + g)(x, y) := f(x, y) + g(x, y)$$
$$(rf)(x, y) := rf(x, y).$$

We define the *convolution* $f * g$ of $f, g \in \mathbb{A}_K(P)$ by

$$(f * g)(x, y) := \sum_{x \leq z \leq y} f(x, z)g(z, y).$$

Notice that the right-hand side is well-defined by the local finiteness of $P$. Obviously, $f * g \in \mathbb{A}_K(P)$ again, where we set $(f * g)(x, y) = 0$ if $x \not\leq y$.

**Definition.** The set $\mathbb{A}_K(P)$ together with the operations addition, scalar multiplication and convolution is called the *incidence algebra* of $P$ over $K$ and its elements the *incidence functions* of $P$.

That $\mathbb{A}_K(P)$ is indeed an algebra is spelled out in our first proposition.

**4.1 Proposition.** *$\mathbb{A}_K(P)$ is an associative $K$-algebra with the Kronecker function $\delta$ as two-sided identity.*

*Proof.* Let us just verify the associative law. For $f, g, h \in \mathbb{A}(P)$ we have

$$(f * (g * h))(x, y) = \sum_{x \leq z \leq y} f(x, z)(g * h)(z, y) = \sum_{x \leq z \leq y} f(x, z)\left(\sum_{z \leq w \leq y} g(z, w)h(w, y)\right)$$
$$= \sum_{x \leq w \leq y}\left(\sum_{x \leq z \leq w} f(x, z)g(z, w)\right)h(w, y)$$
$$= \sum_{x \leq w \leq y} (f * g)(x, w)h(w, y) = ((f * g) * h)(x, y). \quad \square$$

Of special interest are the units in $\mathbb{A}(P)$, i.e., the elements which possess an inverse with respect to the convolution.

**4.2 Proposition.** *An element $f \in \mathbb{A}_K(P)$ is a unit if and only if $f(x, x) \neq 0$ for all $x \in P$. A unit $f$ possesses a unique two-sided inverse $f^{-1}$.*

*Proof*. If $f$ is a unit and $g \in \mathbb{A}(P)$ with $f * g = \delta$ (or $g * f = \delta$) then for all $x \in P$

$$1 = (f * g)(x, x) = (g * f)(x, x) = f(x, x)g(x, x)$$

by the definition of convolution, and thus $f(x, x) \neq 0$.

Conversely, let $f(x, x) \neq 0$ for all $x \in P$. We define the left inverse inductively by

$$f^{-1}(x, x) = \frac{1}{f(x, x)}$$

$$f^{-1}(x, y) = \frac{1}{f(y, y)}\left(- \sum_{x \leq z < y} f^{-1}(x, z) f(z, y)\right).$$

In the same way, the existence of a right inverse is proved, and the two inverses are the same by associativity. $\square$

The convolution of two incidence functions bears an obvious resemblance to the ordinary matrix product. To see this, let us first choose a total extension of $P$ and use this extension to index the elements of $P$, i.e., $x_i \leq_P x_j \Rightarrow i \leq j$. We remark that a total extension of a locally finite poset need not be locally finite. As a matter of fact a total extension is locally finite precisely when it is order-isomorphic to a subset of the chain $\mathbb{Z}$. On the other hand, it is known that any countable locally finite poset can be extended to a locally finite chain.

Now with every $f \in \mathbb{A}_K(P)$ we associate the $K$-matrix $\hat{f}$ whose rows and columns correspond to the indexing set of $P$ with

$$\hat{f}(i, j) := f(x_i, x_j).$$

Any such matrix $\hat{f}$ is upper triangular, and we have for all $f, g \in \mathbb{A}_K(P)$

$$\widehat{f + g} = \hat{f} + \hat{g}, \qquad \widehat{rf} = r\hat{f}.$$

Furthermore,

$$(\widehat{f * g})(i, j) = \sum_{x_i \leq z \leq x_j} f(x_i, z) g(z, x_j),$$

$$(\hat{f} \cdot \hat{g})(i, j) = \sum_k f(x_i, x_k) g(x_k, x_j) = \sum_{i \leq k \leq j} f(x_i, x_k) g(x_k, x_j)$$

$$= \sum_{x_i \leq z \leq x_j} f(x_i, z) g(z, x_j),$$

hence

$$\widehat{f * g} = \hat{f} \cdot \hat{g}.$$

Let us summarize our results in the following proposition.

**4.3 Proposition.** *The mapping $f \to \hat{f}$ is a monomorphism from $\mathbb{A}_K(P)$ into the $K$-algebra of all upper triangular matrices, indexed by a total extension of $P$.* □

For finite posets $P$, $\mathbb{A}_K(P)$ can be viewed as a subalgebra of the algebra of all upper triangular matrices of order $|P|$. **4.2** reduces in this case to the fact that a triangular matrix is invertible if and only if all elements in the main diagonal are non-zero.

**Example.** Let $P$ be an antichain. The mapping $\phi: \mathbb{A}_K(P) \to \prod_{x \in P} K_x$, where $K_x \cong K$ for all $x$, given by $\phi f := (\ldots, f(x, x), \ldots)$ is an isomorphism, i.e., $\mathbb{A}_K(P) \cong \prod_{x \in P} K_x$. It is easy to see that, in fact, antichains are characterized by this property.

We remark that instead of a field any integral domain containing the rationals could be taken as scalar domain. In particular, we may choose the ring $K[x]$ of polynomials over a field of characteristic 0.

We have seen that a locally finite poset $P$ gives rise to a unique $K$-algebra $\mathbb{A}_K(P)$. A basic result of Stanley asserts the converse that the incidence algebra uniquely determines its underlying poset. The proof is not hard but since we shall not need it in the sequel it is left as an exercise.

**4.4 Theorem** (Stanley). *Let $P$ and $Q$ be locally finite posets and $K$ a field of characteristic 0. Then*

$$\mathbb{A}_K(P) \cong \mathbb{A}_K(Q) \Rightarrow P \cong Q. \quad \square$$

## B. Important Incidence Functions

In this section we shall list some incidence functions which are of interest when counting coefficients of the underlying poset. The names allotted to these functions will become clear as we proceed. Let us agree that where $P$ has a unique minimal element 0 and a unique maximal element 1, $f(P)$ shall mean $f(0, 1)$.

**Definition.** Let $P$ be a locally finite poset, $K$ a field of characteristic 0. We set

(a) *Delta-function*

$$\delta(x, y) := \begin{cases} 1 & \text{if } x = y \\ 0 & \text{otherwise} \end{cases}$$

(b) *Zeta-function*

$$\zeta(x, y) := \begin{cases} 1 & \text{if } x \leq y \\ 0 & \text{otherwise} \end{cases}$$

(c) *Lambda-function*

$$\lambda(x, y) := \begin{cases} 1 & \text{if } x = y \text{ or } x <\cdot\, y \\ 0 & \text{otherwise} \end{cases}$$

(d) *chain-function*

$$\eta := \zeta - \delta$$

(e) *cover-function*

$$\kappa := \lambda - \delta$$

(f) *Möbius function*

$$\mu := \zeta^{-1}$$

(g) *length function*

$$\rho(x, y) := l[x, y] \text{ (length of } [x, y]).$$

The following relations are immediate consequences of the definition.

**4.5** (i) $\zeta^n = \sum_{i=0}^n \binom{n}{i}\eta^i$, $\eta^n = \sum_{i=0}^n (-1)^{n-i}\zeta^i$,
(ii) $\lambda^n = \sum_{i=0}^n \binom{n}{i}\kappa^i$, $\kappa^n = \sum_{i=0}^n (-1)^{n-i}\lambda^i$. □

The Zeta-function is invertible by **4.2**, hence the definition of the Möbius function $\mu$ as the inverse of $\zeta$ is valid. Notice that all functions remain the same when we switch from $P$ to its dual poset $P^*$. For instance, $\mu(y, x)$ in $P^*$ is equal to $\mu(x, y)$ in $P$, for all $x, y \in P$. The following proposition is immediate from the definition of the convolution and could alternatively be used as a definition for the Möbius function.

**4.6 Proposition.** *Let $\mu$ be the Möbius function of $P$. For all $a, b \in P$:*

$$\mu(a, a) = 1$$

$$\mu(a, b) = -\sum_{a \le z < b} \mu(a, z) = -\sum_{a < z \le b} \mu(z, b) \quad \textit{if } a < b. \quad \square$$

Let $P$ be the poset of Figure 4.1. The numbers in the diagram are the values $\mu(0, a)$ for all $a \in P$. The incidence functions just defined give us information about the intervals of the poset, which we summarize in the following two propositions.

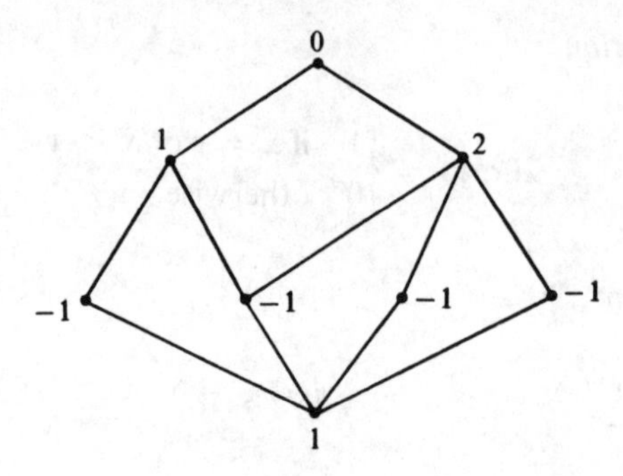


Figure 4.1

**4.7 Proposition.** *Let $P$ be a locally finite poset, $a, b \in P$.*

(i) $\zeta^2(a, b) = |[a, b]|$.
(ii) $(\kappa * \zeta)(a, b) = \#\{$*points in* $[a, b]\}$.
(iii) $(\zeta * \kappa)(a, b) = \#\{$*copoints in* $[a, b]\}$.

*Proof.* (i) is clear from the definition of convolution. Let us verify (ii), (iii) being the dual proposition. We have

$$(\kappa * \zeta)(a, b) = \sum_{a \leq z \leq b} \kappa(a, z)\zeta(a, b) = \sum_{a < \cdot z \leq b} \kappa(a, z). \quad \square$$

Let $f \in \mathbb{A}(P)$. By induction on $k$ it is clear that

$$f^k(a, b) = \sum_{a = z_1 \leq \cdots \leq z_k = b} f(a, z_1) f(z_1, z_2) \ldots f(z_{k-1}, b),$$

where the sum is extended over all $a, b$-chains of length $k$ with repetitions. From this observation the following formulae become apparent.

**4.8 Proposition.** *Let $P$ be a locally finite poset, $a, b \in P$.*

(i) $\eta^k(a, b) = \#\{$*$a, b$-chains of length $k$*$\}$.
(ii) $\kappa^k(a, b) = \#\{$*maximal $a, b$-chains of length $k$*$\}$.
(iii) $\zeta^k(a, b) = \#\{$*$a, b$-chains with repetitions of length $k$*$\}$.
(iv) $\lambda^k(a, b) = \#\{$*maximal $a, b$-chains with repetitions of length $k$*$\}$.
(v) $\sum_{k \geq 0} \eta^k(a, b) = (2\delta - \zeta)^{-1}(a, b) = \#\{$*$a, b$-chains*$\}$.
(vi) $\sum_{k \geq 0} \kappa^k(a, b) = (2\delta - \lambda)^{-1}(a, b) = \#\{$*maximal $a, b$-chains*$\}$. $\square$

**Examples.** Consider the chain $\mathscr{C}(n) \cong \{0 < 1 < \cdots < n\}$. The chains $0 \leq z_1 \leq \cdots \leq z_k = n$ with repetitions correspond bijectively to the monotone words $z_1 z_2 \ldots z_{k-1}$ of length $k - 1$ taken from $\{0, 1, \ldots, n\}$. The chains $0 < z_1 < \cdots < z_k = n$ without repetitions correspond to the $(k - 1)$-subsets of $\{1, \ldots, n - 1\}$, or equivalently, by rewriting the chain as $n = z_1 + (z_2 - z_1) + \cdots + (n - z_{k-1})$, to the ordered $k$-partitions of $n$. Hence, looking at table **3.22** we have

$$\eta^k(\mathscr{C}(n)) = \binom{n-1}{k-1}, \qquad \zeta^k(\mathscr{C}(n)) = \binom{(n+1)+(k-1)-1}{k-1} = \binom{n+k-1}{n},$$

$$\kappa^k(\mathscr{C}(n)) = \delta_{n,k}, \qquad \lambda^k(\mathscr{C}(n)) = \sum_{i=0}^{k} \binom{k}{i} \kappa^i(\mathscr{C}(n)) = \binom{k}{n}.$$

As our next example let $P = \mathscr{B}(S)$, $|S| = n$. The $\varnothing$, $S$-chains $\varnothing < A_1 < \cdots < A_k = S$ correspond bijectively to the ordered $k$-(set)-partitions $S = A_1 | A_2 - A_1 | \cdots | S - A_{k-1}$. By **3.22** this gives

$$\eta^k(\mathscr{B}(n)) = k!S_{n,k}, \qquad \zeta^k(\mathscr{B}(n)) = \sum_{i=0}^{k} [k]_i S_{n,i} = k^n,$$

$$\kappa^k(\mathscr{B}(n)) = \delta_{n,k} k!, \qquad \lambda^k(\mathscr{B}(n)) = \sum_{i=0}^{k} [k]_i \delta_{n,i} = [k]_n.$$

For finite posets $P$ with 0 and 1 the Zeta-function gives rise to an interesting counting polynomial, generalizing the order polynomial of section III.4. Let $P$ be of length $l$, i.e., $l$ is the length of the longest chain in $P$. By **4.8**(i) we know that $\eta^k(P) = 0$ for all $k > l$. Applying **4.5**(i) we have

$$\zeta^n(P) = \sum_{k=0}^{l} \binom{n}{k} \eta^k(P).$$

This means that $\zeta^n(P)$ is a polynomial in $n$ degree $l$, suggesting the following definition.

**Definition.** Let $P$ be a finite poset with 0 and 1. The expression

$$Z(P; x) := \zeta^x(0, 1) = \#\{0, 1\text{-chains with repetitions of length } x\}$$

is called the *Zeta-polynomial* of $P$. Thus $Z(P; n)$ counts the number of 0, 1-chains with repetitions of length $n$, for every $n \in \mathbb{N}$.

Since $\mu = \zeta^{-1}$ we infer

$$Z(P; -n) = \mu^n(P) \qquad (n \in \mathbb{N}).$$

Hence if we are able to give the expression $\mu^n(P)$ a combinatorial meaning we shall have found a combinatorial interpretation of the Zeta-polynomial for negative arguments. The most interesting case where this is indeed possible is the following.

Suppose $P$ has a rank function and suppose further that $\mu(a, b) = 0$ or $\mu(a, b) = (-1)^{r[a,b]}$ (the value depending on the interval $[a, b]$) for all $a, b \in P$. Then

$$\mu^n(P) = \sum_{0 \le z_1 \le \cdots \le z_n = 1} \mu(0, z_1) \ldots \mu(z_{n-1}, 1) = \sum_{0 \le z_1 \le \cdots \le z_n = 1} (-1)^{r(P)},$$

where the last sum runs over all 0, 1-chains $0 \le z_1 \le \cdots \le z_n = 1$ with $\mu(z_{i-1}, z_i) \ne 0$ for all $i$. But this means that $(-1)^{r(P)}\mu^n(P)$ is precisely the number of these chains, and we obtain the following general reciprocity theorem referred to in section III.1.

**4.9 Proposition** (Stanley). *Let $P$ be a finite poset with* 0 *and* 1 *and a rank function such that $\mu(a, b) = 0$ or $\mu(a, b) = (-1)^{r(b)-r(a)}$ for all $a \le b \in P$. Let $Z(P; x)$ be the Zeta-polynomial and $\bar{Z}(P; x)$ be the number of* 0, 1*-chains $0 \le z_1 \le \cdots \le z_x = 1$ with $\mu(z_{i-1}, z_i) \ne 0$ for all $i$. Then*

$$Z(P; -x) = (-1)^{r(P)}\bar{Z}(P; x). \quad \square$$

**Example.** Let us again consider the simplest case, $P = \mathscr{C}(n)$. It follows trivially from **4.6** that

$$\mu(\mathscr{C}(0)) = 1, \qquad \mu(\mathscr{C}(1)) = -1 \quad \text{and} \quad \mu(\mathscr{C}(n)) = 0 \quad \text{for } n \ge 2.$$

Hence the assumption in **4.9** holds and because of $\mu(\mathscr{C}(n)) = 0$ for $n \ge 2$ $\bar{Z}(\mathscr{C}(n); x)$ counts the number of *maximal* 0, 1-chains with repetitions of length $x$. Hence $Z(\mathscr{C}(n); x) = \zeta^x(\mathscr{C}(n)), \bar{Z}(\mathscr{C}(n); x) = \lambda^x(\mathscr{C}(n))$, and we infer from the example after **4.8** that

$$Z(\mathscr{C}(n); x) = \frac{[x]^n}{n!}, \qquad \bar{Z}(\mathscr{C}(n); x) = \frac{[x]_n}{n!}.$$

Thus **4.9** reduces in this case to our reciprocity law **3.5**.

Let us briefly consider posets $P$ for which $\mu(a, b) = (-1)^{r(b)-r(a)}$ always holds. For such a poset $P$, the Zeta-polynomial is self-reciprocal, i.e., $Z(P; -x) = (-1)^{r(P)}Z(P; x)$ which leads to a difference equation for $Z(P; x)$.

**4.10 Proposition.** *Let $P$ be a finite poset with* 0 *and* 1 *and a rank function. The Zeta-polynomial of $P$ is self-reciprocal, i.e.,*

$$Z(P; -x) = (-1)^{r(P)}Z(P; x)$$

*if and only if $\sum_{j=r(a)}^{r(b)} (-1)^j h_j(a, b) = 0$ for all $a < b \in P$ where $h_j(a, b)$ denotes the number of $z \in [a, b]$ of rank $j$.*

*Proof.* We have to show that the condition of the theorem is equivalent to $\mu(a, b) = (-1)^{r(b)-r(a)}$ for all $a \leq b \in P$. Assume $\mu(a, b) = (-1)^{r(b)-r(a)}$ and choose $a < b$. Then by **4.6**

$$0 = \sum_{a \leq z \leq b} \mu(a, z) = \sum_{r(a) \leq j \leq r(b)} (-1)^{j-r(a)} h_j(a, b) = (-1)^{r(a)} \sum_{j=r(a)}^{r(b)} (-1)^j h_j(a, b).$$

The converse is just as easy. □

**Examples.** The Boolean algebras $\mathscr{B}(n)$ satisfy the condition $\mu(a, b) = (-1)^{r(b)-r(a)}$. From the example after **4.8** we see that

$$Z(\mathscr{B}(n); x) = x^n,$$

resulting in the trivial relation $(-x)^n = (-1)^n x^n$.

A very interesting class of posets which satisfy **4.10** are the *face lattices of convex polytopes*. Let $\mathscr{P}$ be a convex polytope and order the faces by inclusion. The resulting lattice $L(\mathscr{P})$ satisfies the condition $\sum_{j=r(a)}^{r(b)} (-1)^j h_j(a, b) = 0$ which for $a = \varnothing$, $b = \mathscr{P}$ is just the classical Euler relation for convex polytopes. (See Grünbaum [1, ch. 8] for definitions and results.) Hence if $Z(L(\mathscr{P}); x)$ denotes the number of $\varnothing$, $\mathscr{P}$-chains with repetitions of length $x$, then

$$Z(L(\mathscr{P}); -x) = (-1)^{d+1} Z(L(\mathscr{P}); x) \qquad (d = \dim \mathscr{P}).$$

Consider, for example, the octahedron $\mathscr{O}$ depicted in Figure 4.2. Since $d = 3$, we must have $Z(L(\mathscr{O}); -x) = Z(L(\mathscr{O}); x)$. As the degree of $Z(L(\mathscr{O}); x)$ is 4 it suffices by the self-reciprocity property to find 3 values of $Z(L(\mathscr{O}); x)$. Now, obviously $Z(L(\mathscr{O}); 0) = 0$, $Z(L(\mathscr{O}); 1) = 1$ and $Z(L(\mathscr{O}); 2) = 28$. By interpolation we easily obtain $Z(L(\mathscr{O}); x) = 2x^4 - x^2$.

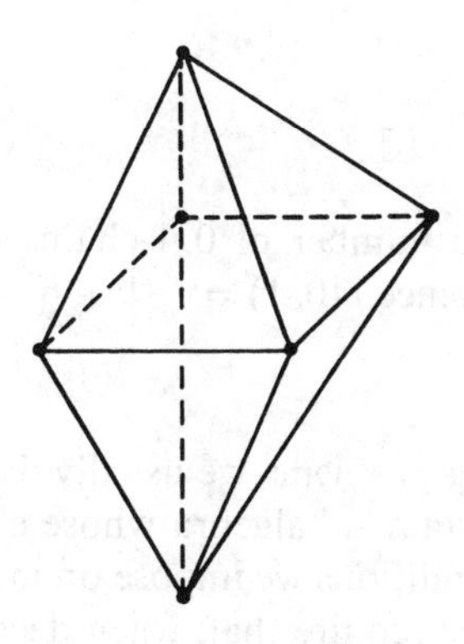

Figure 4.2

If $\mathscr{P}$ is a simplex of dimension $d$, then $L(\mathscr{P}) \cong \mathscr{B}(d+1)$ whence we obtain the trivial relation described above. For simplicial polytopes, the self-reciprocity law is equivalent to the Dehn–Sommerville equations (see Grünbaum [1, p. 145] and Stanley [7] for more details).

We shall have occasion to apply theorem **4.9** to other classes of posets as we go along.

In **4.2** the inverse $f^{-1}$ of a unit $f \in \mathbb{A}(P)$ was computed inductively. For many purposes it is convenient to have an explicit formula for $f^{-1}$ reflecting the interval structure of $P$. Let $f \in \mathbb{A}(P)$ be invertible. We define $\tilde{f} \in \mathbb{A}(P)$ by

$$\tilde{f}(x, y) := \begin{cases} f(x, x) & \text{if } x = y \\ 0 & \text{otherwise.} \end{cases}$$

Obviously, $\tilde{f}$ is invertible again. As examples we have

$$\tilde{\zeta} = \tilde{\lambda} = \delta.$$

**4.11 Proposition.** *Let $f \in \mathbb{A}_K(P)$ be invertible. Then*

$$f^{-1} = \left( \sum_{k \geq 0} (-1)^k ((\tilde{f}^{-1} * f) - \delta)^k \right) * \tilde{f}^{-1}.$$

*Proof.* We have

$$\begin{aligned} f^{-1} &= (\tilde{f} + (f - \tilde{f}))^{-1} = (\tilde{f} * (\delta + (\tilde{f}^{-1} * (f - \tilde{f}))))^{-1} \\ &= (\delta + (\tilde{f}^{-1} * (f - \tilde{f})))^{-1} * \tilde{f}^{-1} = (\delta + ((\tilde{f}^{-1} * f) - \delta))^{-1} * \tilde{f}^{-1}. \end{aligned}$$

Now for any $g \in \mathbb{A}(P)$ with $g(a, a) = 0$ for all $a \in P$, we have

$$(\delta + g)^{-1} = \delta - g + g^2 - g^3 \pm \cdots,$$

where the sum on the right hand side terminates after a finite number of terms for every pair $(a, b)$. Setting $g = (\tilde{f}^{-1} * f) - \delta$ then yields the theorem. □

**4.12 Corollary.** *We have*

(i) $\mu = \sum_{k \geq 0} (-1)^k \eta^k$,
(ii) $\lambda^{-1} = \sum_{k \geq 0} (-1)^k \kappa^k$. □

**Example.** In Figure 4.1, the number of 0, 1-chains of length 0, 1, 2, and 3 are 0, 1, 6, and 5, respectively. Hence $\mu(0, 1) = -1 + 6 - 5 = 0$.

## C. Multiplicative Functions

For applications to counting problems we usually do not need the full incidence algebra $\mathbb{A}(P)$ but only a certain subalgebra whose elements reflect the structure under consideration. The conditions we impose on incidence functions can be put into two categories. First, we require that, when decomposing an interval into a direct product, the function value also decomposes into the product of the values on the two subintervals. Secondly, we introduce an equivalence relation on the set of intervals of $P$ and require that the function be constant on the equivalence classes. In this section we discuss some aspects connecting these questions. Further theorems are contained in V.1.

**4.13 Proposition.** *Let* $\mathbb{S}(P) := \{f \in \mathbb{A}_K(P) \colon [x, y] \cong [u, v] \Rightarrow f(x, y) = f(u, v)\}$. $\mathbb{S}(P)$ *is a subalgebra of* $\mathbb{A}(P)$, *called the* standard algebra *of* $P$. *If* $f \in \mathbb{S}(P)$ *and* $f$ *is invertible in* $\mathbb{A}(P)$, *then* $f$ *is already invertible in* $\mathbb{S}(P)$.

*Proof.* $\mathbb{S}(P)$ is clearly a subspace of the vector space $\mathbb{A}(P)$. Let $f, g \in \mathbb{S}(P)$ and $\phi \colon [x, y] \to [u, v]$ an isomorphism. Then

$$(f * g)(x, y) = \sum_{x \le z \le y} f(x, z)g(z, y) = \sum_{u \le \phi(z) \le v} f(u, \phi(z))g(\phi(z), v) = (f * g)(u, v).$$

The last assertion is verified as in **4.2.** □

The functions $\delta$, $\zeta$, $\lambda$, $\eta$, and $\kappa$ are clearly in $\mathbb{S}(P)$, hence $\mu$ is in $\mathbb{S}(P)$. Also $\rho \in \mathbb{S}(P)$.

**Definition.** Let $P$ be a lattice. An incidence function $f \in \mathbb{A}_K(P)$ is called *multiplicative* if for all $a, b \in P$

$$[a \wedge b, a \vee b] \cong [a \wedge b, a] \times [a \wedge b, b] \Rightarrow f(a \wedge b, a \vee b) = f(a \wedge b, a) f(a \wedge b, b).$$

Examples of multiplicative functions of arbitrary lattices are $\delta$, $\zeta$ and $\zeta^2$ (see **4.7**(i)).

**4.14 Proposition.** *Let* $P$ *be a lattice. The multiplicative invertible functions in* $\mathbb{S}(P)$ *form a group with respect to convolution.*

*Proof.* We have to show that $f * g$ and $f^{-1}$ are multiplicative whenever $f$ and $g$ are. Suppose $[a \wedge b, a \vee b] \cong [a \wedge b, a] \times [a \wedge b, b]$. Then

$$\begin{aligned}(f * g)(a \wedge b, a \vee b) &= \sum_{a \wedge b \le z \le a \vee b} f(a \wedge b, z)g(z, a \vee b) \\ &= \sum_{a \wedge b \le z \le a \vee b} f(a \wedge b, z \wedge a) f(a \wedge b, z \wedge b) g(z, z \vee a) g(z, z \vee b) \\ &= \sum_{a \wedge b \le z \le a \vee b} (f(a \wedge b, z \wedge a) g(z \wedge a, a))(f(a \wedge b, z \wedge b) g(z \wedge b, b))\end{aligned}$$

(since $[z, z \vee a] \cong [z \wedge a, a]$, $[z, z \vee b] \cong [z \wedge b, b]$)

$$\begin{aligned}&= \sum_{a \wedge b \le u \le a} f(a \wedge b, u) g(u, a) \sum_{a \wedge b \le v \le b} f(a \wedge b, v) g(v, b) \\ &= (f * g)(a \wedge b, a) \cdot (f * g)(a \wedge b, b).\end{aligned}$$

To prove that $f^{-1}$ is multiplicative we use induction on the length $l[a \wedge b, a \vee b]$. For $l = 0$ there is nothing to prove. Assume $l[a \wedge b, a \vee b] > 0$. Then

$$
\begin{aligned}
0 &= \delta(a \wedge b, a \vee b) = (f^{-1} * f)(a \wedge b, a \vee b) \\
&= \sum_{a \wedge b \leq z \leq a \vee b} f^{-1}(a \wedge b, z) f(z, a \vee b) \\
&= \sum_{a \wedge b \leq z < a \vee b} f^{-1}(a \wedge b, z \wedge a) f^{-1}(a \wedge b, z \wedge b) f(z, z \vee a) f(z, z \vee b) \\
&\quad + f^{-1}(a \wedge b, a \vee b) \\
&= \left( \sum_{a \wedge b \leq u < a} f^{-1}(a \wedge b, u) f(u, a) \cdot \sum_{a \wedge b \leq v < b} f^{-1}(a \wedge b, v) f(v, b) \right) \\
&\quad + \left( f^{-1}(a \wedge b, a) \sum_{a \wedge b \leq v < b} f^{-1}(a \wedge b, v) f(v, b) \right) \\
&\quad + \left( f^{-1}(a \wedge b, b) \sum_{a \wedge b \leq u < a} f^{-1}(a \wedge b, u) f(u, a) \right) + f^{-1}(a \wedge b, a \vee b) \\
&\qquad \text{(induction!)} \\
&= (-f^{-1}(a \wedge b, a) \cdot - f^{-1}(a \wedge b, b)) + (f^{-1}(a \wedge b, a) \cdot - f^{-1}(a \wedge b, b)) \\
&\quad + (f^{-1}(a \wedge b, b) \cdot - f^{-1}(a \wedge b, a)) + f^{-1}(a \wedge b, a \vee b),
\end{aligned}
$$

whence

$$f^{-1}(a \wedge b, a \vee b) = f^{-1}(a \wedge b, a) \cdot f^{-1}(a \wedge b, b). \quad \square$$

**4.15.** Let us compute the Möbius function for some fundamental lattices.

(i) $\mu(\mathscr{C}(0)) = 1$, $\mu(\mathscr{C}(1)) = -1$ and $\mu(\mathscr{C}(n)) = 0$ for $n \geq 2$.
(ii) From **4.14** we obtain

$$\mu\left(\prod_{i=1}^{t} \mathscr{C}(k_i)\right) = \begin{cases} 1 & \text{if } k_1 = k_2 = \cdots = k_t = 0 \\ (-1)^s & \text{if } s \text{ of the } k_i\text{'s are 1 and the rest 0} \\ 0 & \text{if } k_i \geq 2 \text{ for some } i. \end{cases}$$

(iii) In particular, for the divisor lattice $\mathscr{T}$ this gives (see **2.61**)

$$\mu(l, m) = \begin{cases} 1 & \text{if } \; l = m \\ (-1)^s & \text{if } \; \dfrac{m}{l} = p_1 p_2 \ldots p_s \text{ with } p_i \neq p_j,\ p_i \text{ prime} \\ 0 & \text{if } \; \dfrac{m}{l} = r^2 t \neq 1. \end{cases}$$

In number theory, the Möbius function $\bar{\mu}: \mathbb{N} \to \mathbb{R}$ is defined by

$$\bar{\mu}(n) = \begin{cases} 1 & \text{if } \ n = 1 \\ (-1)^s & \text{if } \ n = p_1 \dots p_s, p_i \neq p_j, p_i \text{ prime} \\ 0 & \text{if } \ n = r^2 t, n \neq 1. \end{cases}$$

Hence we have

$$\mu(l, m) = \bar{\mu}\left(\frac{m}{l}\right) \qquad (l | m \in \mathbb{N}).$$

This relation is the origin of the name Möbius function.

(iv) For the Boolean algebra $\mathscr{B}(S)$, $|S| = n$, we have $\mathscr{B}(n) = [\mathscr{C}(1)]^n$ and thus by (ii)

$$\mu(\mathscr{B}(n)) = (-1)^n \qquad (n \in \mathbb{N}_0).$$

In general,

$$\mu(A, B) = (-1)^{|B-A|} \quad \text{for all } A \subseteq B \subseteq S.$$

As a final application of the multiplicativity of $\mu$ we derive an interesting relation between the Möbius function $\mu_P$ of a locally finite poset $P$ and the Möbius function $\mu_{\text{Int}(P)}$ of its interval poset $\text{Int}(P)$ where

$$\text{Int}(P) := \{[x, y]: x \leq y \in P\} \cup \{\varnothing\}$$

endowed with the inclusion relation.

Let $S = [[a, b], [c, d]]$ be a non-empty interval of $\text{Int}(P)$ with $[a, b] \neq \varnothing$. Clearly, if $[x, y] \in S$ then

$$c \leq x \leq a \leq b \leq y \leq d.$$

The mapping $\phi: S \to [c, a]^* \times [b, d]$ given by

$$\phi[x, y] = (x, y)$$

is easily seen to be an order isomorphism between $S$ and $[c, a]^* \times [b, d]$. Hence $\text{Int}(P)$ is locally finite, and we have by the multiplicativity of $\mu$

$$\mu_{\text{Int}}([a, b], [c, d]) = \mu_P(c, a)\mu_P(b, d).$$

If $[a, b] = \varnothing$, $[c, d] \neq \varnothing$, then by **4.6**

$$\begin{aligned} \mu_{\text{Int}}(\varnothing, [c, d]) &= -\sum_{c \leq x \leq y \leq d} \mu_{\text{Int}(P)}([x, y], [c, d]) = \sum_{c \leq x \leq y \leq d} \mu_P(c, x)\mu_P(y, d) \\ &= -\sum_{y \in P} \delta(c, y)\mu_P(y, d) = -\mu_P(c, d). \end{aligned}$$

In summary we have the following result.

**4.16 Proposition** (Crapo). *Let $P$ be a locally finite poset and* $\operatorname{Int}(P)$ *its interval set (including* $\varnothing$*) with the inclusion relation. Then for a non-empty interval* $[[a, b], [c, d]]$ *we have*

$$\mu_{\operatorname{Int}(P)}([a, b], [c, d]) = \begin{cases} \mu_P(c, a)\mu_P(b, d) & \textit{if} \quad [a, b] \neq \varnothing \\ -\mu_P(c, d) & \textit{if} \quad [a, b] = \varnothing. \end{cases}$$

*If, in particular, $P$ has* 0 *and* 1, *then*

$$\mu(\operatorname{Int}(P)) = -\mu(P). \quad \square$$

EXERCISES IV.1

→ 1. Let $\mathbb{A}_K(P)$ be the incidence algebra of a locally finite poset $P$. Show that $\mathbb{A}(P)$ is commutative if and only if $P$ is an antichain.

2. Prove that a total extension of a locally finite poset is locally finite if and only if it is order-isomorphic to a subset of the chain $\mathbb{Z}$.

3. Show that any countable locally finite poset can be extended to a locally finite chain.

→ 4. Let $\mathbb{A}_K(P)$ be the incidence algebra of a locally finite poset. Prove:
   (i) For all $x \in P$, $J_x(P) := \{f \in \mathbb{A}(P) : f(x, x) = 0\}$ is a maximal two-sided ideal in $\mathbb{A}(P)$ and $\mathbb{A}(P)/J_x(P) \cong K$.
   (ii) $\operatorname{Rad}(P) := \bigcap_{x \in P} J_x(P) = \{f \in \mathbb{A}(P) : f(x, x) = 0 \text{ for all } x \in P\}$ is the *radical* of $\mathbb{A}(P)$. (For the definition see, for instance, Jacobson [1, p. 4].)
   (iii) If $f \in \operatorname{Rad}(P)$ then there exists for all $x, y \in P$ a natural number $k = k(x, y)$ with $f^k(x, y) = 0$. Is the converse also true?
   (iv) $\bigcap_{i=1}^{\infty} R^i = \{0\}$, $R = \operatorname{Rad}(P)$.

5. Let $\mathbb{A}_K(P)$ be as before. For all $u, v \in P$ define $e_{u,v}, e_u \in \mathbb{A}_K(P)$ by

$$e_{u,v}(x, y) := \begin{cases} 1 & \text{if } x = u,\ y = v \\ 0 & \text{otherwise,} \end{cases}$$

$$e_u := e_{u,u}.$$

   Prove:
   (i) The elements $e_x$, $x \in P$, are a system of orthogonal idempotents in $\mathbb{A}(P)$, i.e.,

$$e_x * e_y = e_y * e_x = \begin{cases} e_x & \text{if } x = y \\ 0 & \text{otherwise.} \end{cases}$$

   (ii) For all $x, y \in P$ and $f \in \mathbb{A}(P)$ we have

$$e_x * f * e_y = f(x, y)e_{x,y},$$
$$e_x * \mathbb{A}(P) * e_y \neq \{0\} \Leftrightarrow x \leq_P y.$$

→ 6.* Prove Stanley's theorem **4.4**. (Hint: Use the two previous exercises.)

7.* Let $\mathbb{A}$ be an algebra of upper triangular matrices of order $n$ which contains the identity matrix. Show that $\mathbb{A}$ is isomorphic to $\mathbb{A}(P)$ for some poset $P$ with $|P| = n$ in the sense of **4.3** if and only if $[a_{ij}] \in \mathbb{A} \Leftrightarrow a_{ij}E_{ij} \in \mathbb{A}$ for all $i, j$ where $E_{ij} = \widehat{e_{ij}}$ of ex. 5. (Smith)

8. Define $\nu \in \mathbb{A}(P)$ by $\nu := \lambda^{-1}$ and show:
   (i) $\nu(x, x) = 1$;
   (ii) $\nu(x, y) = -\sum_{z <\cdot y} \nu(x, z) = -\sum_{x <\cdot z} \nu(z, y)$;
   (iii) $\nu = \sum_{i \geq 0} (-1)^i \kappa^i$.

9. Using **4.8** and **4.15** derive some further identities, for instance:
   (i) $\binom{n-1}{r-1} = \sum_{j=0}^{r-1} (-1)^j \binom{r}{j} \binom{n+r-j-1}{n}$;
   (ii) $\binom{r+n-1}{n} = \sum_{j=0}^{r-1} \binom{r}{j} \binom{n-1}{r-j-1}$.

10.* Compute the Zeta polynomial $Z(\mathscr{P}(n); x)$.

→11. By applying **4.12** prove:
   (i) $\mu^r = \sum_{j=1}^{r} (-1)^j \binom{r+j-1}{j} \eta^i$;
   (ii) $\mu^r(\prod_{i=1}^{n} \mathscr{C}(k_i)) = (-1)^{\sum_{i=1}^{n} k_i} \prod_{i=1}^{n} \binom{r}{k_i}$;
   and from this
   (iii) $\sum_j (-1)^{n-j} \binom{r+j-1}{j} j! S_{n,j} = r^n$;
   (iv) $\sum_j (-1)^j \binom{r+j-1}{j} \binom{n-1}{j-1} = (-1)^n \binom{r}{n}$.

→12.* Try to find, for arbitrary incidence algebras $\mathbb{A}(P)$, an analogue of the arithmetical sum function $\sigma(n) = \sum_{d|n} d$, $n \in \mathbb{N}$. (Hint: Let $M: P \to \mathscr{T}$ be a fixed monotone mapping and define $\tau \in \mathbb{A}(P)$ by $\tau(x, y) := M(y)/M(x)$ for $x \leq y$. Now study $\sigma \in \mathbb{A}(P)$, $\sigma := \tau * \zeta$.) (Smith)

13. (Continuation of the previous exercise) Define $\varphi \in \mathbb{A}(P)$ by $\varphi := \mu * \tau$ and prove:
   (i) $\varphi^{-1}(x, y) = \sum_{x \leq z \leq y} \mu(x, z)(M(z)/M(x))$;
   (ii) $M(y)/M(x) = \sum_{x \leq z \leq y} \varphi(z, y)$.
   $\varphi$ is the analogue of the classical Eulerian $\varphi$-function (see **4.24**, example 1).

→14. A function $f \in \mathbb{A}_K(P)$ is called *strongly multiplicative* if $f(x \wedge y, x \vee y) = f(x \wedge y, x) \cdot f(x \wedge y, y)$ for all $x, y \in P$. Let $P$ be a locally finite lattice. Prove:
   (i) $P$ is a chain $\Leftrightarrow \lambda$ is strongly multiplicative;
   (ii) $P$ is distributive $\Leftrightarrow \zeta^2$ is strongly multiplicative $\Leftrightarrow \mu$ is strongly multiplicative $\Leftrightarrow$ the strongly multiplicative invertible functions form a group.
   (Smith)

15. Suppose $P$ is a locally finite lattice and there exists a strongly multiplicative function $f \in \mathbb{A}(P)$ with $f(x, y) \neq 1$ for all $x <\cdot y$. Prove that $P$ is modular.

## 2. Möbius Inversion

In this section we turn to the discrete differential calculus on posets referred to in the introduction. Let $P$ again be a locally finite poset and $K$ a field of characteristic 0 (or more generally an integral domain containing the rationals). By analogy with the indefinite integral in calculus we give the following definition.

**Definition.** Let $f: P \to K$ be any mapping. The operators $S_{\leq}$ and $S_{\geq}$ on Map$(P, K)$ given by

$$(S_{\leq} f)(x) := \sum_{y \leq x} f(y)$$

$$(S_{\geq} f)(x) := \sum_{y \geq x} f(y)$$

are called the *lower* and *upper sum operator* respectively.

The content of the fundamental theorem on Möbius inversion is that for $S_{\leq}$ and $S_{\geq}$ there exist inverse operators $D_{\leq}$ and $D_{\geq}$ which are completely determined by the Möbius function $\mu \in \mathbb{A}_K(P)$.

### A. Difference Operators

We note the following trivial lemma.

**4.17 Lemma.** *Let $f, g, h \in \mathbb{A}_K(P)$ and $h$ be invertible. Then*

$$g = f * h \Leftrightarrow f = g * h^{-1}$$

$$g = h * f \Leftrightarrow f = h^{-1} * g. \quad \square$$

**4.18 Möbius inversion.** *Let $P$ be a locally finite poset, $K$ a field of characteristic* 0 *and $f, g \in$* Map$(P, K)$.

(i) *Inversion from below. Suppose all principal ideals of $P$ are finite. Then*

$$g(x) = \sum_{y \leq x} f(y)\ (x \in P) \Leftrightarrow f(x) = \sum_{y \leq x} g(y)\mu(y, x)\ (x \in P).$$

(ii) *Inversion from above. Suppose all principal filters of $P$ are finite. Then*

$$g(x) = \sum_{y \geq x} f(y)\ (x \in P) \Leftrightarrow f(x) = \sum_{y \geq x} g(y)\mu(x, y)\ (x \in P).$$

*Proof.* We show (i). First adjoin a new 0-element $\bar{0}$ to $P$ and denote the enlarged poset by $\bar{P}$. Now we define $\bar{f}, \bar{g} \in \mathbb{A}_K(\bar{P})$ by

$$\bar{f}(\bar{0}, x) := f(x), \qquad \bar{g}(\bar{0}, x) := g(x) \quad \text{for all } x \in P$$

$$\bar{f}(p, x) = \bar{g}(p, x) = 0 \quad \text{otherwise.}$$

The theorem follows from the relation

$$\bar{g} = \bar{f} * \zeta \Leftrightarrow \bar{f} = \bar{g} * \mu,$$

evaluated at the interval $[\bar{0}, x]$. □

By **4.18**, the difference operator $D_{\leq}$, inverse to $S_{\leq}$, and $D_{\geq}$, inverse to $S_{\geq}$, are given by

$$(D_{\leq} g)(x) = \sum_{y \leq x} g(y)\mu(y, x) \quad \textit{lower difference operator}$$

$$(D_{\geq} g)(x) = \sum_{y \geq x} g(y)\mu(x, y) \quad \textit{upper difference operator.}$$

**4.19.** Let us compute the difference operators for our basic posets.

(i) *Chains.* Consider the chain $\{0 < 1 < \cdots < n\}$. By **4.15**(i)

$$\mu(i, j) = \begin{cases} 1 & \text{if } i = j \\ -1 & \text{if } i = j - 1 \\ 0 & \text{otherwise.} \end{cases}$$

Hence for any $f, g \in \operatorname{Map}(\mathscr{C}(n), K)$

$$g(m) = \sum_{i=0}^{m} f(i) \Leftrightarrow f(m) = g(m) - g(m - 1),$$

$$g(m) = \sum_{i=m}^{n} f(i) \Leftrightarrow f(m) = g(m) - g(m + 1).$$

The difference operators thus have the form

$$(D_{\leq} g)(m) = g(m) - g(m - 1)$$

$$(D_{\geq} g)(m) = g(m) - g(m + 1).$$

Hence $D_{\leq}$ and $D_{\geq}$ are precisely the backward and forward difference operators (apart from the sign) studied in section III.3.

(ii) *Boolean algebras.* Let $S$ be an $n$-set. We found in **4.15**(iv) that

$$\mu(A, B) = (-1)^{|B - A|} \quad \text{for all } A \subseteq B \subseteq S.$$

Hence for any $f, g \in \operatorname{Map}(\mathscr{B}(S), K)$ we have

$$g(B) = \sum_{A \subseteq B} f(A) \Leftrightarrow f(B) = \sum_{A \subseteq B} (-1)^{|B - A|} g(A),$$

and for $|S| < \infty$

$$g(B) = \sum_{A \supseteq B} f(A) \Leftrightarrow f(B) = \sum_{A \supseteq B} (-1)^{|A - B|} g(A).$$

Thus the difference operators take the form

$$(D_{\leq} g)(B) = g(B) - \sum_{A \subseteq B, |B-A|=1} g(A) + \sum_{A \subseteq B, |B-A|=2} g(A) \mp \cdots + (-1)^{|B|} g(\emptyset),$$

$$(D_{\geq} g)(B) = g(B) - \sum_{A \supseteq B, |A-B|=1} g(A) + \sum_{A \supseteq B, |A-B|=2} g(A) \mp \cdots + (-1)^{|S-B|} g(S).$$

These two alternating sums are the basis for the sieve formulae of section $B$ and later for the results on valuations of distributive lattices.

(iii) *Divisor lattice.* **4.15**(iii) yields the well-known Möbius inversion formula in number theory. For any $f, g \in \operatorname{Map}(\mathbb{N}, K)$

$$g(n) = \sum_{k|n} f(k) \Leftrightarrow f(n) = \sum_{k|n} \bar{\mu}\left(\frac{n}{k}\right) g(k).$$

Let us compute the Möbius function for the other two fundamental lattice families, the finite partition lattices $\mathscr{P}(n)$ and the vector space lattices $\mathscr{L}(n, q)$. By **2.65**, **2.67** and the multiplicativity of $\mu$ it suffices to evaluate $\mu(\mathscr{P}(n))$ and $\mu(\mathscr{L}(n, q))$. Our method will be as follows. We denote by $f(x)$ the number of certain mappings, determine the sum function and deduce $f$ by Möbius inversion. $f(0)$ will then be the desired value. As scalar domain we take the polynomial ring $\mathbb{Q}[x]$. Let $S$ and $X$ be finite sets, $|S| = n$, $|X| = x$, and define $f: \mathscr{P}(S) \to \mathbb{Q}[x]$ by

$$f(\pi) := |\{h \in \operatorname{Map}(S, X): \ker(h) = \pi\}|.$$

Clearly,

$$(S_{\geq} f)(\pi) = x^{b(\pi)}$$

and thus

$$f(\pi) = \sum_{\sigma \geq \pi} \mu(\pi, \sigma) x^{b(\sigma)} \qquad (\pi \in \mathscr{P}(S)).$$

For $\pi = 0$, this gives $f(0) = |\operatorname{Inj}(S, X)| = [x]_n$, hence

$$f(0) = x(x-1)\ldots(x-n+1) = \sum_{\sigma \in \mathscr{P}(S)} \mu(0, \sigma) x^{b(\sigma)}.$$

Comparing coefficients for $x$ yields

(iv) $\mu(\mathscr{P}(n)) = (-1)^{n-1}(n-1)!$.

Now let $V$ and $X$ be finite dimensional vector spaces over $GF(q)$ with $\dim(V) = n$ and $|X| = x$, and define $f: \mathscr{L}(V) \to \mathbb{Q}[x]$ by

$$f(U) := |\{h \in \operatorname{Hom}(V, X): \ker(h) = U\}|.$$

We have for $U \in \mathscr{L}(V)$

$$(S_{\geq} f)(U) = x^{n-r(U)}$$
$$f(U) = \sum_{W \supseteq U} \mu(U, W) x^{n-r(W)}.$$

For $U = 0$ this gives $f(0) = g_n(x)$ (Gaussian polynomials of **3.27**). Hence

$$f(0) = (x-1)(x-q)\ldots(x-q^{n-1}) = \sum_{U \in \mathscr{L}(V)} \mu(0, U) x^{n-r(U)},$$

and thus

(v) $\mu(\mathscr{L}(n, q)) = (-1)^n q^{\binom{n}{2}}$.

The same method can be applied to $\mathscr{B}(n)$ yielding our old formulae

$$(x-1)^n = \sum_{A \subseteq S} \mu(\varnothing, A) x^{n-|A|}$$
$$\mu(\mathscr{B}(n)) = (-1)^n.$$

Notice that in all three polynomials $f(0)$, i.e., $[x]_n$, $g_n(x)$ and $(x-1)^n$, the coefficients are grouped together according to their rank. This suggests the following definition.

**Definition.** Let $P$ be a finite poset with 0 and 1 and rank function $r$. The polynomial

$$\chi(P; x) = \sum_{a \in P} \mu(0, a) x^{r(1)-r(a)}$$

is called the *characteristic polynomial* of $P$. The coefficient $w_k = \sum_{r(a)=k} \mu(0, a)$ of $x^{r(1)-k}$ is called the *k-th level number of the first kind*, the cardinality of the $k$-th level $W_k = \sum_{r(a)=k} 1$ the *k-th level number of the second kind*.

The characteristic polynomial gives information about the behavior of $\mu$ on the levels. Obviously, $\chi(P; x)$ is of degree $r(P)$, and we have $w_0 = 1$, $w_{r(P)} = \mu(P)$. We shall develop methods for computing $\chi(P; x)$ in section 3.C. The specifications "first and second kind" are derived from the analogy with the Stirling numbers, spelled out in the following proposition which summarizes our results.

**4.20 Proposition.** *We have*

(i) $\chi(\mathscr{C}(n); x) = x^{n-1}(x-1)$; $w_0 = 1$, $w_1 = -1$ *and* $w_k = 0$ *for* $k \geq 2$; $W_k = 1$ *for all* $k = 0, \ldots, n$.
(ii) $\chi(\mathscr{B}(n); x) = (x-1)^n$; $w_k = (-1)^k \binom{n}{k}$; $W_k = \binom{n}{k}$; $\mu(\mathscr{B}(n)) = (-1)^n$.
(iii) $\chi(\mathscr{L}(n, q); x) = \prod_{i=0}^{n-1} (x - q^i)$; $w_k = (-1)^k q^{\binom{k}{2}} \binom{n}{k}_q$; $W_k = \binom{n}{k}_q$; $\mu(\mathscr{L}(n, q)) = (-1)^n q^{\binom{n}{2}}$.
(iv) $\chi(\mathscr{P}(n); x) = [x-1]_{n-1}$; $w_k = s_{n,n-k}$; $W_k = S_{n,n-k}$; $\mu(\mathscr{P}(n)) = (-1)^{n-1}(n-1)!$. □

As an application we shall rederive some of the inversion formulae of section III.2.

**4.21 Proposition.** *Let $\mathscr{L} = \{L_n : n = 0, 1, 2, \ldots\}$ be a sequence of posets with 0 and 1 and rank function r, where $r(L_n) = n$ for all n. Furthermore, suppose that all intervals of $L_n$ of corank k are isomorphic to $L_k$, for all k and n. If we set $w_{n,k} := w_k(L_n)$, $W_{n,k} := W_k(L_n)$ with $w_{0,0} = W_{0,0} = 1$ and $w_{n,k} = W_{n,k} = 0$ for $n < k$, then the matrices $[w_{n,n-k}]$ and $[W_{n,n-k}]$ are inverses of each other, i.e.,*

$$\sum_{k \geq 0} W_{n,n-k} w_{k,k-s} = \sum_{k \geq 0} w_{n,n-k} W_{k,k-s} = \delta_{n,s} \qquad (n, s \in \mathbb{N}_0).$$

*Proof.* We have

$$\sum_k W_{n,n-k} w_{k,k-s} = \sum_{a \in L_n} \left( \sum_{\substack{b \in L_n \\ b \geq a,\, r(b) = n-s}} \mu(a, b) \right) = \sum_{b \in L_n} \delta_{n,r(b)+s} \sum_{\substack{a \in L_n \\ a \leq b}} \mu(a, b)$$

$$= \sum_{b \in L_n} \delta_{n,r(b)+s} \cdot \delta(0, b) = \delta_{n,s}.$$

The second equation is verified similarly. □

**4.22 Corollary.** *Let $\mathscr{L} = \{L_n : n = 0, 1, 2, \ldots\}$ be a sequence of posets satisfying the conditions of the preceding theorem. Then for all $n \in \mathbb{N}_0$:*

$$\chi(L_n; x) = \sum_{k=0}^{n} w_{n,n-k} x^k,$$

$$x^n = \sum_{k=0}^{n} W_{n,n-k} \chi(L_k; x). \quad \square$$

Clearly, the sequences $\{\mathscr{C}(n)\}$, $\{\mathscr{B}(n)\}$, $\{\mathscr{P}(n)\}$, and $\{\mathscr{L}(n, q)\}$ all satisfy the conditions of **4.21**, thus yielding the binomial, the Stirling, and the Gauss inversion formula, respectively, whereas in the case of $\{\mathscr{C}(n)\}$ the trivial relation

$$u_n = \sum_{k=0}^{n} v_k \Leftrightarrow v_n = u_n - u_{n-1}$$

is obtained.

## B. Sieve Formulae

In this section we demonstrate that Möbius inversion on Boolean algebras is equivalent to a widely used counting method, called the *principle of inclusion-exclusion.* By this we mean the following. Let $S$ be a finite set and $E_1, E_2, \ldots, E_t$ certain properties which the elements of $S$ may or may not possess. Question: How many elements of $S$ do not possess any of the properties?

Purely set-theoretically the problem reads as follows. Suppose $A_1, \ldots, A_t$ are subsets of $S$ ($A_i := \{a \in S: a$ possesses $E_i\}$), what is $|S - \bigcup_{i=1}^t A_i|$? Heuristically, the problem is solved in the following way. We take all elements of $S$, subtract those elements which are in at least one $A_i$, add those which are in at least two $A_i$'s, subtract those appearing in at least three $A_i$'s, etc. This inclusion-exclusion process is justified by formula **2.12** which implies

$$(+)\ \left|S - \bigcup_{i=1}^{t} A_i\right| = |S| - \sum_{i=1}^{t} |A_i| + \sum_{i<j}^{t} |A_i \cap A_j| \mp \cdots + (-1)^t |A_1 \cap \cdots \cap A_t|.$$

Let us rederive $(+)$ and some useful generalizations by means of Möbius inversion.

Let $A_1, \ldots, A_t \subseteq S$ and $T = \{1, \ldots, t\}$. We define $f(I)$, $I \subseteq T$, as the number of elements of $S$ which are precisely in the sets $A_i$ with $i \in I$ and no others. Hence

$$f(I) = \left|\bigcap_{i \in I} A_i \cap \left(S - \bigcup_{j \in T - I} A_j\right)\right|.$$

Clearly,

$$(S_{\geq} f)(I) = \sum_{I \subseteq J \subseteq T} f(J) = \left|\bigcap_{i \in I} A_i\right|,$$

whence we obtain

$$f(I) = \sum_{I \subseteq J \subseteq T} (-1)^{|J-I|} \left|\bigcap_{j \in J} A_j\right|.$$

For $I = \varnothing$ this gives precisely $(+)$ above. (Note that by definition $\bigcap_{i \in I} A_i = S$ for $I = \varnothing$.)

Next, consider the problem of computing the number of elements of $S$ which lie in exactly $p$ of the sets $A_i$. This number is given by

$$\sum_{I \subseteq T, |I|=p} f(I) = \sum_{|I|=p} \sum_{I \subseteq J \subseteq T} (-1)^{|J|-p} \left|\bigcap_{j \in J} A_j\right| = \sum_{|J| \geq p} \sum_{I \subseteq J, |I|=p} (-1)^{|J|-p} \left|\bigcap_{j \in J} A_j\right|$$

$$= \sum_{k=p}^{t} (-1)^{k-p} \binom{k}{p} \sum_{J \subseteq T, |J|=k} \left|\bigcap_{j \in J} A_j\right|.$$

We summarize our results in the following proposition.

**4.23** (Special sieve formula). *Let $S$ be a finite set, $A_1, \ldots, A_t$ subsets of $S$, and $T = \{1, \ldots, t\}$. The number $e_p$ of elements which lie in precisely $p$ of the sets $A_i$ is given by*

$$e_p = \sum_{k=p}^{t} (-1)^{k-p} \binom{k}{p} \sum_{J \subseteq T, |J|=k} \left|\bigcap_{j \in J} A_j\right|. \quad \square$$

Consider again the Boolean algebra $\mathscr{B}(S)$ over a finite set $S$ and a mapping $w: S \to R$ from $S$ onto a commutative ring $R$ with $w(\varnothing) = 0$. We extend $w$ to all of $\mathscr{B}(S)$ by setting

$$w(A) := \sum_{a \in A} w(a) \qquad (A \subseteq S).$$

The enlarged mapping is called a *valuation* of $\mathscr{B}(S)$ in $R$; in particular, if $R = \mathbb{R}$ and $w(a) \geq 0$ for all $a \in S$ then $w$ is called a *measure* on $S$. For example, the cardinality function is the measure $w \equiv 1$. By the same argument as above, **4.23** can be extended to the following formula.

**4.24** (General sieve formula). *Let $S$ be a finite set, $A_1, \ldots, A_t$ subsets of $S$, and $T = \{1, \ldots, t\}$. Let $w$ be a valuation of $\mathscr{B}(S)$ with $w(\varnothing) = 0$ and let $E_p := \{b \in S: b$ belongs to precisely $p$ of the sets $A_i\}$, $e_p = w(E_p)$. Then:*

(i) $w(\bigcup_{i=1}^t A_i) = \sum_{i=1}^t w(A_i) - \sum_{i<j}^t w(A_i \cap A_j) \pm \cdots + (-1)^{t-1} w(A_1 \cap \cdots \cap A_t)$

(ii) $e_p = \sum_{k=p}^t (-1)^{k-p} \binom{k}{p} \sum_{J \subseteq T, |J|=k} w(\bigcap_{j \in J} A_j)$. $\square$

**Example 1** (Euler $\varphi$-function). For $n \in \mathbb{N}$ let $\bar{\varphi}(n)$ be the number of integers between 1 and $n$ which are relatively prime to $n$. We want to determine $\bar{\varphi}(n)$. Suppose $n = p_1^{k_1} \ldots p_t^{k_t}$ and set $S = \{1, \ldots, n\}$, $A_i = \{l \in S: p_i | l\}$ for $i = 1, \ldots, t$, and $T = \{1, \ldots, t\}$. Clearly,

$$\left| \bigcap_{i \in I} A_i \right| = \frac{n}{\prod_{i \in I} p_i} \qquad (I \subseteq T),$$

and from **4.23** we conclude

$$\bar{\varphi}(n) = e_0 = n - \sum_{i=1}^t \frac{n}{p_i} + \sum_{i<j}^t \frac{n}{p_i p_j} \mp \cdots + (-1)^t \frac{n}{p_i \cdots p_t} = n \prod_{i=1}^t \left(1 - \frac{1}{p_i}\right).$$

By the definition of $\bar{\mu}$, this can also be written as

$$\bar{\varphi}(n) = \sum_{d | n} d \bar{\mu}\left(\frac{n}{d}\right),$$

whence we obtain through Möbius inversion over $[1, n] \subseteq \mathscr{T}$

$$n = \sum_{d | n} \bar{\varphi}(d).$$

**Example 2** (Problème des rencontres). Let $S_n$ be the set of permutations of $N = \{1, \ldots, n\}$. We seek the number $e_p(n)$ of permutations having precisely $p$ fixed-points (=rencontres). To find this number, we set $S = S_n$, $A_i = \{f \in S : f(i) = i\}$ for $i = 1, \ldots, n$. Obviously,

$$\left|\bigcap_{i \in I} A_i\right| = (n - |I|)! \qquad (I \subseteq N).$$

Hence we have

$$e_p(n) = \sum_{k=p}^{n} (-1)^{k-p} \binom{k}{p} \binom{n}{k} (n-k)! = \frac{n!}{p!} \sum_{k=p}^{n} \frac{(-1)^{k-p}}{(k-p)!} = \frac{n!}{p!} \sum_{k=0}^{n-p} \frac{(-1)^k}{k!}.$$

For $p = 0$, this gives the number $D_n$ of fixed-point-free permutations (called *derangements*):

$$D_n = n! \sum_{k=0}^{n} \frac{(-1)^k}{k!}.$$

Notice that

$$\frac{D_n}{n!} \to \frac{1}{e} \quad \text{for } n \to \infty,$$

implying the following surprising fact. Suppose that $n$ pages of a manuscript are disarranged by a gust of wind and are reordered arbitrarily. Our formula for $D_n$ states that for large $n$ the probability that not a single page is in its right place is greater than $\frac{1}{3}$.

**Example 3** (Permanent of a matrix). Let $A = [a_{ij}]$ be an $m \times n$-matrix over a commutative ring, $m \leq n$, where we enumerate the rows $R = \{1, \ldots, m\}$ and the columns $C = \{1, \ldots, n\}$. The *permanent* $\operatorname{per}(A)$ is defined by

$$\operatorname{per}(A) := \sum_{f \in \operatorname{Inj}(R, C)} a_{1f(1)} a_{2f(2)} \cdots a_{mf(m)}.$$

Hence $\operatorname{per}(A)$ consists of $[n]_m$ summands. If $m = n$, then the terms in $\operatorname{per}(A)$ are, apart from the sign, just the terms in the expansion of $\det(A)$. For a subset $I \subseteq C$, denote by $A|I$ the $m \times |I|$ matrix which results after deleting the columns in $C - I$. Finally, for any matrix $B = [b_{ij}]$ let $P(B)$ be the product of its row sums, i.e., $P(B) = \prod_i (\sum_j b_{ij})$.

As an example, take

$$A = \begin{bmatrix} 3 & 0 & 1 & -2 \\ 4 & -1 & 3 & 0 \end{bmatrix}.$$

Here

$$\mathrm{per}(A) = 3\cdot(-1) + 3\cdot 3 + 3\cdot 0 + 0\cdot 4 + 0\cdot 3 + 0\cdot 0 + 1\cdot 4 + 1\cdot(-1) + 1\cdot 0 + (-2)\cdot 4 + (-2)\cdot(-1) + (-2)\cdot 3 = -3.$$

$$A\,|\,1,4 = \begin{bmatrix} 3 & -2 \\ 4 & 0 \end{bmatrix}, \qquad P(A\,|\,2,3,4) = (0 + 1 - 2)\cdot(-1 + 3 + 0) = -2.$$

We use the general sieve formula **4.24** to compute the permanent. The basis set is Map$(R, C)$ and we define the valuation $w$ on Map$(R, C)$ by

$$w(f) := \prod_{i=1}^{m} a_{if(i)} \qquad (f \in \mathrm{Map}(R, C)).$$

Finally, let $A_i = \{f \in \mathrm{Map}(R, C)\colon i \notin \mathrm{im}(f)\}$ for $i = 1, \ldots, n$. With these definitions we clearly have in the notation of **4.24**

$$\mathrm{per}(A) = e_{n-m}.$$

Now

$$\bigcap_{i\in I} A_i = \{f \in \mathrm{Map}(R, C)\colon \mathrm{im}(f) \subseteq C - I\}$$

and thus

$$w\left(\bigcap_{i\in I} A_i\right) = \sum_{\mathrm{im}(f)\subseteq C-I} w(f) = P(A\,|\,C - I),$$

as is easily seen by expanding $P(A\,|\,C - I)$.

Hence by **4.24**(ii) we obtain

$$\mathrm{per}(A) = \sum_{k=1}^{m} (-1)^{m-k} \binom{n-k}{n-m} \sum_{I\subseteq C,\,|I|=k} P(A\,|\,I) \qquad \text{(Ryser)}.$$

If, in particular, $A$ is a square matrix, then

$$\mathrm{per}(A) = \sum_{k=1}^{n} (-1)^{n-k} \sum_{I\subseteq C,\,|I|=k} P(A\,|\,I).$$

In our example above we have

$$\mathrm{per}(A) = -\binom{3}{2}(12 + 0 + 3 + 0) + \binom{2}{2}(9 + 28 + 4 + 2 + 2 - 3) = -3.$$

Some interesting identities result upon applying the permanent formula to special matrices. Consider as examples the $n \times n$-matrix $J_n$ all of whose entries are 1 and the matrix $J_n - I_n$ where $I_n$ is the identity matrix. Clearly, $\operatorname{per}(J_n) = n!$ and $\operatorname{per}(J_n - I_n) = D_n$, the number of derangements. By the permanent formula, this gives

$$n! = \sum_{k=1}^{n} (-1)^{n-k} \binom{n}{k} k^n$$

$$D_n = \sum_{k=1}^{n} (-1)^{n-k} \binom{n}{k} k^{n-k}(k-1)^k.$$

C. Some Applications

*Zeta-polynomial.* Let $L$ be a finite distributive lattice. There are two possibilities for $\mu(L)$. If $1 \in L$ is a supremum of points, then $L$ is complemented and thus a geometric lattice by **2.38**, i.e., a Boolean algebra. Hence in this case we have by **4.15**(iv) that $\mu(L) = (-1)^{r(L)}$. If, on the other hand, 1 is not a supremum of points, then $\mu(L) = 0$. This is easily seen using **4.12**(i) and will be proved for arbitrary finite lattices in the next section. The assumptions in **4.9** are thus satisfied, whence we have

$$(+) \qquad Z(L; -x) = (-1)^{r(L)} \bar{Z}(L; x),$$

where $\bar{Z}(L; x)$ counts the number of 0, 1-chains $0 \leq z_1 \leq \cdots \leq z_x = 1$ in which each interval $[z_{i-1}, z_i]$ is a Boolean algebra.

Let us verify that $(+)$ is precisely the reciprocity theorem **3.65**(iii) for the order polynomial $\omega(P; x)$ of a finite poset $P$. The lattice $L = \mathfrak{J}(P)$ of order ideals of $P$ is distributive (see **2.5**). A monotone mapping $f: P \to \mathbb{N}$ corresponds, as in the proof of **3.61**, to a chain $\varnothing \leq J_1 \leq \cdots \leq J_l = P$ via the rule $a \in J_i \Leftrightarrow f(a) \leq i$, and this correspondence is bijective. Hence

$$Z(\mathfrak{J}(P); x) = \omega(P; x).$$

Similarly, the strict monotone mappings $f: P \to \mathbb{N}$ correspond bijectively to the chains $\varnothing \leq J_1 \leq J_2 \leq \ldots$ where all sets $J_i - J_{i-1}$ are antichains in $P$. But this means that all intervals $[J_{i-1}, J_i]$ in $\mathfrak{J}(P)$ are Boolean algebras, thus implying

$$\bar{Z}(\mathfrak{J}(P); x) = \bar{\omega}(P; x).$$

*Waring's Formula.* Let $x_1, \ldots, x_r$ be variables over a commutative ring with unity. A function $f(x_1, \ldots, x_r)$ over this ring is called *symmetric* if

$$f(x_{g(1)}, \ldots, x_{g(r)}) = f(x_1, \ldots, x_r) \quad \text{for all } g \in S_r.$$

The simplest examples of symmetric functions are

$$a_n(x_1, \ldots, x_r) := \sum_{i_1 < \ldots < i_n} x_{i_1} x_{i_2} \ldots x_{i_n},$$

where the sum is extended over all $\binom{r}{n}$ possible products of $n$ of the variables, and

$$s_n(x_1, \ldots, x_r) := \sum_{i=1}^{r} x_i^n.$$

The functions $a_n$ and $s_n$ are called the *elementary symmetric functions* and the *power functions*, respectively.

The main theorem on symmetric function states that any symmetric polynomial can be uniquely written as a polynomial in the elementary symmetric functions $a_1, a_2, \ldots$.

As an example we have $s_4 = a_1^4 - 4a_1^2 a_2 + 2a_2^2 + 4a_1 a_3 - 4a_4$ for any number $r \geq 4$ of variables, as is readily verified by expanding the products on the right-hand side.

We want to show conversely that the polynomials $a_n$ can also be (necessarily uniquely again) represented in terms of the power functions $s_1, s_2, \ldots$.

Let $N$ be an $n$-set, $R = \{x_1, \ldots, x_r\}$. We assign to each mapping $h: N \to R$ the product $x_1^{k_1(h)} \ldots x_r^{k_r(h)}$ with $k_i(h) = |h^{-1}(x_i)|$, $i = 1, \ldots, r$. For $\pi \in \mathscr{P}(N)$ we define

$$f(\pi) := \sum_{\substack{h \in \mathrm{Map}(N, R) \\ \ker(h) = \pi}} x_1^{k_1(h)} \ldots x_r^{k_r(h)}.$$

If we set

$$g(\pi) = (S_{\geq} f)(\pi) = \sum_{\pi \leq \sigma \in \mathscr{P}(N)} f(\sigma),$$

then by Möbius inversion on $\mathscr{P}(N)$

$$(+) \qquad f(0) = \sum_{\pi \in \mathscr{P}(N)} \mu(0, \pi) g(\pi).$$

Let us show that $(+)$ is just the formula we want. By definition,

$$f(0) = \sum_{h \in \mathrm{Inj}(N, R)} x_1^{k_1(h)} \ldots x_r^{k_r(h)} = \sum_{i_s \neq i_t} x_{i_1} \ldots x_{i_n} = n! a_n(x_1, \ldots, x_r).$$

Suppose $\pi \in \mathscr{P}(N)$ has type $1^{b_1} \ldots n^{b_n}$. We claim

$$(++) \qquad g(\pi) = (x_1 + \cdots + x_r)^{b_1} (x_1^2 + \cdots + x_r^2)^{b_2} \cdots (x_1^n + \cdots + x_r^n)^{b_n}.$$

We have $g(\pi) = \sum_{h \in \mathrm{Map}(N, R), \ker(h) \geq \pi} x_1^{k_1(h)} \ldots x_r^{k_r(h)}$. Now $\ker(h) \geq \pi$ means that $h$ is constant on the blocks of $\pi$. Hence

$$x_1^{k_1(h)} \ldots x_r^{k_r(h)} = x_{i_1} \ldots x_{i_{b_1}} x_{j_1}^2 \ldots x_{j_{b_2}}^2 \ldots x_{l_1}^n \ldots x_{l_{b_n}}^n.$$

On the other hand, expansion of the right-hand side of (+ +) yields precisely all terms of this form. By the multiplicativity of $\mu$ and **4.19**(iv) we have

$$\mu(0, \pi) = (-1)^{n - \sum_{i=1}^n b_i} \prod_{i=1}^n (i-1)!^{b_i},$$

and thus the following result.

**4.25 Proposition** (Waring). *Let $s_k = s_k(x_1, \ldots, x_r)$ for all k. Then*

$$a_n(x_1, \ldots, x_r) = \sum_{\substack{b_1, \ldots, b_n \\ \sum i b_i = n}} \frac{(-1)^{n - \sum_{i=1}^n b_i}}{\prod_{i=1}^n (b_i!) \prod_{i=1}^n i^{b_i}} s_1^{b_1} s_2^{b_2} \ldots s_n^{b_n}. \quad \square$$

*Primitive Elements in Finite Fields.* Let $K = GF(p^n)$ be the finite field with $p^n$ elements. A basic theorem in algebra states that $GF(p^n)$ is unique up to isomorphism and that its elements are the roots of the equation $x^{p^n} - x = 0$. (See Van der Waerden [1, p. 129].) Let $\alpha \in K$. Among all polynomials $f(x) \in P[x]$ (where $P = GF(p)$ is the prime field of $K$) which have $\alpha$ as a root there is a unique polynomial $f_\alpha(x)$ of smallest degree and leading coefficient 1. $f_\alpha(x)$ is called the *minimal polynomial* of $\alpha$. $\alpha$ is called a *primitive element* of $K$ over $P$ if $K = P(\alpha)$, which is equivalent to the fact that $\deg(f_\alpha(x)) = n$. We want to compute the number of primitive elements in $K$.

First we claim that $x^{p^n} - x = \prod f(x)$ in $P(x)$, where the product is extended over all irreducible polynomials over $P$ with leading coefficient 1 whose degree divides $n$. To see this let the irreducible polynomial $f(x)$ be a divisor of $x^{p^n} - x$ and let $\alpha$ be a root of $f(x)$. $P(\alpha)$ is a subfield of $K$, hence

$$\deg(f(x)) = [P(\alpha): P] \,|\, [K: P] = n.$$

Conversely, let $f(x)$ be irreducible over $P$ with $\deg(f(x)) = d \,|\, n$ and $\alpha$ a root of $f(x)$. $P(\alpha)$ can be thought of as a subfield of $K$ because $d \,|\, n$, which implies $\alpha \in K$ and thus $f(x) \,|\, x^{p^n} - x$. Since $x^{p^n} - x$ decomposes into distinct linear factors we conclude that every such irreducible polynomial appears exactly once on the right-hand side.

If we denote by $U_n$ the number of irreducible polynomials over $P$ with degree $n$ and leading coefficient 1, then $nU_n$ is clearly the number we seek. From the polynomial identity above we infer that

$$p^n = \sum_{d|n} d U_d,$$

and hence, by Möbius inversion over $[1, n] \subseteq \mathcal{T}$,

$$nU_n = \sum_{d|n} \bar{\mu}\left(\frac{n}{d}\right) p^d.$$

**Example.** Consider $K = GF(2^4)$. The number of primitive elements equals $-2^2 + 2^4 = 12$ corresponding to the three polynomials $x^4 + x^3 + 1, x^4 + x + 1$ and $x^4 + x^3 + x^2 + x + 1$.

EXERCISES IV.2

→ 1. Let $P$ be a finite semilattice with respect to $\vee$ (i.e., all suprema $x \vee y$ exist in $P$), $g \in \text{Map}(P, K)$, and $G$ a $K$-matrix with rows and columns corresponding bijectively to the elements of $P$, $G(x, y) := g(x \vee y)$. Prove:

$$\det G = \prod_{x \in P} (D_{\geq}(g))(x) = \prod_{x \in P} \left( \sum_{y \geq x} g(y)\mu(x, y) \right).$$

(Hint: Set $f = D_{\geq}(g)$, $F$ = diagonal matrix with $F(x, x) = f(x)$ and prove $G = \hat{\zeta} F \hat{\zeta}^{\mathrm{T}}$ where $\hat{\zeta}$ is the matrix corresponding to $\zeta$ as in **4.3**.) (Wilf)

2. Compute $\mu(\mathscr{A}(n, q))$ and $\chi(\mathscr{A}(n, q); x)$ for the affine lattices $\mathscr{A}(n, q)$.

→ 3. Reprove the inversion formulae in **3.38** as follows. Define $f: \mathscr{B}(S) \to \mathbb{R}$ by $f(A) = u_k :\Leftrightarrow |A| = k$. Then we have for an $n$-set $S$

$$g(S) = \sum_{A \subseteq S} f(A) = \sum_{k=0}^{n} \binom{n}{k} u_k.$$

Now apply Möbius inversion.

4. Generalize **4.20**(iii): Let $P_k(x, y) := \prod_{i=0}^{k-1} (x - q^i y)$ and show that

$$P_n(x, y) = \sum_{k=0}^{n} \binom{n}{k}_q P_k(x, z) P_{n-k}(z, y).$$

→ 5. Reprove formula **3.39** for the Stirling numbers $S_{n,k}$ using inclusion-exclusion. (Hint: Set $S = \text{Map}(N, R)$ with $|N| = n$ and $|R| = r$, and $A_i := \{f \in S : i \notin \text{im}(f)\}$.)

6. Given an alphabet $\{a_1, \ldots, a_n\}$, how many words of length $2n$ can be formed containing each letter $a_i$ exactly twice such that no two like letters appear next to each other? For example, for $n = 2$ we have $a_1 a_2 a_1 a_2$ and $a_2 a_1 a_2 a_1$ as the only possible words.

7.* Consider $\{1, \ldots, n\}$ and define the $n \times n$-matrix $M = [m_{ij}]$, $m_{ij} = \gcd(i, j)$. Show that $\det M = \prod_{i=1}^{n} \bar{\varphi}(i)$, where $\bar{\varphi}$ is the Euler $\varphi$-function.

8. Prove that

$$\sum_{\substack{1 \le x \le n \\ \gcd(x,n)=1}} x^2 = \frac{n^2}{3}\bar{\varphi}(n) + (-1)^t \tfrac{1}{6} p_1 p_2 \ldots p_t \bar{\varphi}(n),$$

where $p_1, \ldots, p_t$ are the distinct prime divisors of $n$.

9. Verify the following formulae for the derangement numbers $D_n$:

   (i) $D_1 = 0$;
   (ii) $D_n = nD_{n-1} + (-1)^n$;
   (iii) $D_n = (n-1)(D_{n-1} + D_{n-2})$;
   (iv) $\sum_{n \ge 0} D_n(t^n/n!) = e^{-t}(1-t)^{-1}$.

→10.* Problème des ménages. Given $n$ ladies and their husbands and a round table. In how many ways can we place these $2n$ people around the table such that no couple ever sits together? (Hint: We seek the number of permutations $f = f(1) \ldots f(n)$ such that $i \ne f(i), \ne f(i-1), i = 2, \ldots, n$, and $1 \ne f(1), \ne f(n)$.) (Answer: $M_n = \sum_{k=0}^{n} (-1)^k (2n/(2n-k))\binom{2n-k}{k}(n-k)!$.)

11. Prove $\sum_{k=0}^{n} (-1)^k \binom{n}{k} = 0$ by means of inclusion-exclusion. Deduce some other binomial formulae by the same method.

→12.* Consider $t$ subsets $A_1, \ldots, A_t$ of a set $S$ some of which may be empty and set $T = \{1, \ldots, t\}$. The *algebra* Boole($\mathfrak{A}$) generated by $\mathfrak{A} = \{A_1, \ldots, A_t\}$ consists of all subsets of $S$ which are obtained from the $A_i$'s by a finite sequence of the operations $\cup, \cap$ and complementation $A \to A^c := S - A$. The elements of Boole($\mathfrak{A}$) are called *Boolean functions*. Example: $f_1 = (A_1 \cap A_2) \cup A_3^c$, $f_2 = ((A_1 \cup A_2) \cap (A_3 \cup A_4^c))^c$. A *complete product* $f \in$ Boole($\mathfrak{A}$) is any of the $2^t$ functions of the form

$$f_I = \left(\bigcap_{i \in I} A_i\right) \cap \left(\bigcap_{j \in T-I} A_j^c\right) \qquad (I \subseteq T).$$

For instance, the 8 complete products generated by $A_1, A_2, A_3$ are

$$\begin{array}{llll} A_1 \cap A_2 \cap A_3, & A_1 \cap A_2 \cap A_3^c, & A_1 \cap A_2^c \cap A_3, & A_1^c \cap A_2 \cap A_3, \\ A_1 \cap A_2^c \cap A_3^c, & A_1^c \cap A_2 \cap A_3^c, & A_1^c \cap A_2^c \cap A_3, & A_1^c \cap A_2^c \cap A_3^c. \end{array}$$

Prove the *disjunctive normal form* for Boolean functions: For every $f \in$ Boole($\mathfrak{A}$) there exists a unique subset $M \subseteq 2^T$ such that

$$f = \bigcup_{I \in M} f_I.$$

In particular, $|\text{Boole}(\mathfrak{A})| = 2^{2^t}$.

13. Show that the number $c_n$ of connected simple graphs with vertices $\{1, \ldots, n\}$ is given by

$$c_n = \sum_{\substack{(b_1, \ldots, b_n) \\ \sum i b_i = n}} (-1)^{\sum_{i=1}^n b_i - 1} \left( \sum_{i=1}^n b_i - 1 \right)! \, 2^{\sum_{i=1}^n b_i \binom{i}{2}}.$$

(Hint: Every graph induces, via the connected components, a partition of $\{1, \ldots, n\}$. Denote by $f(\pi)$ the number of graphs whose induced partition is just $\pi$. Then $c_n = f(1)$.)

14. Deduce from **4.25** the identity

$$\sum_{n \geq 0} (-1)^n a_n z^n = \exp\left( -\sum_{k \geq 1} \frac{s_k}{k} z^k \right).$$

→15. An $n \times n$-matrix $H = [h_{ij}]$ over $\mathbb{Z}$ is called a *Hadamard matrix* if $h_{ij} = \pm 1$ and if the inner product of any two distinct rows is 0; hence $HH^T = nI_n$. It follows that $|\det H| = n^{n/2}$ and a theorem of Hadamard's states that for every real matrix $M = [m_{ij}]$ with $|m_{ij}| \leq 1$, we have $|\det M| \leq n^{n/2}$ with equality if and only if $M$ is a Hadamard matrix. We construct a class of matrices whose determinant comes close to the upper bound. Let $k \in \mathbb{N}$ and write $k = \sum_{i=0}^n a_i 2^i$ in its binary representation and identify $k$ with $(a_0, a_1, a_2, \ldots)$. Let $q(k) := \sum_{i \geq 0} a_i$ and $b_{ij} = i \cdot j$ (i.e., $b_{ij}$ is the number of ones common to the binary representations of $i$ and $j$). Prove:
(i) For $M = [m_{ij}]$, $0 \leq i, j \leq n$, where $m_{ij} = (-1)^{b_{ij}}$, we have

$$|\det M| = 2^{\sum_{k=0}^n q(k)}.$$

(ii) Compute $M$ for $n = 2, 3, 4$. Which are Hadamard matrices? (Hint: Use the dual form of ex. 1.) (Wilf)

## 3. The Möbius Function

After having demonstrated the applicability of Möbius inversion we now want to develop methods for computing the Möbius function of an arbitrary poset. We have already seen that $\mu$ is multiplicative. Hence we may confine ourselves to indecomposable posets. To obtain some deeper results we proceed as follows. Suppose $P$ and $L$ are posets with Möbius functions $\mu_P$ and $\mu_L$ respectively. We wish to determine the relationship between $\mu_P$ and $\mu_L$ when the posets $P$ and $L$ are connected by some order function. In general, one of the Möbius functions, say $\mu_P$, is known and our goal is to compute the unknown function $\mu_L$ by means of $\mu_P$ and the connecting function. Consider as an example a finite geometric lattice $L$ and the Boolean algebra $P = \mathcal{B}(S)$ of its set $S$ of points with the closure operator $A \in P \to \bar{A} \in L$ as the connecting function. Here $\mu_P$ is known and we want to derive $\mu_L$. To continue our analogy with calculus we could interpret this method as the discrete counterpart of the substitution of variables.

We are going to discuss in detail two types of connecting functions, *closure* and *Galois connection*, and conclude this section with a brief study of the characteristic polynomial introduced in section 2.A.

## A. Closure

Let us generalize the concept of a closure operator from Boolean algebras to arbitrary posets.

**Definition.** Let $P$ be a poset. A mapping $H: P \to P$ is called a *closure operator* (or simply *closure*) *on* $P$ if for all $a, b \in P$:

(i) $a \leq Ha$
(ii) $a \leq b \Rightarrow Ha \leq Hb$ (monotone)
(iii) $HHa = Ha$ (idempotent).

As before, we usually set $\bar{a} := Ha$ and call an element $a$ *closed* if $\bar{a} = a$. The set $Q$ of closed elements is the *quotient* of $P$ relative to the closure $H$ and we sometimes write $Q = P/H$.

The following proposition is readily verified.

**4.26 Proposition.** *Let $P$ be a lattice. Then every quotient $Q$ of $P$ is a complete lattice and infima in $Q$ coincide with infima in $P$.* □

Let $x \to \bar{x}$ be a closure on $P$ with quotient $Q$. We regard $Q$ as a subposet of $P$ and the incidence algebra $\mathbb{A}(Q)$ as a subset of $\mathbb{A}(P)$ by defining

$$f(x, y) = 0 \quad \text{if } x \notin Q \text{ or } y \notin Q \qquad (f \in \mathbb{A}(Q)).$$

Denote by $\mu$ and $\mu_Q$ the Möbius functions of $P$ and $Q$, respectively, and similarly $\zeta, \zeta_Q$ and $\delta, \delta_Q$. The following theorem is basic.

**4.27 Theorem** (Rota). *Let $P$ be a locally finite poset and $x \to \bar{x}$ a closure on $P$ with quotient $Q$. Then for all $x, y \in P$*

$$\sum_{z \in P, \bar{z} = \bar{y}} \mu(x, z) = \begin{cases} \mu_Q(\bar{x}, \bar{y}) & \text{if } x = \bar{x} \\ 0 & \text{if } x < \bar{x}. \end{cases}$$

*Proof.* We have

$$\begin{aligned} \sum_{z \in P, \bar{z} = \bar{y}} \mu(x, z) &= \sum_{z} \mu(x, z) \delta_Q(\bar{z}, \bar{y}) = \sum_{z, \bar{w}} \mu(x, z) \zeta_Q(\bar{z}, \bar{w}) \mu_Q(\bar{w}, \bar{y}) \\ &= \sum_{z, \bar{w}} \mu(x, z) \zeta(z, \bar{w}) \mu_Q(\bar{w}, \bar{y}) \quad (\text{since } z \leq \bar{w} \Leftrightarrow \bar{z} \leq \bar{w}) \\ &= \sum_{\bar{w} \in Q} \delta(x, \bar{w}) \mu_Q(\bar{w}, \bar{y}). \quad \square \end{aligned}$$

As a corollary, we note the relationship between the difference operators $D_{\leq}$ of $P$ and $D_{\leq}^{(Q)}$ of $Q$.

**4.28 Corollary.** *Let $P$ be a poset all of whose principal ideals are finite and $x \to \bar{x}$ a closure on $P$ with quotient $Q$, $K$ a field of characteristic 0. For a mapping $g: P \to K$ we have*

$$(D_{\leq} g)(\bar{y}) = \sum_{z \in P, \bar{z} = \bar{y}} (D_{\leq} g)(z). \quad \square$$

**4.27** and **4.28** yield a variety of interesting results corresponding to the choice of the closure or the mapping $g$. We give an example of each.

**4.29 Proposition.** *Let $S$ be a finite set, $A \to \bar{A}$ a closure on $\mathscr{B}(S)$ with $\bar{\varnothing} = \varnothing$ and $Q$ its quotient. If $r_k$ denotes the number of $k$-subsets $A$ of $S$ with $\bar{A} = S$, then*

$$\sum_{k \geq 0} (-1)^k r_k = \mu_Q(0, 1).$$

*Proof.* In **4.27** choose $P = \mathscr{B}(S)$, $x = \varnothing$, $y = S$ and apply **4.15**(iv). $\square$

**Examples.** Let $P$ be a finite lattice and set $S = P - \{0\}$. The mapping

$$A \to \bar{A} = \{a \in S: a \leq \sup A\} \qquad (\varnothing \neq A \subseteq S)$$
$$\varnothing \to \bar{\varnothing} = \varnothing$$

is a closure on $\mathscr{B}(S)$ with $Q \cong P$. Hence we obtain the formula

$$\mu(0, 1) = \sum_{k \geq 0} (-1)^k r_k,$$

where $r_k = |\{A \subseteq P - \{0\}: |A| = k, \sup A = 1\}|$.

Now assume $P$ is a point lattice with point set $S$. If we choose the closure

$$A \to \bar{A} = \{p \in S: p \leq \sup A\} \qquad (\varnothing \neq A \subseteq S)$$
$$\varnothing \to \bar{\varnothing} = \varnothing$$

then again $Q \cong P$, and thus

$$\mu(0, 1) = \sum_{k \geq 0} (-1)^k r_k,$$

where $r_k = |\{A \subseteq S: |A| = k, \sup A = 1\}|$.

The dual propositions hold similarly.

As an application of **4.28**, consider an arbitrary finite poset $P$ with 0 and 1 and define $g: P \to \mathbb{Q}$ by

$$g(z) := \zeta^2(0, z) = |[0, z]| \qquad (\text{cf. } \mathbf{4.7}(\text{i})).$$

If we set $y = 1$ then the right-hand side of **4.28** becomes $\sum_{z \in P, \bar{z} = 1} \zeta(0, z) = |\{z \in P: \bar{z} = 1\}|$ whence we obtain the following result.

**4.30 Proposition.** *Let $P$ be a finite poset with 0 and 1, and let $x \to \bar{x}$ be a closure on $P$ with quotient $Q$. Then*

$$|\{z \in P: \bar{z} = 1\}| = \sum_{x \in P} |[0, x]| \mu_Q(x, 1). \quad \square$$

The case $P = \mathscr{B}(S)$ yields the following useful companion formula to **4.29**.

**4.31 Proposition.** *Let $S$ be a finite set, $A \to \bar{A}$ a closure on $\mathscr{B}(S)$ with $\bar{\varnothing} = \varnothing$ and $Q$ the quotient. If $r_k$ denotes the number of $k$-subsets $A$ of $S$ with $\bar{A} = S$, then*

$$\sum_{k \geq 0} r_k = \sum_{A \subseteq S} 2^{|A|} \mu_Q(A, 1). \quad \square$$

**Example.** Consider the $n$-dimensional vector space $V(n, q)$. The linear closure on the set $\mathscr{B}(V(n, q))$ of vectors has quotient $Q \cong \mathscr{L}(n, q)$. Since a $k$-dimensional subspace contains $q^k$ vectors we have from **4.20**(iii) the following formulae:

$$\sum_{k \geq 0} (-1)^k r_k = (-1)^n q^{\binom{n}{2}}$$

$$\sum_{k \geq 0} r_k = \sum_{l \geq 0} (-1)^{n-l} \binom{n}{l}_q 2^{q^l} q^{\binom{n-l}{2}},$$

where $r_k = |\{A \subseteq V(n, q): \bar{A} = V(n, q)\}|$.

The following application involving the connection between the Möbius function of a lattice and the properties of being a point lattice or of being complemented is most important from a theoretical point of view.

**4.32 Proposition** (Hall). *Let $P$ be a finite lattice. Then*

$$\mu(0, 1) = 0$$

*if 0 is not the infimum of copoints or if 1 is not the supremum of points.*

*Proof.* We choose the closure

$$x \to \bar{x} = \inf\{c \in P: x \leq c \lessdot 1\}$$

and apply **4.27** and its dual form. $\square$

**4.32** determines the values of $\mu$ for finite distributive and for finite modular lattices of rank $n \geq 4$. (The cases $n = 2$ or $3$ are trivial.) The condition $\mu(0, 1) \neq 0$ implies in both cases that $P$ is geometric whence we may use the structure theorems **2.53** and **2.55**:

$$P \textit{ distributive} \Rightarrow \mu(0, 1) = 0 \textit{ or } (-1)^{r(P)},$$

$$P \textit{ modular indecomposable} \Rightarrow \mu(0, 1) = 0 \textit{ or } (-1)^{r(P)} q^{\binom{r(P)}{2}} \textit{ for a prime power } q.$$

**4.32** states that $\mu(0, 1) \neq 0$ implies that 1 is a supremum of points. We are now going to prove the stronger result that $\mu(0, 1) \neq 0$ even implies that $P$ is complemented. Hence if $\mu(a, b) \neq 0$ for all $a \leq b \in P$ then $P$ is relatively complemented. The converse of this is false as is shown by the relatively complemented lattice of Figure 4.3. But we shall prove as a central result that in finite geometric lattices we indeed have $\mu(a, b) \neq 0$ for all $a \leq b$.

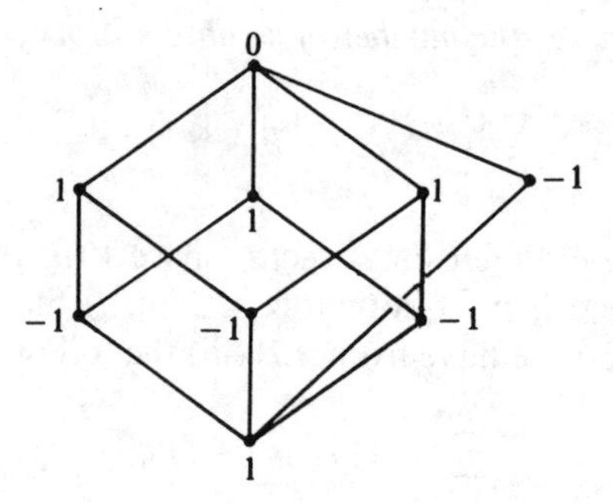


Figure 4.3

**4.33 Theorem** (Crapo). *Let $P$ be a finite lattice. For $a \in P$ denote by $a^{\perp}$ the set of complements of $a$. Then for all $a \in P$:*

$$\mu(0, 1) = \sum_{w, z \in a^{\perp}} \mu(0, w)\zeta(w, z)\mu(z, 1).$$

*Proof.* We may assume $a < 1$. Let $z \in P$ with $z \vee a = 1$. The mapping

$$w \to \bar{w} := \begin{cases} w & \text{if } w \vee a < 1 \\ z & \text{if } w \vee a = 1 \end{cases} \qquad (w \leq z)$$

is clearly a closure on the principal ideal generated by $z$, with quotient $Q_z = \{w \in P: w = z \text{ or } w < z, w \vee a < 1\}$. Denote by $\mu_z$ the Möbius function of $Q_z$. By **4.27** we have

$$\sum_{w \leq z,\, w \vee a = 1} \mu(0, w) = \mu_z(0, z).$$

If we regard $\mu(0, z)$ and $\mu_z(0, z)$ as functions of $z$ only and apply Möbius inversion on the poset $F = \{z \in P: z \vee a = 1\}$ then (noting that $\mu_F(z, 1) = \mu(z, 1)$ for all $z \in F$)

$$\mu(0, 1) = \sum_{z \in P, z \vee a = 1} \mu_z(0, z)\mu(z, 1) = \sum_{z \in P, z \vee a = 1} \left( \sum_{w \in P, w \vee a = 1} \mu(0, w)\zeta(w, z) \right) \mu(z, 1).$$

Since $z \wedge a = 0$, $w \leq z$ imply $w \wedge a = 0$, it remains to be shown that $\mu_z(0, z) = 0$ for all $z \in P$ with $z \vee a = 1$, $z \wedge a > 0$. Clearly, $u, v \in Q_z$ implies $u \wedge v \in Q_z$. Hence $Q_z$ with the induced order is a lattice. Furthermore if $w \in Q_z$, then $w \vee (z \wedge a) \leq z$ and $(w \vee (z \wedge a)) \vee a = w \vee a < 1$ if $w \vee (z \wedge a) < z$, thus $w \vee (z \wedge a) \in Q_z$. Since $w \in Q_z, w < z \Rightarrow w \vee (z \wedge a) < z$, this implies that $w \vee (z \wedge a) = w$, i.e., $z \wedge a \leq w$ for all copoints $w$ of $Q_z$. Hence $z \wedge a$ is a lower bound for all copoints of $Q_z$, and we conclude by **4.32** that $\mu_z(0, z) = 0$ whenever $z \wedge a > 0$. □

As a corollary we now have a fairly complete picture of the range of the Möbius function for our fundamental lattice classes.

**4.34 Corollary.** *Let $P$ be a finite lattice.*

(i) *$P$ not complemented $\Rightarrow \mu(0, 1) = 0$.*
(ii) *$P$ modular $\Rightarrow \mu(0, 1) = \mu(0, a) \sum_{z \in a^\perp} \mu(0, z)$ for all $a \in P$.*
(iii) *$P$ semimodular $\Rightarrow \mu(0, 1) = \mu(0, a) \sum_{z \in a^\perp} \mu(0, z)$ for all modular elements $a \in P$.*
(iv) *$P$ geometric $\Rightarrow \mu(0, 1) \neq 0$, where $\mu(0, 1) > 0$ if $r(P)$ is even and $\mu(0, 1) < 0$ if $r(P)$ is odd.*

*Proof.* (i) is a direct consequence of **4.33**. **2.22**(ii) implies that the complements of a modular element $a$ form an antichain whence by **4.33** we infer $\mu(0, 1) = \sum_{z \in a^\perp} \mu(0, z)\mu(z, 1)$. (ii) and (iii) follow now from **2.17** which says $[z, 1] \cong [0, a]$ if $z \in a^\perp$ and $a$ is modular. Finally, we use induction to prove (iv). For $r(P) = 1$ or 2 the assertion is trivial. Let $r(P) \geq 2$ and $p$ a point of $P$. All complements of $p$ must be copoints in $P$ and by (iii) we have $\mu(0, 1) = -\sum_{z \in p^\perp} \mu(0, z)$. Now from the induction hypothesis, all values $\mu(0, z)$ are non-zero and have the same sign. Hence $\mu(0, 1) \neq 0$ with the opposite sign. □

## B. Galois Connection

Let us now discuss the most interesting connection principle as described in the introduction.

**Definition.** Let $P$ and $L$ be posets. A pair $(\sigma, \tau)$ of maps $\sigma: P \to L$, $\tau: L \to P$ is called a *Galois connection* between $P$ and $L$ if

(i) $\sigma$ and $\tau$ are antitone,
(ii) $\tau\sigma x \geq x$ for all $x \in P$, $\sigma\tau z \geq z$ for all $z \in L$.

**4.35 Lemma.** *Let $(\sigma, \tau)$ be a Galois connection between $P$ and $L$. Then*

$$\sigma\tau\sigma = \sigma, \qquad \tau\sigma\tau = \tau.$$

*Proof.* Let $x \in P$. By (ii), $\tau\sigma x \geq x$ and thus $\sigma\tau\sigma x \leq \sigma x$. On the other hand, $\sigma x \in L$ and hence $\sigma\tau\sigma x \geq \sigma x$. Now reverse the rôles of $\sigma$ and $\tau$. □

**4.36 Proposition.** *Let $(\sigma, \tau)$ be a Galois connection between $P$ and $L$.*

(i) *$\tau\sigma$ and $\sigma\tau$ are closure operators on $P$ and $L$, respectively.*
(ii) *$P/\tau\sigma = \tau L$, $L/\sigma\tau = \sigma P$, i.e., the quotient of $P$ relative to $\tau\sigma$ is precisely the image $\tau L$, and similarly for $L$.*
(iii) *$P/\tau\sigma$ and $L/\sigma\tau$ are antiisomorphic with $\sigma$ and $\tau$ as inverse mappings.*

*Proof.* By symmetry it suffices to verify the first assertion in (i) and (ii). The first two properties of a closure operator are immediate consequences of the definition of a Galois connection, whereas the idempotency follows from **4.35** which also implies part (ii). The restrictions of $\sigma$ and $\tau$ to the quotients $P/\tau\sigma$ and $L/\sigma\tau$, respectively, are inverse mappings, which proves (iii). □

**Example.** The classical example of a Galois connection from which its name is derived occurs in the theory of fields. Consider a commutative field $K$ and its automorphism group $G$. To each subfield $L$ of $K$ we associate its fixed-group, i.e., the subgroup of $G$ consisting of all those automorphisms which leave $L$ pointwise fixed. Conversely, to every subgroup of $G$ we associate its fixed-field in $K$. The correspondence fixed-group-fixed-field is a Galois connection between the lattice of subfields of $K$ and the lattice of subgroups of $G$. The main theorem of Galois theory states that when $G$ is finite every subgroup of $G$ is closed, whereas the closed subfields of $K$ are precisely those relative to which $K$ is a Galois extension.

For many applications it is advantageous to use monotone instead of antitone mappings. To accomplish this we merely turn one of the posets $P$ or $L$ upside down and translate the definition of a Galois connection into this new situation.

First let us call a mapping $x \to \bar{x}$ a *coclosure* on a poset $P$ if for all $a, b \in P$:

(i) $a \geq \bar{a}$
(ii) $a \leq b \Rightarrow \bar{a} \leq \bar{b}$
(iii) $\bar{\bar{a}} = \bar{a}$.

**Definition.** Let $P$ and $L$ be posets. A mapping $\sigma: P \to L$ is called a *Galois function* if there exists a function $\sigma^+: L \to P$ such that

(i) $\sigma, \sigma^+$ are monotone,
(ii) $\sigma^+\sigma x \geq x$ for all $x \in P$, $\sigma\sigma^+ z \leq z$ for all $z \in L$.

**4.37 Proposition.** *Let $\sigma: P \to L$ be a Galois function. Then*

(i) *$\sigma^+\sigma$ is a closure on $P$; $\sigma\sigma^+$ is a coclosure on $L$.*
(ii) *$P/\sigma^+\sigma = \sigma^+L$, $L/\sigma\sigma^+ = \sigma P$.*
(iii) *$P/\sigma^+\sigma \cong L/\sigma\sigma^+$.* □

**Example.** Let $L$ be any lattice and $a, b \in L$. The map $\phi_b: z \to z \vee b$ is a Galois function from $[a \wedge b, a]$ into $[b, a \vee b]$ with $\phi_b^+: w \to w \wedge a$. **4.35** reduces in this case to proposition **2.16** whereas **2.17** reads: A lattice $L$ is modular if and only if all elements of $[a \wedge b, a]$ and $[b, a \vee b]$ are closed, for all $a, b \in L$.

The proof of the following statement is an easy exercise.

**4.38 Proposition.** *Let $P$ and $L$ be complete lattices. Any Galois function $\sigma: P \to L$ is supremum-preserving and we have*

$$\sigma^+ z = \sup\{x \in P: \sigma x \leq z\} \qquad (z \in L).$$

*Conversely, any supremum-preserving mapping from $P$ to $L$ is a Galois function where $\sigma^+$ is defined as above.* □

Whenever we study complete lattices we shall replace Galois function by supremum-preserving function. The next two propositions are the central theorems of this section and are expressed in terms of both Galois connections and Galois functions. Of course, we need only prove one assertion each time.

**4.39 Theorem** (Rota). *Let $(\sigma, \tau)$ be a Galois connection between the locally finite posets $P$ and $L$, and $Q = P/\tau\sigma$. Then for all $x \in P$, $y \in L$:*

$$\sum_{z \in P,\, \sigma z = y} \mu_P(x, z) = \sum_{u \in L,\, \tau u = x} \mu_L(y, u) = \mu_Q(x, \tau y).$$

*Let $\sigma: P \to L$ be a Galois function. Then for all $x \in P$, $y \in L$:*

$$\sum_{z \in P,\, \sigma z = y} \mu_P(x, z) = \sum_{u \in L,\, \sigma^+ u = x} \mu_L(u, y) = \mu_Q(x, \sigma^+ y).$$

*Proof.* If either of $x$ or $y$ is not closed, then both sums are 0. Otherwise, we have by **4.27** and **4.36**(iii)

$$\sum_{z \in P,\, \sigma z = y} \mu_P(x, z) = \mu_{P/\tau\sigma}(x, \tau y) = \mu_{L/\sigma\tau}(y, \sigma x) = \sum_{u \in L,\, \tau u = x} \mu_L(y, u). \quad \square$$

In analogy to **4.28** we obtain the following formula for "substitution of variables."

**4.40 Theorem.** *Let $(\sigma, \tau)$ be a Galois connection between $P$ and $L$ and $Q = P/\tau\sigma$, where we assume that all principal ideals of $P$ and $L$ are finite. Further, let $f: P \to K$ and $g: L \to K$ be functions into a field $K$ of characteristic* 0. *Then*

$$\sum_{z \in P} g(\sigma z)(D_{\leq}^{(P)} f)(z) = \sum_{u \in L} f(\tau u)(D_{\leq}^{(L)} g)(u) = \sum_{z, u \in Q} f(z)\mu_Q(z, u) g(\sigma u).$$

*Let $\sigma: P \to L$ be a Galois function, $Q = P/\sigma^+\sigma$, where in $P$ all principal ideals, and in $L$ all principal filters, are finite. Then*

$$\sum_{z \in P} g(\sigma z)(D_{\leq}^{(P)} f)(z) = \sum_{u \in L} f(\sigma^+ u)(D_{\geq}^{(L)} g)(u) = \sum_{z, u \in Q} f(z)\mu_Q(z, u) g(\sigma u). \quad \square$$

The main advantage of the notion of Galois connection is that we are able to compute the Möbius function $\mu_L$ of a lattice $L$ by looking at an arbitrary subset $P$ and the Boolean algebra $\mathscr{B}(P)$ it generates, whereas up to now special sets $P$ (e.g., the point set as in the example after **4.29**) had to be chosen.

Consider a finite lattice $L$ and an arbitrary non-empty subset $M \subseteq L$. The mapping $\sigma: \mathscr{B}(M) \to L$ given by

$$\sigma A = \sup A \qquad (\varnothing \neq A \subseteq M)$$
$$\sigma\varnothing = 0$$

is clearly supremum-preserving with

$$\sigma^+ u = M_u = \{p \in M : p \leq u\} \qquad (u \in L).$$

Letting $x = \varnothing \in \mathscr{B}(M)$, $y = 1 \in L$ and applying **4.39** the following proposition results.

**4.41 Proposition.** *Let $L$ be a finite lattice, $M \subseteq L$ and $M_u := \{p \in M : p \leq u\}$ for all $u \in L$. If $r_k$ denotes the number of $k$-subsets $A \subseteq M$ with* $\sup A = 1$, *then*

$$\sum_{k \geq 0} (-1)^k r_k = \sum_{u \in L, M_u = \varnothing} \mu_L(u, 1).$$

*The dual statement holds similarly.* $\square$

Of special interest is the case when $M_u = \varnothing$ is only satisfied for $u = 0$. Such a set $M$ possesses the following properties:

(i) $0 \notin M$;
(ii) for all $0 \neq u \in L$ there is an element $p \in M$ with $p \leq u$.

Let us rewrite (ii) briefly as: $M < u$ for $0 \neq u \notin M$.

**Definition.** A subset $M$ of a finite lattice $L$ is called a *lower cross-cut* if $0 \notin M$ and $M < u$ for all $0 \neq u \notin M$. Dually, $M$ is called *upper cross-cut* if $1 \notin M$ and $u < M$ for all $1 \neq u \notin M$. $M$ is called a *cross-cut* if $0, 1 \notin M$ and if further for all $u \notin M$ precisely one of $u < M$ or $M < u$ holds and if every maximal chain contains at least one element of $M$.

**4.41** specialized to cross-cuts yields the following important theorem.

**4.42 First cross-cut theorem** (Rota). *Let $L$ be a finite lattice, $M$ a lower (upper) cross-cut and $r_k$ the number of $k$-sets $A \subseteq M$ with* $\sup A = 1$ ($\inf A = 0$). *Then*

$$\mu(0, 1) = \sum_{k \geq 0} (-1)^k r_k. \quad \square$$

Any lower cross-cut clearly contains the point set $S$ of $L$ and similarly any upper cross-cut contains the copoint set $C$. Since we are often interested in as small a cross-cut as possible the cases $M = S$ and $M = C$ warrant special attention.

**4.43 Proposition.** *Let $L$ be a finite lattice with point set $S$ and copoint set $C$. If $r_k$ denotes the number of $k$-sets $A \subseteq S$ with* $\sup A = 1$ *or the number of $k$-sets $A \subseteq C$ with* $\inf A = 0$, *then in each case*

$$\mu(0, 1) = \sum_{k \geq 0} (-1)^k r_k. \quad \square$$

We have already shown **4.43** as a corollary to **4.29** under the additional assumption that $L$ is a point lattice. Comparison of these two results leads to the conclusion that we may confine ourselves to the sublattice generated by the points when computing the Möbius function. For semimodular lattices, in particular, the geometric lattice induced by its point set suffices to determine $\mu$.

As in **4.30** and **4.31** we may now apply **4.39** to obtain an expression for $\sum_{k \geq 0} r_k$ (see the exercises).

As a final application of **4.39** let us derive a symmetric formula for cross-cuts. For the moment, let $M$ be any subset of the finite lattice $L$. We define $\sigma: \mathscr{B}(M) \to \operatorname{Int} L$ (the interval lattice of $L$ including $\varnothing$) by

$$\sigma A = [\inf A, \sup A] \qquad (\varnothing \neq A \subseteq M)$$
$$\sigma \varnothing = \varnothing.$$

$\sigma$ is clearly supremum-preserving, and by **4.38** we have

$$\sigma^+[x, y] = \bigcup \{A \subseteq M : x \leq \inf A \leq \sup A \leq y\} \qquad (\varnothing \neq [x, y] \in \operatorname{Int} L)$$
$$\sigma^+ \varnothing = \varnothing.$$

This is easily seen to be equivalent to

$$\sigma^+[x, y] = M \cap [x, y]$$
$$\sigma^+ \varnothing = \varnothing.$$

From **4.39** and **4.16** we deduce the following result.

**4.44 Proposition.** *Let $L$ be a finite lattice, $M \subseteq L$ and $r_k$ the number of $k$-sets $A \subseteq M$ with* $\inf A = 0$, $\sup A = 1$. *Then*

$$\sum_{k \geq 0} (-1)^k r_k = \sum_{u, v \in L, [u, v] \cap M = \varnothing} \mu(0, u) \zeta(u, v) \mu(v, 1) - \mu(0, 1). \quad \square$$

**4.45 Second cross-cut theorem** (Rota). *Let $L$ be a finite lattice, $M$ a cross-cut of $L$ and $r_k$ the number of $k$-sets $A \subseteq M$ with* $\inf A = 0$, $\sup A = 1$. *Then*

$$\mu(0, 1) = \sum_{k \geq 0} (-1)^k r_k.$$

*Proof.* According to the definition of a cross-cut there are two types of non-empty intervals $[u, v]$ with $[u, v] \cap M = \emptyset$. Either $u \leq v < M$ or $M < u \leq v$, and exactly one of these possibilities holds. Hence we have

$$\sum_{u, v \in L, [u, v] \cap M = \emptyset} \mu(0, u)\zeta(u, v)\mu(v, 1) = \sum_{v \in L, v < M} \delta(0, v)\mu(v, 1) + \sum_{u \in L, M < u} \mu(0, u)\delta(u, 1) = 2\mu(0, 1),$$

and the theorem follows from **4.44**. □

A particularly important class of cross-cuts are antichains which have a non-empty intersection with every maximal chain in $L$. Let us note this case separately.

**4.46 Corollary.** *Let $L$ be a finite lattice, $M$ an antichain in $L$ with $0, 1 \notin M$ which has a non-empty intersection with every maximal chain. If $r_k$ is the number of $k$-sets $A \subseteq M$ with* $\inf A = 0$, $\sup A = 1$, *then*

$$\mu(0, 1) = \sum_{k \geq 0} (-1)^k r_k. \quad \square$$

Consider a lattice with a rank function. Here all levels are cross-cuts for which **4.46** applies. **4.46** provides a convenient method for computing $\mu$ if the lattice contains small cross-cuts $M$. For instance, if $|M| = 1$ then $\mu = 0$. For $|M| = 2$ or 3 the possible values are $\mu = 0, 1$ and $\mu = 0, \pm 1, 2$ respectively. (See the exercises for bounds on the range of $\mu$ when the lattice contains a cross-cut of cardinality $k$.)

## C. The Characteristic Polynomial

The theorems of sections $A$ and $B$ permit a closer analysis of the characteristic polynomial $\chi(P; x)$ of a poset with a rank function. Our method involves decomposing $\chi(P; x)$ into a product of characteristic polynomials of posets with smaller rank.

Let us first define the *characteristic function* of $P$. Suppose the poset $P$ satisfies the JD-chain condition and consider the algebra $\mathbb{A}_{K[x]}(P)$ over the polynomial ring $K[x]$ where $K$ is a field of characteristic 0. Let $\rho \in \mathbb{A}(P)$ be the length function, given by $\rho(a, b) = l[a, b]$ for $a \leq b \in P$, and define $\bar{\rho} \in \mathbb{A}(P)$ by

$$\bar{\rho}(a, b) := x^{\rho(a, b)}.$$

Clearly

$$\bar{\rho}(a, c) = \bar{\rho}(a, b)\bar{\rho}(b, c) \quad \text{for } a \leq b \leq c \in P;$$

thus $\bar{\rho} \in \mathbb{S}(P)$ is a multiplicative invertible function.

**Definition.** Let $P$ a poset which satisfies the JD-chain condition. The *characteristic function* char of $P$ is defined by

$$\text{char} := \mu * \bar{\rho} \in \mathbb{A}_{K[x]}(P).$$

Comparing this definition with that of the characteristic polynomial $\chi(P; x) = \sum_{a \in P} \mu(0, a) x^{r(1)-r(a)}$ we see that for finite posets with 0 and 1 and with a rank function

$$\text{char}(0, 1) = \chi(P; x).$$

Applying **4.14** we note a first decomposition theorem.

**4.47 Proposition.** *Let $P$ be a finite poset with 0 and 1 and a rank function. If $P = P_1 \times P_2$, then*

$$\chi(P; x) = \chi(P_1; x) \cdot \chi(P_2; x). \quad \square$$

**Example.** Consider $P = \prod_{i=1}^{t} \mathscr{C}(k_i)$. We have

$$\chi(P; x) = \prod_{i=1}^{t} \chi(\mathscr{C}(k_i); x) = \prod_{i=1}^{t} (x^{k_i} - x^{k_i - 1}) = x^{r(P)-t}(x-1)^t$$

and, in particular,

$$\chi(\mathscr{B}(n); x) = (x-1)^n.$$

**4.48 Theorem** (Stanley). *Let $P$ be a finite lattice with rank function $r$ and let $a \in P$ be a modular element (in the sense of* **2.42**(iii)). *Then*

$$\chi(P; x) = \chi([0, a]; x) \cdot \sum_{z \in P,\, z \wedge a = 0} \mu(0, z) x^{r(1)-r(a)-r(z)}.$$

*Proof.* Let $a$ be a modular element of $P$. It is a trivial exercise to show that $a \wedge w$ is modular in $[0, w]$ for all $w \in P$. Applying **4.34**(ii) we have

$$\begin{aligned}
\chi(P; x) &= \sum_{w \in P} \mu(0, w) x^{r(1)-r(w)} \\
&= \sum_{w \in P} \mu(0, a \wedge w) \sum_{z \wedge (a \wedge w) = 0,\, z \vee (a \wedge w) = w} \mu(0, z) x^{r(1)-r(w)} \\
&= \sum_{w \in P} \mu(0, a \wedge w) \sum_{z \wedge a = 0,\, z \leq w \leq a \vee z} \mu(0, z) x^{r(1)-r(w)} \quad \text{(since } a \text{ is modular)} \\
&= \sum_{z \in P,\, z \wedge a = 0} \mu(0, z) x^{r(1)-r(a)-r(z)} \sum_{w \in [z, a \vee z]} \mu(z, w) x^{r(a)+r(z)-r(w)} \\
&\qquad \text{(since } [0, a \wedge w] \cong [z, w]).
\end{aligned}$$

Since $[0, a] \cong [z, a \vee z]$ the last sum is precisely $\chi([0, a]; x)$. $\square$

**Examples.** The characteristic polynomials of finite modular lattices $P$ are completely determined by **4.48**. Let $0 <\cdot a_1 <\cdot \cdots <\cdot a_n = 1$ be any maximal chain and $\alpha_i$ the number of points which are below $a_i$ but not below $a_{i-1}$, $i = 1, \ldots, n$. Let $z \in P$ with $z \wedge a_{n-1} = 0$. Since $P$ is modular we infer that $z = 0$ or $z$ is a point. Hence by the theorem just proved $\chi(P; x) = \chi([0, a_{n-1}]; x) \cdot (x - \alpha_n)$. Now consider the interval $[0, a_{n-1}]$, and so on. Finally, we obtain

$$\chi(P; x) = \prod_{i=1}^{n} (x - \alpha_i).$$

This result implies, in particular, that in any maximal chain the numbers $\alpha_1, \ldots, \alpha_n$ must appear in some order. For $P = \mathscr{L}(n, q)$ we again obtain **4.20**(iii) using **3.11**:

$$\chi(\mathscr{L}(n, q); x) = \prod_{i=0}^{n-1} (x - q^i).$$

In an arbitrary finite lattice $P$ with a rank function **4.48** gives the result $\chi(P; x) = \prod_{i=1}^{n} (x - \alpha_i)$ as long as *one* maximal chain $0 <\cdot a_1 <\cdot \cdots <\cdot a_n = 1$ exists in which $a_{i-1}$ is modular in $[0, a_i]$ for all $i$. It is readily seen that this condition is equivalent to requiring that a maximal chain of modular elements exists. For the partition lattices $\mathscr{P}(n)$ such a "modular" chain is readily constructed. It is an easy exercise to show that in $\mathscr{P}(n)$ the modular partitions are precisely those which consist of at most one non-trivial block and otherwise just singleton blocks. Hence if $\pi_i := \{1, \ldots, i\} | i + 1 | \cdots | n$, then

$$0 = \pi_1 <\cdot \pi_2 <\cdot \cdots <\cdot \pi_n = 1$$

is such a modular chain where $\alpha_i = i$ for $i = 1, \ldots, n - 1$. **4.48** thus again yields our result **4.20**(iv):

$$\chi(\mathscr{P}(n); x) = \prod_{i=1}^{n-1} (x - i).$$

Let us summarize all the information we have gained about the coefficients of the characteristic polynomial.

**4.49 Proposition.** *Let $P$ be a finite poset with 0 and 1 and a rank function. Then*

(i) $\chi(P; x) = \sum_{i=0}^{n} w_i x^{n-i}$ *has degree* $n = r(P)$.
(ii) $w_0 = 1$, $-w_1 = \#\{\text{points}\}$, $w_n = \mu(P)$.
(iii) *If $P$ is a geometric lattice, then* $(-1)^i w_i > 0$. $\square$

EXERCISES IV.3

→ 1. Use **4.27** to prove Weisner's formula: Let $a > 0$ in a finite lattice $L$. Then

(i) $\sum_{x \vee a = 1} \mu(0, x) = 0$; and more generally
(ii) $\sum_{x \vee a = b} \mu(0, x) = 0$ for all $b \in L$.

2. Derive from ex. 1 the formulae for the Möbius functions in $\mathscr{B}(n)$, $\mathscr{L}(n, q)$, and $\mathscr{P}(n)$. (Hint: Use induction.)

→ 3. Apply **4.34** to prove the following:

(i) Let $U \in \mathscr{L}(n, q)$, $r(U) = k$. Then the number of complements of $U$ is given by $|U^{\perp}| = q^{k(n-k)}$.
(ii) Deduce a similar formula for the lattice $\mathscr{P}(n)$.
(iii) Derive, conversely, $\mu(\mathscr{L}(n, q))$ and $\mu(\mathscr{P}(n))$ from (i) and (ii).

→ 4. Let $S$ be a set and $K$ a field. The set $\mathrm{Map}(S, K)$ together with the usual addition and scalar multiplication is a vector space $V$ over $K$. Let $W \in \mathscr{L}(V)$ and define for $A \subseteq S$, $U \subseteq W$:

$$h(A) := \{f \in W : f(x) = 0 \text{ for all } x \in A\}$$
$$k(U) := \{a \in S : g(a) = 0 \text{ for all } g \in U\}.$$

Prove that $(h, k)$ is a Galois connection between $2^S$ and $2^W$ and that the closure $kh: 2^S \to 2^S$ satisfies the exchange axiom.

5. Let $P$ and $L$ be finite posets with 0, and $\sigma: P \to L$ monotone, such that $\sigma^{-1}[0, b]$ is an interval of $P$ for all $b \in L$. Prove that $\sigma$ is a Galois function whose quotient $Q$ on $P$ is isomorphic to $L$ and that

$$\sum_{\sigma x = b} \mu_P(0, x) = \mu_Q(0, b) \quad \text{for all } b \in L.$$

(Rota)

6. Let $L$ be a locally finite lattice, $x, y \in L$. Prove:

(i) $\phi_y: [x \wedge y, x] \to [y, x \vee y]$ given by $\phi_y(z) = z \vee y$ is a Galois function with $\phi_y^+ = \psi_x$, $\psi_x(w) = w \wedge x$ (see **2.16**).
(ii) For the quotient $Q$ on $[x \wedge y, x]$, we have $Q = [x \wedge y, x]$ if and only if $x$ and $y$ are a modular pair.

→ 7. Let $L$ be a finite lattice, $a \in L$, and let $L_y$ denote the image of $[0, a]$ under the mapping $z \to z \vee y$. Show:

(i) $\mu_L(0, 1) = \sum_{y \in a^{\perp}} \mu_L(0, y)\mu_{L_y}(y, 1)$;
(ii) if $L$ is semimodular and $(x, y)M$ then

$$|\mu_L(0, 1)| \leq |\mu_L(0, a)| \sum_{\substack{y \in a^{\perp} \\ (y, a)M}} |\mu_L(0, y)|$$

with equality if and only if $a$ is modular.
(Greene)

8. Prove **4.38**.

9. Let $R$ be a binary relation, $R \subseteq S \times T$. Define $\sigma: \mathcal{B}(S) \to \mathcal{B}(T)$, $\tau: \mathcal{B}(T) \to \mathcal{B}(S)$ by

$$\sigma A := \begin{cases} \{z \in T: (x, z) \in R \text{ for all } x \in A\} & \text{if } A \neq \varnothing \\ T & \text{if } A = \varnothing; \end{cases}$$

similarly $\tau$. Prove that $(\sigma, \tau)$ is a Galois connection.

→10. Deduce from ex. 9 the following proposition: Let $Q$ be a finite lattice and $S$ and $T$ the set of supremum-irreducible elements $\neq 0$ and infimum-irreducible elements $\neq 1$, respectively. We define for $z \in Q$

$$s(z) := |\{p \in S: p \leq z\}|$$

$$t(z) := |\{q \in T: q \geq z\}|.$$

Verify the polynomial identity:

$$\sum_{A \subseteq S} (x-1)^{|A|} y^{t(\sup A)} = \sum_{B \subseteq T} x^{s(\inf B)} (y-1)^{|B|} = \sum_{u, v \in Q} x^{s(u)} \mu_Q(u, v) y^{t(v)}.$$

(Hint: Define $R \subseteq S \times T$ by $(a, b)R \Leftrightarrow a \leq b$ and apply **4.40**.)

11. Let $L$ be a finite lattice and $M \subseteq L$ a lower cross-cut. Define $M_u$ and $r_k$ as in **4.41**. Show that

$$\sum_{k \geq 0} r_k = \sum_{u \in L} 2^{|M_u|} \mu_L(u, 1) = \sum_{A \in Q} 2^{|A|} \mu_Q(A, 1),$$

where as usual $Q$ is the quotient of $\mathcal{B}(M)$ under the mapping $\sigma: A \to \sup A$.

→12. Generalize **4.45** as follows. Let $L$ be a finite lattice and $R, S \subseteq L$. Let $r_k$ be the number of $k$-sets $A$ in $R$ with $A \cap S = \varnothing$, $\inf A = 0$, $\sup A = 1$. Prove:

$$\sum_{k \geq 0} (-1)^k r_k = \sum_{u, v \in L} \zeta(R \cap [u, v], S) \mu(0, u) \zeta(u, v) \mu(v, 1) - \mu(0, 1).$$

(Crapo)

13. Deduce from the formula in ex. 12 the complementation theorem **4.33**. (Hint: $R = L$, $S = a^{\perp}$.)

14.* Let the finite lattice $L$ have a cross-cut $M$ which is an antichain with $|M| = m$. Prove that

$$\binom{m-1}{2\left\lfloor \frac{m+1}{4} \right\rfloor} \leq \mu(L) \leq \binom{m-1}{2\left\lfloor \frac{m-1}{4} \right\rfloor + 1}.$$

(Scheid)

15. Let $\alpha_1, \ldots, \alpha_n$ be arbitrary non-negative integers. Construct a finite lattice $L$ with a rank function such that $\chi(L; x) = \prod_{i=1}^{n} (x - \alpha_i)$.

# 4. Valuations

In this section only a few new results are presented. Instead, the emphasis is on interpreting the concepts of Möbius function and Möbius inversion from a new algebraic point of view. Roughly speaking, most of the known formulae for the Möbius function will be rederived as identities between coefficients of two linear expressions. The main advantage of this approach is a better understanding of the algebraic basis for the sieve formulae of section 2.B and their generalizations to arbitrary distributive lattices.

## A. The Möbius Algebra

Let $P$ be a locally finite poset with 0 and $K$ a field of characteristic 0 (or more generally an integral domain containing $\mathbb{Q}$). We regard $P$ as a basis of a vector space $V(P)$ over $K$, defining $V(P)$ as the set of all finite linear combinations of elements of $P$ with coefficients in $K$ and coordinatewise addition and scalar multiplication. To distinguish better between $P$ and the basis of $V(P)$ let us denote by $\varepsilon_p$ the basis element of $V(P)$ corresponding to $p \in P$. Hence

$$V(P) = \left\{ \sum_{p \in P} a_p \varepsilon_p : a_p \in K \text{ and } a_p = 0 \text{ for all but finitely many } p \in P \right\}.$$

$V(P)$ is called the *free vector space* over $K$ generated by $P$. It will sometimes be convenient to regard an element $\sum_{p \in P} a_p \varepsilon_p \in V(P)$ as a vector $(\ldots, a_p, \ldots) \in K^P$. For example, we can write

$$\varepsilon_p = (\ldots, a_q, \ldots) \quad \text{with } a_q = \begin{cases} 1 & \text{if } q = p \\ 0 & \text{if } q \neq p. \end{cases}$$

For $p \in P$ we define $\iota_p \in V(P)$ by

$$\iota_p := \sum_{q \leq p} \varepsilon_q.$$

In vector form,

$$\iota_p = (\ldots, b_q, \ldots) \quad \text{with } b_q = \begin{cases} 1 & \text{if } q \leq p \\ 0 & \text{if } q \not\leq p. \end{cases}$$

**4.50 Proposition.** *Let $\mu$ be the Möbius function of $P$. Then*

$$\varepsilon_p = \sum_{q \in P} \mu(q, p) \iota_q.$$

*Furthermore, the $\iota_p$'s are linearly independent and thus form a basis of $V(P)$.*

*Proof.* The expression for $\varepsilon_p$ follows from Möbius inversion of $\iota_p$ where we regard $\mu(q, p)$ as an element of $\mathbb{Z}$ and $V(P)$ as a $\mathbb{Z}$-module. It is easily verified that the inversion formulae of section 2 can be generalized to this situation. To prove the linear independence assume that $\sum_{i=1}^{t} a_i \iota_{q_i} = 0$ with $a_i \neq 0, i = 1, \ldots, t$. Suppose $q_m$ is maximal in $\{q_1, \ldots, q_t\}$. For suitable $a_i', b_r$ we have

$$\iota_{q_m} = \sum_{i=1, i\neq m}^{t} a_i' \iota_{q_i}$$

$$\varepsilon_{q_m} + \sum_{p\in P, p<q_m} \varepsilon_p = \sum_{i=1, i\neq m}^{t} \sum_{r\leq q_i} b_r \varepsilon_r.$$

But this implies $q_m \leq q_i$ for some $i$, contradicting the hypothesis. □

Suppose now $P$ is a finite lattice. For $p \in P$ we define $\kappa_p \in V(P)$ by

$$\kappa_p := \sum_{q\in P, q\vee p=1} \varepsilon_q.$$

**4.51 Proposition.** *Let $P$ be a finite lattice, $p \in P$. Then*

(i) $\kappa_p = \sum_{q\in P, q\geq p} \mu(q, 1)\iota_q$.

(ii) $\mu(p, 1)\iota_p = \sum_{q\in P, q\geq p} \mu(p, q)\kappa_q$.

*If, in particular, $\mu(p, 1) \neq 0$ for all $p \in P$, then $\{\kappa_p : p \in P\}$ is a basis of $V(P)$ with*

(iii) $\varepsilon_p = \sum_{t\in P} \nu(p, t)\kappa_t$ $(p \in P)$ *where* $\nu(p, t) = \sum_{q\in P, q\leq p\wedge t} \mu(q, p)\mu(q, t)/\mu(q, 1)$.

*Proof.* We have

$$\sum_{q\geq p} \mu(q, 1)\iota_q = \sum_{q\geq p} \mu(q, 1) \sum_{r\leq q} \varepsilon_r = \sum_{r\in P} \left( \sum_{q\geq p\vee r} \mu(q, 1) \right) \varepsilon_r = \kappa_p \quad (\text{by } \mathbf{4.6}).$$

(ii) is proved by Möbius inversion and (iii) by substituting **4.50** into (ii). □

From the fact that $\{\kappa_p : p \in P\}$ is a basis, two very interesting results on the level numbers of a geometric lattice can be deduced. First we note the following useful proposition.

**4.52 Proposition** (Dowling–Wilson). *Let $P$ be a finite lattice with $\mu(p, 1) \neq 0$ for all $p \in P$. There exists a permutation $\phi: P \to P$ such that*

$$p \vee \phi(p) = 1 \quad \textit{for all } p \in P.$$

*Proof.* Write $\kappa_p$ as a vector, i.e.,

$$\kappa_p = (\ldots, c_q, \ldots) \quad \text{with } c_q = \begin{cases} 1 & \text{if } p \vee q = 1 \\ 0 & \text{otherwise.} \end{cases}$$

Since $\{\kappa_p: p \in P\}$ is a basis of $V(P)$, the matrix $A = [a_{p,q}]$ (indexed by $P$), where

$$a_{p,q} = \begin{cases} 1 & \text{if } p \vee q = 1 \\ 0 & \text{otherwise,} \end{cases}$$

is nonsingular. Hence at least one term in the determinant expansion of $A$ must be nonzero, proving the theorem. □

**4.53 Theorem** (Dowling–Wilson). *Let $P$ be a finite geometric lattice of rank $n$ and $\{W_k: k = 0, \ldots, n\}$ the sequence of level numbers of $P$. For all $k$ between 0 and $n$ we have*

$$W_0 + W_1 + \cdots + W_k \leq W_n + W_{n-1} + \cdots + W_{n-k}.$$

*Proof.* The assumption $\mu(p, 1) \neq 0$ for all $p$ is satisfied according to **4.34**(iv) and **2.39**. Hence there is a permutation $\phi: P \to P$ with $p \vee \phi(p) = 1$ for all $p$. From the semimodular inequality we infer that for all $p \in P$

$$r(\phi(p)) \geq n + r(p \wedge \phi(p)) - r(p) \geq n - r(p)$$

and thus

$$\phi\{p \in P: r(p) \leq k\} \subseteq \{q \in P: r(q) \geq n - k\}$$

whence the theorem. □

**4.54 Corollary.** *Let $P$ be a finite geometric lattice of rank $n$. Then $W_1 \leq W_{n-1}$ and $W_1 = W_{n-1}$ if and only if $P$ is modular.*

*Proof.* Since the dual lattice of a modular geometric lattice is again geometric we have $W_1 = W_{n-1}$ by applying **4.53** twice. Now assume $W_1 = W_{n-1}$. We define $V_1 \subseteq V(P)$ as the subspace of $V(P)$ spanned by $\{\varepsilon_p: p \in P, r(p) \leq 1\}$. Obviously,

$$\dim V_1 = 1 + W_1.$$

Let the linear transformation $T: V(P) \to V_1$ be given by

$$T(\varepsilon_p) = \begin{cases} \varepsilon_p & \text{if } r(p) \leq 1 \\ 0 & \text{otherwise.} \end{cases}$$

This gives

$$T(\kappa_p) = \sum_{q \vee p = 1} T(\varepsilon_q) = \sum_{q \vee p = 1, r(q) \leq 1} \varepsilon_q \qquad (p \in P)$$

and thus

$$T(\kappa_p) = 0 \quad \text{for all } p \text{ with } r(p) < n - 1.$$

For $\varepsilon_p$ with $r(p) \leq 1$ we have by **4.51**(iii)

$$\varepsilon_p = T(\varepsilon_p) = \sum_{r(t) \geq n-1} v(p, t) T(\kappa_t).$$

It follows that the set $\{T(\kappa_t): r(t) \geq n - 1\}$ spans the subspace $V_1$ and thus forms a basis of $V_1$ because $W_1 = W_{n-1}$. To prove the modularity of $P$ we use the characterization **2.43**. Let $g$ be a line of $P$. Then

$$0 = T(\varepsilon_g) = \sum_{r(t) \geq n-1} v(g, t) T(\kappa_t)$$

hence $v(g, t) = 0$ for all copoints $t$. By **4.51**(iii) this gives

$$\sum_{q \leq g \wedge t} \frac{\mu(q, g)\mu(q, t)}{\mu(q, 1)} = 0 \quad \text{for all copoints } t.$$

Suppose $g \wedge t = 0$. In this case, the left-hand side contains only one summand, namely $\mu(0, g)\mu(0, t)/\mu(0, 1) \neq 0$, which is a contradiction. We conclude $g \wedge t > 0$ for all lines $g$ and copoints $t$, thus proving the theorem. □

**Definition** (Solomon). Let $P$ be a locally finite poset with 0 and $V(P)$ the free vector space (over some field $K$ of characteristic 0). The *Möbius algebra* $\text{Möb}_K(P)$ is the vector space $V(P)$ over $K$ together with the coordinatewise product:

$$\text{If} \quad \alpha = \sum_{p \in P} a_p \varepsilon_p, \ \beta = \sum_{p \in P} b_p \varepsilon_p \quad \text{then} \quad \alpha \cdot \beta = \sum_{p \in P} (a_p b_p) \varepsilon_p.$$

Möb($P$) could equivalently be defined by the system of orthogonal idempotents $\{\varepsilon_p: p \in P\}$ and linear extension to all of $V(P)$:

$$\varepsilon_p \cdot \varepsilon_q = \begin{cases} \varepsilon_p & \text{if } p = q \\ 0 & \text{if } p \neq q. \end{cases}$$

Consider the basis $\{\iota_p: p \in P\}$. **4.50** shows how the multiplication in Möb($P$) can be defined in terms of the $\iota_p$'s.

**4.55 Proposition.** *Let $P$ be a locally finite poset with* 0. *The multiplication in* Möb($K$) *is given by linear extension of the relations*

$$\iota_p \cdot \iota_q = \sum_{t \in P} \left( \sum_{u \leq p, u \leq q} \mu(t, u) \right) \iota_t.$$

*If, in particular, p and q have a unique lower bound $p \wedge q$ then*

$$\iota_p \cdot \iota_q = \iota_{p \wedge q}. \quad \square$$

In the sequel we shall use both definitions to compute products in Möb($K$). For $\wedge$-semilattices, i.e., for posets in which a unique infimum $p \wedge q$ exists for all $p$ and $q$, **4.55** is especially convenient and suggests the following definition.

**Definition.** Let $P$ be a locally finite $\wedge$-semilattice with 0. The vector space $V(P)$ together with the product $\wedge$ is called the *semigroup algebra* $A(P, \wedge)$. Hence

$$A(P, \wedge) = \left\{ \sum_{p \in P} a_p p : a_p = 0 \text{ for all but finitely many } p \right\}$$

and

$$\sum_{p \in P} a_p p \cdot \sum_{q \in P} b_q q := \sum_{p, q \in P} a_p b_q (p \wedge q).$$

The dual concept is the semigroup algebra $A(P, \vee)$ for locally finite $\vee$-semilattices with 1.

**4.56 Proposition.** *Let $P$ be a locally finite $\wedge$-semilattice with 0. Then*

$$A(P, \wedge) \cong \text{Möb}(P)$$

*by means of $\phi: p \to \iota_p$.*

*Proof.* $\phi$ maps the basis $\{p\}$ of $A(P, \wedge)$ onto the basis $\{\iota_p\}$ of Möb($P$) and preserves products by **4.55**. $\square$

To emphasize the connection between $A(P, \wedge)$ and Möb($P$) let us from now on identify $p$ with $\iota_p$ for all $p \in P$ (but retaining the notation $\varepsilon_p$ and $\kappa_p$). What is the importance of the Möbius algebra? Instead of order relations we are now considering *linear identities*. Our method consists of establishing identities $\sum a_p p = \sum b_p p$ in Möb($P$), thus enabling us to deduce the equation $a_p = b_p$ for all $p$. We give one example to illustrate this approach; a few others are contained in the exercises.

**Example.** Let $P$ be a finite lattice with copoint set $C$. We want to prove **4.43** again. Using the identification $\iota_p \equiv p$, for any copoint $c$ we have

$$1 - c = \sum_{p \not\leq c} \varepsilon_p,$$

and thus

$$\prod_{c \in C} (1 - c) = \varepsilon_1$$

since $\varepsilon_1$ is the only $\varepsilon_p$ which appears in all factors $1 - c$, $c \in C$. If we express $\prod (1 - c)$ and $\varepsilon_1$ in terms of the basis $\{p \in P\}$ and compare the coefficients of $0 \in P$, we obtain

$$\sum_{k \geq 0} (-1)^k r_k = \mu(0, 1),$$

where $r_k$ is the number of $k$-sets $A \subseteq C$ with $\inf A = 0$.

The most important result in this section is the following factorization theorem.

**4.57 Theorem** (Greene). *Let $P$ be a finite poset with 1 and $a$ an element of $P$ for which all suprema $a \vee p$, $p \in P$, exist. The following identity holds in* Möb($P$):

$$\sum_{p \in P} \mu(p, 1)p = \sum_{q \in P, q \geq a} \mu(q, 1)q \cdot \sum_{r \in P, r \vee a = 1} \mu(r, 1)r.$$

*Proof.* The left-hand side is just $\varepsilon_1$ by **4.50**. The factorization on the right is made plausible by the following argument. Let $C \subseteq P$ be the set of copoints, $C_1 := \{c \in C : c \geq a\}$, $C_2 := C - C_1$. As in the example above, we have

$$\varepsilon_1 = \prod_{c \in C} (1 - c) = \prod_{c' \in C_1} (1 - c') \cdot \prod_{c'' \in C_2} (1 - c'').$$

Now clearly,

$$\prod_{c' \in C_1} (1 - c') = \sum_{p \not\leq c' (c' \in C_1)} \varepsilon_p = \sum_{p \vee a = 1} \varepsilon_p = \kappa_a.$$

By **4.51**(i), $\kappa_a$ is also equal to the first factor on the right-hand side of our equation. The second factor, $\sum_{r \vee a = 1} \mu(r, 1)r$, is in general different from $\prod_{c'' \in C} (1 - c'')$ but the coefficients of all $r$ with $r \vee a < 1$ cancel out when multiplied with $\kappa_a$. Indeed, we have

$$\sum_{r \vee a = 1} \mu(r, 1)r = \sum_{r \vee a = 1} \mu(r, 1) \sum_{q \leq r} \varepsilon_q = \sum_{q \in P} \left( \sum_{r \geq q, r \vee a = 1} \mu(r, 1) \right) \varepsilon_q,$$

thus obtaining for the right-hand side of the asserted equation

$$\sum_{p \vee a = 1} \varepsilon_p \cdot \sum_{q \in P} \left( \sum_{r \geq q, r \vee a = 1} \mu(r, 1) \right) \varepsilon_q = \sum_{u \vee a = 1} \left( \sum_{r \geq u} \mu(r, 1) \right) \varepsilon_u = \varepsilon_1. \quad \square$$

**4.57** permits very short proofs of several decomposition theorems, for instance, of the formula **4.48** concerning the characteristic polynomial as shown below; a few others are contained in the exercises.

**Example.** Let $P$ be a finite lattice with rank function $r$, $a \in P$. Let us first note the formula dual to **4.57** in the algebra $A(P, \vee)$:

$$(+) \qquad \sum_{p \in P} \mu(0, p)p = \sum_{q \leq a} \mu(0, q)q \cdot \sum_{s \wedge a = 0} \mu(0, s)s.$$

Set $P_1 := [0, a]$, $P_2 := \{s \in P : s \wedge a = 0\}$. If $a$ is modular in $P$, then clearly

$$r(q \vee s) = r(q) + r(s) \quad \text{for all } q \in P_1, s \in P_2.$$

Define the polynomials $\bar{p}$, $\bar{q}$ and $\bar{s} \in K[x]$ by

$$\bar{p} = x^{r(1) - r(p)} \qquad (p \in P)$$

$$\bar{q} = x^{r(a) - r(q)} \qquad (q \in P_1)$$

$$\bar{s} = x^{r(1) - r(a) - r(s)} \qquad (s \in P_2).$$

Since for all $q \in P_1$, $s \in P_2$

$$\bar{q} \cdot \bar{s} = x^{r(1) - r(q) - r(s)} = x^{r(1) - r(q \vee s)} = \overline{q \vee s} = \overline{q \cdot s},$$

we conclude that $(+)$ is preserved by the substitution $p \to \bar{p}$, $q \to \bar{q}$, $s \to \bar{s}$, thus yielding **4.48**.

## B. The Valuation Ring

Content of this section is a generalization of the inclusion-exclusion principle from Boolean algebras to arbitrary distributive lattices. To this end we introduce the notion of a valuation and describe all valuations of a distributive lattice in the setting of a new algebraic structure called the valuation ring. The main theorem links the valuation ring to the Möbius algebra described in the preceding section.

**Definition.** Let $P$ be a lattice and $G$ an abelian group. A mapping $w: P \to G$ is called a *valuation* of $P$ in $G$ if for all $a, b \in P$:

$$w(a \wedge b) + w(a \vee b) = w(a) + w(b).$$

Let $V(P)$ again denote the free vector space over some field $K$ of characteristic 0 (or, more generally, over some integral domain containing $\mathbb{Q}$). By $M(P)$ we denote the subspace of $V(P)$ generated by all elements in $V(P)$ of the form $a \wedge b + a \vee b - a - b$, where as in section $A$ we use the identification $\iota_p \equiv p$ for all $p \in P$.

**Definition.** The vector space $W(P) = V(P)/M(P)$ regarded as an abelian group is called the *valuation module* of $P$.

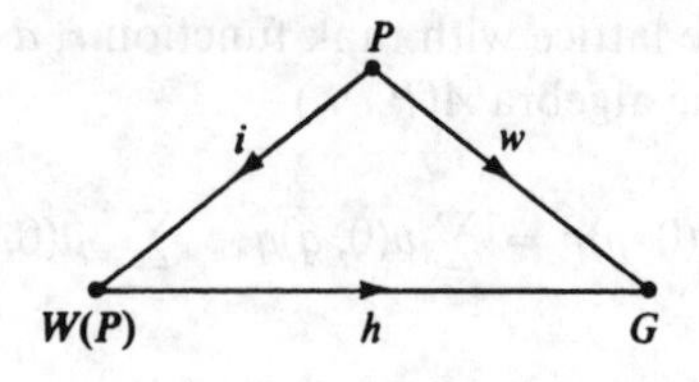


Figure 4.4

**4.58 Proposition.** *Let $P$ be a lattice and $G$ an abelian group. The canonical injection $i: P \to W(P)$ given by $i(p) = p + M(P)$ is a valuation of $P$ in $W(P)$. Conversely, for every valuation $w: P \to G$ there exists a unique homomorphism $h: W(P) \to G$ such that the diagram in Figure 4.4 commutes, and vice versa.*

*Proof.* The definition of $M(P)$ immediately implies that $i$ is a valuation. Let $w: P \to G$ be any valuation in $G$. We extend $w$ linearly to all of $V(P)$. This extension is identically 0 on $M(P)$ whence $h: W(P) \to G$ defined by $h(p + M(P)) := w(p)$ is the desired homomorphism. The converse is clear. □

Because of **4.58** we may identify the additive group of valuations in $G$ with $\mathrm{Hom}(W(P), G)$.

**Examples.** Every constant function $w: P \to G$, $w \equiv c \in G$, is a valuation. This implies that $p \notin M(P)$ for all $p \in P$ since otherwise the homomorphism $h_c$ corresponding to some $c \neq 0$ maps $p \in M(P)$ onto 0.

Let $P$ be a locally finite distributive lattice with 0, and $p$ an irreducible element. That the function $w_p: P \to \mathbb{Z}$ given by

$$w_p(a) := \begin{cases} 1 & \text{if } a \geq p \\ 0 & \text{if } a \ngeq p \end{cases}$$

is a valuation follows directly from **2.5**. By the isomorphism there we may assume w.l.o.g. that for $a \neq b \in P$ there exists an irreducible element $p$ with $p \leq a$, $p \nleq b$. This means for $w_p$ that $w_p(a) = 1$, $w_p(b) = 0$. It follows, in particular, that for locally finite distributive lattices $P$ with 0 the embedding $i: P \to W(P)$ is *injective*. The converse is easily seen. If $P$ is not distributive, then by **2.22**(i) there exist distinct elements $a, b, c \in P$ with $a \wedge c = b \wedge c$ and $a \vee c = b \vee c$ which implies $i(a) = i(b)$.

Let us introduce the multiplication $pq := p \wedge q$ as in section A.

**4.59 Proposition.** *Let $P$ be a locally finite distributive lattice with 0. Then $M(P)$ is an ideal of* Möb($P$) *(regarded as a ring).*

*Proof.* We have to show that for any triple $a, b, c \in P$ the expression $c \wedge (a \wedge b + a \vee b - a - b)$ lies in $M(P)$. By distributivity,

$$\begin{aligned} c \wedge (a \wedge b + a \vee b - a - b) \\ &= (c \wedge a \wedge b) + ((c \wedge a) \vee (c \wedge b)) - (c \wedge a) - (c \wedge b) \\ &= ((c \wedge a) \wedge (c \wedge b)) + ((c \wedge a) \vee (c \wedge b)) - (c \wedge a) - (c \wedge b) \in M(P). \end{aligned}$$

□

**Definition.** Let $P$ be a locally finite distributive lattice with 0. The factor ring $W(P) = \text{Möb}(P)/M(P)$ is called the *valuation ring* of $P$.

The following result generalizes **4.58** to the present situation and is proved by an identical argument.

**4.60 Proposition.** *Let $P$ be a locally finite distributive lattice with* 0 *and $R$ any commutative ring. The canonical injection $i: P \to W(P)$, $i(p) = p + M(P)$, is a valuation of $P$ in $W(P)$ with $i(p \wedge q) = i(p)i(q)$. Conversely, for any valuation $w: P \to R$ with $w(p \wedge q) = w(p)w(q)$ there exists a unique ring homomorphism $h: W(P) \to R$ such that $w = hi$, and vice versa.* □

Let us assume from now on that $w$ is a valuation with $w(p \wedge q) = w(p)w(q)$. According to **4.60** we may identify $p \in P$ with $i(p) \in W(P)$, and further the valuation $w$ with the homomorphism $h \in \text{Hom}(W(P), R)$ where $w = hi$. The following proposition is the generalization of **2.12** and the sieve formulae of section 2.B referred to above.

**4.61 Proposition** (Rota). *Let $P$ be a locally finite distributive lattice with* 0 *and $p_1, \ldots, p_t \in P$. We have in $W(P)$*

$$\bigvee_{i=1}^{t} p_i = \sum_{i=1}^{t} p_i - \sum_{i<j} (p_i \wedge p_j) \pm \cdots + (-1)^{t-1}(p_1 \wedge \cdots \wedge p_t).$$

*Equivalently, for $q \geq \bigvee_{i=1}^{t} p_i$ we have in $W(P)$*

$$\prod_{i=1}^{t} (q - p_i) = q - \bigvee_{i=1}^{t} p_i.$$

*Proof.* For $t = 1$ or $t = 2$ there is nothing to prove. The induction step is immediate from the distributive property. □

**4.62** (General sieve formula). *Let $P$ be a locally finite distributive lattice with* 0 *and $w$ a valuation. For any elements $p_1, \ldots, p_t \in P$*

$$w\left(\bigvee_{i=1}^{t} p_i\right) = \sum_{i=1}^{t} w(p_i) - \sum_{i<j} w(p_i \wedge p_j) \pm \cdots + (-1)^{t-1} w(p_1 \wedge \cdots \wedge p_t). \quad \square$$

**Example.** Consider the chain $\mathbb{N}_0$ and the identity valuation $w: \mathbb{N}_0 \to \mathbb{Z}$, $w(n) = n$. Then for $n_1, \ldots, n_t \in \mathbb{N}_0$

$$\max(n_1, \ldots, n_t) = \sum_{i=1}^{t} n_i - \sum_{i<j}^{t} \min(n_i, n_j) \pm \cdots + (-1)^{t-1} \min(n_1, \ldots, n_t).$$

The main theorem in this section gives a complete description of all valuations of a distributive lattice.

**4.63 Theorem** (Davis–Rota). *Let $L$ be a locally finite distributive lattice with 0 and $P(L)$ the poset of irreducible elements (including 0).*

(i) *$W(L) \cong$ Möb$(P(L))$ with the isomorphism $\phi$: Möb$(P(L)) \to W(L)$, $\phi(p) = p + M(L)$. In particular, every valuation is uniquely determined by its values on $P(L)$, and these values can be arbitrarily prescribed.*
(ii) *We have in $W(L)$, i.e., modulo $M(L)$:*
  (a) *$\{\varepsilon_p: p \in P(L)\}$ is a basis of orthogonal idempotents.*
  (b) $q = \sum_{p \in P(L),\, p \le q} \varepsilon_p \quad (q \in L)$.
  (c) $\varepsilon_p = \sum_{r \in P(L)} \mu_{P(L)}(r, p) r \quad (p \in P(L))$.
  (d) $q = \sum_{r,\, p \in P(L),\, p \le q} \mu_{P(L)}(r, p) r \quad (q \in L)$.

*Proof.* Let us first show that $P(L)$ mod $M(L)$ spans $W(L)$. In other words, for $q \in L$ we have to find $p_1, p_2, \ldots \in P(L)$ with $q = \sum a_i p_i$ mod $M(L)$. For $q = 0$ there is nothing to prove. Suppose the assertion is true for all $r < q$. Since for $q \in P(L)$ our claim is trivial we may assume there are $p_1, \ldots, p_t \in P(L)$ with

$$q = \bigvee_{i=1}^{t} p_i.$$

By **4.61**, this implies

$$q = \sum_i p_i - \sum_{i<j} (p_i \wedge p_j) \pm \cdots.$$

Every one of the summands on the right-hand side is $< q$, and the assertion follows by induction. Our next claim is that $P(L)$ mod $M(L)$ is a basis of $W(L)$. Let $p_1, \ldots, p_t \in P(L)$ and suppose $p_1$ is maximal among the $p_i$'s. If $w_{p_1}$ denotes the valuation of the example after **4.58**, then

$$w_{p_1}(p_i) = \begin{cases} 1 & \text{if } i = 1 \\ 0 & \text{if } i \ne 1. \end{cases}$$

We infer that $p_1$ is linearly independent of $\{p_2, \ldots, p_t\}$ in $W(L)$. Repeating this argument for the set $\{p_2, \ldots, p_t\}$ and so on, we conclude that $\{p_1, \ldots, p_t\}$ is linearly independent in $W(L)$. Before showing that the mapping $\phi$ preserves products we prove (ii). It suffices to verify (b). For $q = 0$ we have $0 = \varepsilon_0$ by definition. Suppose (b) holds for all $r < q$. If $q \in P(L)$ then there exists a unique element $q' \in L$ with

$q' <\cdot q$. Hence $q = q' + \varepsilon_{q'}$ and we are finished. If, on the other hand, $q \notin P(L)$ then there are elements $a < q, b < q$ with $q = a \vee b$. By induction, we have

$$a = \sum_{s \in P(L), s \leq a} \varepsilon_s, \qquad b = \sum_{t \in P(L), t \leq b} \varepsilon_t,$$

and hence

$$a \wedge b = ab = \sum_{u \in P(L), u \leq a \wedge b} \varepsilon_u.$$

Applying **2.5** this gives

$$q = a \vee b = a + b - a \wedge b = \sum_{p \in P(L), p \leq q} \varepsilon_p.$$

Finally, by **4.55**, we have for $p, q \in P(L)$

$$pq = \sum_{t \in P(L)} \left( \sum_{u \in P(L), u \leq p \wedge q} \mu_{P(L)}(t, u) \right) t,$$

and thus

$$\begin{aligned} \phi(pq) &= \sum_{t, u \in P(L), u \leq p \wedge q} \mu_{P(L)}(t, u) t \bmod M(L) \\ &= p \wedge q = pq \bmod M(L) \quad \text{(by (ii)(d))}. \quad \square \end{aligned}$$

**Examples. 4.63**(i) says that to define a valuation on $L$ we just have to give its values on $P(L)$ (or on some other basis of $W(L)$). We obtain the two fundamental examples by using the bases $\{\varepsilon_p\}$ and $\{p\}$ of Möb($P(L)$) respectively. Define the valuation $\rho: L \to \mathbb{Z}$ by

$$\rho(\varepsilon_p) := \begin{cases} 1 & \text{if } p \in P(L), p \neq 0, \\ 0 & p = 0. \end{cases}$$

By **4.63**(ii)(b), we have for $a \in L$

$$\rho(a) = |\{p \in P(L): p \neq 0, p \leq a\}|,$$

i.e., $\rho$ is just the rank function of $L$.

The valuation $\chi: L \to \mathbb{Z}$ defined by

$$\chi(p) := \begin{cases} 1 & \text{if } p \in P(L), p \neq 0, \\ 0 & p = 0 \end{cases}$$

is called the *characteristic* of $L$ and is the subject of the next section. Figure 4.5 shows a distributive lattice and its characteristic.

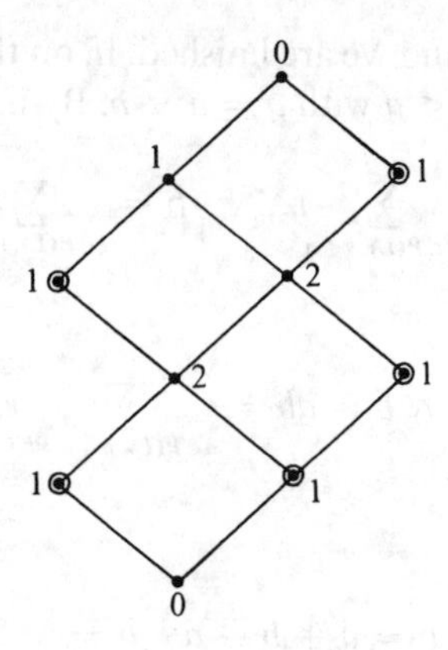


Figure 4.5

## C. The Characteristic

$L$ is again a locally finite distributive lattice with 0. We want to express the characteristic $\chi$ using the Möbius function of the poset $P(L)$ of irreducible elements.

**4.64 Proposition** (Rota). *Let $L$ be a locally finite distributive lattice with 0 and $P(L)$ the poset of irreducible elements of $L$ (including 0). The characteristic $\chi$ of $L$ is given by*

$$\chi(0) = 0$$
$$\chi(a) = -\sum_{p \in P(L),\, 0 \neq p \leq a} \mu_{P(L)}(0, p) \qquad (0 \neq a \in L).$$

*Proof.* By **4.63**(ii)(d) we have in $W(L)$

$$a = \sum_{r,\, p \in P(L),\, p \leq a} \mu_{P(L)}(r, p) r.$$

Applying the unique homomorphism $h \in \operatorname{Hom}(W(L), \mathbb{Z})$ with $\chi = hi$, we deduce that

$$\chi(a) = \sum_{p \in P(L),\, p \leq a} \; \sum_{r \in P(L),\, 0 \neq r \leq p} \mu_{P(L)}(r, p) = -\sum_{p \in P(L),\, 0 \neq p \leq a} \mu_{P(L)}(0, p). \quad \square$$

**4.65 Proposition.** *Let $L$ be a locally finite distributive lattice with 0 and $P(L)$ the poset of irreducible elements of $L$ including 0. For $a \in L$ let $J(a) = \{p \in P(L) : p \leq a\}$ and define the poset $J'(a) = J(a) \cup \{\hat{a}\}$ by setting $p < \hat{a}$ for all $p \in J(a)$. Then*

$$\chi(a) = \mu_{J'(a)}(0, \hat{a}) + 1 \qquad (a \in L).$$

*In particular, we have*

$$\chi(p') = \mu_{P(L)}(0, p) + 1 \qquad (0 \neq p \in P(L))$$

*where $p'$ is the unique element in $L$ with $p' <\cdot\, p$.*

*Proof.* By **4.6** and **4.64**

$$\mu_{J'(a)}(0, \hat{a}) = - \sum_{p \in P(L),\, p \leq a} \mu_{P(L)}(0, p) = \chi(a) - \mu_{P(L)}(0, 0) = \chi(a) - 1. \quad \square$$

We can now use our knowledge of the Möbius function to derive sieve-like expressions for the characteristic. Two examples are contained in the following result.

**4.66 Corollary.** *Let $L$ be a locally finite distributive lattice with 0 and $a \in L$.*

(i) *If $r_k$ denotes the number of totally ordered $k$-subsets $\{p_1, \ldots, p_k\} \subseteq P(L)$ with $0 < p_1 < p_2 < \cdots < p_k \leq a$, then*

$$\chi(a) = \sum_{k \geq 1} (-1)^{k-1} r_k.$$

(ii) *Let $P(a)$ be the set of maximal irreducible elements $\leq a$, and $s_k$ the number of $k$-sets $\{q_1, \ldots, q_k\} \subseteq P(a)$ with $\bigwedge_{i=1}^{k} q_i = 0$. Then*

$$\chi(a) = \sum_{k \geq 1} (-1)^k s_k + 1. \; \square$$

**Example.** Let us compute the characteristic of the divisor lattice $\mathscr{T}$. **4.64** yields for $n = p_1^{k_1} \ldots p_t^{k_t} \in \mathbb{N}$

$$\chi(n) = - \sum_{i=1}^{t} \sum_{s=1}^{k_i} \bar{\mu}(p_i^s) = - \sum_{i=1}^{t} (-1) = t.$$

Thus the characteristic is the number of distinct prime divisors. For Boolean algebras, one immediately sees that $\chi(A) = |A|$.

To conclude, let us derive a useful result on simplicial complexes. Recall that a simplical complex on a set $S$ is an ideal of the Boolean algebra $\mathscr{B}(S)$. To avoid computational difficulties one usually defines simplicial complexes as ideals of the lattice $\mathscr{B}(S) - \{\emptyset\}$.

**4.67 Proposition.** *The characteristic of a simplicial complex $\mathfrak{C} \in \mathfrak{J}(\mathscr{B}(n) - \{0\})$ is given by*

$$\chi(\mathfrak{C}) = \sum_{k \geq 1} (-1)^{k+1} r_k$$

*where $r_k$ is the number of $k$-sets in $\mathfrak{C}$.*

*Proof.* We have by **4.64**

$$\chi(\mathfrak{C}) = -\sum_{\varnothing \neq A \in \mathfrak{C}} \mu_{\mathscr{B}(n)}(\varnothing, A) = -\sum_{k \geq 1} \sum_{A \in \mathfrak{C}, |A| = k} (-1)^k. \quad \square$$

In combinatorial topology the sets $A \in \mathfrak{C}$ are called the *faces* of the complex $\mathfrak{C}$. The face $A$ is said to have dimension $k$ if $|A| = k + 1$. **4.67** thus reads: The characteristic of a simplicial complex equals the alternating sum $f_0 - f_1 + f_2 \mp \cdots$ where $f_k$ is the number of faces in $\mathfrak{C}$ of dimension $k$.

Similar formulae can be derived for complexes in partition lattices and in vector space lattices. The reader is referred to the exercises for more details.

EXERCISES IV.4

→ 1. Prove Weisner's theorem in ex. IV.3.1 by comparing coefficients in **4.51**.

2.* Strengthen **4.52**: Let $P$ be a finite lattice with $\mu(0, p)\mu(p, 1) \neq 0$ for all $p \in P$. Then there exists a bijection $\phi: P \to P$ such that $\phi(p)$ is a complement of $p$, for all $p \in P$. (Dowling)

3. Strengthen **2.43**: A geometric lattice $L$ is modular if and only if there is some $k$ such that for all $x, y \in L$ with $r(x) = k + 1$ and $r(y) = r(L) - k, x \wedge y > 0$.

4. Deduce from ex. 3: A finite geometric lattice of rank $n$ is modular if and only if $\sum_{i=0}^{k} W_i = \sum_{i=0}^{k} W_{n-i}$ for some $k$ between 1 and $n - 1$. (Dowling–Wilson)

→ 5. Prove by using the Möbius algebra: Let $L$ be a finite lattice, $x, y, z \in L$. Then

$$\sum_{t \wedge x = z} \mu(t, y) = \begin{cases} \mu(z, y) & \text{if } x \geq y \\ 0 & \text{otherwise.} \end{cases}$$

→ 6. Prove **4.27** with the aid of the algebra $A(P, \vee)$.

7. Prove the multiplicativity of $\mu$ by a suitable evaluation in **4.57**.

→ 8. Let $P$ be a finite lattice, and $a \in P$. For $y \in P$ denote by $P_y$ the image of $[0, a]$ under the map $x \to x \vee y$. Then

$$\sum_{x \in P} \mu(0, x)x = \sum_{y \wedge a = 0} \mu(0, y) \sum_{s \in P_y} \mu_{P_y}(y, s)s.$$

(Greene)

9. Use ex. 8 to prove the following analogue of **4.33**. Let $P$ be a finite lattice, $a \in L$. Then

$$\mu(0, 1) = \sum_{y \in a^{\perp}} \mu(0, y)\mu_{P_y}(y, 1).$$

10. Let $P$ be a finite semilattice with respect to $\wedge$, $\{a_1, \ldots, a_k\}$ and $\{b_1, \ldots, b_l\}$ subsets of $P$. Let $u \in P$ with $u \geq a_i, u \geq b_j$ for all $i, j$. Show that in Möb$(P)$ the following identity is valid:

$$\prod_i (u - a_i) + \prod_j (u - b_j) - \prod_{i,j} (u - a_i \wedge b_j) = \prod_i (u - b_i) \prod_j (u - b_j).$$

→11. Use ex. 10 to prove the following: Let $P_c(m, n)$ be the number of binary relations $R \subseteq \mathbb{N}_m \times \mathbb{N}_n$ with $|R| = c$ such that $R(a) \neq \emptyset$ *and* $R(b) \neq \emptyset$ for every $a \in \mathbb{N}_m$ and $b \in \mathbb{N}_n$. Show that

$$\sum_c (-1)^c P_c(m, n) = (-1)^{m+n}.$$

(Hint: Choose a semilattice and elements $a_i$, $b_j$ in which all expressions $a_{i_1} \wedge \cdots \wedge a_{i_k} \wedge b_{j_1} \wedge \cdots \wedge b_{j_l}$ are distinct.) (Klee)

→12.* Let $P$ be an indecomposable lattice of rank $n$ and $w: P \to R$ a valuation. Prove:

(i) $w$ is constant on all levels and $w(x) = w(0) + k(w(p) - w(0))$ for all $x, p \in P$ with $r(x) = k$, $r(p) = 1$;
(ii) if $P$ is not modular, then $w$ must be constant;
(iii) if $P$ is modular, then $w = ar + b$ for some $a, b \in R$, where $r$ is the rank function.

(Hint: Use ex. II.3.14 and induction on the rank.)

13. Using ex. 12, determine the structure of the valuation ring of a finite geometric lattice.

14. Compute the characteristic of the Young lattice.

→15. Prove **4.66** and use other identities involving $\mu$ to derive formulae for the characteristic. In particular, establish the results analogous to **4.67** for complexes in $\mathscr{L}(n, q)$ and $\mathscr{P}(n)$.

## Notes

The notion of an incidence algebra and general inversion on posets seem to have appeared first in Ward [1, 2] and Weisner [1]; see also Carlitz [4], Smith [1], Rota [1], and the bibliographies listed there. Rota's influential paper [1] led to a renewed activity and a thorough investigation of the Möbius function. The main theorems of sections 2 and 3 are contained in Rota [1] and Crapo [2, 5]. For a comprehensive account of the inclusion-exclusion principle and its applications, the reader is referred to Comtet [1, ch. 4], Ryser [1, ch. 2], and Riordan [1, ch. 2]. The concept of a Möbius algebra was introduced by Solomon [1] and studied further by Greene [4], Rota [4], Davis [2], Geissinger [1], and others.

Chapter V

# Generating Functions

In the course of our investigation of counting problems we have encountered many functions $c = c(k)$ depending on an integral parameter $k = 0, 1, 2, \ldots$. Most of the time the parameter had something to do with the type classification of the underlying incidence algebra. Our goal is now to find the solution $c(0)$, $c(1)$, $c(2), \ldots$ in a closed form instead of having to evaluate each term $c(k)$ individually. To accomplish this we regard $c(k) = c_k$ as *coefficient* of a formal power series and develop methods to compute this series, called the *generating function* for the coefficients $c_k$.

Let us illustrate this idea with two examples. Let $S$ be an $n$-set. We want to count the $k$-subsets of $S$, $k = 0, 1, 2, \ldots$, hence $c_k = \binom{n}{k}$. We know by the binomial theorem

$$\sum_{k=0}^{n} \binom{n}{k} t^k = (1 + t)^n.$$

Accordingly, we say $(1 + t)^n$ is the generating function for the binomial coefficients $\binom{n}{k}$, for a fixed $n$.

Consider next the number $c_n$ of solutions of $x^2 = id$ in the symmetric group $S_n$. We know from **3.15**(i) that

$$c_n = \sum_{i+2j} \frac{n!}{i!\,2^j j!}.$$

If we take $c_n$ as a coefficient of an *exponential series* we obtain

$$\sum_{n \geq 0} \left( \sum_{i+2j=n} \frac{n!}{i!\,2^j j!} \right) \frac{t^n}{n!} = \sum_{i,j \geq 0} \frac{t^i}{i!} \frac{(t^2/2)^j}{j!} = \exp\left(t + \frac{t^2}{2}\right).$$

Hence we say $\exp(t + \frac{1}{2}t^2)$ is the generating function for the number of solutions of $x^2 = id$ in $S_n$.

These two examples raise the following questions:

1. How does one compute the generating function of a given sequence of coefficients?

2. Which type of generating function is appropriate for the given sequence? (Apart from the types in the examples above, called ordinary and exponential power series, we shall encounter some others later.)

We consider an enumeration problem as solved in the sense of generating functions if this function can be computed from the given data, but it should be kept in mind that the evaluation of some or all coefficients of the generating function may still present considerable difficulties. Historically, generating functions and their uses have always played a prominent role in the analysis of combinatorial counting procedures as witnessed by the book of MacMahon [1].

The first two sections of this chapter form a bridge between the notion of an incidence algebra in the last chapter and the notion of generating functions. We shall see that computing certain incidence functions is essentially equivalent to solving certain enumeration problems concerning ordered objects (section 1) and unordered objects (section 2). The last two sections present an account of the enumeration of $G$-patterns and $G, H$-patterns, culminating in the theory of Polya–deBruijn.

# 1. Ordered Structures

In this section, we study the enumeration of ordered things, for instance, ordered number-partitions, ordered set-partitions, etc. Recall that in **4.8** these numbers were seen to be equal to $\sum_{k \geq 0} \eta^k$ evaluated over $\mathscr{C}(n)$ and $\mathscr{B}(n)$, respectively. Hence we want to first obtain some deeper knowledge of the corresponding incidence algebras.

## A. Reduced Algebras

The definition of the standard algebra $\mathbb{S}(P)$ of a locally finite poset $P$ can be rephrased as follows. We consider the partition of the interval set Int($P$) of $P$ into its isomorphism classes. $\mathbb{S}(P)$ is then the incidence algebra on the $\cong$-classes (regarded as new elements) with the corresponding convolution. Let us slightly generalize this concept.

**Definition.** Let $P$ be a locally finite poset and Int($P$) be the set of non-empty intervals of $P$. An equivalence relation $\approx$ on Int($P$) is said to be *compatible* with the convolution $*$ if for any $f, g \in \mathbb{A}(P)$ the convolution $f * g$ is constant on the $\approx$-classes whenever $f$ and $g$ are. The classes of a compatible relation $\approx$ are called the *types* of $\approx$.

The isomorphism relation $\cong$ is, of course, compatible for any poset $P$. This is precisely the content of **4.13**. Another important example is the relation $\approx$ in the divisor lattice $\mathscr{T}$ discussed in **2.61**. To see this recall that $\approx$ is finer than isomorphism. Suppose $f, g \in \mathbb{A}(\mathscr{T})$ are constant on the $\approx$-classes, i.e.,

$$f(k, l) = f(m, n), \qquad g(k, l) = g(m, n) \quad \text{whenever} \quad \frac{l}{k} = \frac{n}{m}.$$

The mapping $i \to im/k$ is an isomorphism from $[k, l]$ onto $[m, n]$ whence we have

$$(f * g)(k, l) = \sum_{k|i|l} f(k, i)g(i, l) = \sum_{k|i|l} f\left(m, \frac{im}{k}\right) g\left(\frac{im}{k}, n\right)$$

$$= \sum_{m|j|n} f(m, j)g(j, n) = (f * g)(m, n).$$

**5.1 Proposition.** *Let $\approx$ be a compatible relation on the locally finite poset $P$. Then:*

(i) *Intervals of the same type have the same cardinality.*
(ii) *Intervals of the same type possess the same number of chains of length $k$, for all $k$, and also the same number of maximal chains of length $k$, for all $k$.*

*Proof.* The Zeta-function $\zeta$ is trivially constant ($= 1$) on the classes of any equivalence relation $\approx$. Hence if $\approx$ is compatible, $\zeta^2$ must be constant, proving (i) by **4.7**(i). This implies that $\delta$ is constant on $\approx$-classes being 1 on all one-element intervals and 0 elsewhere, and thus that $\eta = \zeta - \delta$ is constant, too. Since by **4.8**(i) $\eta^k$ counts the number of chains of length $k$, the first part of (ii) follows. Part two of (ii) is proved similarly by **4.8**(ii) and the observation that $\kappa$ and hence $\kappa^k$ are constant on $\approx$-classes. □

The strong conditions spelled out in **5.1** suggest that two intervals of the same type are in fact isomorphic, or in other words, that the isomorphism is always the coarsest compatible relation. An example to show that this is not true in general is contained in the exercises.

**Definition.** Let $\approx$ be a compatible equivalence relation with respect to the incidence algebra $\mathbb{A}(P)$ over $K$, and $T = \{\alpha, \beta, \gamma, \ldots\}$ the types of $\approx$. We define $\mathbb{F}(P, \approx)$ as the set of functions $f : T \to K$ from the types in $T$ into $K$. Let $\alpha, \beta, \gamma \in T$ and $[x, y] \in \alpha$. The numbers

$$\binom{\alpha}{\beta\ \gamma} := |\{z \in [x, y] : [x, z] \in \beta \text{ and } [z, y] \in \gamma\}|$$

are called the *incidence coefficients* of $\mathbb{F}(P, \approx)$.

We have to show that $\binom{\alpha}{\beta\ \gamma}$ is independent of the choice of $[x, y] \in \alpha$. Suppose $[u, v]$ is another interval of type $\alpha$. For each $\omega \in T$ we define $f_\omega \in \mathbb{A}(P)$ by

$$f_\omega(x, y) := \begin{cases} 1 & \text{if } [x, y] \in \omega \\ 0 & \text{otherwise.} \end{cases}$$

Then by the compatibility of $\approx$

$$\binom{\alpha}{\beta\ \gamma}_{[x,y]} = (f_\beta * f_\gamma)(x, y) = (f_\beta * f_\gamma)(u, v) = \binom{\alpha}{\beta\ \gamma}_{[u,v]}.$$

**5.2 Theorem** (Scheid–Smith). *Let $P$ be a locally finite poset and $\approx$ a compatible equivalence relation. The set $\mathbb{F}(P, \approx)$ together with the usual addition and scalar multiplication of functions and the product $h = f * g$ defined by*

$$h(\alpha) := \sum_{\beta, \gamma \in T} \binom{\alpha}{\beta \ \gamma} f(\beta) g(\gamma) \qquad (\alpha \in T)$$

*is an associative algebra with unity, called the* reduced algebra modulo $\approx$. *$\mathbb{F}(P, \approx)$ is isomorphic to a subalgebra of $\mathbb{A}(P)$ and is closed with respect to inverses. In particular, $\delta$, $\zeta$, $\eta$, $\kappa$, and $\mu$ are elements of any reduced algebra.*

*Proof.* We define $\phi: \mathbb{F}(P, \approx) \to \mathbb{A}(P)$ by $\phi f = \bar{f}$ with

$$\bar{f}(x, y) := f(\alpha) \quad \text{if } [x, y] \in \alpha.$$

$\phi$ is clearly injective and we have, for $h = f * g$ and $[x, y] \in \alpha$,

$$\bar{h}(x, y) = h(\alpha) = \sum_{\beta, \gamma \in T} \binom{\alpha}{\beta \ \gamma} f(\beta) g(\gamma) = \sum_{x \leq z \leq y} \bar{f}(x, z) \bar{g}(z, y) = (\bar{f} * \bar{g})(x, y).$$

$\mathbb{F}(P, \approx)$ possesses a unity by **5.1**. That $\mathbb{F}(P, \approx)$ is closed with respect to taking inverses follows from **4.11**, since all terms on the right-hand side there lie in $\mathbb{F}(P, \approx)$. □

**5.3 Corollary.** $\mathbb{F}(P, \approx)$ *is commutative if and only if* $\binom{\alpha}{\beta \ \gamma} = \binom{\alpha}{\gamma \ \beta}$ *for all incidence coefficients.* □

Let us add the obvious remark that the reduced algebra modulo the isomorphism relation is isomorphic to the standard algebra $\mathbb{S}(P)$ introduced in section IV.1, whence we may regard from now on the elements of $\mathbb{S}(P)$ as functions $f: T \to K$.

## B. The Fundamental Series

Following our program set out in the introduction let us now look at the fundamental lattices and the reduced algebras corresponding to the natural type classification described in section II.4. It is convenient to take the lattices $\mathbb{N}_0$, $\mathcal{B}(\infty)$, $\mathcal{T}$ and $\mathcal{L}(\infty, q)$ and regard chains $\mathcal{C}(n)$, Boolean algebras $\mathcal{B}(n)$, etc., as sublattices of them. The reader is once again referred to II.4 where all the necessary results are collected.

*Chains.* Consider the isomorphism relation on $\text{Int}(\mathcal{C}(\infty))$. The types $T = \{0, 1, 2, \ldots\}$ correspond bijectively to the natural numbers (**2.59**). Hence any element $f \in \mathbb{S}(\mathbb{N}_0)$ is uniquely determined by the sequence

$$a_0, a_1, a_2, \ldots \quad \text{with} \quad f(n) = a_n \text{ for all } n \in T.$$

The incidence coefficients trivially are

$$\binom{n}{k\ l} = \begin{cases} 1 & \text{if } n = k + l \\ 0 & \text{otherwise.} \end{cases}$$

Suppose $f$ and $g \in \mathbb{S}(\mathbb{N}_0)$ are given by the sequences $\{a_n = f(n): n = 0, 1, 2, \ldots\}$ and $\{b_n = g(n): n = 0, 1, 2, \ldots\}$ respectively. By **5.2**, the coefficients $c_n = (f * g)(n)$ of the product $f * g$ are determined by

$$c_n = \sum_{k=0}^{n} a_k b_{n-k} \qquad (n \in \mathbb{N}_0).$$

We notice that this is precisely the rule by which the coefficients of the product of two ordinary power series $\sum_{n \geq 0} a_n t^n$, $\sum_{n \geq 0} b_n t^n$ are computed. Hence we obtain the following result.

**5.4 Proposition.** *The standard algebra $\mathbb{S}(\mathbb{N}_0)$ is isomorphic to the algebra of ordinary power series $\sum_{n \geq 0} a_n t^n$ under the isomorphism $\phi$ where*

$$\phi f = \sum_{n \geq 0} f(n) t^n. \quad \square$$

**Examples.** The Zeta function $\zeta$ corresponds to the power series $\sum_{n \geq 0} t^n = (1 - t)^{-1}$; hence the Möbius function $\mu$ corresponds to $1 - t$, a result which, of course, also follows from **4.15**(i).

**5.5.** *The number $c_n$ of ordered number-partitions of $n \in \mathbb{N}_0$ is*

$$c_0 = 1$$

$$c_n = 2^{n-1} \qquad (n > 0).$$

*Proof.* Let $c \in \mathbb{S}(\mathbb{N}_0)$ be defined by $c(n) = c_n$ for $n \in T$. According to **4.8**(v) we have

$$c = (2\delta - \zeta)^{-1}$$

and thus

$$\sum_{n \geq 0} c_n t^n = \left(2 - \frac{1}{1-t}\right)^{-1} = \frac{1-t}{1-2t} = (1-t) \sum_{n \geq 0} (2t)^n = 1 + \sum_{n \geq 1} 2^{n-1} t^n. \quad \square$$

We proceed in analogous fashion for the other lattice types. The set of types is always associated bijectively with $\mathbb{N}_0$ whence any function $f$ is determined by the sequence $a_0, a_1, a_2, \ldots$ with $a_n = f(n)$. The incidence coefficients will then tell us which class of power series corresponds to the given lattice.

*Boolean Algebras.* Let $S$ be a countable set and $\mathscr{B}(S)$ the lattice of finite subsets of $S$. The types $T$ of $\operatorname{Int}(\mathscr{B}(S))$ correspond bijectively to $\mathbb{N}_0$ by

$$[A, B] \in (n) \Leftrightarrow |B - A| = n \qquad (A \subseteq B \subseteq S).$$

The incidence coefficients of $\mathbb{S}(\mathscr{B}(\infty))$ are clearly given by

$$\begin{pmatrix} n \\ k \;\; l \end{pmatrix} = \begin{cases} \dbinom{n}{k} & \text{if } n = k + l \\ 0 & \text{otherwise.} \end{cases}$$

Hence, if $f, g \in \mathbb{S}(\mathscr{B}(\infty))$ are determined by the sequences $\{a_n = f(n) : n = 0, 1, 2, \ldots\}$ and $\{b_n = g(n) : n = 0, 1, 2, \ldots\}$, then $c_n = (f * g)(n)$ is given by

$$c_n = \sum_{k=0}^{n} \binom{n}{k} a_k b_{n-k}.$$

Written differently, we have

$$\frac{c_n}{n!} = \sum_{k=0}^{n} \frac{a_k}{k!} \frac{b_{n-k}}{(n-k)!}$$

whence we obtain the following result.

**5.6 Proposition.** *The standard algebra $\mathbb{S}(\mathscr{B}(\infty))$ is isomorphic to the algebra of exponential power series $\sum_{n \ge 0} a_n(t^n/n!)$ under the isomorphism $\phi$ where*

$$\phi f = \sum_{n \ge 0} f(n) \frac{t^n}{n!}. \quad \square$$

**Examples.** We have $\phi\zeta = \sum_{n \ge 0} (t^n/n!) = e^t$, $\phi\mu = e^{-t}$. If $c_n$ counts the number of ordered set-partitions of an $n$-set, then the function $c \in \mathbb{S}(\mathscr{B}(\infty))$ defined by $c(n) = c_n$ is given by $c = (2\delta - \zeta)^{-1}$ (see **4.8**(v)). Hence we have

$$\sum_{n \ge 0} c_n \frac{t^n}{n!} = \frac{1}{2 - e^t}.$$

As a further simple example consider $\zeta^2$. We know that $\zeta^2$ counts the cardinality of intervals. This yields in our case

$$\sum_{n \ge 0} |\mathscr{B}(n)| \frac{t^n}{n!} = (e^t)^2 = e^{2t} = \sum_{n \ge 0} \frac{(2t)^n}{n!} = \sum_{n=0}^{\infty} 2^n \frac{t^n}{n!},$$

thus

$$|\mathscr{B}(n)| = 2^n.$$

*The Divisor Lattice.* The types $T$ of $\operatorname{Int}(\mathscr{T})$ correspond to $\mathbb{N}$ by the rule

$$[k, l] \in (n) \Leftrightarrow \frac{l}{k} = n \qquad (k \mid l \in \mathbb{N}).$$

The resulting reduced algebra $\mathbb{F}(\mathscr{T}, \approx)$ contains $\mathbb{S}(\mathscr{T})$ as the subalgebra of all those functions $f$ which do not distinguish between prime numbers, i.e., for which $f(m) = f(n)$ whenever $m = p_1^{k_1} \ldots p_t^{k_t}$, $n = q_1^{k_1} \ldots q_t^{k_t}$ with the same $t$ and $k_1, \ldots, k_t$.

The incidence coefficients of $\mathbb{F}(\mathscr{T}, \approx)$ clearly are

$$\binom{n}{k \; l} = \begin{cases} 1 & \text{if } n = kl \\ 0 & \text{otherwise.} \end{cases}$$

Hence if $f, g \in \mathbb{F}(\mathscr{T}, \approx)$ are determined by the sequences $\{a_n = f(n) : n = 0, 1, 2, \ldots\}$ and $\{b_n = g(n) : n = 0, 1, 2, \ldots\}$ then $c_n = (f * g)(n)$ is given by

$$c_n = \sum_{i \mid n} a_i b_{n/i}.$$

**5.7 Proposition.** *The reduced algebra* $\mathbb{F}(\mathscr{T}, \approx)$ *with* $[k, l] \approx [m, n]$ *if and only if* $l/k = n/m$ *is isomorphic to the algebra of Dirichlet series*[1] $\sum_{n \geq 1} a_n n^{-s}$ *under the isomorphism* $\phi$ *given by*

$$\phi f = \sum_{n \geq 1} f(n) n^{-s}. \quad \square$$

**Examples.** The Zeta-function $\zeta$ is mapped onto $\phi\zeta = \sum_{n \geq 1} n^{-s}$ which is the classical *Riemannian* $\zeta$*-function*, usually denoted by $\zeta(s)$. This function is the origin of the name Zeta-function in arbitrary incidence algebras. From $\phi\mu = \sum_{n \geq 1} \bar{\mu}(n) n^{-s}$ we obtain the well-known formula in number theory

$$\zeta(s)^{-1} = \sum_{n \geq 1} \bar{\mu}(n) n^{-s}.$$

Here and elsewhere convergence questions will be disregarded, i.e., power series will be thought of as sequences of their coefficients which are manipulated in certain ways.

The function $c$ analogous to that of the previous examples counts the number of *ordered factorizations* of $n$ into numbers $\geq 2$. By **4.8** (v) we have

$$\sum_{n \geq 1} c_n n^{-s} = \frac{1}{2 - \zeta(s)}.$$

[1] It is a convention to denote the variable of a Dirichlet series by $-s$.

The covering function $\kappa$ is mapped onto $\phi\kappa = \sum_{p \text{ prime}} p^{-s}$. Thus if $d_n$ counts the number of *ordered factorizations of n into primes* then

$$\sum_{n \geq 1} d_n n^{-s} = \frac{1}{1 - \sum_{p \text{ prime}} p^{-s}}.$$

*Vector Space Lattices.* Let $V(\infty, q)$ be a vector space over $GF(q)$ with countable basis and $\mathscr{L}(\infty, q)$ the lattice of all finite dimensional subspaces of $V(\infty, q)$. The types of $\operatorname{Int}(\mathscr{L}(\infty, q))$ with respect to isomorphism correspond bijectively to $\mathbb{N}_0$ by means of

$$[U, W] \in (n) \Leftrightarrow \dim W - \dim U = n \qquad (U \subseteq W \subseteq V).$$

Hence we have for the incidence coefficients

$$\binom{n}{k\ l} = \begin{cases} \binom{n}{k}_q & \text{if } n = k + l \\ 0 & \text{otherwise.} \end{cases}$$

If $f, g \in \mathbb{S}(\mathscr{L}(\infty, q))$ are determined by the sequences $\{a_n = f(n): n = 0, 1, 2, \ldots\}$, $\{b_n = g(n): n = 0, 1, 2, \ldots\}$ then $c_n = (f * g)(n)$ is given by

$$c_n = \sum_{k=0} \binom{n}{k}_q a_k b_{n-k}.$$

By **3.12**, this can be written as

$$\frac{c_n}{\prod_{i=1}^{n} (q^i - 1)} = \sum_{k=0}^{n} \frac{a_k}{\prod_{i=1}^{k} (q^i - 1)} \frac{b_{n-k}}{\prod_{i=1}^{n-k} (q^i - 1)}.$$

Hence we obtain the following result, where, as usual, the products in the denominator are multiplied by $-1$.

**5.8 Proposition.** *The standard algebra* $\mathbb{S}(\mathscr{L}(\infty, q))$ *is isomorphic to the algebra of Eulerian series* $\sum_{n \geq 0} a_n(t^n / \prod_{i=1}^{n} (1 - q^i))$ *under the isomorphism* $\phi$ *where*

$$\phi f = \sum_{n \geq 0} f(n) \frac{t^n}{\prod_{i=1}^{n} (1 - q^i)}. \qquad \square$$

**Examples.** The functions $\zeta$, $\mu$, and $\kappa$ are mapped onto

$$\phi\zeta = \sum_{n \geq 0} \frac{t^n}{\prod_{i=1}^{n} (1 - q^i)}, \qquad \phi\mu = \sum_{n \geq 0} (-1)^n q^{\binom{n}{2}} \frac{t^n}{\prod_{i=1}^{n} (1 - q^i)} \quad \text{and}$$

$$\phi\kappa = \frac{t}{1 - q},$$

whence we obtain the formula

$$\left(\sum_{n\geq 0}\frac{t^n}{\prod_{i=1}^n(1-q^i)}\right)^{-1}=\sum_{n\geq 0}(-1)^n q^{\binom{n}{2}}\frac{t^n}{\prod_{i=1}^n(1-q^i)}.$$

For the Eulerian series with coefficients $c_n$ and $d_n$ analogous to those in the previous examples we have

$$\sum_{n\geq 0}c_n\frac{t^n}{\prod_{i=1}^n(1-q^i)}=\frac{1}{2-\sum_{n\geq 0}(t^n/\prod_{i=1}^n(1-q^i))},$$

$$\sum_{n\geq 0}d_n\frac{t^n}{\prod_{i=1}^n(1-q^i)}=\frac{1-q}{1-q-t}.$$

A final example involves the Galois numbers $G_{n,q}$. Let

$$z(t,q)=\sum_{n\geq 0}\frac{t^n}{\prod_{i=1}^n(1-q^i)}.$$

Then

$$z(t,q)^2=\sum_{n\geq 0}\frac{G_{n,q}t^n}{\prod_{i=1}^n(1-q^i)}.$$

## C. Partition Lattices

Let $S$ be a countable set and $\mathscr{P}(S)$ the lattice of all partitions of $S$ into a finite number of blocks. We know from **2.67** that the types of $\mathrm{Int}(\mathscr{P}(S))$ induced by isomorphism correspond bijectively to the sequences $(k_1, k_2, \ldots)$ where

$$[\pi,\sigma]\in(k_1,k_2,\ldots)\Leftrightarrow[\pi,\sigma]\cong\prod_{i\geq 1}\mathscr{P}(i)^{k_i}\Leftrightarrow\operatorname{type}[\pi,\sigma]=1^{k_1}2^{k_2}\ldots,$$

with $b(\pi)=\sum_{i\geq 1}ik_i$, $b(\sigma)=\sum_{i\geq 1}k_i$.

We see that the types are now determined by *sequences* of natural numbers $\geq 1$ rather than $\mathbb{N}_0$ as in the previous examples. This suggests using the semigroup $\mathbb{M}(\mathscr{P}(\infty))$ of *multiplicative* (but not necessarily invertible) *functions* in $\mathbb{S}(\mathscr{P}(\infty))$ (see **4.14**). Each $f\in\mathbb{M}(\mathscr{P}(\infty))$ is uniquely determined by the sequence $a_1, a_2, a_3, \ldots$ with

$$f[\pi,\sigma]=a_1^{k_1}a_2^{k_2}\ldots:\Leftrightarrow[\pi,\sigma]\in(k_1,k_2,\ldots).$$

In particular, we have $f[\pi,\sigma]=a_n$ if $[\pi,\sigma]\cong\mathscr{P}(n)$ so that we may simply set $f(n):=a_n$.

**Example.** The functions $\delta$, $\zeta$, and $\mu$ are represented by the sequences $(1, 0, 0, \ldots)$, $(1, 1, 1, \ldots)$ and $(\ldots, (-1)^{n-1}(n-1)!, \ldots)$, respectively.

**5.9 Proposition.** *The semigroup $\mathbb{M}(\mathscr{P}(\infty))$ is antiisomorphic to the set $\sum_{n\geq 1} a_n(t^n/n!)$ of exponential series with constant term 0 where the product is the composition of power series. The antiisomorphism is given by $\phi$ where*

$$\phi f = \sum_{n\geq 1} f(n)\frac{t^n}{n!}.$$

*Thus we have*

$$\phi(f * g) = \sum_{n\geq 1} g(n)\frac{(\phi f)^n}{n!}.$$

*Proof.* Let $f$ and $g$ correspond to the sequences $a_n = f(n)$ and $b_n = g(n)$, respectively. The coefficient $c_n = (f * g)(n)$ is then

$$c_n = \sum_{\substack{(k_1,\ldots,k_n)\\ \Sigma ik_i = n}} P(k_1,\ldots,k_n)a_1^{k_1}\ldots a_n^{k_n} b_{\Sigma k_i}$$

where $P(k_1,\ldots,k_n)$ is the number of partitions in $\mathscr{P}(n)$ of type $1^{k_1}\ldots n^{k_n}$. Hence we have to show that for all $n$

$$\sum_{\substack{(k_1,\ldots,k_n)\\ \Sigma ik_i = n}} P(k_1,\ldots,k_n)a_1^{k_1}\ldots a_n^{k_n} b_{\Sigma k_i} = n!\sum_{j\geq 1}\frac{b_j}{j!}\cdot\left(\text{coeff. of } t^n \text{ in } \left(\sum_{i\geq 1} a_i\frac{t^i}{i!}\right)^j\right).$$

This last coefficient is given by

$$\sum_{\substack{(i_1,\ldots,i_j)\\ \Sigma i_l = n}} \frac{a_{i_1}}{i_1!}\cdots\frac{a_{i_j}}{i_j!} = \sum_{\substack{(k_1,\ldots,k_n)\\ \Sigma ik_i = n,\ \Sigma k_i = j}} \frac{j!}{k_1!\ldots k_n!}\frac{a_1^{k_1}\ldots a_n^{k_n}}{1!^{k_1}\ldots n!^{k_n}}$$

since there are $j!/k_1!\ldots k_n!$ ordered number-partitions of $n$ of type $1^{k_1}\ldots n^{k_n}$ and $\sum k_i = j$ (see ex. III.1.7). Altogether we obtain for the right-hand side

$$n!\sum_{j\geq 1}\frac{b_j}{j!}\sum_{\substack{(k_1,\ldots,k_n)\\ \Sigma ik_i = n,\ \Sigma k_i = j}} \frac{j!}{k_1!\ldots k_n!}\frac{a_1^{k_1}\ldots a_n^{k_n}}{1!^{k_1}\ldots n!^{k_n}}$$

$$= \sum_{\substack{(k_1,\ldots,k_n)\\ \Sigma ik_i = n}} \frac{n!}{k_1!\ldots k_n!}\frac{a_1^{k_1}\ldots a_n^{k_n}}{1!^{k_1}\ldots n!^{k_n}} b_{\Sigma k_i}$$

which is equal to the left hand side by **3.15**(ii). □

**Examples.** $\delta$ and $\zeta$ are mapped onto

$$\phi\delta = t \quad\text{and}\quad \phi\zeta = e^t - 1.$$

This implies that $\mu$ corresponds to

$$\phi\mu = \log(1 + t).$$

Looking at the series expansion $\log(1+t) = t - (t^2/2) + (t^3/3) \mp \cdots = \sum_{n\geq 1}(-1)^{n-1}(n-1)!(t^n/n!)$ we have another proof of our old formula $\mu(\mathscr{P}(n)) = (-1)^{n-1}(n-1)!$. $\zeta^2$ counts the Bell numbers whence

$$\sum_{n\geq 0} B_n \frac{t^n}{n!} = e^{e^t - 1} \qquad (B_0 := 1).$$

Let us interpret the Möbius inversion in the light of **5.9**. We call a function $g: \mathscr{P}(n) \to K$ *multiplicative* if there exists a sequence $a_1, a_2, \ldots$ such that

$$\begin{aligned} g(\pi) = a_1^{k_1} a_2^{k_2} \ldots a_n^{k_n} &\Leftrightarrow [0, \pi] \in (k_1, \ldots, k_n) \\ &\Leftrightarrow \text{type}(\pi) = 1^{k_1} \ldots n^{k_n}. \end{aligned}$$

The function $f: \mathscr{P}(n) \to K$ defined by

$$f(\sigma) = \sum_{\pi \leq \sigma} g(\pi)\mu(\pi, \sigma)$$

is also multiplicative and by extending the definition of $f$ and $g$ as in the proof of **4.18** we have in $\mathbb{M}(\mathscr{P}(n))$

$$f = g * \mu, \quad g = f * \zeta.$$

Therefore, if $f$ and $g$ are represented by the sequences $c_1, c_2, \ldots$ and $a_0 = 1, a_1, a_2, \ldots$, respectively, then

$$\sum_{n\geq 1} c_n \frac{t^n}{n!} = \log\left(1 + \sum_{n\geq 1} a_n \frac{t^n}{n!}\right)$$

$$\sum_{n\geq 0} a_n \frac{t^n}{n!} = \exp\left(\sum_{n\geq 1} c_n \frac{t^n}{n!}\right).$$

Thus Möbius inversion over the partition lattice corresponds to the logarithm applied to the algebra of exponential series. Let us illustrate this relationship with examples from graph theory.

We want to determine the number $c_n$ of simple *connected labeled graphs* on $n$ vertices. By a labeled graph we mean that the vertices are numbered 1 to $n$ regarding two labeled graphs as indistinguishable if there exists a graph isomorphism which preserves all labels. Graphs in the ordinary sense, i.e., unlabeled graphs, will be considered in the next section.

**Example.** The graphs of Figure 5.1 represent the same labeled graph. If in the right graph we exchange the numbers 1 and 2 we obtain two graphs which are the same unlabeled graph but are different as labeled graphs.

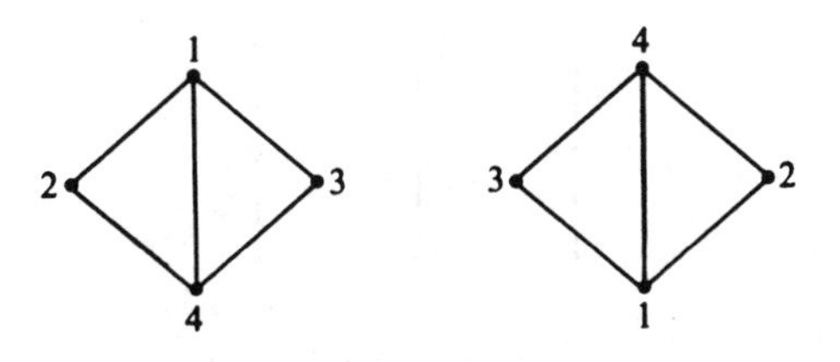


Figure 5.1

Back to our problem. The total number of simple labeled graphs on $n$ vertices is clearly $2^{\binom{n}{2}}$. Let $V$ be the vertex-set and $\pi \in \mathscr{P}(V)$. If $f(\pi)$ denotes the number of labeled graphs on $V$ whose connected components are precisely the blocks of $\pi$ then

$$\sum_{\pi \leq \sigma} f(\pi) = (2^{\binom{2}{2}})^{k_2}(2^{\binom{3}{2}})^{k_3} \dots (2^{\binom{n}{2}})^{k_n}, \quad \text{where type}(\sigma) = 1^{k_1} \dots n^{k_n},$$

and thus

$$c_n = f(1) = \sum_{\pi \in \mathscr{P}(V)} (2^{\binom{1}{2}})^{k_1} \dots (2^{\binom{n}{2}})^{k_n} \mu(\pi, 1), \quad \text{with type}(\pi) = 1^{k_1} \dots n^{k_n}.$$

Setting $a_n = 2^{\binom{n}{2}}$ in the expression above we obtain the following result.

**5.10 Proposition.** *Let $c_n$ be the number of simple connected labeled graphs on $n$ vertices. Then*

$$\sum_{n \geq 1} c_n \frac{t^n}{n!} = \log\left(1 + \sum_{n \geq 1} 2^{\binom{n}{2}} \frac{t^n}{n!}\right). \quad \square$$

The same method can be applied to subclasses of graphs such as colored graphs (see the exercises).

A classical problem concerns the enumeration of labeled *trees*. Let us write $b_n$ for the number of such trees on $n$ vertices. By analogy with **5.10** we can derive a formula connecting $b_n$ with the number of labeled *forests*, but a direct recursion is obtained by looking at rooted trees. We call a labeled tree $T$ *rooted at the vertex $i$* simply by distinguishing $i$ from the remaining vertices. Two rooted trees are considered equal if there exists a label-isomorphism which carries root onto root. For the number $u_n$ of rooted labeled trees on $n$ vertices we have trivially

$$u_n = nb_n.$$

**Example.** All labeled trees on 3 vertices are shown in Figure 5.2. Each one induces three rooted labeled trees by taking 1, 2, and 3 as root, in turn.

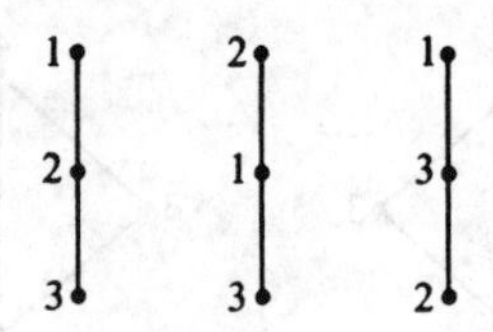


Figure 5.2

More generally, we speak of a *rooted forest* if each component is a rooted tree. Let $v_n$ be the number of labeled rooted forests on $n$ vertices. Now let $T$ be an arbitrary rooted tree on $n$ vertices. By removing the vertex labeled $n$ and distinguishing its neighbors as roots of this new graph we obtain a rooted forest $\phi T$ on $n - 1$ vertices (see Figure 5.3).

For any rooted forest $F$ on $n - 1$ vertices there are $n$ trees $T$ with $\phi T = F$ since we may choose any vertex of the resulting tree as root. Hence we have

$$v_{n-1} = \frac{u_n}{n} \qquad (n \geq 1),$$

and thus

$$\exp\left(\sum_{n\geq 1} u_n \frac{t^n}{n!}\right) = \sum_{n\geq 1} v_{n-1} \frac{t^{n-1}}{(n-1)!} = \frac{1}{t} \sum_{n\geq 1} u_n \frac{t^n}{n!}.$$

**5.11 Proposition** (Polya). *Let $u_n$ be the number of labeled rooted trees on n vertices. The generating function $u(t) = \sum_{n\geq 1} u_n(t^n/n!)$ satisfies the functional equation*

$$u(t) = t \exp(u(t))$$

*with the initial condition $u_1 = 1$.* □

Differentiating both sides of **5.11** we have

$$u'(t) = \frac{u(t)}{t} + u(t)u'(t)$$

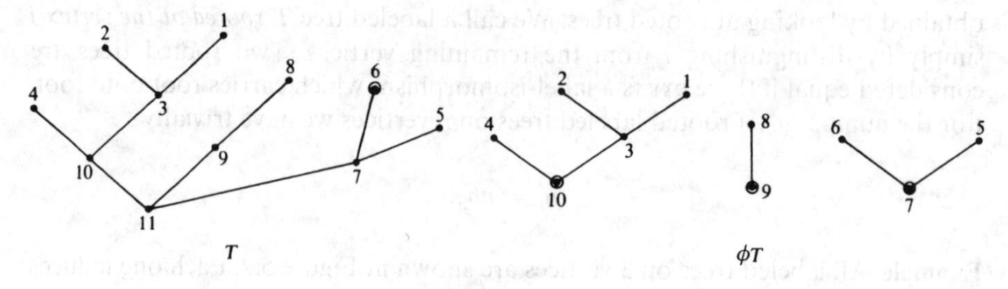


Figure 5.3

whence by comparing coefficients the recursion

$$u_n = \frac{n}{n-1} \sum_{k=1}^{n-1} \binom{n-1}{k} u_k u_{n-k} \qquad (n \in \mathbb{N})$$

results. From this one easily derives by induction $u_n = n^{n-1}$ and hence the following classical formula.

**5.12 Proposition** (Cayley). *The number of labeled trees on n vertices is*

$$b_n = n^{n-2}. \quad \square$$

As a final example consider the symmetric group $S_n$. For $g \in S_n$ of type $1^{b_1}2^{b_2} \ldots n^{b_n}$ (see section I.3) we set

$$t(g) := t_1^{b_1} t_2^{b_2} \ldots t_n^{b_n},$$

where $t_1, \ldots, t_n$ are $n$ variables. The polynomial

$$Z(S_n) = Z(S_n; t_1, \ldots, t_n) := \frac{1}{n!} \sum_{g \in S_n} t(g)$$

is called the *cycle indicator* of $S_n$. We shall take up this notion for arbitrary permutation groups in the next section. Our purpose now is to derive a formula for the generating function of $Z(S_n)$. Each permutation $g$ on $N = \{1, \ldots, n\}$ induces a partition $\phi g$ on $N$, namely its cycle decomposition. Let us set

$$f(\pi) := \sum_{\phi(g) = \pi} t(g).$$

In particular, this gives

$$f(1) = (n-1)!\, t_n,$$

since there are precisely $(n-1)!$ permutations in $S_n$ consisting of just one cycle of length $n$. For $\sigma \in \mathscr{P}(N)$ of type $1^{b_1} \ldots n^{b_n}$ we have

$$g(\sigma) = \sum_{\pi \leq \sigma} f(\pi) = (1!Z(S_1)^{b_1})(2!Z(S_2)^{b_2}) \ldots (n!Z(S_n)^{b_n})$$

and thus the following result.

**5.13 Proposition.** *Let $Z(S_n; t_1, \ldots, t_n)$ be the cycle indicator of $S_n$. Then*

$$\sum_{n \geq 0} Z(S_n; t_1, \ldots, t_n) u^n = \exp\left( \sum_{n \geq 1} t_n \frac{u^n}{n} \right). \quad \square$$

EXERCISES V.1

1. Let $P_<$ be a locally finite poset and $C(P)$ the set of compatible equivalence relations of Int($P$) in the sense of **5.1**. Prove that $C(P)$ with the refinement relation is a complete lattice in which suprema coincide with suprema in $\mathscr{P}(\mathrm{Int}(P))$.

→ 2. Let $L_1$ and $L_2$ be the lattices of flats of two non-isomorphic projective planes and identify the maximal element of $L_1$ with the minimal element of $L_2$. Call the resulting lattice $L$ and define $[x, y] \approx [u, v]$ in Int($L$) if and only if $[x, y] \cong [u, v]$ or $[x, y] \cong L_1$, $[u, v] \cong L_2$. Prove that $\approx$ is a compatible equivalence relation which is coarser than isomorphism.

→ 3. Let $P$ be a locally finite poset with 0 and rank function $r$. We call $[x, y] \in \mathrm{Int}(P)$ an *m,n-interval* if $r(x) = m$, $r(y) = n$. $P$ is said to satisfy the *strong* JD-*condition* if all $m,n$-intervals possess the same number of maximal chains. Show that $[x, y] \approx_r [u, v] :\Leftrightarrow [x, y]$ and $[u, v]$ are $m,n$-intervals for some $m, n \in \mathbb{N}_0$ is a compatible equivalence relation.

4. Let the locally finite poset $P$ satisfy the strong JD-condition and let $\mathbb{F}(P, \approx_r)$ be the reduced algebra with respect to $\approx_r$. Show that $\phi: \mathbb{F}(P, \approx_r) \to D_{n+1}(K)$ given by $\phi f = [f(i, j)/c(i, j)]$, $0 \le i, j \le n$, is an isomorphism onto the algebra of all upper triangular matrices of order $n + 1$, where $c(i, j) = \#\{\text{maximal chains in an } i, j\text{-interval}\}$ and $n = r(P)$.

→ 5. Converse of ex. 4: Let $P$ be a locally finite poset with 0 and let $\mathbb{F}(P, \approx)$ be a reduced algebra whose types $T$ are ordered pairs $(m, n)$, $m \le n \in \mathbb{N}_0$, such that $(m, n) \in T$ and $m \le m' \le n' \le n$ imply $(m', n') \in T$. Suppose there are integers $c_{i,j}$ such that the mapping $\phi$ in ex. 4 is an isomorphism. Prove:

   (i) $P$ satisfies the strong JD-condition.
   (ii) $c_{n,n} = 1$ if $(n, n) \in T$.
   (iii) If $c_{n,n+1} = 1$ for all $(n, n + 1) \in T$, then $c_{ij} = \#\{\text{maximal chains in any interval of type } (i, j)\}$.
   
   (Doubilet–Rota–Stanley)

6.* Let $L$ be a locally finite lattice with 0 satisfying the strong JD-condition and $T(n) = c_{n,n+2} - 1$ with $c_{i,j}$ as in ex. 5. Prove:
   (i) $T(n) \ne 0$ for all $(n, n + 2) \in T \Leftrightarrow L$ is a point lattice.
   (ii) If $L$ is a semimodular point lattice, then $T(n) \ne 0$ for all $(n, n + 2) \in T$.
   (iii) $L$ is semimodular $\Leftrightarrow c_{m,n}/c_{m+1,n} = 1 + T(m)T(m + 1) + \cdots + T(m)T(m + 1) \ldots T(n - 2)$ for all $(m, n) \in T$.
   
   (Doubilet–Rota–Stanley)

→ 7. Show that $\mathscr{C}(n)$, $\mathscr{B}(n)$, $\mathscr{L}(n, q)$, and $\mathscr{A}(n, q)$ all satisfy the strong JD-condition and compute $T(k)$.

8.* Determine all finite modular lattices which satisfy the strong JD-condition.

9. Let $f(n)$ be the number of pairs $(p, c)$ in $\mathscr{L}(n, q)$ with $0 \lessdot p \le c \lessdot 1$. Determine the Eulerian generating function for the $f(n)$'s.

10. Denote by $f_n^{(k)}$ and $c_n^{(k)}$ the number of $k$-colored labeled graphs and of $k$-colored connected labeled graphs on $n$ vertices, respectively. Show

$$\exp\left(\sum_{n\geq 1} c_n^{(k)} \frac{t^n}{n!}\right) = \sum_{n\geq 0} f_n^{(k)} \frac{t^n}{n!}.$$

11. Fill in the details in the derivation of Cayley's formula **5.12**.

→12. Let $a_{n,k} := \#\{g \in S_n : g^k = id\}$. Prove that

$$1 + \sum_{n\geq 1} a_{n,k} \frac{t^n}{n!} = \exp\left(\sum_{i|k} \frac{t^i}{i}\right).$$

(Hint: Use **5.13**.)

→13. For $g \in S_n$ set $t(g) = t_1^{b_1} \ldots t_n^{b_n}$ as usual and $z(g) = z^k t_1^{b_1} \ldots t_n^{b_n}$ with $k = \sum_{i=1}^n b_i$. Prove that

(i) $\sum_{n\geq 0} \sum_{g \in S_n} z(g)(u^n/n!) = \exp(z \sum_{n\geq 1} t_n(u^n/n)$.
(ii) $(n+1)! = \sum_{g \in S_n} 2^{b(g)}$, $b(g) = \#\{\text{cycles in } g\}$.

→14. Using ex. 13 prove that

$$|s_{n,k}| = \#\{g \in S_n : b(g) = k\},$$

where $s_{n,k}$ is the Stirling number of the first kind.

15. Find the generating function of the permutations in $S_n$ which have precisely $k$ cycles of which none is trivial.

## 2. Unordered Structures

The emphasis in this section is on explaining the rôle of the variable in the generating function. We shall see that in certain instances a simple compatibility condition will determine the type of generating function suitable for a given problem. This will prove particularly useful in connection with the enumeration of unordered objects. In this sense the present section is a counterpart to the previous one.

### A. Weighted Structures

Consider the generating function $\sum_{k=0}^n \binom{n}{k} t^n$ of the number of $k$-subsets of an $n$-set. We may interpret this function as collecting all subsets of the same cardinality $k$, assigning them the *weight* $t^k$ and then *summing* all these weights.

Take as another example factorizations of integers. If we assign to the factorization $n_1 n_2 \ldots n_t$ the weight $(\prod_{i=1}^t n_i)^{-s}$ then the generating function $\sum_{n\geq 1} F_n n^{-s}$ of factorizations of $n$ is again the sum of the weights of all factorizations.

Hence we are led to the following definition.

**Definition.** Let $S$ be a set (which is to be enumerated) and $W$ a field of characteristic 0. A mapping $w: S \to W$ is called a *weight function* of $S$ in $W$ and

$$\gamma(S) := \sum_{a \in S} w(a)$$

the *enumerator* of $S$ (with respect to $w$).

According to this definition we can say that the variable corresponds roughly to the weight and the generating function to the enumerator.

Next we notice that in our two examples the objects, i.e., subsets or factorizations, are composed of smallest subobjects. In the first case any set is composed of its one-element subsets, in the second case any factorization is composed of "smallest" subfactorizations consisting of a single integer. If $A = \{a_1, \ldots, a_k\}$ is a $k$-subset, then

$$w(A) = t^k = t \,.\, t \ldots t = w(a_1)w(a_2) \ldots w(a_k).$$

Similarly, we have for the factorization $n_1 n_2 \ldots n_t$,

$$w(n_1 n_2 \ldots n_t) = \left(\prod_{i=1}^{t} n_i\right)^{-s} = n_1^{-s} \ldots n_t^{-s} = w(n_1)w(n_2) \ldots w(n_t).$$

In both cases the weight function is compatible with the composition, suggesting the following definition.

**Definition** (Bender–Goldman). Let $S$ be a set, $\circ$ a binary operation on $S$, called *composition*, and $w: S \to W$ a *weight function*. The triple $(S, \circ, w)$ is called a *weighted structure*[(2)] (or to be more precise a weighted composition structure) if the following holds:

(i) $\circ$ is associative, commutative and possesses a two-sided neutral element $e$. An element $p \in S$, $p \neq e$, is called *prime* if $p = a \circ b$ implies $a = e$ or $b = e$.

(ii) Every $a \in S$ possesses a unique factorization into primes,

$$a = p_1^{k_1} \circ p_2^{k_2} \circ \cdots \circ p_t^{k_t}.$$

(iii) The weight $w$ is compatible with $\circ$, i.e., $w(a \circ b) = w(a)w(b)$ for all $a, b \in S$.

The expression $\gamma(S) := \sum_{a \in S} w(a)$ is called the *enumerator* of $S$. It is convenient to extend this definition to arbitrary subsets $A \subseteq S$, writing $\gamma(A) := \sum_{a \in A} w(a)$. For subsets $A, B \subseteq S$ we set $A \circ B := \{a \circ b : a \in A, b \in B\}$. Elements $a, b \in S$ are called *relatively prime* if they have no common prime factors.

The fact that $\circ$ is commutative makes it clear that we are dealing with unordered objects; for generalizations see Bender–Goldman [1] and the exercises.

(2) Bender and Goldman call this a *prefab*.

**5.14 Theorem** (Bender–Goldman). *Let $(S, \circ, w)$ be a weighted structure and $A, B \subseteq S$ such that any two elements $a \in A$ and $b \in B$ are relatively prime. Then*

$$\gamma(A \circ B) = \gamma(A)\gamma(B).$$

*Proof.* We first show that each $c \in A \circ B$ can be written *uniquely* as a product $c = a \circ b$ with $a \in A$ and $b \in B$. Suppose $a \circ b = a' \circ b'$ and let the prime factorizations be $a = \prod p_i^{k_i}$, $b = \prod q_j^{l_j}$, $a' = \prod r_i^{m_i}$ and $b' = \prod s_j^{n_j}$. Hence

$$\prod p_i^{k_i} \circ \prod q_j^{l_j} = \prod r_i^{m_i} \circ \prod s_j^{n_j}.$$

Since any two elements of $A$ and $B$ are relatively prime it follows from the uniqueness property that the $p_i^{k_i}$'s and $r_i^{m_i}$'s are the same up to order and similarly the $q_j^{l_j}$'s and $s_j^{n_j}$'s. Hence $a = a'$ and $b = b'$ and we conclude

$$\gamma(A \circ B) = \sum_{c \in A \circ B} w(c) = \sum_{a \in A} \sum_{b \in B} w(a \circ b) = \sum_{a \in A} w(a) \sum_{b \in B} w(b) = \gamma(A)\gamma(B). \quad \square$$

**5.15 Corollary.** *Let $(S, \circ, w)$ be a weighted structure and $P$ the set of primes. If we set $P^i = \{p^i : p \in P\}$, then*

(i) $\gamma(S) = \prod_{p \in P} (1 - w(p))^{-1}$

(ii) $\gamma(S) = \exp(\sum_{i \geq 1} (\gamma(P^i)/i))$.

*Proof.* Let $P_k = \{e, p_k, p_k^2, \ldots\}$ for any $p_k \in P$. Any two sets $P_k \neq P_l$ are trivially relatively prime whence

$$S = \prod_{\circ} P_k \quad \text{over all primes } p_k \in P.$$

Notice that the products on the right hand side contain only a finite number of factors $\neq e$. By **5.14**,

$$\gamma(S) = \prod_{\circ} \gamma(P_k) = \prod_{p \in P} (1 + w(p) + w(p^2) + \cdots) = \prod_{p \in P} \frac{1}{1 - w(p)}.$$

Taking logarithms of both sides we have

$$\log \gamma(S) = \log\left( \prod_{p \in P} \frac{1}{1 - w(p)} \right) = \sum_{p \in P} (-\log(1 - w(p)) = \sum_{p \in P} \left( w(p) + \frac{w(p^2)}{2} + \cdots \right)$$

and thus

$$\gamma(S) = \exp\left( \sum_{p \in P} \left( w(p) + \frac{w(p^2)}{2} + \cdots \right) \right) = \exp\left( \sum_{i \geq 1} \frac{\gamma(P^i)}{i} \right). \quad \square$$

**Examples.** Let us first derive for unordered partitions and factorizations the formulae analogous to those of section 1. Let $S$ be the set of unordered number-partitions $\{n_1 + \cdots + n_t : n_i \in \mathbb{N}\}$. We define the composition of two partitions $\alpha : n_1 + \cdots + n_r$ and $\beta : m_1 + \cdots + m_s$ by juxtaposition, i.e., $\alpha \circ \beta : n_1 + \cdots + n_r + m_1 + \cdots + m_s$. This composition is clearly associative and commutative with the empty partition as neutral element. The unique factorization property is satisfied with the primes being the partitions consisting of a single term $n$. The weight $w : S \to \mathbb{Q}(t)$ defined by

$$w(e) = 1$$

$$w(n_1 + \cdots + n_r) = t^{\sum_{i=1}^{r} n_i}$$

is clearly compatible with the composition whence we obtain the following result.

**5.16 Proposition.** *Let $P_n$ be the number of (unordered) partitions of $n \in \mathbb{N}_0$ where we set $P_0 = 1$. Then*

$$\sum_{n \geq 0} P_n t^n = \prod_{n \geq 1} \frac{1}{1 - t^n}. \quad \square$$

Now let $S$ be the set of factorizations $\{n_1 n_2 \ldots n_t : n_i \geq 2\}$ of integers into numbers $\geq 2$. Again we define the composition of two factorizations by adjoining the factors. The empty factorization is the neutral element and the primes are the factorizations consisting of a single term $n \geq 2$. To satisfy the compatibility condition we define the weight $w$ by

$$w(e) = 1$$

$$w(n_1 \ldots n_t) = \left(\prod_{i=1}^{t} n_i\right)^{-s},$$

obtaining the formula

$$\sum_{n \geq 1} F_n n^{-s} = \prod_{n \geq 2} \frac{1}{1 - n^{-s}},$$

where $F_n$ is the number of factorizations of $n$ into integers $\geq 2$, $F_1 = 1$.

It is clear that **5.15** and **5.16** remain valid if we restrict ourselves to subsets $T \subseteq S$ which are closed with respect to composition. Let us note two interesting examples.

Denote by $S^{[k]}$ the set of number-partitions $n_1 + \cdots + n_r$ with $n_j \leq k$ for all $j$. $S^{[k]}$ is closed and contains as primes the singleton partitions $1, 2, \ldots, k$. Thus we infer that

$$\sum_{n \geq 0} P_n^{[k]} t^n = \prod_{i=1}^{k} \frac{1}{1 - t^i},$$

where $P_n^{[k]}$ is the number of partitions of $n$ into summands $\leq k$.

In particular, this formula yields the generating function of partitions of $n$ whose greatest summand equals $k$. Now according to **3.18** this number is precisely $P_{n,k}$, the number of partitions of $n$ into $k$ summands. Hence we obtain

$$\sum_{n\geq 0} P_{n,k}t^n = \prod_{i=1}^{k} \frac{1}{1-t^i} - \prod_{i=1}^{k-1} \frac{1}{1-t^i} = t^k \prod_{i=1}^{k} \frac{1}{1-t^i}.$$

In our next example let $S$ again be the weighted structure of factorizations of integers and denote by $S_{\text{prime}}$ the subset of factorizations into *prime numbers.* $S_{\text{prime}}$ is obviously closed and contains as primes the factorizations consisting of a single prime number. Since by the Fundamental Theorem of Arithmetic each natural number $>1$ possesses a unique prime number factorization we have proved the classical result:

$$\zeta(s) = \sum_{n\geq 1} n^{-s} = \prod_{p \text{ prime}} \frac{1}{1-p^{-s}}.$$

Another variant of **5.15** is obtained by restricting the composition to the set $\bar{S} \subseteq S$ whose elements decompose into *distinct* primes. The proofs of **5.14** and **5.15** can be carried over to this situation except that we now use the set $\{e, p_k\}$ instead of $P_k = \{e, p_k, p_k^2, \ldots\}$.

**5.17 Proposition.** *Let $(S, \circ, w)$ be a weighted structure and $\bar{S}$ the subset of $S$ consisting of all elements which decompose into distinct primes. If $P$ denotes the set of primes, then*

(i) $\gamma(\bar{S}) = \prod_{p\in P} (1 + w(p))$

(ii) $\gamma(\bar{S}) = \exp(\sum_{i\geq 1} (-1)^{i+1}(\gamma(P^i)/i))$. $\square$

**Examples.** The generating functions of partitions and factorizations into unequal summands and unequal factors, respectively, are

$$\prod_{n\geq 1}(1+t^n) \quad \text{and} \quad \prod_{n\geq 2}(1+n^{-s}).$$

We have seen in **3.19** that the number $\bar{P}_n$ of self-conjugate partitions of $n$ equals the number of partitions containing only unequal odd terms. Hence their generating function is

$$\sum_{n\geq 0} \bar{P}_n t^n = \prod_{k\geq 0}(1+t^{2k+1}).$$

## B. Applications to Graphs

We consider ordinary finite graphs, i.e., unlabeled graphs as explained in section 1.C. Let $S$ be the set of non-isomorphic simple finite graphs, i.e., for each class of isomorphic graphs we choose a representative and put it in $S$. For $G, H \in S$ we define

$G \circ H$ to be the graph whose connected components are precisely the components of $G$ and $H$. This composition obviously satisfies all requirements with the empty graph as neutral element and the connected graphs as primes. The function $w: S \to \mathbb{Q}(t)$ defined by

$$w(G) := t^n, \qquad n = \#\{\text{vertices of } G\},$$

is a compatible weight, whence we obtain the following result.

**5.18 Proposition.** *Denote by $g_n$ and $c_n$ the number of simple unlabeled graphs and simple connected unlabeled graphs, respectively. Then*

(i) $\sum_{n \geq 0} g_n t^n = \prod_{n \geq 1} (1 - t^n)^{-c_n}$
(ii) $\sum_{n \geq 0} g_n t^n = \exp(\sum_{i \geq 1} (c(t^i)/i))$,

*where* $c(t) = \sum_{n \geq 1} c_n t^n$. $\square$

To enumerate trees we proceed as in **5.11**. A *rooted tree* is a tree in which one vertex has been distinguished as the root. Two rooted trees are considered equal if there exists a graph isomorphism between them which maps root onto root. Figure 5.4 depicts all rooted trees on 4 vertices. More generally, we speak of a *rooted forest* when each component tree is rooted. Let $u_n$, $v_n$ denote the number of rooted trees and rooted forests on $n$ vertices, respectively. If $T$ is a rooted tree with $n$ vertices then, by deleting the root and declaring its neighbors to be the new roots, we obtain a rooted forest $\phi T$ on $n - 1$ vertices (compare Figure 5.3). $\phi$ is clearly bijective, whence

$$v_{n-1} = u_n \qquad (n \geq 1).$$

The set consisting of all rooted forests is trivially closed with respect to graph composition and contains the rooted trees as primes. Hence we obtain the following companion formula to **5.11**.

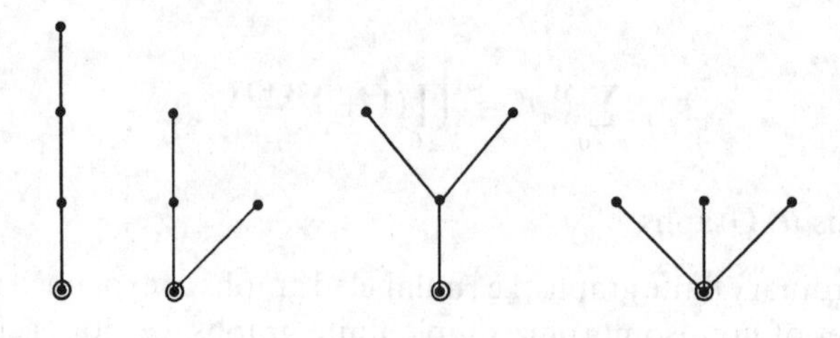

Figure 5.4

**5.19 Proposition** (Cayley–Polya). *Let* $u(t) = \sum_{n\geq 1} u_n t^n$ *be the generating function of unlabeled rooted trees. Then*

(i) $u(t) = t \prod_{t\geq 1} (1 - t^n)^{-u_n}$,
(ii) $u(t) = t \exp(\sum_{i\geq 1} (u(t^i)/i))$. □

The second formula can be thought of as the defining functional equation for $u(t)$.

Just as we distinguish a single vertex as the root, we may also distinguish a single edge as the root-edge of a tree $T$, then calling $T$ an *edge-rooted tree.* Two edge-rooted trees are considered equal if there exists a graph isomorphism between them which preserves the root-edge. Let $u^{(2)}(t) = \sum_{n\geq 1} u_n^{(2)} t^n$ be their generating function, where $n$ is the number of vertices. Let $T$ be a tree with $n$ vertices rooted at the edge $e = \{a, b\}$. We define $\phi^{(2)}(T)$ as that vertex-rooted forest which results after deletion of $e$ and distinguishing $a$ and $b$ as roots. Clearly, $\phi^{(2)}(T)$ has always two connected components (see Figure 5.5). $\phi^{(2)}$ is a bijection between the set of

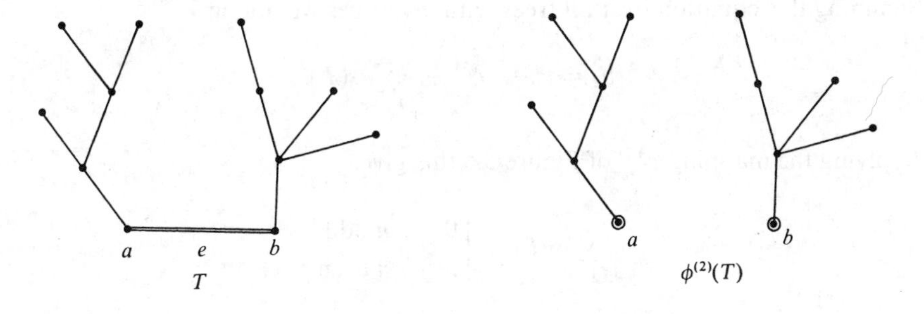


Figure 5.5

all edge-rooted trees on $n$ vertices and the set of all unordered pairs of rooted trees (including pairs with equal components) whose vertex-numbers add up to $n$. Hence we have (with $u_0 = 0$)

$$u_n^{(2)} = \begin{cases} \dfrac{1}{2}\sum_{i=0}^{n} u_i u_{n-i} & \text{for odd } n \\ \dfrac{1}{2}\sum_{i=0}^{n} u_i u_{n-i} + \frac{1}{2}u_{n/2} & \text{for even } n, \end{cases}$$

and thus

$$u^{(2)}(t) = \tfrac{1}{2}((u(t))^2 + u(t^2)).$$

Finally, let us consider the number $b_n$ of unlabeled trees on $n$ vertices and their generating function $b(t) = \sum_{n \geq 1} b_n t^n$. How many different rooted trees are generated by a given tree $T$? Two rooted trees $(T, a)$ and $(T, b)$ are apparently isomorphic as rooted trees precisely when there is an automorphism of $T$ carrying $a$ into $b$. Hence if $\mathscr{V}(T)$ is the permutation group on $V$ induced by $\operatorname{Aut}(T)$ and $v(T)$ the number of distinct rooted trees derived from $T$, then

$$v(T) = \#\{\mathscr{V}(T)\text{-orbits}\}.$$

The corresponding formula holds for the permutation group $\mathscr{E}(T)$ induced by $\operatorname{Aut}(T)$ on the edges and the number $e(T)$ of different edge rooted-trees:

$$e(T) = \#\{\mathscr{E}(T)\text{-orbits}\}.$$

Now we have the following result (see the exercises): Let $s(T)$ be the number of edges of $T$ which connect vertices of the same $\mathscr{V}(T)$-orbit. Then

$$1 = v(T) - e(T) + s(T).$$

Summing this equation over all trees with $n$ vertices we obtain

$$b_n = u_n - u_n^{(2)} + \sum_{|T|=n} s(T).$$

Applying the mapping $\phi^{(2)}$ of Figure 5.5 this gives

$$\sum_{|T|=n} s(T) = \begin{cases} 0 & n \text{ odd} \\ u_{n/2} & n \text{ even.} \end{cases}$$

Hence we have

$$\sum_{n \geq 1} \left( \sum_{|T|=n} s(T) \right) t^n = u(t^2),$$

and thus the following result.

**5.20 Proposition** (Otter). *Let $b(t) = \sum_{n \geq 1} b_n t^n$ and $u(t) = \sum_{n \geq 1} u_n t^n$ be the generating functions of unlabeled trees and unlabeled rooted trees, respectively. Then*

$$b(t) = u(t) - \tfrac{1}{2}((u(t))^2 - u(t^2)). \quad \square$$

EXERCISES V.2

1. Generalize the definition of a weighted structure as follows. Let $S$ and $w: S \to W$ be as before. The composition $\circ$ is allowed to be multivalued, i.e., we regard $a \circ b$ as a subset of $S$, satisfying the following axioms:

(i) If $c = \prod_i p_i^{k_i} \circ \prod_j q_j^{l_j}$ where the $p_i$'s and $q_j$'s are prime, then there exist unique elements $a \in \prod_i p_i^{k_i}$ and $b \in \prod_j q_j^{l_j}$ with $c \in a \circ b$.

(ii) There is a function $f: S \to W$ such that $f(c_1) = f(c_2)$ if $c_1, c_2 \in a \circ b$; hence $f(a \circ b)$ can be uniquely defined and we require

$$|a \circ b| = \frac{f(a \circ b)}{f(a)f(b)}.$$

(iii) $w(c) = w(a)w(b)$ whenever $c \in a \circ b$.

Let $\gamma(A) := \sum_{a \in A} w(a)/f(a)$ and show that **5.14** and **5.15** are valid. (Bender–Goldman)

→ 2. Solve the following problem by using ex. 1. We seek the number of words of length $n$ in $\{a, b\}$ in which either letter occurs an odd number of times. (Hint: Choose $S$ to be the set of words in $\{a, b\}$ and define $\alpha \circ \beta$ as the set of words which can be obtained from $\alpha$ and $\beta$ by taking the union of $\alpha$ and $\beta$ (as multisets) preserving the order of the letters within either word. Example: $aab \circ ab = \{aabab, aaabb, abaab\}$.)

→ 3. Prove **5.10** using ex. 1. (Hint: Let $S$ be the set of simple labeled graphs on $n$ vertices and define $G \circ H$ as follows: $V(G) = \mathbb{N}_m$, $V(H) = \mathbb{N}_n$. To every partition of $\mathbb{N}_{m+n}$ into an $m$-set $\{v_1, \ldots, v_m\}$ and an $n$-set $\{w_1, \ldots, w_n\}$ associate the graph which consists of all connected components of $G$ and $H$ and denote $i \in V(G)$ by $v_i$ and $j \in V(H)$ by $w_j$ for all $i, j$. $G \circ H$ shall consist of all graphs that are obtained in this way. Now prove $|G \circ H| = \binom{m+n}{n}$ and choose $f(G) := |V(G)|!$, $w(G) = x^{|V(G)|}$.)

→ 4. Let $(S, \circ, w, f)$ be a weighted structure in the sense of ex. 1 and suppose $|p^k| = f(p^k)/f(p)^k k!$ for all primes $p \in P$. Prove that

$$\gamma(S) = \exp(\gamma(P)).$$

5. Verify the formula for the generating function of the Bell numbers, **5.11**, and **5.13** by means of ex. 3 and 4.

6.* Try to find a concept analogous to the partitions of a set for subspaces of a vector space $V(n, q)$. In particular, try to derive a formula analogous to the example after **5.9**.

7. What is the generating function for number-partitions into odd summands?

→ 8. Show that there are equally many partitions of $n$ into odd summands as there are into distinct summands by applying ex. 7 and **5.17**.

9. Using **5.17**, prove the identity

$$\prod_{n \geq 0} (1 + zt^{2n+1}) = 1 + \sum_{n \geq 1} z^n \frac{t^{n^2}}{\prod_{i=1}^{n} (1 - t^{2i})}.$$

→10. Let $P_n$ be the number of partitions of $n$. Show that $P_n^* = P_n - P_{n-1}$ is the number of partitions of $n$ into summands greater than 1. Deduce from this that $P_{n+2} - 2P_{n+1} + P_n \geq 0$ for all $n$. When do we have equality?

11. Define $c_n$ and $g_n$ as in **5.18**. Show: If the numbers $d_n$ are given by

$$\log\left(\sum_{n\geq 0} g_n t^n\right) = \sum_{n\geq 1} d_n t^n$$

then

$$c_n = \sum_{k|n} \frac{\bar{\mu}(k)}{k} d_{n/k}.$$

Compute from this $c_1$ through $c_5$.

12. Let $u_n$ be the number of unlabeled rooted trees on $n$ vertices. Show that

$$u_{n+1} \leq \sum_{i=1}^{n} u_i u_{n-i+1}$$

and from this

$$u_n \leq \frac{1}{2}\binom{2n-2}{n-1}.$$

(Otter)

→13.* Verify the details in the proof of Otter's formula **5.20**, in particular, the equality $1 = v(T) - e(T) + s(T)$. (Hint: $s(T) = 0$ or 1; characterize the vertices which are joined by this one possible edge.)

14. Compute the first values of $b(t)$ in **5.20**.

15. A simple unlabeled graph $G(V, E)$ is called an *identity graph* if $G$ admits only the trivial automorphism. Let $a(t) = \sum_{n\geq 0} a_n t^n$ be the generating function of identity trees, $n = \#\{\text{vertices}\}$, and $A(t) = \sum_{n\geq 0} A_n t^n$ the generating function of rooted identity trees. Prove:

(i) $A(t) = t\exp(\sum_{n\geq 1}(-1)^{n+1}(A(t^n)/n))$.
(ii) $a(t) = A(t) - \frac{1}{2}((A(t))^2 + A(t^2))$.

## 3. $G$-patterns

After our detailed discussion of ordered and unordered structures, with corresponding permutation groups $G = E_n$ and $G = S_n$, respectively, we now study $G$-patterns with respect to arbitrary groups $G$. Let us repeat the main problem: Given Map($N$, $R$) and a permutation group $G$ on $N$, count the number of $G$-patterns

($=E(R)^G$-orbits of Map($N$, $R$)). More generally, we assign *weights* $w$ to the $G$-patterns $M$ and try to calculate the *enumerator* $\sum_M w(M)$.

The solution, provided by the results of deBruijn-Polya, combines in an elegant fashion the concepts of generating function, cycle decomposition and weighted structures, opening the way to a variety of interesting applications.

## A. The Problem

Throughout this section we make the following assumptions: $N$ and $R$ are sets, $N$ finite and $R$ at most countable with $|N| = n$ and $|R| = r$; $G$ is a permutation group on $N$ (this fact shall henceforth be denoted by $G \leq S(N)$); $\mathscr{U}(G)$ is the subgroup lattice of $G$ and $\mathscr{P}(N)$ the partition lattice of $N$.

$G$ induces an equivalence relation $\approx_G$ on Map($N$, $R$) by

$$f \approx_G f' :\Leftrightarrow \exists g \in G \text{ with } f' = fg.$$

The equivalence classes under $\approx_G$ are called *G-patterns* or just patterns when it is clear which group $G$ is meant. Hence the $G$-patterns correspond to the different sets $fG$ where $f \in$ Map($N$, $R$).

In many situations we do not need the full set Map($N$, $R$) but only a part of it. In order for $G$ to induce an equivalence relation on $\mathscr{F} \subseteq$ Map($N$, $R$) we have to make the following assumption:

**Definition.** A subset $\mathscr{F} \subseteq$ Map($N$, $R$) is called *closed* with respect to $G \leq S(N)$ if

$$f \in \mathscr{F}, g \in G \Rightarrow fg \in \mathscr{F}.$$

Hence $\mathscr{F}$ contains with any $f \in$ Map($N$, $R$) the whole pattern $fG$. Let $\mathfrak{M}$ denote the set of all patterns and $\mathfrak{M}_{\mathscr{F}}$ the patterns in $\mathscr{F}$.

We now introduce a weight function on Map($N$, $R$). To every element $j \in R$ we assign a variable $x_j$ (over $\mathbb{Q}$), the *weight* of $j$, and take as weight domain the ring of all rational functions in the variables $x_j, j \in R$. The weight of a mapping $f: N \to R$ is defined by

$$w(f) := \prod_{i \in N} x_{f(i)} \qquad (f \in \text{Map}(N, R)).$$

If $f' = fg$, then

$$w(f') = \prod_{i \in N} x_{f'(i)} = \prod_{i \in N} x_{fg(i)} = \prod_{i \in N} x_{f(i)} = w(f),$$

since $g(i)$ runs through all of $N$ as $i$ does. Hence all mappings of a $G$-pattern have the same weight and we may therefore define the weight of a pattern $M$ by

$$w(M) := w(f) \quad \text{for } f \in M.$$

Now we are ready to state the main problem.

**Main Problem.** *Given sets $N$ and $R$, the permutation group $G \leq S(N)$ and the closed set $\mathscr{F} \subseteq \mathrm{Map}(N, R)$. Determine the enumerator*

$$\gamma(\mathscr{F}; G) := \sum_{M \in \mathfrak{M}_{\mathscr{F}}} w(M).$$

Evaluation of $\gamma(\mathscr{F}; G)$ at $x_j = 1$ for all $j \in R$ clearly yields the number $|\mathfrak{M}_{\mathscr{F}}|$ of different $G$-patterns in $\mathscr{F}$. By substituting other values for the $x_j$'s we shall be able to solve a variety of other counting problems.

## B. The Main Theorem

We continue the notation of section A. Let the mappings $\phi\colon \mathscr{U}(G) \to \mathscr{P}(N)$, $\psi\colon \mathscr{P}(N) \to \mathscr{U}(G)$ be defined as follows:

$\phi H :=$ partition of $N$ into $H$-orbits,
$\psi\pi :=$ subgroup of all permutations in $G$ which leave the blocks of $\pi$ invariant.

**5.21 Proposition.** *$\phi$ is a Galois function from $\mathscr{U}(G)$ to $\mathscr{P}(N)$ with $\phi^+ = \psi$.*

*Proof.* Let $A \leq B \in \mathscr{U}(G)$. Any $A$-orbit is contained in some $B$-orbit, hence $\phi A \leq \phi B \in \mathscr{P}(N)$. On the other hand, if $\pi \leq \sigma \in \mathscr{P}(N)$, then any permutation which leaves the blocks of $\pi$ invariant leaves those of $\sigma$ invariant too, since the blocks of $\sigma$ are disjoint unions of $\pi$-blocks. Thus $\psi\pi \leq \psi\sigma \in \mathscr{U}(G)$. The conditions $\psi\phi H \geq H$ and $\phi\psi\pi \leq \pi$ are similarly verified. □

Let $\mathscr{P}(N, G)$ denote the lattice of coclosed partitions in $\mathscr{P}(N)$ under the coclosure $\phi\psi$. We know from **4.36**(ii) that $\pi \in \mathscr{P}(N, G)$ if and only if $\pi = \phi H$ for some $H \in \mathscr{U}(G)$.

**Example.** If $H = \langle g \rangle$ is a cyclic subgroup of $G$ then $\phi(\langle g \rangle)$ is just the cycle partition of $N$ induced by $g$. Hence $\psi\phi(\langle g \rangle)$ is the subgroup of all permutations in $G$ which leave the cycles of $g$ invariant. We write $\phi(g) := \phi(\langle g \rangle)$ for short.

**5.22 Lemma.** *Let $f \in \mathrm{Map}(N, R)$. Then*

(i) $\psi(\ker f) = \{g \in G : fg = f\}$,
(ii) $\ker f \geq \phi(g) \Leftrightarrow fg = f$.

*Proof.* A permutation $g \in G$ is in $\psi(\ker f)$ if and only if $f$ is constant on the cycles of $g$ which is the case if and only if $fg = f$ or, equivalently, $\ker f \geq \phi(g)$. □

**5.23 Proposition.** *Let $f \in \mathrm{Map}(N, R)$. The cardinality of the pattern $fG$ is given by*

$$|fG| = \frac{|G|}{|\psi(\ker f)|}.$$

*Proof.* We have

$$fg_1 = fg_2 \Leftrightarrow fg_1g_2^{-1} = f \Leftrightarrow g_1g_2^{-1} \in \psi(\ker f) \quad \text{by } \mathbf{5.22}$$
$$\Leftrightarrow \psi(\ker f)g_1 = \psi(\ker f)g_2.$$

Hence different elements in $fG$ correspond bijectively to different cosets of $\psi(\ker f) \leq G$. □

**5.23** is analogous to proposition **1.6** and was proved in the same way. Yet the situation is not quite the same since in the present case $G$ only *acts* on Map$(N, R)$ with $E(R)^G$ being the concrete permutation group.

**Definition.** Let $\mathscr{P}(N, G)$ be the lattice of coclosed partitions. The *Euler function* $\varphi: \mathscr{P}(N, G) \to \mathbb{N}_0$ is defined by

$$\varphi(\pi) := |\{g \in G: \phi(g) = \pi\}|.$$

Hence $\varphi(\pi)$ counts the number of permutations whose cycle decomposition is precisely $\pi$.

**Example.** Let $N = \{1, \ldots, n\}$ and $G = \langle g \rangle$ be cyclic where $g = (1, 2, \ldots, n)$ is a cycle of length $n$. We proved in section I.2 that

$$\mathscr{U}(\langle g \rangle) \cong [1, n] \subseteq \mathscr{T}.$$

$G$ contains a unique subgroup of order $k$, generated by $g^{n/k}$, for every $k|n$. Clearly, the permutations $g^i$ which leave all cycles of $g^{n/k}$ invariant are just the members of the subgroup $\langle g^{n/k} \rangle$. Thus every subgroup of $G$ is closed, whence we deduce from **5.21** that

$$\mathscr{P}(N, G) \cong [1, n] \in \mathscr{T}.$$

Let $\pi = \phi(g^{n/k}) \in \mathscr{P}(N, G)$. $\pi$ consists of $n/k$ blocks each of cardinality $k$. The permutations in $G$ which possess precisely these blocks as cycles are the generators of the subgroup $\langle g^{n/k} \rangle$, i.e., all permutations of the form $g^{in/k}$ with $i$ relatively prime to $k$. Recalling example 1 after **4.24** this means

$$\varphi(\pi) = \bar{\varphi}(k)$$

where $\bar{\varphi}$ is the Euler function in number theory.

We come to the main theorem.

**5.24 Theorem** (Polya–de Bruijn). *Let $N$ be a finite set and $R$ at most countable. Let $G$ be a permutation group on $N$ with $\mathscr{P}(N, G)$ the lattice of coclosed partitions in $\mathscr{P}(N)$ and $\varphi$ the Euler function on $\mathscr{P}(N, G)$. Let $\mathscr{F} \subseteq$ Map$(N, R)$ be closed with* *respect to G and* $K(\pi) = \sum_{f \in \mathscr{F}, \ker f \geq \pi} w(f)$ *for* $\pi \in \mathscr{P}(N)$. *The enumerator* $\gamma(\mathscr{F}; G)$ *is then given by*

(i) $\gamma(\mathscr{F}; G) = \frac{1}{|G|} \sum_{g \in G} \left( \sum_{f \in \mathscr{F}, fg = f} w(f) \right)$

(ii) $\gamma(\mathscr{F}; G) = \frac{1}{|G|} \sum_{\pi \in \mathscr{P}(N, G)} \varphi(\pi) K(\pi)$.

*Proof.* By definition $\gamma(\mathscr{F}; G) = \sum_{M \in \mathfrak{M}_{\mathscr{F}}} w(M)$. Hence

$$\begin{aligned}\gamma(\mathscr{F}; G) &= \sum_{f \in \mathscr{F}} \frac{w(f)}{|fG|} = \frac{1}{|G|} \sum_{f \in \mathscr{F}} w(f) |\psi(\ker f)| \quad (\text{by } \mathbf{5.23}) \\ &= \frac{1}{|G|} \sum_{f \in \mathscr{F}} w(f) |\{g \in G: fg = f\}| \quad (\text{by } \mathbf{5.22}(\text{i})) \\ &= \frac{1}{|G|} \sum_{g \in G} \left( \sum_{f \in \mathscr{F}, fg = f} w(f) \right).\end{aligned}$$

To prove (ii) we have

$$\begin{aligned}\frac{1}{|G|} \sum_{f \in \mathscr{F}} w(f) |\psi(\ker f)| &= \frac{1}{|G|} \sum_{f \in \mathscr{F}} w(f) |\{g \in G: \ker f \geq \phi(g)\}| \quad (\text{by } \mathbf{5.22}(\text{ii})) \\ &= \frac{1}{|G|} \sum_{\pi \in \mathscr{P}(N, G)} |\{g \in G: \phi(g) = \pi\}| \sum_{f \in \mathscr{F}, \ker f \geq \pi} w(f) \\ &= \frac{1}{|G|} \sum_{\pi \in \mathscr{P}(N, G)} \varphi(\pi) K(\pi). \quad \square\end{aligned}$$

By setting $x_j = 1$ for all $j \in R$, or in other words $w(f) = 1$ for all $f \in \mathscr{F}$, we obtain the number of different patterns in $\mathscr{F}$.

**5.25 Corollary.** *Let N, R, G, and* $\mathscr{F}$ *be as in* **5.24**. *The number of different G-patterns in* $\mathscr{F}$ *is*

$$|\mathfrak{M}_{\mathscr{F}}| = \frac{1}{|G|} \sum_{g \in G} |\{f \in \mathscr{F}: fg = f\}|. \quad \square$$

What is the significance of **5.24**? Consider part (ii). The expression $K(\pi)$ is totally independent of the group $G$ and $\varphi(\pi)$ is only $\neq 0$ when $\pi$ is the cycle partition of some $g \in G$, and hence is in $\mathscr{P}(N, G)$. In summary: The coclosed partitions in $\mathscr{P}(N)$ which are in addition cycle partitions of elements of $G$ already determine the enumerator $\gamma(\mathscr{F}; G)$. Hence the only data of the group $G$ that we need are the *cycle types* of the permutations; this suggests the use of the cycle indicator of a permutation group already hinted at in **5.13**. Indeed, in the case $\mathscr{F} = \text{Map}(N, R)$ this indicator will enable us to present the theorem of Polya–deBruijn very concisely.

**Example.** Consider a roulette wheel with $n$ boxes. We color each box with one of $r$ colors and seek the number of differently colored wheels. Two colored wheels are considered different when it is not possible to transform one wheel into the other by some rotation. Reformulated in terms of **5.24** this reads as follows. Let $N$ be the set of $n$ positions $1, 2, \ldots, n$ on the wheel and $R$ the set of colors. Each coloring corresponds bijectively to a map $f: N \to R$. Let $G = \langle g \rangle$ be the group generated by $g = (1, 2, \ldots, n)$. Then the colored wheels correspond precisely to the different $G$-patterns of Map($N$, $R$). Applying **5.24**(ii) and the result of our previous example we obtain for the number $h(n, r)$ of colored wheels

$$h(n, r) = \frac{1}{n} \sum_{k|n} \bar{\varphi}(k)[K(\pi)]_{x_j=1}$$

where $\pi$ consists of $n/k$ blocks each of cardinality $k$. For such $\pi$ we clearly have

$$[K(\pi)]_{x_j=1} = |\{f \in \text{Map}(N, R): \ker f \geq \pi\}| = r^{n/k},$$

whence we conclude

$$h(n, r) = \frac{1}{n} \sum_{k|n} \bar{\varphi}(k) r^{n/k}.$$

For $n = 5$, $r = 2$ this gives $h(5, 2) = 8$ wheels depicted in Figure 5.6.

**Example.** Let $N$ and $R$ be finite sets with $|N| = n$, $|R| = r$ and $n = n_1 + \cdots + n_r$ with $n_j \geq 0$. The set $\mathfrak{M}_{n_1, \ldots, n_r} := \{f \in \text{Map}(N, R): |f^{-1}(j)| = n_j \text{ for all } j \in R\}$ is

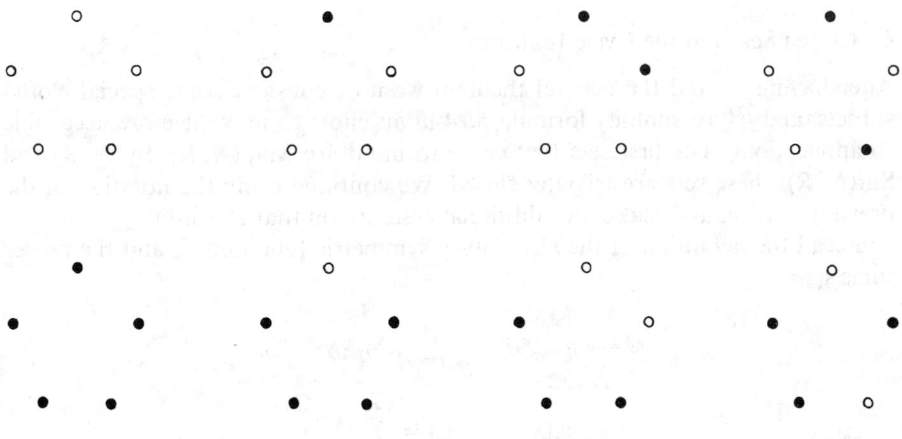

Figure 5.6

closed with respect to any group $G \leq S(N)$, since it is precisely the set of all functions with weight $x_1^{n_1} \dots x_r^{n_r}$. Hence the number of $G$-patterns in $\mathfrak{M}_{n_1,\dots,n_r}$ is

$$|\mathfrak{M}_{n_1,\dots,n_r}| = \frac{1}{|G|} \sum_{\substack{(b_1,\dots,b_n)\\ \Sigma i b_i = n}} a_G(b_1,\dots,b_n) p_{n_1,\dots,n_r}(b_1,\dots,b_n)$$

where $a_G(b_1,\dots,b_n) = \#\{\text{permutations in } G \text{ of type } 1^{b_1}\dots n^{b_n}\}$ and $p_{n_1,\dots,n_r}(b_1,\dots,b_n) = \#\{\text{functions in } \mathfrak{M}_{n_1,\dots,n_r}$ which are constant on the blocks of some arbitrary, but fixed, partition of $N$ of type $1^{b_1}\dots n^{b_n}\}$.

Continuing our example from above, let us see how many colored wheels exist with exactly $n_j$ boxes colored by $j$, for all $j \in R$. Since $G$ is cyclic the types of the permutations in $G$ are of the form $k^{n/k}$ with $k \mid n$. Hence $p_{n_1,\dots,n_r}(k^{n/k}) = 0$ if one of the $n_j$'s is not a multiple of $k$. Otherwise it is readily seen that

$$p_{n_1,\dots,n_r}(k^{n/k}) = \binom{n/k}{n_1/k,\dots,n_r/k} \quad \text{(multinomial coefficient).}$$

Thus the desired number is

$$\frac{1}{n} \sum_{k \mid \gcd(n_1,\dots,n_r)} \bar{\varphi}(k) \binom{n/k}{n_1/k,\dots,n_r/k}.$$

For $n = 5$, $r = 3$ and $n_1 = n_2 = 2$, $n_3 = 1$ this gives

$$\frac{1}{5}\binom{5}{2,2,1} = 6$$

wheels shown in Figure 5.7 with the colors ∘, • and ⊙ in this order.

## C. Closed Sets and the Cycle Indicator

After having proved the general theorem we now consider some special closed subsets, and try to simplify formula **5.24** in an effort to make it more accessible to applications. The first sets that come to mind are Map($N$, $R$), Inj($N$, $R$) and Sur($N$, $R$); these sets are trivially closed. We continue using the notation of the previous section and make the additional assumption that $R$ is finite.

Recall the definition of the elementary symmetric functions $a_n$ and the power sums $s_k$:

$$a_n(x_1,\dots,x_r) = \sum_{j_1 < \dots < j_n} x_{j_1} x_{j_2} \dots x_{j_n},$$

$$s_k(x_1,\dots,x_r) = \sum_{i=1}^{r} x_i^k.$$

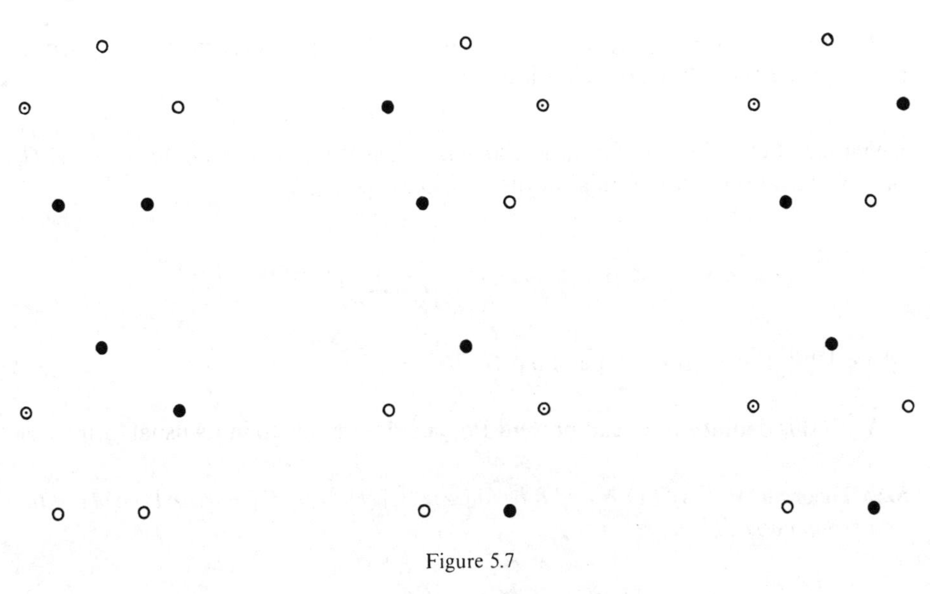

Figure 5.7

**5.26 Theorem** (Polya). *Let $N$ and $R$ be finite sets, $|N| = n, |R| = r$ and $G$ a permutation group on $N$. Then*

$$\gamma(\mathrm{Map}(N, R); G) = \frac{1}{|G|} \sum_{g \in G} s_1^{b_1(g)} s_2^{b_2(g)} \dots s_n^{b_n(g)},$$

*where $1^{b_1(g)} \dots n^{b_n(g)}$ is the type of $g \in G$ and $s_k = s_k(x_1, \dots, x_r)$ for $k = 1, \dots, n$.*

*Proof*. We have, by **5.24**,

$$\gamma(\mathrm{Map}(N, R); G) = \frac{1}{|G|} \sum_{g \in G} \left( \sum_{\ker f \geq \phi(g)} w(f) \right).$$

Let $g \in G$ be of type $1^{b_1} \dots n^{b_n}$ and $A_1, \dots, A_m$ the blocks of $\phi(g)$, $m = \sum_{i=1}^n b_i(g)$. Clearly, if $f$ is constant on the $A_i$'s, then

$$w(f) = x_{f(A_1)}^{|A_1|} x_{f(A_2)}^{|A_2|} \dots x_{f(A_m)}^{|A_m|}.$$

Since we consider the full set Map($N$, $R$), all $m$-tuples of subscripts must appear on the right-hand side, whence

$$\begin{aligned} \sum_{\ker f \geq \phi(g)} w(f) &= \sum_{(k_1, \dots, k_m)} \prod_{j=1}^m x_{k_j}^{|A_j|} = \prod_{j=1}^m \sum_{k=1}^r x_k^{|A_j|} \\ &= \left( \sum_{k=1}^r x_k \right)^{b_1(g)} \left( \sum_{k=1}^r x_k^2 \right)^{b_2(g)} \dots \left( \sum_{k=1}^r x_k^n \right)^{b_n(g)}. \quad \square \end{aligned}$$

Comparison of the polynomial in **5.26** with the cycle indicator of $S_n$ used in **5.13** suggests the following definition.

**Definition.** Let $G$ be a permutation group of degree $n$. The *cycle indicator* $Z(G)$ of $G$ is the rational polynomial in the variables $t_1, \dots, t_n$

$$Z(G) = Z(G; t_1, \dots, t_n) := \frac{1}{|G|} \sum_{g \in G} t_1^{b_1(g)} t_2^{b_2(g)} \dots t_n^{b_n(g)}$$

where $1^{b_1(g)} \dots n^{b_n(g)}$ is the type of $g \in G$.

With this definition we can present Polya's theorem **5.26** in its usual form.

**5.27 Theorem** (Polya). *Let $N$ and $R$ be finite sets, $|N| = n, |R| = r$, and $G$ a permutation group on $N$. Then*

$$\gamma(\mathrm{Map}(N, R); G) = Z(G; s_1, \dots, s_n),$$

*where $s_k = s_k(x_1, \dots, x_r)$, $k = 1, \dots, n$. In particular, the total number of $G$-patterns is given by*

$$|\mathfrak{M}| = Z(G; r, r, \dots, r). \quad \square$$

**Example.** Let $N$ be an $n$-set and $G \leq S(N)$. We say that two subsets $A, B \subseteq N$ are *$G$-equivalent* if $B = \{g(a) : a \in A\}$ for some $g \in G$. How many non-equivalent subsets are there? To fit this problem into our framework we choose $R = \{0, 1\}$ and identify $A \subseteq N$ with its characteristic function $f_A : N \to \{0, 1\}$. Then $A \approx_G B$ holds if and only if $f_A \approx_G f_B$ in the sense of our theorem. If we choose as weights

$$w(1) = x,$$
$$w(0) = 1,$$

then

$$w(f_A) = x^{|A|},$$

implying the following result.

**5.28 Corollary.** *Let $N$ be an $n$-set and $G$ a permutation group on $N$. The generating function of the number $m_k$ of $G$-patterns of $k$-subsets in $N$ is given by*

$$\sum_{k=0}^{n} m_k x^k = Z(G; 1 + x, 1 + x^2, \dots, 1 + x^n).$$

*In particular, the total number of subset patterns is*

$$\sum_{k=0}^{n} m_k = Z(G; 2, \ldots, 2) = \frac{1}{|G|} \sum_{g \in G} 2^{b(g)},$$

*where $b(g)$ is the number of cycles of $g \in G$.* □

As convenient short notation we usually write

$$Z(G; 1 + x) := Z(G; 1 + x, 1 + x^2, \ldots, 1 + x^n)$$

when there is no danger of confusion.

It is clear from the correspondence: subset ↔ characteristic function that for $r = 2$ any $G$-pattern can be interpreted as a subset pattern.

**Examples.** For $G = E(N)$ any two subsets are non-equivalent, hence $m_k = \binom{n}{k}$ and we obtain $(1 + x)^n$ as the generating function of the binomial coefficients. In the case $G = S(N)$, any two subsets of the same cardinality are equivalent, i.e., $m_k = 1$ for all $k$, whence we infer

$$\sum_{k=0}^{n} x^k = Z(S_n; 1 + x).$$

In particular, for $x = 1$, this yields

$$(n + 1)! = \sum_{g \in S_n} 2^{b(g)}.$$

Before turning to the sets Inj($N$, $R$) and Sur($N$, $R$) let us compute the cycle indicator of some important groups. We denote by $E_n$, $S_n$, $A_n$ and $C_n$ as usual the identity group, the symmetric group, the alternating group, and the cyclic group of degree $n$.

**5.29 Proposition.** *We have*

(i) $$Z(E_n) = t_1^n,$$

(ii) $$Z(S_n) = \sum_{\substack{(b_1, \ldots, b_n) \\ \sum ib_i = n}} \frac{1}{b_1! \ldots b_n! 1^{b_1} \ldots n^{b_n}} t_1^{b_1} \ldots t_n^{b_n},$$

(iii) $$Z(A_n) = \sum_{\substack{(b_1, \ldots, b_n) \\ \sum ib_i = n}} \frac{1 + (-1)^{n - \sum_{i=1}^n b_i}}{b_1! \ldots b_n! 1^{b_1} \ldots n^{b_n}} t_1^{b_1} \ldots t_n^{b_n},$$

(iv) $$Z(C_n) = \frac{1}{n} \sum_{k|n} \varphi(k) t_k^{n/k}.$$

*Proof.* We already know (i), (ii), and (iv). If $g \in A_n$ is of type $1^{b_1} \ldots n^{b_n}$, then $n - \sum_{i=1}^{n} b_i$ is even by **1.12**. Hence every permutation in $S_n$ is counted twice on the right-hand side of (iii), thereby accounting for $|A_n| = |S_n|/2$. □

**Example.** Let us generalize our problem of the colored wheels. We now suppose that the wheel can also be turned upside down. In other words we consider two colored wheels as equal if they can be carried into one another by some *rotation* and *reflexion*. Another interpretation is that of *colored necklaces*. For instance, the configurations in Figure 5.8 are inequivalent as wheels but they can be superimposed by a reflexion about the 1,4-axis followed by a rotation.

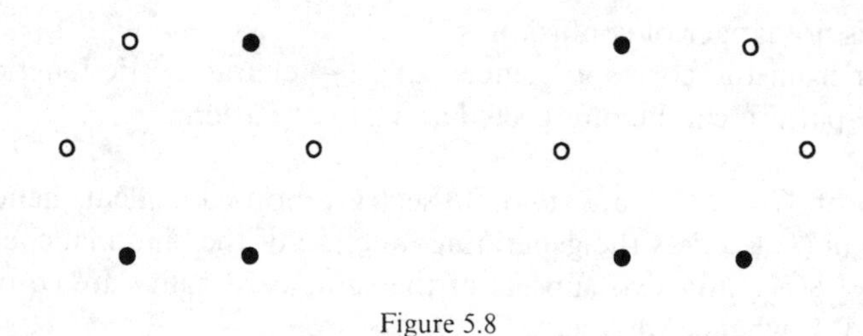

Figure 5.8

The corresponding permutation group is called the *dihedral group* $D_n$. $D_n$ is generated by the permutations $(1, 2, \ldots, n)$ and $(1, n)(2, n-1)(3, n-2)\ldots$. The reader is invited to verify the following formula for the cycle indicator of $D_n$.

**5.29**

$$\text{(v)} \qquad Z(D_n) = \begin{cases} \frac{1}{2}Z(C_n) + \frac{1}{4}(t_2^{n/2} + t_1^2 t_2^{n/2-1}) & n \text{ even} \\ \frac{1}{2}Z(C_n) + \frac{1}{2}t_1 t_2^{(n-1)/2} & n \text{ odd.} \end{cases} \qquad \square$$

To conclude this section let us compute the enumerator of the sets $\mathrm{Inj}(N, R)$ and $\mathrm{Sur}(N, R)$.

**5.30 Proposition.** *Let $N$ and $R$ be finite sets, $|N| = n$, $|R| = r$, and $G$ a permutation group on $N$. Then*

$$\gamma(\mathrm{Inj}(N, R); G) = \frac{n!}{|G|} a_n(x_1, \ldots, x_r)$$

$$= \frac{n!}{|G|} Z(S_n; s_1, -s_2, s_3, \mp \cdots).$$

*Proof.* According to **5.24**

$$\gamma(\mathrm{Inj}(N, R); G) = \frac{1}{|G|} \sum_{f \in \mathrm{Inj}(N,R)} w(f).$$

Now clearly,

$$\sum_{f \in \mathrm{Inj}(N, R)} w(f) = \sum x_{j_1} \dots x_{j_n},$$

where $(j_1, \dots, j_n)$ runs through all $[r]_n$ strict words of length $n$ in $R$. Hence by the definition of $a_n(x_1, \dots, x_r)$

$$\sum_{f \in \mathrm{Inj}(N, R)} w(f) = n!\,a_n(x_1, \dots, x_r).$$

Comparing Waring's formula **4.25** for the functions $a_n(x_1, \dots, x_r)$ with the cycle indicator $Z(S_n)$ we conclude that

$$a_n(x_1, \dots, x_r) = Z(S_n; s_1, -s_2, s_3, \mp \cdots). \quad \square$$

**Examples.** Set $\gamma(n, r) := \gamma(\mathrm{Map}(N, R); S_n) = Z(S_n; s_1, \dots, s_n)$ and $\gamma_r(t) := \sum_{n \geq 0} \gamma(n, r)t^n$. We know from **5.13** that

$$\gamma_r(t) = \exp\left(\sum_{n \geq 1} s_n \frac{t^n}{n}\right) = \exp\left(\sum_{n \geq 1}\left(\frac{(x_1 t)^n}{n} + \cdots + \frac{(x_r t)^n}{n}\right)\right)$$
$$= \prod_{j=1}^{r} \exp\left(\sum_{n \geq 1} \frac{(x_j t)^n}{n}\right) = \prod_{j=1}^{r} \exp(-\log(1 - x_j t)),$$

and thus

$$\gamma_r(t) = \sum_{n \geq 0} \gamma(n, r)t^n = \prod_{j=1}^{r} \frac{1}{1 - x_j t}.$$

By setting $x_j = 1$ for all $j$ we obtain the generating function for the number $m(n, r)$ of monotone words of length $n$ taken from a chain with $r$ elements (see **1.13**):

$$\sum_{n \geq 0} m(n, r)t^n = \frac{1}{(1 - t)^r}.$$

From the expansion $(1 - t)^{-r} = \sum \binom{-r}{n}(-1)^n t^n$, we deduce $m(n, r) = (-1)^n \binom{-r}{n} = [r]^n/n!$, verifying our old formula **3.6**.

Another interesting identity is obtained by setting $x_j = j \in \mathbb{N}$. In this case the weight of a function $f \in \mathrm{Map}(N, R)$ is given by

$$w(f) = 1^{|f^{-1}(1)|} 2^{|f^{-1}(2)|} \dots r^{|f^{-1}(r)|}.^{(3)}$$

(3) Strictly speaking the symbol $j \in R$ appearing in the exponent is different from $j \in \mathbb{N}$, but this should cause no confusion.

Since the $S_n$-patterns are determined by the numbers $|f^{-1}(1)|, \ldots, |f^{-1}(r)|$ and since all possible $r$-tuples summing to $n$ occur, we have

$$\gamma(\mathrm{Map}(N, R); S_n)_{x_j=j} = \sum_{1 \le k_1 \le k_2 \le \cdots \le k_n \le r} k_1 k_2 \ldots k_n.$$

On the other hand we have

$$S_{n,r} = \sum_{1 \le k_1 \le \cdots \le k_{n-r} \le r} k_1 k_2 \ldots k_{n-r},$$

as is easily seen by applying the recursion **3.29**(ii) or by a direct combinatorial argument. Hence we conclude $\gamma(n, r)|_{x_j = j (j \in R)} = S_{n+r,r}$ and thus

$$\sum_{n \ge 0} S_{n,r} t^n = \frac{t^r}{(1-t)(1-2t)\cdots(1-rt)}.$$

Now set $\bar{\gamma}(n, r) := \gamma(\mathrm{Inj}(N, R); S_n)$ and $\bar{\gamma}_r(t) := \sum_{n \ge 0} \bar{\gamma}(n, r) t^n$. From **5.30** and the definition of $a_n(x_1, \ldots, x_r)$ we immediately have

$$\bar{\gamma}_r(t) = \sum_{n \ge 0} a_n(x_1, \ldots, x_r) t^n = \prod_{j=1}^{r} (1 + x_j t).$$

Together with the expression for $\gamma_r(t)$, this gives the rational analogue of the reciprocity theorem **3.5**

$$\bar{\gamma}_r(-t) = \gamma_r(t)^{-1}.$$

Let us again consider the identity resulting from the substitution $x_j = j$ for all $j$. In this case

$$\bar{\gamma}_r(t)|_{x_j=j} = \sum_{n \ge 0} \left( \sum_{1 \le k_1 < \cdots < k_n \le r} k_1 k_2 \ldots k_n \right) t^n = (1+t)(1+2t)\cdots(1+rt).$$

The right-hand side can be written as

$$t^{r+1} \frac{1}{t}\left(\frac{1}{t} + 1\right)\left(\frac{1}{t} + 2\right) \cdots \left(\frac{1}{t} + r\right) = t^{r+1} \left[\frac{1}{t}\right]^{r+1} = t^{r+1} \sum_{i=0}^{r+1} |s_{r+1,i}| \frac{1}{t^i}$$

by **3.26**(i), whence we obtain the formula

$$|s_{r+1,r+1-n}| = \sum_{1 \le k_1 < \cdots < k_n \le r} k_1 k_2 \ldots k_n.$$

For the closed subset $\mathrm{Sur}(N, R)$ no comparatively simple formula is known. By inclusion-exclusion one readily proves the following result.

**5.31 Proposition.** *Let $N$ and $R$ be finite sets, $|N| = n, |R| = r$, and $G$ a permutation group on $N$. Then*

$$\gamma(\mathrm{Sur}(N, R); G) = \frac{1}{|G|} \sum_{g \in G} \left( s_1^{b_1(g)} \dots s_n^{b_n(g)} - \sum_i (s_1 - x_i)^{b_1(g)} \dots (s_n - x_i^n)^{b_n(g)} + \sum_{i<j} (s_1 - x_i - x_j)^{b_1(g)} \dots (s_n - x_i^n - x_j^n)^{b_n(g)} \mp \dots \right). \quad \square$$

**Example.** For $G = E(N)$ we obtain the formula **3.39** for the Stirling numbers of the second kind. For $G = S(N)$ we clearly have

$$\sigma(n, r) := \gamma(\mathrm{Sur}(N, R); S_n) = \sum_{\substack{\sum b_j = n \\ b_j \geq 1}} x_1^{b_1} x_2^{b_2} \dots x_n^{b_n}.$$

Let $\sigma_r(t) := \sum_{n \geq 1} \sigma(n, r) t^n$. Then

$$\sigma_r(t) = \left( \sum_{i \geq 1} (x_1 t)^i \right) \left( \sum_{j \geq 1} (x_2 t)^j \right) \cdots \left( \sum_{k \geq 1} (x_r t)^k \right),$$

hence

$$\sigma_r(t) = \sum_{n \geq 1} \sigma(n, r) t^n = \prod_{j=1}^{t} \frac{x_j}{1 - x_j t} t^r.$$

Substituting $x_j = 1$ for all $j$ yields the generating function for the number $p(n, r)$ of ordered $r$-partitions of $n$ (see **1.13**):

$$\sum_{n \geq 1} p(n, r) t^n = \left( \frac{t}{1 - t} \right)^r.$$

The expression

$$\sum_{r \geq 1} \left( \frac{t}{1 - t} \right)^r = \frac{t}{1 - t} \sum_{r \geq 0} \left( \frac{t}{1 - t} \right)^r = \frac{t}{1 - 2t}$$

therefore counts *all* ordered partitions of $n$, in accordance with **5.5**.

## EXERCISES V.3

1. Let $N = \mathbb{N}_6$ and consider the dihedral group $D_6$ on $N$. Determine the lattice $\mathscr{P}(N, D_6)$. Which subgroups of $D_6$ are closed? Compute the Euler function $\varphi$.

2. Study the similarity of **5.23** to Burnside's lemma **1.7** Deduce **1.7** from **5.24**.

→ 3. Let $S_n$ act on $\mathrm{Map}(N, R)$, $|N| = n$, $|R| = r$. Show that $f' \approx_{S_n} f \Leftrightarrow w(f') = w(f)$ and deduce from this formula **5.13** by using the following steps:

   (i) $\gamma(n, r) := \gamma(\mathrm{Map}(N, R); S_n) = \sum_{(b_1, \dots, b_r), \sum ib_i = n} \prod_{j=1}^{r} x_j^{b_j}$,
   (ii) $\gamma_r(t) := \sum_{n \geq 0} \gamma(n, r) t^n = \exp(\sum_{n \geq 1} s_n(t^n/n)) = \sum_{n \geq 0} Z(S_n; s_1, \dots, s_n) t^n$.
   (iii) Since the $x_j$'s may assume arbitrary values, deduce **5.13**.

4. Prove Waring's formula **4.25** by means of **5.30**.

→ 5. Reprove **5.19** by the following approach. Let $v$ be the root with $\gamma(v) = k$. Let $N = \mathbb{N}_k$, $R = \{\text{unlabeled trees}\}$ and $w(T) = x^n$ if $T \in R$ has $n$ vertices. Compute $\gamma(\text{Map}(N, R); S_k)$ and deduce **5.19** by letting $k = 1, 2, 3, \ldots$. (Note: $R$ is infinite, but this plays no role here.)

→ 6.* A *planted tree* is a rooted unlabeled tree whose root has degree 1. Consider the planted trees as embedded in the real Euclidean plane and define $T$ and $T'$ to be equal if $T$ can be superimposed upon $T'$ by a continuous transformation. For instance,

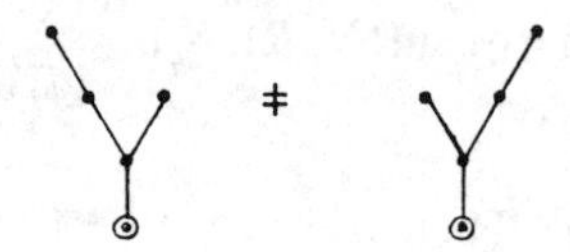

Let $k_n$ be the number of planted trees with $n$ edges ($= n + 1$ vertices) whose vertices all have degree 1 or 3, and set $k(t) = \sum_{n \geq 1} k_n t^n$. Prove:

(i) $k(t) = t + t(k(t))^2$,
(ii) $k_{2n} = 0$, $k_{2n+1} = (2n)!/n!(n + 1)!$ for all $n$.

(deBruijn)

7. Verify formula **5.29**(v) for $Z(D_n)$.

8. For some $k \in \mathbb{N}$ we have $Z(C_n; 2, \ldots, 2) = Z(D_n; 2, \ldots, 2)$ for all $n \leq k$ and $Z(C_n; 2, \ldots, 2) \neq Z(D_n; 2, \ldots, 2)$ for all $n > k$. Find $k$.

9. Let $G \leq S(N)$, $H \leq S(R)$ be permutation groups where $N$ and $R$ are disjoint. The *product* $G \cdot H$ is the permutation group on $N \cup R$, $G \cdot H := \{g \cdot h : g \in G, h \in H\}$, where

$$(g \cdot h)(a) := \begin{cases} ga & \text{if } a \in N \\ ha & \text{if } a \in R. \end{cases}$$

Prove that

$$Z(G \cdot H) = Z(G) \cdot Z(H).$$

→ 10. Let $N, R, G$, and $H$ be as in ex. 9. The *cartesian product* $G \times H$ is the permutation group on $N \times R$, $G \times H := \{(g, h) : g \in G, h \in H\}$, where

$$(g, h)(a, b) := (ga, hb) \qquad (a \in N, b \in R).$$

Prove that

$$Z(G \times H) = \frac{1}{|G||H|} \sum_{(g,h)} \prod_{k=1,l=1}^{n,r} t_{\text{lcm}(k,l)}^{\gcd(k,l) b_k(g) b_l(h)}.$$

→11. Consider the colorings of a cube $W$. Suppose we color the faces of $W$ red or blue. Two colorings are considered indistinguishable if and only if one can be transformed into the other by a rotation of the cube. Determine the group $G$ of rotations of the cube ($|G| = 24$) and find the generating function for face colorings, edge colorings, and vertex colorings with 2 colors.

→12. How many face colorings (edge colorings, vertex colorings) of a cube exist in which the two colors are used the same number of times?

13. Compute $Z(G)$ where $G$ is the group of rotations of the octahedron.

14. Show that there are precisely 3 colorings of the vertices of the octahedron such that 3 vertices are colored red, 2 are colored blue, and 1 is colored green.

→15. Consider $\sigma(n, r)$ as in the last example of this section. What does $\sigma(n, r)|_{x_j = j,\, j=1,\ldots,r}$ count?

## 4. $G, H$-patterns

To conclude our exposition of combinatorial counting let us now study the enumeration of $G, H$-patterns in full generality. As in section 3 we are going to reduce the problem of finding the pattern enumerator to the determination of the cycle structure of the groups $G$ and $H$. After proving the main theorem in section $B$ we discuss some important sets of mappings in detail, closing with applications to a variety of interesting counting problems.

### A. Patterns Invariant Under a Permutation

As the main step towards proving the principal theorem let us apply **5.24** to the following situation. $N$ and $R$ are finite sets, $G$ a subgroup of $S(N)$ and $\mathscr{F} \subseteq \mathrm{Map}(N, R)$ closed with respect to $G$. In addition, we are given a permutation $h$ on $R$, and we set

$$\mathscr{F}_h := \{f \in \mathscr{F} : h(fG) = fG\}.$$

Hence $\mathscr{F}_h$ is the set of all mappings $f \in \mathscr{F}$ whose pattern $fG$ is invariant under the permutation $h$. We claim that $\mathscr{F}_h$ is closed. To see this notice that $f \in \mathscr{F}_h$ is equivalent to $hf \in fG$, i.e., $hf = fk_f$ for some $k_f \in G$. Let $f \in \mathscr{F}_h$ and $g \in G$. Then

$$h(fg) = fk_f g = (fg)(g^{-1}k_f g),$$

hence $fg \in \mathscr{F}_h$ with $k_{fg} = g^{-1}k_f g$, proving our assertion.

**5.32 Proposition** (de Bruijn). *Let $N$ and $R$ be finite sets, $G$ a permutation group on $N$ and $\mathscr{F} \subseteq \mathrm{Map}(N, R)$ closed with respect to $G$. Further, let $h$ be a permutation on $R$. Then $\mathscr{F}_h = \{f \in \mathscr{F} : h(fG) = fG\}$ is closed and has as enumerator*

$$\gamma(\mathscr{F}_h; G) = \frac{1}{|G|} \sum_{g \in G} \left( \sum_{f \in \mathscr{F},\, hf = fg} w(f) \right).$$

*Proof*. Applying **5.24** we have

$$\gamma(\mathcal{F}_h; G) = \frac{1}{|G|} \sum_{g \in G} \left( \sum_{f \in \mathcal{F}_h, fg = f} w(f) \right).$$

A term $w(f)$, $f \in \mathcal{F}_h$, appears in the inner sum $|G(f)|$ times where $G(f) := \{g \in G: fg = f\}$. Set $hf = fk_f$ with $k_f \in G$ as above. Since for $g \in G$

$$hf = fg \Leftrightarrow gk_f^{-1} \in G(f) \Leftrightarrow g \in G(f)k_f$$

we see that

$$|G(f)| = |\{g \in G: hf = fg\}|$$

whence the theorem follows. □

**Example**. Consider $\mathcal{F} = \text{Map}(N, R)$. To determine $\sum_{hf=fg} w(f)$ for $g \in G$ let $a \in N$ with $f(a) = b \in R$ and assume $hf = fg$. Suppose $a$ lies in a $k$-cycle of $g$ (i.e., a cycle of length $k$) and $b$ in a $j$-cycle of $h$. Applying $f$ to the cycle of $g$ containing $a$ we infer

$$(+) \qquad \begin{array}{c} a \;\; ga \;\; g^2a \;\cdots\; g^{k-1}a \\ f \downarrow \\ b \;\; hb \;\; h^2b \;\cdots\; h^{k-1}b. \end{array}$$

It follows that

$$f(g^k a) = f(a) = b = h^k(b),$$

hence $j|k$. If, conversely, we choose to every $k$-cycle of $g$ a $j$-cycle of $h$ with $j|k$ and determine $f$ according to $(+)$, then $f$ lies in $\text{Map}(N, R)_h$ with $hf = fg$. Since we may pick as the image of $a$ any of the $j$ elements of a $j$-cycle of $h$ we obtain

$$\sum_{hf=fg} w(f) = \lambda_1^{b_1(g)} \lambda_2^{b_2(g)} \ldots \lambda_n^{b_n(g)}$$

with

$$\lambda_k = \sum_{j|k} j \sum_{j\text{-cycles of } h} (x_i x_{hi} \ldots x_{h^{j-1}i})^{k/j},$$

and thus we have the following result.

**5.33 Proposition.** *Let $N$ and $R$ be finite sets, $|N| = n$, $|R| = r$, $G$ a permutation group on $N$, and $h$ a permutation on $R$. Then*

$$\gamma(\text{Map}(N, R)_h; G) = Z(G; \lambda_1, \ldots, \lambda_n)$$

*where*

$$\lambda_k = \sum_{j|k} j \sum_{j\text{-cycles of } h} (x_i x_{hi} \ldots x_{h^{j-1}i})^{k/j} \qquad (k = 1, \ldots, n).$$

*As a corollary, the total number* $|\mathfrak{M}_h|$ *of G-patterns invariant under h is given by*

$$|\mathfrak{M}_h| = Z(G; \lambda_1(h), \ldots, \lambda_n(h))$$

*where*

$$\lambda_k(h) = \sum_{j|k} j b_j(h) \qquad (k = 1, \ldots, n). \quad \square$$

**Example.** Let us determine all necklaces colored with at most three colors which are invariant under the color permutation $1 \to 2 \to 3 \to 1$. By definition of $\lambda_k$ we have

$$\lambda_k = \begin{cases} 0 & \text{for } 3 \nmid k \\ 3x_1^{k/3}x_2^{k/3}x_3^{k/3} & \text{for } 3 | k. \end{cases}$$

Using **5.29**(v) we obtain the answer

$$Z(D_n; \lambda_1, \ldots, \lambda_n) = \frac{1}{2n} \sum_{3|k|n} \bar{\varphi}(k)\lambda_k^{n/k} = \frac{1}{2n} \sum_{3|k|n} \bar{\varphi}(k)3^{n/k}x_1^{n/3}x_2^{n/3}x_3^{n/3}.$$

For $n = 9$ this gives 4 colored necklaces; they are depicted in Figure 5.9.

B. The Main Theorem

Let $N$ and $R$ be finite sets and $G \leq S(N)$, $H \leq S(R)$ permutation groups. As explained in section I.3 the groups $G$ and $H$ induce the *power group* $H^G = \{h^g : g \in G, h \in H\}$ on Map$(N, R)$ where

$$h^g(f) = hfg \qquad (f \in \text{Map}(N, R)).$$

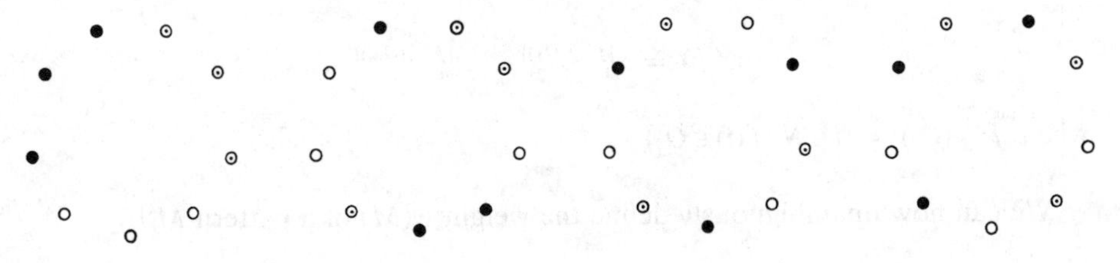

Figure 5.9

The $H^G$-orbits are called *G, H-patterns* and it is now our aim to enumerate these patterns. Clearly,

$$f \approx_G f' \Rightarrow f \approx_{H^G} f' \qquad (f, f' \in \mathrm{Map}(N, R));$$

thus $\approx_G$ is a finer relation than $\approx_{H^G}$. Let us define the *weight* of a mapping $f$ as before by

$$w(f) = \prod_{i \in N} x_{f(i)},$$

where the $x_j$'s are rational variables assigned to the elements $j \in R$. In contrast to section 3, i.e., $H = E(R)$, equivalent mappings may now have different weights when $H$ is an arbitrary subgroup of $S(R)$. More precisely, we have for $f' = hfg$

$$w(f') = x_{h(1)}^{|f^{-1}(1)|} \dots x_{h(r)}^{|f^{-1}(r)|} = w(f) = x_1^{|f^{-1}(1)|} \dots x_r^{|f^{-1}(r)|}$$

if and only if

$$|f^{-1}(j)| = |f^{-1}(h^{-1}(j))| \quad \text{for all } j \in R.$$

Thus $w(f) = w(f')$ if and only if $|f^{-1}(j)|$ is constant on the $h$-cycle containing $j$, for all $j$. Letting $h$ run through $H$ we infer that

$$(+) \quad \left(f \underset{H^G}{\approx} f' \Rightarrow w(f) = w(f')\right) \Leftrightarrow |f^{-1}(j)| \text{ is constant on the } H\text{-orbits } O(j), j \in R, \text{ for all } f \in \mathrm{Map}(N, R).$$

To give a definition of a $G, H$-pattern which is independent of the chosen representative we therefore have to restrict the mappings under consideration to those satisfying $(+)$, or we must alter the definition of the weight. The first alternative would, in general, leave us with too few mappings to be useful, hence we choose a new definition of the weight motivated by $(+)$.

**Definition.** Let $G$ and $H$ be permutation groups on the finite sets $N$ and $R$, respectively. To each $H$-orbit $O_j$, $j = 1, \dots, m$, we assign a variable $x_j$ and then define the *weight* of $f \in \mathrm{Map}(N, R)$ by

$$w(f) := x_1^{|f^{-1}(O_1)|} \dots x_m^{|f^{-1}(O_m)|}$$

where $f^{-1}(O_j) = \{i \in N : f(i) \in O_j\}$.

We can now unambiguously define the weight $w(M)$ of a pattern $M$ by

$$w(M) := w(f) \quad \text{for } f \in M.$$

Notice that for $H = E(R)$ all orbits consist of a single element; in this case the definition of the weight reduces to the one given in section 3.

**Definition.** A subset $\mathcal{F} \subseteq \mathrm{Map}(N, R)$ is called *closed* with respect to $G \leq S(N)$ and $H \leq S(R)$ if

$$f \in \mathcal{F}, g \in G, h \in H \Rightarrow hfg \in \mathcal{F}.$$

Again we denote by $\mathfrak{M}$ the set of $G$, $H$-patterns and by $\mathfrak{M}_{\mathcal{F}}$ the subset of patterns contained in $\mathcal{F}$. The central result computes the *enumerator*

$$\gamma(\mathcal{F}; G, H) := \sum_{M \in \mathfrak{M}_{\mathcal{F}}} w(M).$$

**5.34 Theorem** (de Bruijn). *Let $N$ and $R$ be finite sets and $G$ and $H$ permutation groups on $N$ and $R$, respectively. Let $\mathcal{F} \subseteq \mathrm{Map}(N, R)$ be closed with respect to $G$ and $H$. If $\mathcal{F}_h$ denotes the closed set as in* **5.32** *then*

$$\gamma(\mathcal{F}; G, H) = \frac{1}{|H|} \sum_{h \in H} \gamma(\mathcal{F}_h; G) = \frac{1}{|G||H|} \sum_{g \in G, h \in H} \left( \sum_{f \in \mathcal{F}, hf = fg} w(f) \right).$$

*Proof.* The set $fG$ is contained in the $G$, $H$-pattern $H(fG)$. By the same argument as in **5.23** we have

$$|H(fG)| = \frac{|H|}{|\{h \in H : h(fG) = fG\}|}.$$

Hence we conclude as in **5.24**

$$\gamma(\mathcal{F}; G, H) = \sum_{\substack{fG \\ f \in \mathcal{F}}} \frac{w(fG)}{|H(fG)|} = \frac{1}{|H|} \sum_{\substack{fG \\ f \in \mathcal{F}}} w(fG) |\{h \in H : h(fG) = fG\}|,$$

and thus

$$\gamma(\mathcal{F}; G, H) = \frac{1}{|H|} \sum_{h \in H} \left( \sum_{\substack{fG \\ f \in \mathcal{F}, h(fG) = fG}} w(fG) \right) = \frac{1}{|H|} \sum_{h \in H} \gamma(\mathcal{F}_h; G)$$

$$= \frac{1}{|G||H|} \sum_{g \in G, h \in H} \left( \sum_{f \in \mathcal{F}, hf = fg} w(f) \right). \quad \square$$

**5.35 Corollary.** *Let $N$ and $R$ be finite sets, $G \leq S(N)$ and $H \leq S(R)$, and $\mathcal{F} \subseteq \mathrm{Map}(N, R)$ closed. The number of different $G$, $H$-patterns in $\mathcal{F}$ is given by*

$$|\mathfrak{M}_{\mathcal{F}}| = \frac{1}{|G||H|} \sum_{g \in G, h \in H} |\{f \in \mathcal{F} : hf = fg\}|. \quad \square$$

Corresponding to section 3 we may now apply **5.34** to various closed sets; see the exercises. Here we deal only with the most important case $\mathscr{F} = \mathrm{Map}(N, R)$.

**5.36 Theorem.** *Let $N$ and $R$ be finite sets, $|N| = n$, $|R| = r$, and $G \leq S(N)$, $H \leq S(R)$. Denote the $H$-orbits of $R$ by $O_1, \ldots, O_m$ and assign a variable $x_j$ to $O_j$, $j = 1, \ldots, m$. Any $h \in H$ thus decomposes uniquely into a product $h = h_1 h_2 \ldots h_m$ where $h_i \in S(O_i)$. The enumerator of* $\mathrm{Map}(N, R)$ *is then given by*

$$\gamma(\mathrm{Map}(N, R); G, H) = \frac{1}{|H|} \sum_{h \in H} Z(G; \lambda_1(h), \ldots, \lambda_n(h))$$

*where*

$$\lambda_k(h) = \sum_{i=1}^{m} \left( \sum_{j|k} j b_j(h_i) \right) x_i^k \qquad (k = 1, \ldots, n). \quad \square$$

**5.37 Corollary.** *Let $G$ and $H$ be permutation groups on the finite sets $N$ and $R$, respectively, $|N| = n$ and $|R| = r$. The total number of $G$, $H$-patterns is*

$$|\mathfrak{M}| = \frac{1}{|H|} \sum_{h \in H} Z(G; \lambda_1(h), \ldots, \lambda_n(h))$$

*where*

$$\lambda_k(h) = \sum_{j|k} j b_j(h) \qquad (k = 1, \ldots, n). \quad \square$$

**5.38 Examples.** To illustrate **5.37** let us look at the table of patterns in **1.13**.

(i) If $H = E_r$ then **5.36** and **5.37** reduce to **5.27**.

(ii) Let $G = E_n$ and denote by $m(E_n, H)$ the number of $E_n$, $H$-patterns. Then

$$m(E_n, H) = \frac{1}{|H|} \sum_{h \in H} (b_1(h))^n.$$

If $H = S_r$ then $m(E_n, S_r)$ counts the number of partitions of an $n$-set into at most $r$ blocks. In this case $m(E_n, S_r) = (1/r!) \sum_{h \in S_r} (b_1(h))^n$ and thus

$$\sum_{n \geq 0} m(E_n, S_r) \frac{t^n}{n!} = \frac{1}{r!} \sum_{n \geq 0} \sum_{h \in S_r} \frac{(b_1(h)t)^n}{n!} = \frac{1}{r!} \sum_{h \in S_r} \left( \sum_{n \geq 0} \frac{(b_1(h)t)^n}{n!} \right)$$

$$= \frac{1}{r!} \sum_{h \in S_r} e^{t b_1(h)} = Z(S_r; e^t, 1, \ldots, 1).$$

Applying **5.13** we have

$$\sum_{r\geq 0} Z(S_r; e^t, 1, \ldots, 1)u^r = \exp\left(e^t u + \frac{u^2}{2} + \frac{u^3}{3} + \cdots\right)$$
$$= \exp(u(e^t - 1) - \log(1 - u))$$

and thus

$$\sum_{n,r\geq 0} m(E_n, S_r)\frac{t^n u^r}{n!} = \frac{\exp(u(e^t - 1))}{1 - u}.$$

The expression $m(E_n, S_r) - m(E_n, S_{r-1})$ counts the number of partitions of an $n$-set into precisely $r$ blocks, whence

$$\sum_{n,r\geq 0} S_{n,r}\frac{t^n u^r}{n!} = \exp(u(e^t - 1)).$$

For $u = 1$ the coefficient of $t^n/n!$ counts all partitions of an $n$-set yielding the generating function for the Bell numbers already derived after **3.59** and **5.9**:

$$\sum_{n\geq 0} B_n \frac{t^n}{n!} = e^{e^t - 1}.$$

(iii) Let $G = S_n$ and denote by $m(S_n, H)$ the number of $S_n,H$-patterns. By **5.13** we have

$$\sum_{n\geq 0} m(S_n, H)t^n = \frac{1}{|H|}\sum_{h\in H}\sum_{n\geq 0} Z(S_n; \lambda_1(h), \ldots, \lambda_n(h))t^n$$
$$= \frac{1}{|H|}\sum_{h\in H}\exp\left(\sum_{k\geq 1}\lambda_k(h)\frac{t^k}{k}\right).$$

Now $\lambda_k(h) = \sum_{j|k} jb_j(h)$, hence

$$\sum_{k\geq 1}\lambda_k(h)\frac{t^k}{k} = \sum_{j\geq 1} jb_j(h)\left(\frac{t^j}{j} + \frac{t^{2j}}{2j} + \cdots\right)$$
$$= \sum_{j\geq 1} b_j(h)(-\log(1 - t^j)) = \log\prod_{j=1}^{r}\left(\frac{1}{1 - t^j}\right)^{b_j(h)}.$$

Thus we obtain

$$\sum_{n\geq 0} m(S_n, H)t^n = Z\left(H; \frac{1}{1 - t}, \frac{1}{1 - t^2}, \ldots, \frac{1}{1 - t^r}\right).$$

For $H = E_r$ this again gives $(1 - t)^{-r}$ as the generating function of monotone words of length $n$ taken from an $r$-set. The case $H = S_r$ counts the number of number-partitions of $n$ into at most $r$ summands. Applying **5.13** once more we have

$$\sum_{r \geq 0} Z\left(S_r; \frac{1}{1-t}, \ldots, \frac{1}{1-t^r}\right) u^r = \exp\left(\sum_{k \geq 1} \frac{1}{1-t^k} \frac{u^k}{k}\right)$$

$$= \exp\left(\sum_{k \geq 1} \frac{u^k}{k}(1 + t^k + t^{2k} + \cdots)\right)$$

$$= \exp(-\log(1-u) - \log(1-tu) - \cdots),$$

hence

$$\sum_{n, r \geq 0} m(S_n, S_r) t^n u^r = \prod_{n \geq 0} \frac{1}{1 - t^n u}.$$

By definition, $m(S_n, S_r) - m(S_n, S_{r-1})$ counts the number of partitions of $n$ into precisely $r$ summands, whence

$$\sum_{n, r \geq 0} P_{n,r} t^n u^r = \prod_{n \geq 1} \frac{1}{1 - t^n u}$$

$$\sum_{n \geq 0} P_n t^n = \prod_{n \geq 1} \frac{1}{1 - t^n},$$

in accordance with **5.16**.

(iv) Finally, let $H = S_r$ and $G$ arbitrary. In this case, the patterns may be thought of as *partition patterns*. For $r = 2$, for instance, this gives partition patterns containing at most two blocks. Let $m(G, S_2)$ be their number. By **5.37**,

$$m(G, S_2) = \tfrac{1}{2}(Z(G; 2, 2, \ldots, 2) + Z(G; 0, 2, 0, 2, \ldots)).$$

As in **5.28**, we identify the mapping $f: N \to \{0, 1\}$ with the subset $f^{-1}(1) \subseteq N$. The expression $m(G, S_2)$ then counts all subset patterns, where, in addition, every subset $A \subseteq N$ is equivalent to its complement. Since, by **5.28**, $Z(G; 2, \ldots, 2)$ counts *all* subset patterns of $N$, we conclude that

$$Z(G; 0, 2, 0, 2, \ldots) = \#\{\text{self-equivalent subset patterns, meaning that } A \approx_G N - A, A \subseteq N\}.$$

**Example.** Let $N = \{1, 2, \ldots, 8\}$. We want to find the self-equivalent subset patterns of $N$ with respect to $C_8$ and to $D_8$. Applying **5.29** we have

$$Z(C_8; 0, 2, 0, 2, 0, 2, 0, 2) = 4$$

$$Z(D_8; 0, 2, 0, 2, 0, 2, 0, 2) = 6.$$

The self-equivalent patterns have as representatives

$$\{1, 2, 3, 4\} \approx_{C_8} \{5, 6, 7, 8\}, \qquad \{1, 2, 5, 6\} \approx_{C_8} \{3, 4, 7, 8\},$$
$$\{1, 2, 4, 7\} \approx_{C_8} \{3, 5, 6, 8\}, \qquad \{1, 3, 5, 7\} \approx_{C_8} \{2, 4, 6, 8\},$$

and further

$$\{1, 2, 3, 5\} \approx_{D_8} \{4, 6, 7, 8\}, \qquad \{1, 2, 4, 6\} \approx_{D_8} \{3, 5, 7, 8\}.$$

## C. Applications

The relationship between the enumerator of a set of patterns and the cycle indicator of its corresponding permutation group can be fruitfully exploited in both directions. On the one hand it enables us to compute the enumerator from a known cycle indicator, and on the other hand to determine the cycle structure of a group from the solution of a counting problem. We shall present some typical examples illustrating both aspects of this relation.

*Composition of Permutation Groups.* The cycle indicator $Z(G)$ of a finite permutation group $G$ is a rational polynomial and, as such, can be algebraically manipulated. In particular, we may ask the following question: Suppose $G$ and $H$ are permutation groups on the disjoint sets $N$ and $R$, respectively. Is there a "natural" group $G \cdot H$ such that $Z(G \cdot H) = Z(G) \cdot Z(H)$? The answer to this is almost trivial. Define the *product* $G \cdot H$ as the permutation group on $N \cup R$, $G \cdot H := \{g \cdot h : g \in G, h \in H\}$ with

$$(g \cdot h)(a) := \begin{cases} ga & \text{if } a \in N \\ ha & \text{if } a \in R. \end{cases}$$

It is immediately verified that $Z(G \cdot H) = Z(G) \cdot Z(H)$.

A more intricate problem involves the composition of polynomials. Let $G$ and $H$ be as before. Is there a natural group—which we call the *composition* $G[H]$—such that $Z(G[H]) = Z(G; Z(H))$? By $Z(G; Z(H))$ we mean that the variable $t_k$ is replaced by a suitable expression involving $Z(H)$, to be made precise below. Such a group indeed exists and is the key to the solution of important counting problems that were unaccessible up to now.

**Definition.** Let $G$ and $H$ be permutation groups on the finite sets $N$ and $P$, respectively, $|N| = n$ and $|P| = p$. We set $N = \{a_1, \ldots, a_n\}$ and $P = \{b_1, \ldots, b_p\}$. The *composition* $G[H]$ is the permutation group on $N \times P$ whose elements are all $(n + 1)$-tuples

$$[g; h_1, \ldots, h_n] \text{ with } g \in G, h_1, \ldots, h_n \in H$$

where

$$[g; h_1, \ldots, h_n](a_i, b_j) = (ga_i, h_i b_j) \qquad (i = 1, \ldots, n;\ j = 1, \ldots, p).^{(4)}$$

It is easily seen that different $(n + 1)$-tuples induce different permutations on $N \times P$ and that $G[H]$ is indeed a group. The order of $G[H]$ is $|G||H|^n$ and the degree is $np$.

**5.39 Proposition** (Polya). *Let $G$ and $H$ be permutation groups on the finite sets $N$ and $P$, respectively, $|N| = n$ and $|P| = p$. Then*

$$Z(G[H]; t_1, t_2, \ldots, t_{np}) = Z(G; Z(H; t_1, t_2, \ldots, t_p), Z(H; t_2, t_4, \ldots, t_{2p}), \ldots),$$

*where on the right-hand side $t_k$ is replaced by $Z(H; t_k, t_{2k}, \ldots, t_{pk})$.*

*Proof*. To establish the asserted identity we proceed as follows. We represent both sides as the enumerator of a certain set of patterns (after the substitution $t_k = \sum x_j^k$) and then construct a weight-preserving bijection between the pattern sets. Since the $x_j$'s are variables the two polynomials must then be identically equal.

Let $N = \{a_1, \ldots, a_n\}$, $P = \{b_1, \ldots, b_p\}$ and let $R$ be an $r$-set disjoint to both $N$ and $P$. To each $j \in R$ we assign a variable $x_j$. The enumerator of the $G[H]$-patterns $\mathfrak{M}$ of Map$(N \times P, R)$ is, by **5.27**,

$$\gamma(\text{Map}(N \times P, R); G[H]) = Z(G[H]; \textstyle\sum x_j, \sum x_j^2, \ldots, \sum x_j^{np}).$$

To see what the right-hand side of **5.39** counts, first we consider the enumerator $\gamma(\text{Map}(N, R); H)$ of all $H$-patterns $\mathfrak{P}$ on Map$(P, R)$:

$$\gamma(\text{Map}(N \times P, R); H) = Z(H; \textstyle\sum x_j, \sum x_j^2, \ldots, \sum x_j^p).$$

Next we compute the enumerator $\gamma(\text{Map}(N, \mathfrak{P}); G)$ of all $G$-patterns $\mathfrak{N}$ of mappings $N \to \mathfrak{P}$ where the weight is given by $w(D) = \prod_{i=1}^n w(D(a_i))$ for $D \in \text{Map}(N, \mathfrak{P})$:

$$\gamma(\text{Map}(N, \mathfrak{P}); G) = Z\left(G; \sum_{F \in \mathfrak{P}} w(F), \sum_{F \in \mathfrak{P}} w^2(F), \ldots, \sum_{F \in \mathfrak{P}} w^n(F)\right).$$

Now clearly

$$\sum_{F \in \mathfrak{P}} w^k(F) = Z\left(H; \sum x_j^k, \sum x_j^{2k}, \ldots, \sum x_j^{pk}\right)$$

(4) In group theory $G[H]$ is known as the *wreath product* $H \sim G$.

(just assign the weight $x_j^k$ to $j \in R$) and thus

$$\gamma(\mathrm{Map}(N, \mathfrak{P}); G) = Z(G; Z(H; \sum x_j, \sum x_j^2, \ldots), Z(H; \sum x_j^2, \sum x_j^4, \ldots), \ldots)$$

which is precisely the right-hand side of **5.39** after the substitution $t_k = \sum x_j^k$ for all $k$.

To conclude the proof we define a mapping $\phi\colon \mathfrak{M} \to \mathfrak{N}$ as follows: For $M \in \mathfrak{M}$ with representative $f$ we define $\hat{f}\colon N \to \mathrm{Map}(P, R)$ by

$$(\hat{f}(a_i))(b_j) := f(a_i, b_j) \in R \qquad (i = 1, \ldots, n; j = 1, \ldots, p).$$

The mapping $\hat{f}$ induces another mapping $\hat{F}\colon N \to \mathfrak{P}$ by setting

$$\hat{F}(a_i) := \hat{f}(a_i)H \in \mathfrak{P},$$

and now finally we define

$$\phi M := \hat{F}G \in \mathfrak{N}.$$

It is readily seen that the definition of $\phi$ is independent of the chosen representative and that $\phi$ is in fact a weight-preserving bijection between $\mathfrak{M}$ and $\mathfrak{N}$. □

**Example.** The typical situation for the composition arises when we are given $n$ copies of a structure on which a group $G$ acts. $S_n[G]$ is then the group of symmetries induced by permutations of the $n$ copies followed by permutations in $G$ applied to each individual structure. Consider, for instance, three wheels $A$, $B$, $C$ with 4 boxes each of which is colored red or blue. Two colorings $F$ and $F'$ of the triple $\{A, B, C\}$ are called equivalent if $F'$ arises from $F$ by some permutation of $\{A, B, C\}$ together with some rotations of the wheels. Reformulated in terms of our problem it reads as follows: $N = \{1, 2, 3\}$ is the set of places for the wheels, $P = \{1, 2, 3, 4\}$ is the set of positions of the boxes and $R = \{\text{red}, \text{blue}\}$. The colorings are then all mappings $F\colon N \times P \to R$. Equivalence of $F$ and $F'$ now means that there are permutations $g \in S(N)$ and $h_1, h_2, h_3 \in C_4$ with

$$F'(i, j) = F(gi, h_i j) \qquad (i = 1, 2, 3; j = 1, 2, 3, 4).$$

Hence $F$ and $F'$ are equivalent if and only if $F \approx_{S_3[C_4]} F'$ and we obtain for the number of different colorings

$$Z(S_3[C_4]; 2, 2, \ldots, 2) = Z(S_3; Z(C_4; 2, \ldots, 2), Z(C_4; 2, \ldots, 2), \ldots).$$

By **5.29**,

$$Z(S_3) = \tfrac{1}{6}(t_1^3 + 3t_1t_2 + 2t_3),\ Z(C_4) = \tfrac{1}{4}(t_1^4 + t_2^2 + 2t_4),\ Z(C_4)_{t_j=2} = 6$$

giving $\frac{1}{6}(6^3 + 3.6^2 + 2.6) = 56$ colorings.

*Boolean Functions.* Just as $G \leq S(N)$ and $H \leq S(R)$ induce on $\mathrm{Map}(N, R)$ the power group $H^G$, the composition $G[H] \leq S(N \times R)$ induces a permutation group $[H]^G$ on $\mathrm{Map}(N, R)$ called the *exponentiation group.* The elements of $[H]^G$ are all $(n + 1)$-tuples $[g; h_1, \ldots, h_n]$ such that

$$[g; h_1, \ldots, h_n]f = f' \in \mathrm{Map}(N, R)$$

where $f'$ is defined by

$$f'(a_i) = h_i\, fg(a_i) \qquad (a_i \in N).$$

The degree of $[H]^G$ is $r^n$, the order is $|G||H|^n$, and we have as abstract groups $G[H] \cong [H]^G$.

An interesting example where the exponentiation group acts as a group of symmetries are *Boolean functions* (or *truth functions*). Let $E_1, \ldots, E_n$ be variables with truth values 1 (=true) or 0 (=false). The operations $\wedge$ and $\vee$ are defined as usual:

| $\wedge$ | 1 | 0 |
|---|---|---|
| 1 | 1 | 0 |
| 0 | 0 | 0 |

| $\vee$ | 1 | 0 |
|---|---|---|
| 1 | 1 | 1 |
| 0 | 1 | 0 |

To every well-formed sentence $F(E_1, \ldots, E_n; \wedge, \vee)$ consisting only of the $E_i$'s, parentheses (,) and $\wedge$, $\vee$, we assign a function $f: \{0, 1\}^n \rightarrow \{0, 1\}$ by setting

$$f(a_1, \ldots, a_n) := F(E_1, \ldots, E_n; \wedge, \vee)|_{E_i = a_i} \quad \text{for all } i.$$

For example, if $F = (E_1 \vee E_2) \wedge E_3$ then

$$f(1, 0, 1) = f(0, 1, 1) = f(1, 1, 1) = 1$$

and

$$f(a_1, a_2, a_3) = 0 \quad \text{otherwise.}$$

We call $f$ the *Boolean function* corresponding to $F$ (see ex. IV.2.12). It is plausible to regard $F$ and $F'$ as equal whenever the corresponding Boolean functions $f, f'$ are identically equal. Furthermore, we do not distinguish between $F$ and $F'$ if $F'$ can be obtained from $F$ by permuting variables and/or replacing some of the $E_i$'s by their negations *non-*$E_i$. We want to find the number of different Boolean functions. Set $N = \{1, \ldots, n\}$ and $R = \{0, 1\}$. The domain of the Boolean functions is then $R^N$ and the symmetry group is $[S_2]^{S_n}$ with $S_2$ accounting for the possible negations $E_i \rightarrow$ *non-*$E_i$. Hence we have the following result.

**5.40 Proposition.** *The number of Boolean functions $f: \{0, 1\}^n \to \{0, 1\}$ with precisely $k$ true-values is the coefficient of $x^k$ in*

$$Z([S_2]^{S_n}; 1 + x). \quad \square$$

**Example.** For $n = 3$ one calculates

$$Z([S_2]^{S_3}; t_1, \ldots, t_8) = \tfrac{1}{48}[t_1^8 + 7t_2^4 + 6t_1^4 t_2^2 + 6t_2^4 + 12t_4^2 + 8t_1^2 t_3^2 + 8t_2 t_6],$$

whence

$$Z([S_2]^{S_3}; 1 + x) = x^8 + x^7 + 3x^6 + 3x^5 + 6x^4 + 3x^3 + 3x^2 + x + 1.$$

There are, for instance, 6 different sentences $F(E_1, E_2, E_3; \wedge, \vee)$ with precisely 4 true-values.

*Enumeration of Graphs.* For our final applications let us return to finite graphs. Counting problems arising in the context of graph theory were the original ones considered by Polya and have motivated many important advances of the theory in later years.

We remarked in section 1 that there are $2^{\binom{n}{2}}$ simple labeled graphs on $n$ vertices; more generally, there are clearly

$$\binom{\binom{n}{2}}{k}$$

simple labeled graphs with $n$ vertices and $k$ edges. Let us discuss the corresponding problem for unlabeled graphs. We denote by $V = \{1, \ldots, n\}$ the set of vertices and by $V^{(2)}$ the set of all distinct unordered pairs of vertices. Every simple graph $G(V, E)$ corresponds uniquely to the subset $E \subseteq V^{(2)}$ of its edges, and two graphs are isomorphic if there exists a permutation $g$ of the vertices which carries one edge-set onto the other. Hence the symmetry group is $S_n^{(2)} = \{g^*: g \in S_n\}$ on $V^{(2)}$ where

$$g^*\{i, j\} = \{gi, gj\} \qquad (g \in S_n).$$

$S_n^{(2)}$ is called the *pair group*.

**5.41 Theorem** (Polya). *Let $g_{n,k}$ be the number of simple unlabeled graphs with $n$ vertices and $k$ edges. Then*

$$g_n(x) = \sum_{k=0}^{\binom{n}{2}} g_{n,k} x^k = Z(S_n^{(2)}; 1 + x). \quad \square$$

To compute $Z(S_n^{(2)})$ one proceeds as follows. For any term $t(g) = t_1^{b_1(g)} \dots t_n^{b_n(g)}$ of $Z(S_n)$, one calculates the corresponding term $t(g^*)$ of $Z(S_n^{(2)})$ according to whether in $g^*\{i, j\} = \{gi, gj\}$ the vertices $i$ and $j$ are in the same cycle of $g$ or not.

**Example.** Let $n = 4$. Then:

| terms in $Z(S_4)$ | terms in $Z(S_4^{(2)})$ |
|---|---|
| $t_1^4$ | $t_1^6$ |
| $t_1^2 t_2$ | $t_1^2 t_2^2$ |
| $t_1 t_3$ | $t_3^2$ |
| $t_2^2$ | $t_1^2 t_2^2$ |
| $t_4$ | $t_2 t_4$ |

hence

$$Z(S_4^{(2)}) = \tfrac{1}{24}(t_1^6 + 9t_1^2 t_2^2 + 8t_3^2 + 6t_2 t_4),$$

and thus

$$g_n(x) = 1 + x + 2x^2 + 3x^3 + 2x^4 + x^5 + x^6.$$

Figure 5.10 shows all simple graphs on 4 vertices.

The number $\bar{g}_n$ of self-complementary graphs $G$ (i.e., $G(V, E) \cong G(V, V^{(2)} - E)$) is, according to **5.38**(iv), given by

$$\bar{g}_n = Z(S_n^{(2)}; 0, 2, 0, 2, \dots).$$

For $n = 4$ this gives one such graph, the path of length 3.

**5.41** can be readily generalized to the enumeration of *subgraphs* of a given graph $G$. Let $G = G(V, E)$ be a simple unlabeled graph on $n$ vertices. The group $\mathscr{V}(G)$ of edge-preserving permutations of $V$ is called the *vertex-group* of $G$. $\mathscr{V}(G)$ induces

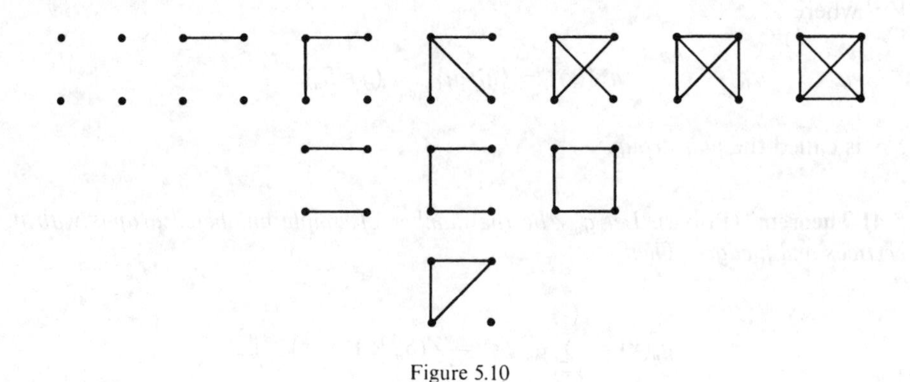

Figure 5.10

a group $\mathscr{E}(G) = \{g^*: g \in \mathscr{V}(G)\}$ on the edge-set $E$, called the *edge-group*, just as above $S_n$ induced $S_n^{(2)}$, i.e.,

$$g^*\{i, j\} = \{gi, gj\} \quad \text{for } \{i, j\} \in E.$$

Hence in the case $G = K_n$ we have $\mathscr{E}(K_n) = S_n^{(2)}$. The subset-patterns of $E$ under the group $\mathscr{E}(G)$ correspond bijectively to the non-isomorphic subgraphs of $G$ (on $n$ vertices). Notice that two subgraphs of $G$ may be isomorphic graphs but non-isomorphic as *subgraphs* of $G$.

**Example.** The subgraphs $G_1$ and $G_2$ of $G$ shown in Figure 5.11 are clearly isomorphic but there is no automorphism of $G$ carrying the edges of $G_1$ onto those of $G_2$, since any automorphism must fix the edge $a$.

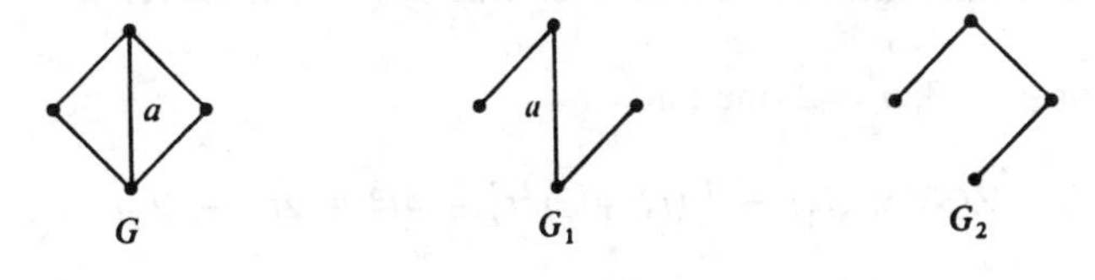


Figure 5.11

**5.42 Proposition.** *Let $G$ be a simple graph and $g_k(G)$ the number of non-isomorphic subgraphs of $G$ with $k$ edges. The generating function of $g_k(G)$ is then given by*

$$g(G, x) = \sum_{k \geq 0} g_k(G)x^k = Z(\mathscr{E}(G); 1 + x).$$

*The number $\bar{g}(G)$ of self-complementary subgraphs of $G$ is*

$$\bar{g}(G) = Z(\mathscr{E}(G); 0, 2, 0, 2, \ldots). \quad \square$$

We now use this result to count 2-*colored graphs*. Recall that $G(V, E)$ is called 2-colored if there exists a partition of $V$ into two blocks $R$ and $S$, called color classes, such that every edge of $G$ has one vertex in $R$ and the other in $S$. Thus 2-colored graphs are just the subgraphs of the complete bipartite graphs $K_{m,n}$ with the defining vertex-sets corresponding to $R$ and $S$, $|R| = m$ and $|S| = n$. Hence if $b_{m,n}(x)$ is the generating function of all 2-colored graphs with precisely $m$ vertices of one color and $n$ vertices of the other color, then

$$b_{m,n}(x) = Z(\mathscr{E}(K_{m,n}); 1 + x).$$

Let us determine $\mathscr{E}(K_{m,n})$. Denote by $R$ and $S$ the defining vertex-sets with $|R| = m$ and $|S| = n$.

*Case* a: $m \neq n$. Clearly, $\mathscr{V}(K_{m,n})$ has $R$ and $S$ as its only orbits and induces the symmetric group on each. Hence

$$\mathscr{E}(K_{m,n}) \cong S_m \times S_n,$$

where $S_m \times S_n = \{(g, h): g \in S_m, h \in S_n\}$ is the direct product of $S_m$ and $S_n$ defined by

$$(g, h)(i, j) = (gi, hj) \quad \text{for } (i, j) \in R \times S,$$

and thus

$$b_{m,n}(x) = Z(S_m \times S_n; 1 + x).$$

It is not hard to compute the cycle indicator $Z(G \times H)$ of two arbitrary groups by examining the contribution of $g$ and $h$ separately (see the exercises).

**Example.** For $m = 3, n = 2$ one finds

$$Z(S_3 \times S_2) = \tfrac{1}{12}(t_1^6 + 3t_1^2t_2^2 + 4t_2^3 + 2t_3^2 + 2t_6)$$

whence

$$b_{3,2}(x) = 1 + x + 3x^2 + 3x^3 + 3x^4 + x^5 + x^6$$

(see Figure 5.12).

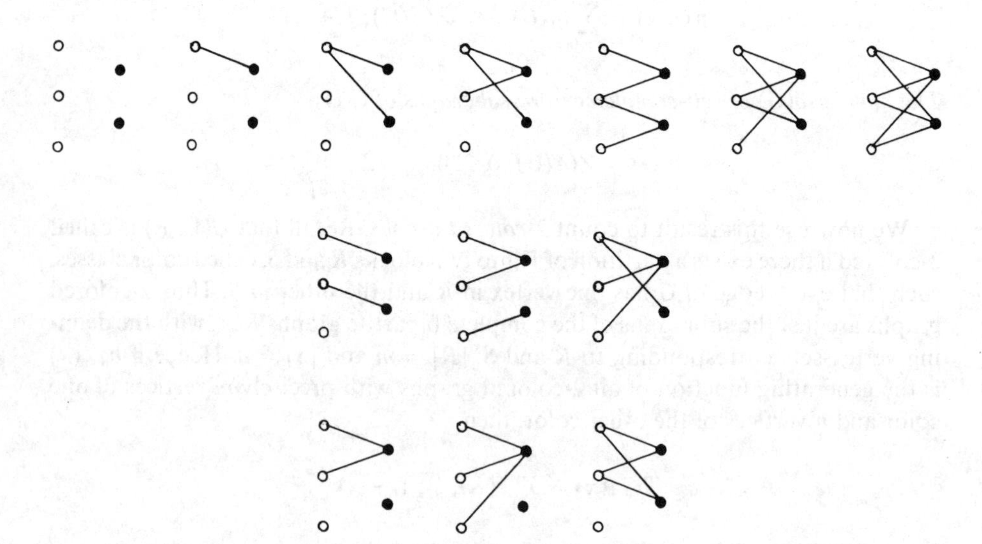

Figure 5.12. All 2-colored graphs with 3 vertices colored white and 2 vertices colored black, where we distinguish between the colors.

*Case* b: $m = n$. In this case it may happen that vertices of $R$ are mapped into $S$ by some $g \in \mathscr{V}(K_{m,n})$ but then all of $R$ is mapped by $g$ onto $S$. Let $R = \{a_1, \ldots, a_n\}$, $S = \{b_1, \ldots, b_n\}$. We let the edge-set $E$ correspond to the set of all functions $f\colon \{R, S\} \to \{1, \ldots, n\}$ by means of

$$\{a_i, b_j\} \in E \leftrightarrow f(R) = i, f(S) = j.$$

Since an arbitrary element $g \in \mathscr{V}(K_{m,n})$ either leaves $R$ and $S$ invariant or interchanges them it follows from the definition of the exponentiation that

$$\mathscr{E}(K_{n,n}) \cong [S_n]^{S_2}.$$

Hence

$$b_{n,n}(x) = Z([S_n]^{S_2}; 1 + x).$$

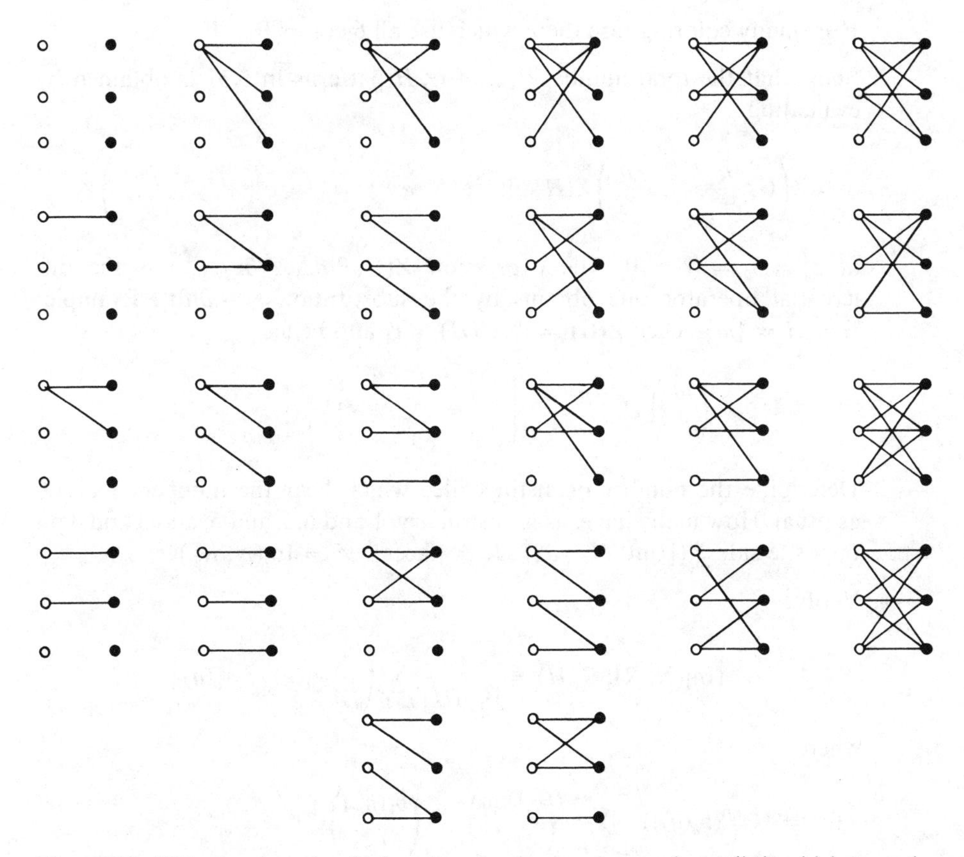

Figure 5.13. All 2-colored graphs with 3 vertices of each color where we do not distinguish between the colors.

**Example.** For $n = 3$ we have

$$Z([S_3]^{S_2} = \tfrac{1}{72}(t_1^9 + 12t_1^3t_2^3 + 8t_3^3 + 9t_1t_2^4 + 18t_1t_4^2 + 24t_3t_6),$$

thus

$$b_{3,3}(x) = 1 + x + 2x^2 + 4x^3 + 5x^4 + 5x^5 + 4x^6 + 2x^7 + x^8 + x^9.$$

(see Figure 5.13).

EXERCISES V.4

→ 1. Suppose we color the faces of a cube $W$ with 6 colors $F_1, \ldots, F_6$. Two colorings are considered equal if they can be transformed into each other by the permutation

$$\begin{pmatrix} F_1 F_2 F_3 F_4 F_5 F_6 \\ F_2 F_1 F_4 F_3 F_6 F_5 \end{pmatrix}.$$

How many colorings are there which use all 6 colors?

2. Show that the total number $|\mathfrak{M}|$ of $G, H$-patterns in **5.37** is obtained by evaluating

$$Z\left(G; \frac{\partial}{\partial z_1}, \ldots, \frac{\partial}{\partial z_n}\right) Z(H; e^{z_1 + z_2 + \cdots}, e^{2(z_2 + z_4 + \cdots)}, e^{3(z_3 + z_6 + \cdots)}, \ldots)$$

at $z_1 = z_2 = \cdots = 0$. The expression $Z(G; \partial/\partial z_1, \partial/\partial z_2, \ldots)$ is the differential operator one obtains by the substitution $t_i \to \partial/\partial z_i$. Example: $G = H = \{id\}$. Then $Z(G) = t_1^n$, $Z(H) = t_1^r$ and hence

$$\left(\frac{d}{dz_1}\right)^n e^{r(z_1 + z_2 + \cdots)}\bigg|_{z_i = 0} = r^n e^{r(z_1 + z_2 + \cdots)}\bigg|_{z_i = 0} = r^n.$$

→ 3. Determine the number of distinct dice which bear the numbers $1, \ldots, 6$ as usual. How many have, as is customary, 1 and 6, 2 and 5, and 3 and 4 on opposite sides? (Hint: $N = \mathbb{N}_6$, $R = \{\text{faces}\}$, $\mathcal{F} = \mathrm{Inj}(N, R)$.)

→ 4. Verify:

$$\gamma(\mathrm{Inj}(N, R); G, H) = \frac{1}{|G||H|} \sum_{h \in H} \left( \sum_{g \in G} v_1(g) \ldots v_n(g) \right)$$

where

$$v_k(g) := k^{b_k(g)}(b_k(g)!) \sum_{\substack{(r_1, \ldots, r_m) \\ \sum r_i = b_k(g)}} \binom{b_k(h_1)}{r_1} \cdots \binom{b_k(h_m)}{r_m} x_1^{r_1 k} \ldots x_m^{r_m k}$$

$$(k = 1, \ldots, n).$$

5. Prove, by analogy to ex. 2, that the number $|\mathfrak{M}_{\mathrm{Inj}}|$ of injective $G$, $H$-patterns is obtained by evaluating

$$Z\left(G;\frac{\partial}{\partial z_1},\ldots,\frac{\partial}{\partial z_n}\right)Z(H;1+z_1,1+2z_2,\ldots,1+rz_r)$$

at $z_1=\cdots=z_r=0$. Show that for $n=r$ this can be reduced to

$$|\mathfrak{M}_{\mathrm{Bij}}|=Z\left(G;\frac{\partial}{\partial z_1},\ldots,\frac{\partial}{\partial z_n}\right)Z(H;z_1,2z_2,\ldots,nz_n)\quad\text{at } z_1=\cdots=z_n=0.$$

6. Compute $|\mathfrak{M}_{\mathrm{Inj}}|$ for $G=S_n$ and $H$ arbitrary. (Answer: This is the coefficient of $x^n$ in $Z(H;1+x)$. Can you prove this directly?)

7. Find the number of self-equivalent subset patterns of $\{1,\ldots,12\}$ where $G=C_{12}$ and $G=D_{12}$.

→ 8. Let $\pi\in\mathscr{P}(N)$, $\sigma\in\mathscr{P}(R)$ with $\operatorname{type}(\pi)=1^{k_1}\ldots n^{k_n}$, $\operatorname{type}(\sigma)=1^{l_1}\ldots r^{l_r}$. Show that the number of $\pi$, $\sigma$-distributions of $N$ in $R$ is given by

$$d(\pi,\sigma)=\frac{1}{\prod_{j=1}^{r}(j!^{l_j})}\sum_{h\in H(\sigma)}Z\left(\prod_{i=1}^{n}(S_i)^{k_i};\lambda_1(h),\ldots,\lambda_n(h)\right)$$

with $\lambda_k(h)=\sum_{j|k}jb_j(h)$ and $H(\sigma)=\prod_{j=1}^{r}(S_j)^{l_j}$. The cycle indicators $Z(\prod S_i^{k_i})$ can now be computed as in ex.V.3.10. Find $d(\pi,\sigma)$ for $\operatorname{type}(\pi)=1^2 2^2$, $\operatorname{type}(\sigma)=1^1 2$.

→ 9. Complete the proof of **5.39**.

→ 10. We are given $n$ cubes whose faces are colored red or blue. Equivalence of colorings is defined by permutations of the cubes and rotations of individual cubes. How many color patterns are there? (Answer: $(n+9)!/n!9!$.) How many color patterns are invariant when the colors are interchanged? (Answer: E.g., 27 for $n=4$.)

11. Fill in the details in the proof **5.40**.

12. Determine the 6 different sentences $F(E_1,E_2,E_3;\wedge,\vee)$ with precisely 4 true-values.

13. Determine $Z(S_5^{(2)})$ and $g_5(x)$ (in the notation of **5.41**).

→ 14. Find the group of symmetries for the enumeration of unlabeled simple digraphs, i.e., there are no loops and at most one arrow $a\to b$ for each pair $a\neq b$ of vertices. Compute the number of these graphs on $n$ vertices.

15.* Let $s_n$ be the number of self-complementary graphs on $n$ vertices and $\vec{s}_n$ the corresponding number for directed graphs, where the complement of a digraph has an arrow $a\to b$ if and only if the arrow $a\to b$ does not appear in the original digraph. Prove that $s_{4n}=\vec{s}_{2n}$. (Read)

## Notes

Generating functions have been a standard tool of combinatorists since Euler and Laplace and have always occupied a central position in the subject. In particular, the connection between the analytical derivation of an identity and its combinatorial interpretation as a bijection between sets has been a prime source of combinatorial activity; see, e.g., Riordan [1, 2], Andrews [2], and Foata [2]. Section 1 is based on Doubilet–Rota–Stanley [1], Smith [1], Scheid [1], and section 2 on Bender–Goldman [1]. For a comprehensive account of graphical enumeration, the reader is referred to Harary–Palmer [1]. Polya's paper [1] and de Bruijn's survey article [3] and [1, 4] provide the basis for sections 3 and 4; see also Kerber [1].

Chapter VI

# Matroids: Introduction

Matroids were introduced in the early 1930's in an attempt to axiomatize and generalize basic notions in linear algebra such as dependence, basis and span. The importance of matroids came to be appreciated with the discovery of new classes of matroids so that today we may rightly consider them as a unifying concept for a large part of combinatorics opening up basic combinatorial questions to algebraic ideas and methods. One of these fields is graph theory; in fact, it was precisely this correspondence between concepts in linear algebra and concepts in graph theory which set the theory of matroids on its way. Since then, other branches of combinatorics such as transversal theory, incidence structures and combinatorial lattice theory have been brought successfully into the realm of matroid theory. Indeed, it is this exchange of ideas from various fields which is one of the most gratifying aspects of matroid theory and also one measure of its success.

We begin this chapter with a discussion of the basic notions of matroid theory and their use in axiomatizing arbitrary matroids. Section 2 introduces the most important classes of matroids: vector spaces, graphs, transversal systems, and incidence geometries. Apart from yielding interesting results, these examples will also motivate much of the further development of the subject. After a discussion of construction methods, we conclude our account of the fundamentals of matroid theory with a study of two of the most fruitful concepts: duality and connectivity.

A word regarding the terminology is in order. It has already been pointed out in section II.3 that the names matroid and pregeometry are both used in the literature; various other terms such as independence space can also be found. Similarly, duality and orthogonality are names for the same concept. Usually, it is the individual approach of the author that prompts the choice of terminology. Pregeometry is preferred by those who emphasize the geometric origin via theorem **2.29**. In this book, the set-theoretical point of view has been taken from the beginning, whence the term matroid and the definition through the closure operator have been adopted. The reader should have no difficulty in equating corresponding notions when consulting other books on the subject.

# 1. Fundamental Concepts

After repeating the definition of a matroid and its associated lattice from section II.3 and presenting some simple examples, we begin the theory by discussing various ways to axiomatize matroids. Apart from providing interesting comparisons, these axiom systems will prove very useful later on; in most examples one or the other system will suggest itself in a natural way.

## A. Definition and Examples

**Definition.** A *matroid* $\mathbf{M}(S)$ is a set $S$ together with a closure $A \to \bar{A}$ such that for all $p, q \in S$, $A \subseteq S$:

(i) $p \notin \bar{A}$, $p \in \overline{A \cup q} \Rightarrow q \in \overline{A \cup p}$ (exchange axiom)
(ii) $\exists B \subseteq A$, $|B| < \infty$ with $\bar{B} = \bar{A}$ (finite basis).

$\mathbf{M}(S)$ is called a *simple matroid* or a *combinatorial geometry* (or just *geometry*) if
(iii) $\bar{\varnothing} = \varnothing$ and $\bar{p} = p$ for all $p \in S$.

$A \subseteq S$ is called *closed* if $A = \bar{A}$. To every matroid $\mathbf{M}(S)$ we associate the lattice $L(S)$ of closed subsets, ordered by inclusion. The lattice operations in $L(S)$ are given by

$$A \wedge B = A \cap B, \qquad A \vee B = \overline{A \cup B} \qquad (A, B \in L(S)).$$

Clearly, for any $A, B \subseteq S$

$$\overline{A \cap B} \subseteq \bar{A} \cap \bar{B}, \qquad \overline{A \cup B} = \overline{\bar{A} \cup \bar{B}}.$$

The members of $L(S)$ are more often called the *flats* of $\mathbf{M}(S)$. Another name is *subspace*, which we shall use when discussing geometric structures.

Two matroids $\mathbf{M}_1(S_1)$, $\mathbf{M}_2(S_2)$ are *isomorphic* if there exists a bijection $\Phi$: $S_1 \to S_2$ such that $p \in \bar{A}$ if and only if $\Phi p \in \overline{\Phi A}$. We then write $\mathbf{M}_1(S_1) \cong \mathbf{M}_2(S_2)$. It may be useful to repeat at this point the fundamental theorem **2.29**.

**6.1 Theorem.**
(i) *The lattice of flats of a matroid is geometric.*
(ii) *If L is a geometric lattice with point set S, then the closure*

$$\bar{A} = \{p \in S : p \leq \sup A\}$$

*induces a simple matroid* $\mathbf{M}(S)$ *on S whose lattice of flats L(S) is isomorphic to L where the isomorphism* $\phi$ *is given by*

$$\phi(A) = \sup A \qquad (A \subseteq S)$$
$$\phi^{-1}(x) = \{p \in S : p \leq x\} \qquad (x \in L). \quad \square$$

Since geometric lattices have a rank function we can now define the *points*, *lines* and *planes* of a matroid $\mathbf{M}(S)$ as the flats of rank 1, 2 and 3, respectively, and similarly *copoints*, *colines*, and *coplanes* as the flats of corank 1, 2, and 3. The *rank of the matroid* is the rank of $S$. A *loop* is an element $p \in S$ with $p \in \bar{\varnothing}$; $p$ and $q$ are called *parallel* if $\bar{p} = \bar{q}$. All these definitions have already been given in section II.3. The reader is advised to go through the relevant parts of section II.3 as the subject is developed to see the correspondence between the set-theoretical notions and their order-theoretical counterparts.

Whenever possible we shall draw a geometry $\mathbf{M}$ as embedded in the real Euclidean space of dimension $\leq 3$. If there are 3 or more points on a line in $\mathbf{M}$ then we pass a curve through these points, if possible a straight line. If there are 4 or more points on a plane in $\mathbf{M}$ then we represent this plane as a surface in the Euclidean space, if possible as a Euclidean plane. Trivial lines through 2 points or trivial planes through 3 points will often be omitted. Figure 6.1 depicts all geometries with at most 4 points in their Euclidean representation.

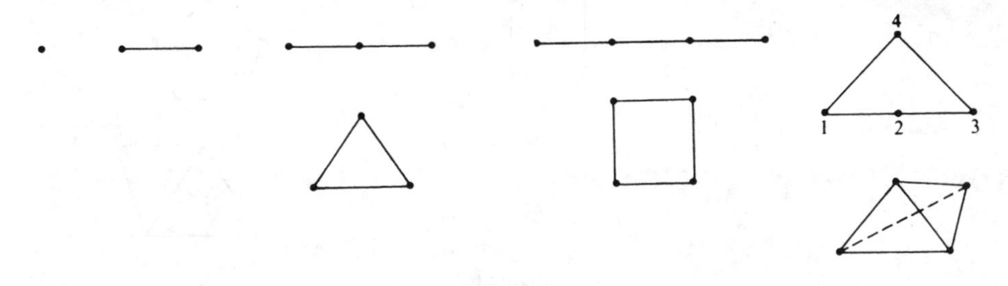


Figure 6.1

Take, for instance, the geometry in the upper right-hand corner of Figure 6.1. The drawing means that the geometry has 4 points and is of rank 3. The only nontrivial line is $\{1, 2, 3\}$, the trivial lines are $\{1, 4\}$, $\{2, 4\}$ and $\{3, 4\}$. It is easy to extend these representations in order to include loops and parallel elements. The loops are drawn separately and encircled, the parallel elements are placed next to the Euclidean point corresponding to their 1-rank closure.

We now describe briefly some matroids suggested by the standard geometric lattices of section II.3.

*Free Matroids and Uniform Matroids.* The *free matroid* $\mathbf{FM}(S)$ on $S$ is the matroid in which every set is closed. By the finite basis property, $S$ must be finite and it is clear that two free matroids are isomorphic if and only if their point sets have the same cardinality. Hence we may define $\mathbf{FM}(n)$ as the *free matroid of rank n*, for $n < \infty$. By definition, the lattice of flats in $\mathbf{FM}(n)$ is isomorphic to the *Boolean algebra* $\mathscr{B}(n)$. The matroids $\mathbf{FM}(1)$ through $\mathbf{FM}(4)$ are shown in Figure 6.1.

We slightly generalize this concept. Let $n \geq 2$. By deleting in $\mathscr{B}(n)$ all levels from rank $k$ up to rank $n$, $0 \leq k \leq n$, and replacing them by a new maximal element we obtain a new lattice $\mathscr{B}_k(n)$, called the *uniform lattice* with parameters $n$ and $k$.

$\mathscr{B}_k(n)$ is of rank $k$ and it is almost immediate that $\mathscr{B}_k(n)$ is again a geometric lattice. In fact, we shall see in section 3.D that this "truncation" yields a geometric lattice whenever the initial lattice is geometric. The corresponding matroid is called the *uniform matroid* $\mathbf{U}_k(n)$. It is defined by the closure on $S$:

$$\bar{A} = \begin{cases} A & \text{if } |A| \leq k-1 \\ S & \text{if } |A| \geq k. \end{cases}$$

Notice that $\mathbf{U}_0(n)$ consists of $n$ loops and $\mathbf{U}_1(n)$ of $n$ parallel non-loops. By definition, $\mathbf{U}_n(n) = \mathbf{FM}(n)$. Figure 6.2 gives the smallest uniform matroids $\mathbf{U}_k(n)$ with $2 \leq k < n \leq 5$ in their Euclidean representation.

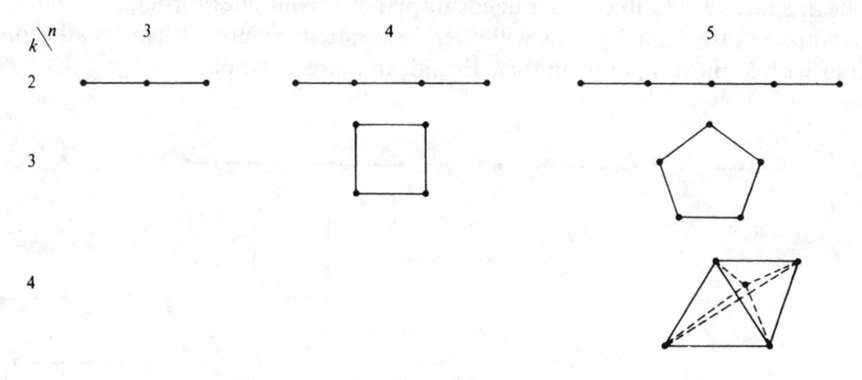


Figure 6.2

*Covering Matroids.* In a uniform matroid $\mathbf{U}_k(n)$ all copoints have the same cardinality $k-1$, which implies trivially that every $(k-1)$-subset of $S$ is in a unique copoint. We generalize this fact by allowing copoints of varying cardinality.

**Definition.** Let $S$ be a set and $t \in \mathbb{N}$. A collection $\mathfrak{H}$ of subsets of $S$ is called a *t-partition* if

(i) $A \in \mathfrak{H} \Rightarrow |A| \geq t$,
(ii) each $t$-subset of $S$ is contained in a unique member of $\mathfrak{H}$.

The name $t$-partition derives from the fact that 1-partitions are just partitions in the ordinary sense.

**6.2 Proposition.** *Let $\mathfrak{H}$ be a t-partition of S with $|\mathfrak{H}| \geq 2$. Then there is a matroid* $\mathbf{M}(S, \mathfrak{H})$ *on S whose copoints are precisely the sets in $\mathfrak{H}$.*

*Proof.* Define a closure on $S$ by

$$\bar{A} = \begin{cases} A & \text{if } |A| \leq t-1 \\ H & \text{if } |A| \geq t \text{ and } A \text{ is contained in the (necessarily unique) set } H \in \mathfrak{H} \\ S & \text{otherwise.} \end{cases}$$

Clearly, the closed sets are all sets $A$ with $|A| \leq t - 1$, the sets in $\mathfrak{H}$, and $S$. From this it follows easily that the lattice of flats satisfies the semimodular law and is thus geometric with the copoints being precisely the sets in $\mathfrak{H}$. □

A matroid $\mathbf{M}$ is called a *covering matroid* (by some authors, paving matroid) if $\mathbf{M} \cong \mathbf{M}(S, \mathfrak{H})$ for some $t$-partition $\mathfrak{H}$ with $|\mathfrak{H}| \geq 2$. Notice that the covering matroid induced by a $t$-partition has rank $t + 1$. Uniform matroids of rank $\geq 2$ are therefore special examples of covering matroids and it is clear that every simple matroid of rank 2 or 3 is a covering matroid. Special as these covering matroids may appear, they predominate in an enumeration of geometries of small rank and it seems probable that this may be the case in general.

*Vector Space Matroids.* Any vector space $V(n, K)$ of rank $n$ over a division ring $K$ is, together with the linear closure, a matroid, denoted by $\mathbf{M}(V(n, K))$. We discussed in **2.32** the relationship between vector space matroids and projective geometries. The simple matroid underlying $\mathbf{M}(V(n, K))$ is the *projective geometry* $\mathbf{PG}(n - 1, K)$ *of g-dimension* $n - 1$. Of particular interest are the projective geometries $\mathbf{PG}(n - 1, q)$ over the finite field $GF(q)$. The projective plane $\mathbf{PG}(2, 2)$ is called the *Fano plane*; its Euclidean representation is shown in Figure 6.3. Because of its importance we denote it simply by the symbol $\mathbf{F}$.

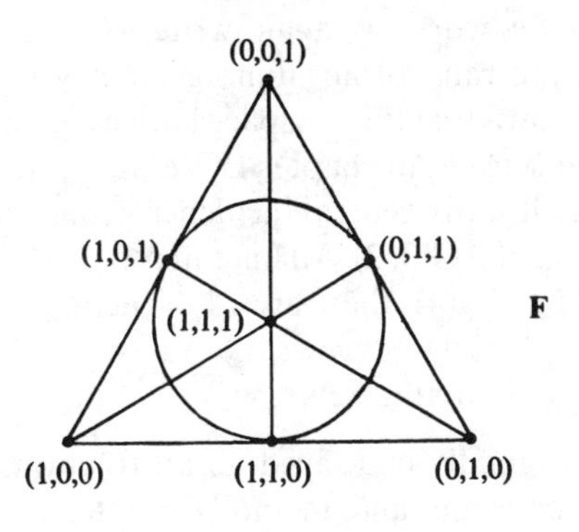


Figure 6.3

*Matroids Induced by Graphs.* We have seen in **2.31** that the simple matroid corresponding to the partition lattice $\mathscr{P}(n)$ can be described as the matroid defined on the edges of the complete graph $K_n$ with a set $A$ of edges being closed if and only if all connected components in the graph generated by $A$ are complete subgraphs of $K_n$. We call this the *polygon matroid* $\mathbf{P}(K_n)$ for reasons to be made clear soon. Hence an edge in $K_n$ corresponds to a point in $\mathbf{P}(K_n)$, a triangle corresponds to a 3-point line, a complete quadrangle to a 4-point plane, etc. Figure 6.4 gives a Euclidean representation of $\mathbf{P}(K_4)$. $\mathbf{P}(K_5)$ is shown in Figure 6.10. Notice that $\mathbf{P}(K_3) \cong \mathbf{PG}(1, 2) \cong \mathbf{U}_2(3)$.

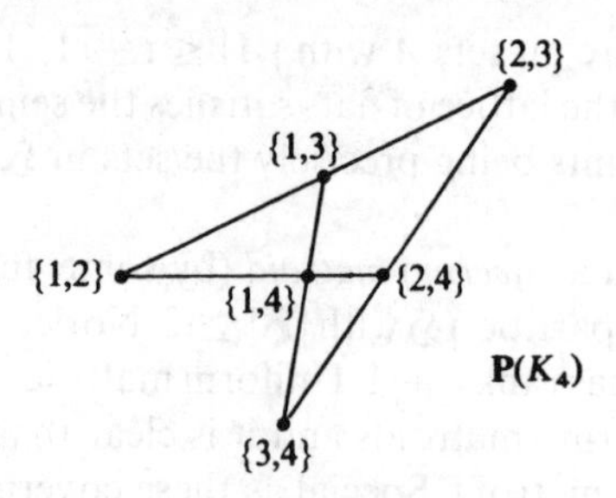


Figure 6.4

*Affine Geometries.* Let $V(n, K)$ be a vector space over the division ring $K$. Recall from linear algebra that the *affine closure* of a set $A \subseteq V$ is the set of all vectors which are affinely dependent on $A$, i.e.,

$$\text{aff } A := \left\{ v \in V : v = \sum_{i=1}^{t} \lambda_i v_i \text{ for some } v_1, \ldots, v_t \in A, \sum_{i=1}^{t} \lambda_i = 1 \right\}.$$

Since $V$ is finite-dimensional the affine closure has the finite basis property. If $q \notin \text{aff } A$, $q \in \text{aff}(A \cup p)$, then $q = \lambda p + \sum \lambda_i v_i$ with $\lambda + \sum \lambda_i = 1$ and $\lambda \neq 0$. Hence $p = \lambda^{-1} q - \sum (\lambda^{-1}\lambda_i) v_i, \lambda^{-1} - \sum (\lambda^{-1}\lambda_i) = 1$, and thus $p \in \text{aff}(A \cup q)$. Any single vector is trivially closed whence we obtain a geometry, the *affine geometry* $\mathbf{AG}(n, K)$ *of g-dimension n*, where we again write $\mathbf{AG}(n, q)$ for $K = GF(q)$. As for projective geometries, the rank of an affine geometry is one higher than its $g$-dimension. It is easily shown that the subspace lattices of $\mathbf{AG}(n, K)$ are precisely the affine lattices $\mathscr{A}(n, K)$ discussed in chapter I. We shall return to these questions in section 2.D when we shall study general incidence geometries. Figure 6.5 depicts a Euclidean representation of $\mathbf{AG}(3, 2)$. All lines in $\mathbf{AG}(3, 2)$ are trivial. All planes are isomorphic to $\mathbf{AG}(2, 2) \cong \mathbf{U}_3(4)$; there are 14 of them.

## B. Independent Sets and Spanning Sets

Since matroids were originally conceived as abstractions of vector spaces it is natural to ask which concepts and methods can be carried over to arbitrary matroids.

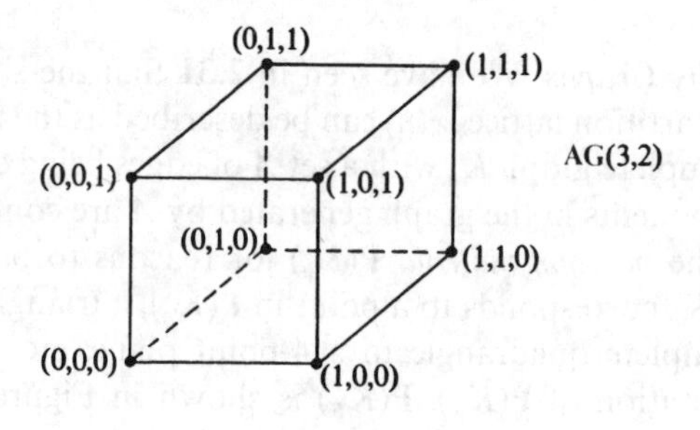


Figure 6.5

**Definition.** A set $A \subseteq S$ in a matroid $\mathbf{M}(S)$ is called *independent* if $p \notin \overline{A - p}$ for all $p \in A$. $A$ is called *dependent* if it is not independent.

**6.3 Proposition.** *Let* $\mathbf{M}(S)$ *be a matroid. Then*:

(i) *Any subset of an independent set is independent.*
(ii) *Any independent set is finite.*
(iii) *If $A$ is independent and $p \notin A$, then $A \cup p$ is independent if and only if $p \notin \bar{A}$.*

*Proof.* (i) is an immediate consequence of the definition. If $A$ is an infinite independent set then by the finite basis axiom there exists a finite subset $B \subseteq A$ with $\bar{B} = \bar{A}$. Hence $p \in \bar{B} \subseteq \overline{A - p}$ for all $p \in A - B$, contradicting the definition of an independent set. In (iii), the implication "$A \cup p$ independent $\Rightarrow p \notin \bar{A}$" is clear. If, conversely, $p \notin \bar{A}$ then $q \notin \overline{A - q}$ for all $q \in A$ and thus by the exchange axiom $q \notin \overline{(A - q) \cup p}$. Hence $A \cup p$ is independent. □

**Definition.** The maximal independent subsets $B$ of $A \subseteq S$ are called the *bases* of $A$. In particular, we call the maximal independent subsets of $S$ the *bases of the matroid* $\mathbf{M}(S)$. The minimal dependent sets are called the *circuits* of $\mathbf{M}(S)$.

To justify these definitions, notice that every subset $A \subseteq S$ contains maximal independent sets. If $B \subseteq A$ is finite with $\bar{B} = \bar{A}$ and $B_0 \subseteq B$ maximally independent in $B$, then we claim that $B_0$ is also maximally independent in $A$. Indeed, by **6.3**(iii), we have $p \in \bar{B}_0$ for all $p \in B - B_0$, thus $\bar{B}_0 = \bar{B} = \bar{A}$ and $B_0$ cannot be enlarged in $A$ to a larger independent set.

The following theorem is the key result that carries over from vector spaces to arbitrary matroids.

**6.4 Basis Theorem.** *Let* $\mathbf{M}(S)$ *be a matroid, and* $A \subseteq S$. *Then*:

(i) *All bases of $A$ have the same (finite) cardinality.*
(ii) *For any basis $B$ of $A$ we have $\bar{B} = \bar{A}$.*
(iii) *Any independent set in $A$ can be enlarged to a basis of $A$.*

*Proof.* We have already seen (ii). Let $B$ be a basis of $A$ and $I$ an arbitrary independent subset of $A$. If we can show $|I| \leq |B|$, (i) will follow. In the case $I \subseteq B$ there is nothing to prove. Let $p_1 \in I - B$. Then $p_1 \notin \overline{I - p_1}$, $p_1 \in \bar{B}$ (because of (ii)); hence $B \nsubseteq \overline{I - p_1}$. Therefore there exists $q_1 \in B$ with $q_1 \notin \overline{I - p_1}$ and we infer from **6.3**(iii) that $I_1 = (I - p_1) \cup q_1$ is independent. For $I_1$ we have $|I_1| = |I|$ and $|I_1 \cap B| > |I \cap B|$. Continuing in this way we arrive finally at a set $I_t \subseteq B$ with $|I| = |I_t| \leq |B|$. (iii) is now trivial since all bases have the same cardinality. □

The basis theorem **6.4** suggests the introduction of a rank function for arbitrary subsets in a matroid.

**Definition.** Let $\mathbf{M}(S)$ be a matroid and $A \subseteq S$. The *rank* $r(A)$ of $A$ is the common cardinality of all bases of $A$; $r(\mathbf{M}) = r(S)$ is called the *rank of the matroid* $\mathbf{M}(S)$. The number $r(S) - r(A)$ is the *corank* of $A$.

**6.5 Proposition.** *Let* $\mathbf{M}(S)$ *be a matroid and* $\mathbf{L}(S)$ *the lattice of flats. Then for* $A \subseteq S$:

(i) $r(A) = |B|$ *where* $B$ *is any basis of* $A$.
(ii) $r(A) \leq |A|$, *and* $r(A) = |A| \Leftrightarrow A$ *is independent.*
(iii) $r(A) = r(\bar{A}) =$ *lattice rank of* $\bar{A}$ *in* $L(S)$.

*Proof.* (i) is just the definition of the rank; (ii) is a trivial consequence of it. Let $B = \{b_1, \ldots, b_k\}$ be a basis of $A$. Then $\bar{B} = \bar{A}$, hence $B$ is by **6.3**(iii) also a basis of $\bar{A}$, implying $r(A) = r(\bar{A})$. Since $B$ is independent it follows from **2.28** that $\varnothing <\!\cdot\, \overline{b_1} <\!\cdot\, \overline{b_1 \cup b_2} <\!\cdot \cdots <\!\cdot\, \overline{b_1 \cup b_2 \cup \cdots \cup b_k} = \bar{A}$ is a maximal $0, \bar{A}$-chain in $L(S)$, proving the last assertion in (iii). □

**Definition.** Let $\mathbf{M}(S)$ be a matroid, $A = \bar{A}$ a flat. A subset $H \subseteq A$ is said to *span* $A$ if $\bar{H} = A$. $H$ is called a *spanning set of the matroid* if $\bar{H} = S$.

The terms independent set, basis and spanning set are well-known from linear algebra. The name circuit is borrowed from graph theory where, as we shall see, the circuits correspond precisely to the circuits of the graph.

The following assertions are immediate.

**6.6 Proposition.** *Let* $\mathbf{M}(S)$ *be a matroid. Then*:

(i) $A$ *independent*, $B \subseteq A \Rightarrow B$ *independent.*
(ii) $H$ *spanning*, $K \supseteq H \Rightarrow K$ *spanning.* □

Hence the families of independent sets and non-spanning sets form an (order) ideal in the lattice $2^S$, the families of spanning sets and dependent sets form a filter in $2^S$. The connection between the concepts "independent" and "spanning" is implied by the basis theorem **6.4**(ii).

**6.7 Proposition.** *We have*:

(i) *The minimal spanning sets of a matroid are precisely the bases of the matroid. More generally, the minimal spanning sets of a flat* $A$ *are the bases of* $A$.
(ii) *The maximal non-spanning sets are the copoints of the matroid.* □

Let us denote by $\mathfrak{H}$, $\mathfrak{B}$, $\mathfrak{I}$ and $\mathfrak{K}$ the families of copoints, bases, independent sets and circuits, respectively. Then **6.7** says that any one of these families uniquely determines the other three.

**6.8 Proposition.** *Let* $\mathbf{M}(S)$ *be a matroid with rank function* $r$, *and* $\mathfrak{I}$ *and* $\mathfrak{K}$ *the families of independent sets and circuits. Then for* $A \subseteq S$, $p \in S$:

(i) $p \in \bar{A} \Leftrightarrow (p \in A$ *or* $\exists I \subseteq A$ *with* $I \in \mathfrak{I}, I \cup p \notin \mathfrak{I})$
(ii) $p \in \bar{A} \Leftrightarrow (p \in A$ *or* $\exists C \subseteq \mathfrak{K}$ *with* $p \in C \subseteq A \cup p)$
(iii) $p \in \bar{A} \Leftrightarrow r(A \cup p) = r(A)$.

*Proof.* Let $p \in \bar{A} - A$ and $B$ a basis of $A$. By **6.5**, $r(\bar{A}) = r(A) = |B| < |B \cup p|$ hence $B \cup p \notin \mathfrak{I}$. If, on the other hand, $p \notin \bar{A}$, then $p \notin \bar{B}$ and thus $B \cup p \in \mathfrak{I}$ by **6.3**(iii) for all $B \subseteq A$ with $B \in \mathfrak{I}$. The equivalence of (i) and (ii) is now clear from the definition of the circuits. To see (iii), notice that $p \in \bar{A} \Leftrightarrow \overline{A \cup p} = \bar{A} \Leftrightarrow r(A \cup p) = r(A)$. □

**6.8** states that the closure operator, and hence the matroid, is fully determined by the family $\mathfrak{I}$ as well as by $\mathfrak{K}$, and hence, by our remark above, by $\mathfrak{B}$ and $\mathfrak{H}$, too. It is the aim of this section to answer the converse question. Let $\mathfrak{F} \subseteq 2^S$. Under what conditions is $\mathfrak{F}$ the family of independent sets, bases, etc., of a matroid on $S$? Since the rank function also determines the matroid we can ask similarly when a function $f: 2^S \to \mathbb{N}_0$ is the rank of a matroid $\mathbf{M}(S)$. Since the proofs of these axiomatizations are quite similar and soon become routine we shall not verify all the statements, leaving part of the proofs to the exercises.

**6.9 Independence axioms.** *Let* $\mathbf{M}(S)$ *be a matroid and* $\mathfrak{I}$ *the family of independent sets. Then*:

(i) $\varnothing \in \mathfrak{I}$; $I \in \mathfrak{I}, J \subseteq I \Rightarrow J \in \mathfrak{I}$.
(ii) $I, J \in \mathfrak{I}, |I| < |J| \Rightarrow \exists p \in J - I$ *with* $I \cup p \in \mathfrak{I}$.
(ii′) *If* $A \subseteq S$, *all maximal subsets of* $A$ *which lie in* $\mathfrak{I}$ *have the same cardinality.*
(iii) *There exists* $n \in \mathbb{N}_0$ *with* $|I| \leq n$ *for all* $I \in \mathfrak{I}$.

*Conversely, if* $\mathfrak{I} \subseteq 2^S$ *satisfies* (i), (ii), (iii) *or equivalently* (i), (ii′), (iii) *then* $\mathfrak{I}$ *is the family of independent sets of a unique matroid* $\mathbf{M}(S)$ (*whose closure is defined according to* **6.8**(i)).

*Proof.* We have already seen properties (i) and (iii) for $\mathfrak{I}$. Condition (ii′) is just the basis theorem **6.4**(i), and applied to $I \cup J$, where $I, J \in \mathfrak{I}$, yields (ii). Let $\mathfrak{I} \subseteq 2^S$ satisfy (i) and (iii). It is immediately clear that (ii) and (ii′) are then equivalent whence we may assume all four conditions. We call the common cardinality of the maximal $\mathfrak{I}$-subsets of $A$ the *rank* $r(A)$ of $A$. Let us define the mapping $A \to \bar{A}$ as in **6.8**(i). The properties "$A \subseteq \bar{A}$" and "$A \subseteq B \Rightarrow \bar{A} \subseteq \bar{B}$" are trivially satisfied. To verify the idempotency $\bar{\bar{A}} = \bar{A}$, we show first $r(\bar{A}) = r(A)$ for all $A \subseteq S$. Suppose on the contrary there are sets $I, J \in \mathfrak{I}$, $I \subseteq A$ and $J \subseteq \bar{A}$ with $|I| = r(A) < r(\bar{A}) = |J|$. Then, by (ii), there exists $p \in J$ with $I \cup p \in \mathfrak{I}$ where $p \in \bar{A} - A$ because of the maximality of $I$. Choose, according to **6.8**(i), a maximal set $I' \subseteq A$ with $I' \in \mathfrak{I}$, $I' \cup p \notin \mathfrak{I}$. Then $r(A) = |I'| < |I \cup p|$ whence, by (ii), $I' \cup p \in \mathfrak{I}$, which is impossible. Now if $p \in \bar{\bar{A}} - \bar{A}$ then $I \cup p \in \mathfrak{I}$ for all $I \in \mathfrak{I}$, $I \subseteq A$, and we obtain the contradiction $r(\bar{\bar{A}}) \geq \max_{I \in \mathfrak{I}, I \subseteq A} |I \cup p| = r(A) + 1 = r(\bar{A}) + 1$. The finite basis property is now easily deduced from (iii). It remains to be shown that $A \to \bar{A}$ satisfies the exchange axiom. Let $p, q \in S$, $A \subseteq S$ and $p \notin \bar{A}$, $p \in \overline{A \cup q}$. By the definition of the closure, there exists $I \in \mathfrak{I}$, $I \subseteq A \cup q$ with $I \cup p \notin \mathfrak{I}$. Since $p \notin \bar{A}$ we must have $q \in I$ and $(I - q) \cup p \in \mathfrak{I}$. Hence we conclude that $I' = (I - q) \cup p \in \mathfrak{I}$, $I' \cup q \notin \mathfrak{I}$, and thus $q \in \overline{A \cup p}$. Finally, it is evident that $A \in \mathfrak{I}$ if and only if $p \notin \overline{A - p}$ for all $p \in A$, i.e., the matroid induced in this way by $\mathfrak{I}$ has precisely $\mathfrak{I}$ as its family of independent sets. □

**Example.** Consider the uniform matroid $\mathbf{U}_k(n)$ on the $n$-set $S$. The independent sets are all $A \subseteq S$ with $|A| \leq k$, and it is clear that uniform matroids are characterized by this condition. In the vector space matroids $\mathbf{M}(V(n, K))$ and affine geometries $\mathbf{AG}(n, K)$ independence coincides with linear independence and affine independence, respectively.

**6.10 Basis axioms.** *Let $\mathbf{M}(S)$ be a matroid and $\mathfrak{B}$ the family of bases of the matroid. Then:*

(i) $B \neq B' \in \mathfrak{B} \Rightarrow B \nsubseteq B', B' \nsubseteq B$.
(ii) *Let $B \neq B' \in \mathfrak{B}$. For every $b \in B$ there is a $b' \in B'$ with $(B - b) \cup b' \in \mathfrak{B}$.*
(iii) *There exists $n \in \mathbb{N}_0$ with $|B| \leq n$ for all $B \in \mathfrak{B}$.*

*Conversely, if $\mathfrak{B} \subseteq 2^S$ satisfies (i), (ii) and (iii) then $\mathfrak{B}$ is the family of bases of a unique matroid.*

*Proof.* Construct $\mathfrak{J}$ as the ideal generated by $\mathfrak{B}$ and apply **6.9**. □

Before giving the axiomatization in terms of circuits let us note some properties relating bases and circuits.

**6.11 Proposition.** *Let $C$ and $C'$ be distinct circuits of a matroid $\mathbf{M}(S)$. Suppose $p \in C \cap C'$. Then there is another circuit $D$ with $D \subseteq (C \cup C') - p$.*

*Proof.* Choose $a \in C - C'$. Then $a \in \overline{C - a}$, $p \in \overline{C' - p}$, and thus

$$a \in \overline{(C - p - a) \cup p} \subseteq \overline{(C - p - a) \cup (C' - p)} = \overline{((C \cup C') - p) - a}.$$

This means $(C \cup C') - p$ is a dependent set and therefore contains a circuit. □

**6.12 Corollary.** *Let $B$ be a basis of $\mathbf{M}(S)$ and $p \notin B$. Then there is a unique circuit $C$ such that $p \in C \subseteq B \cup p$.*

*Proof.* $B \cup p$ is dependent and hence contains at least one circuit, and we have $p \in C$ for every such circuit $C$. If $C'$ is another circuit in $B \cup p$ different from $C$ then $p \in C \cap C'$ whence **6.11** implies the existence of a circuit $D$ with $D \subseteq (C \cup C') - p \subseteq B$, which is impossible as $B$ is independent. □

**6.13 Circuit axioms.** *Let $\mathbf{M}(S)$ be a matroid and $\mathfrak{K}$ the family of circuits. Then:*

(i) $\varnothing \notin \mathfrak{K}$; $C \neq C' \in \mathfrak{K} \Rightarrow C \nsubseteq C', C' \nsubseteq C$.
(ii) $C \neq C' \in \mathfrak{K}, p \in C \cap C' \Rightarrow \exists D \in \mathfrak{K}$ *with* $D \subseteq (C \cup C') - p$.
(ii′) $C \neq C' \in \mathfrak{K}, p \in C \cap C', q \in C - C' \Rightarrow \exists D \in \mathfrak{K}$ *with* $q \in D \subseteq (C \cup C') - p$.
(iii) *There exists $n \in \mathbb{N}_0$ such that $|A| \leq n$ whenever $A \nsupseteq C$ for all $C \in \mathfrak{K}$.*

*If, conversely, $\mathfrak{K} \subseteq 2^S$ satisfies (i), (ii), (iii) or equivalently (i), (ii′), (iii) then $\mathfrak{K}$ is the family of circuits of a unique matroid.*

*Proof.* The family of circuits in a matroid trivially satisfies (i) and (iii); (ii) was shown in **6.11**. We prove next that, for any family $\mathfrak{K}$ of finite subsets of a set $S$ satisfying (i), the properties (ii) and (ii') are equivalent. (Notice that (i) and (iii) imply that all members of $\mathfrak{K}$ must be finite.) We have to show that (ii) $\Rightarrow$ (ii'). Assume otherwise. Among all counterexamples to (ii'), choose $C \neq C' \in \mathfrak{K}$ such that $|C \cup C'|$ is minimal, where $p \in C \cap C'$, $q \in C - C'$. Now there exists $D \in \mathfrak{K}$ with $D \subseteq C \cup C'$ and $p \notin D$, $q \notin D$. Obviously, $D \neq C$, $D \neq C'$ and further $|D \cup C'| < |C \cup C'|$. Let $r \in D - C$. Condition (ii') applied to $r \in C' \cap D$, $p \in C' - D$ yields, by the minimality of $|C \cup C'|$, a set $D' \in \mathfrak{K}$ with $p \in D' \subseteq (C' \cup D) - r$. The pair $C \neq D'$ is now another counterexample with $p \in C \cap D'$, $q \in C - D'$, in contradiction to $|C \cup D'| < |C \cup C'|$.

The family of circuits thus satisfies all the conditions of the theorem. Take, conversely, a family $\mathfrak{K} \subseteq 2^S$ satisfying (i) to (iii). We call $I \subseteq S$ *independent* if and only if $I \not\supseteq C$ for all $C \in \mathfrak{K}$; otherwise it is dependent. Let us just check axiom **6.9**(ii) for the family of independent sets thus defined. Suppose there are sets $I, J \in \mathfrak{I}$ violating **6.9**(ii) and choose among all such pairs $I, J$ one with $|I \cup J|$ minimal. Let $|I| < |J|$ and $J - I = \{p_1, \ldots, p_k\}$. If $k = 1$ then $J \supseteq I$ and **6.9**(ii) is trivially true; hence assume $k \geq 2$. By our hypothesis we have $I \cup p_i \notin \mathfrak{I}$ for all $i$. Thus there exists $C \in \mathfrak{K}$ with $p_1 \in C \subseteq I \cup p_1$ and $C$ must contain some element $q_1 \in I - J$. Consider $(I - q_1) \cup p_1$. If $(I - q_1) \cup p_1 \notin \mathfrak{I}$ then there is some $C' \in \mathfrak{K}$ with $p_1 \in C' \subseteq (I - q_1) \cup p_1$ whence we infer from (ii) the existence of $D \in \mathfrak{K}$ with $D \subseteq (C \cup C') - p_1 \subseteq I$, in contradiction to $I \in \mathfrak{I}$. If, on the other hand, $(I - q_1) \cup p_1 \in \mathfrak{I}$ then because of $|((I - q_1) \cup p_1) \cup J| < |I \cup J|$ there must be some $p_i$, $i \geq 2$, such that $(I - q_1) \cup p_1 \cup p_i \in \mathfrak{I}$. Now since $I \cup p_i \notin \mathfrak{I}$ there exists $C'' \in \mathfrak{K}$ with $p_i \in C'' \subseteq I \cup p_i$. Obviously $C \neq C''$, $q_1 \in C \cap C''$ whence by (ii) there is $D'' \in \mathfrak{K}$ with $D'' \subseteq (C \cup C'') - q_1 \subseteq (I - q_1) \cup p_1 \cup p_i \in \mathfrak{I}$, a contradiction. □

It is, of course, equally possible to derive the closure operator directly from the circuit axioms via proposition **6.8**(ii).

Summarizing our axiomatizations so far, we notice that, roughly speaking, condition (i) corresponds to the closure operator, (ii) to the exchange axiom, and (iii) to the finite basis axiom. There is another axiomatization of matroids, in terms of copoints. Since we shall not need this set of axioms in the sequel and since we shall return to copoints and their relationship to circuits in the section on duality we defer a discussion until then.

### C. Rank Function and Semimodular Functions

We know from **6.8**(iii) that a matroid is also uniquely determined by its rank function. The corresponding axiomatization of matroids is of particular interest since it will enable us to construct some very interesting new classes of matroids.

**6.14 Rank axioms.** *Let* $\mathbf{M}(S)$ *be a matroid with rank function* $r$. *Then for all* $A, B \subseteq S$:

(i) $A \subseteq B \Rightarrow r(A) \leq r(B)$ *(monotone)*
(ii) $r(A \cap B) + r(A \cup B) \leq r(A) + r(B)$ *(semimodular)*
(iii) $0 \leq r(B) \leq |B|$ *for all finite subsets* $B$, *and to every* $A \subseteq S$ *there exists a finite subset* $B \subseteq A$ *with* $r(B) = r(A)$ *(finite basis).*

*Conversely, if $r$ is a function from $2^S$ to $\mathbb{N}_0$ which satisfies (i), (ii), (iii) then there exists a unique matroid on $S$ which has $r$ as its rank function.*

*Proof.* Conditions (i) and (iii) were proved for a matroid in **6.5**. To prove the semimodularity we choose a basis $C$ of $A \cap B$ and extend $C$ to a basis $D$ of $A \cup B$. Then $C = D \cap (A \cap B)$, $D = (D \cap A) \cup (D \cap B)$, and thus

$$\begin{aligned} r(A \cap B) + r(A \cup B) &= |C| + |D| = |D \cap (A \cap B)| + |(D \cap A) \cup (D \cap B)| \\ &= |D \cap A| + |D \cap B| \leq r(A) + r(B). \end{aligned}$$

Now conversely, let $r: 2^S \to \mathbb{N}_0$ satisfy the conditions of the theorem. We call $I \subseteq S$ *independent* if $r(I) = |I|$ and verify the axioms **6.9** for the family $\mathfrak{I}$ of these independent sets. Clearly, $\varnothing \in \mathfrak{I}$. If $I \in \mathfrak{I}$ and $J \subseteq I$ then by (ii) and (iii)

$$|I| = r(I) \leq r(J) + r(I - J) \leq |J| + |I - J| = |I|,$$

hence $r(J) = |J|$, i.e., $J \in \mathfrak{I}$. Since by (i) $r(A) \leq r(S) < \infty$ we infer that the cardinalities of the sets in $\mathfrak{I}$ are bounded above by $r(S)$. It remains to verify **6.9**(ii). Let $I, J \in \mathfrak{I}$, $|I| < |J|$ and $J - I = \{p_1, \ldots, p_k\}$. Suppose $I \cup p_i \notin \mathfrak{I}$ and hence $r(I \cup p_i) < |I| + 1$ for all $p_i$. Then we have

$$|I| = r(I) \leq r(I \cup p_1) < |I| + 1,$$

i.e., $r(I \cup p_1) = |I|$. Suppose we already know $r(I \cup p_1 \cup \cdots \cup p_i) = |I|$. Then

$$\begin{aligned} r(I \cup p_1 \cup \cdots \cup p_i \cup p_{i+1}) &\leq r(I \cup p_1 \cup \cdots \cup p_i) + r(I \cup p_{i+1}) - r(I) \\ &= |I| + |I| - |I| = |I|, \end{aligned}$$

thus, by induction, $r(I \cup J) = |I|$, contradicting $|I| < |J| = r(J) \leq r(I \cup J)$. We conclude that $\mathfrak{I}$ induces a matroid $\mathbf{M}(S)$ and it remains to be shown that $r$ is indeed the rank function of $\mathbf{M}(S)$, i.e., we have to show that $r(A) = |B|$ for some basis $B$ of $A$. The finite basis property implies $r(A) = r(C)$ for some finite set $C \subseteq A$. Now for $C$ we infer by the same argument as above that $r(C) = |B|$ where $B$ is a basis of $C$, and hence $r(A) = |B|$. Since $r$ is monotone $B$ must be a basis of $A$, and the proof is complete. □

Notice that we did not make use of the finite basis property in constructing the matroid $\mathbf{M}(S)$. (iii) is, however, indispensable in proving that $r$ is indeed the *rank function* of $\mathbf{M}(S)$ as the following example demonstrates. Let $S$ be countable and let $r: 2^S \to \mathbb{N}_0$ be defined by

$$r(A) = \begin{cases} 0 & \text{if } A = \varnothing \\ 1 & \text{if } |A| < \infty \\ 2 & \text{otherwise.} \end{cases}$$

It is easy to see that $r$ satisfies the conditions of **6.14** except (iii). The matroid induced by $r$ as in the proof consists wholly of parallel elements and is thus of rank 1 whereas $r(S) = 2$.

A closer analysis of the proof of **6.14** shows that the monotonicity and semimodularity already suffice to prove the exchange axiom **6.9**(ii) for $\mathfrak{I}$. This suggests that it might be possible to obtain matroids by retaining just these two properties and a suitable finiteness condition.

**Definition.** A monotone semimodular function $f: 2^S \to \mathbb{Z}$ is said to satisfy the *finite basis property* if to every $A \subseteq S$ there is a finite subset $B \subseteq A$ with $f(B) = f(A)$.

**6.15 Lemma.** *Let $f: 2^S \to \mathbb{N}_0$ be a function. If $f$ is monotone semimodular with $f(\emptyset) = 0$ then the same holds for the function $\hat{f}$ defined by*

$$\hat{f}(A) := \min_{B \subseteq A}(f(B) + |A - B|) \qquad (A \subseteq S)$$

*with the agreement that $|B| > k$ and $|B| + k = |B|$ for all $k \in \mathbb{N}_0$ whenever $B$ is an infinite set.*

*Proof.* The monotonicity of $\hat{f}$ is easy. Now for all $A_1 \subseteq A$, $B_1 \subseteq B$

$$|A - A_1| + |B - B_1| = |(A \cup B) - (A_1 \cup B_1)| + |(A \cap B) - (A_1 \cap B_1)|,$$

hence

$$\begin{aligned}(f(A_1) + |A - A_1|) &+ (f(B_1) + |B - B_1|) \\ &\geq (f(A_1 \cup B_1) + |(A \cup B) - (A_1 \cup B_1)|) \\ &\quad + (f(A_1 \cap B_1) + |(A \cap B) - (A_1 \cap B_1)|).\end{aligned}$$

It follows that

$$\begin{aligned}\hat{f}(A) + \hat{f}(B) &= \min_{A_1 \subseteq A,\, B_1 \subseteq B} (f(A_1) + |A - A_1| + f(B_1) + |B - B_1|) \\ &\geq \min_{C \subseteq A \cup B,\, D \subseteq A \cap B} (f(C) + |(A \cup B) - C| + f(D) + |(A \cap B) - D|) \\ &= \hat{f}(A \cup B) + \hat{f}(A \cap B). \quad \square\end{aligned}$$

**6.16 Corollary.** *Let $f$ be as in* **6.15**. *Then*

$$\hat{f}(A) = |A| \Leftrightarrow |B| \leq f(B) \quad \textit{for all } B \subseteq A. \quad \square$$

**6.17 Theorem** (Edmonds). *Let $f: 2^S \to \mathbb{N}_0$ be a monotone semimodular function with $f(\emptyset) = 0$ which satisfies the finite basis property. The family $\mathfrak{I} \subseteq 2^S$ defined by*

$$A \in \mathfrak{I} \Leftrightarrow |B| \leq f(B) \quad \textit{for all } B \subseteq A$$

*is the family of independent sets of a matroid* $\mathbf{M}_f(S)$. *Furthermore, the rank of an arbitrary subset* $A \subseteq S$ *is given by*

$$r(A) = \min_{B \subseteq A}(f(B) + |A - B|).$$

*Proof.* We consider the family $\mathfrak{A} = \{A \subseteq S : f(A) < |A|\}$ and show that the minimal members in $\mathfrak{A}$ satisfy the circuit axioms **6.13**. That any $A \in \mathfrak{A}$ contains a minimal set of $\mathfrak{A}$ follows easily from the finite basis property for $f$. **6.13**(i) and (iii) are clear, the cardinalities of $\mathfrak{I}$-sets being bounded by $f(S)$. Let $C \neq C'$ be minimal sets in $\mathfrak{A}$ and $p \in C \cap C'$. Then $|C| \geq 2$ and hence

$$|C| - 1 = |C - p| \leq f(C - p) \leq f(C) < |C|,$$

i.e., $f(C) = |C| - 1, f(D) \geq |D|$ for all $D \subsetneqq C$, and similarly $f(C') = |C'| - 1$. In particular, we have $|C \cap C'| \leq f(C \cap C')$ and thus

$$\begin{aligned} f((C \cup C') - p) &\leq f(C \cup C') \leq f(C) + f(C') - f(C \cap C') \\ &\leq |C| - 1 + |C'| - 1 - |C \cap C'| \\ &= |C \cup C'| - 2 < |(C \cup C') - p|. \end{aligned}$$

$(C \cup C') - p$ is therefore in $\mathfrak{A}$ and contains a minimal set of $\mathfrak{A}$.

The family $\mathfrak{I}$ as defined in the theorem induces, by **6.13**, a matroid $\mathbf{M}_f(S)$ and it remains to be shown that $\hat{f}$ is the rank function in $\mathbf{M}_f(S)$. We know from **6.15** that $\hat{f}$ is monotone and semimodular. Furthermore, $0 \leq \hat{f}(A) \leq |A|$ clearly holds for all finite subsets $A \subseteq S$. If we can show that $\hat{f}$ satisfies the finite basis property, the proof will be complete since, by **6.14**, $\hat{f}$ will then be the rank function of some matroid, which by the definition of $\mathfrak{I}$ and **6.16** must be precisely $\mathbf{M}_f(S)$. The proof of the finite basis property for $\hat{f}$ poses certain technical difficulties. Since it is not important for the subject as a whole we have deferred the details of the proof to the exercises. (See Pym–Perfect [1], Aigner [1, vol. II, p. 33].) □

We remark that even functions $f$ which are not semimodular may induce a matroid by the construction of **6.17**. As an example, take a finite set $S$ with $|S| \geq 3$ and define $f: 2^S \to \mathbb{N}_0$ by $f(\emptyset) = 0, f(A) = |S|$ for $\emptyset \neq A \neq S$ and $f(S) = 2|S|$. Then $f$ induces the free matroid on $S$ but is not semimodular.

Let us note a slight but very useful generalization of **6.17** where $f(\emptyset)$ is allowed to be any integer. The proof of **6.17** can be carried over verbatim.

**6.18 Theorem.** *Let* $f: 2^S \to \mathbb{Z}$ *be a monotone semimodular function with* $f(A) \geq 0$ *for all* $A \neq \emptyset$ *which satisfies the finite basis property. The family* $\mathfrak{I} \subseteq 2^S$ *defined by*

$$A \in \mathfrak{I} \Leftrightarrow |B| \leq f(B) \quad \textit{for all } \emptyset \neq B \subseteq A$$

*is then the family of independent sets of a matroid* $\mathbf{M}_f(S)$. *The rank function in* $\mathbf{M}_f(S)$ *is given by*

$$r(A) = \min_{\varnothing \neq B \subseteq A} (|A|, f(B) + |A - B|). \quad \square$$

We shall apply **6.17** in our discussion of transversal matroids in the next section, the sum of matroids in section 3.B and the Dilworth completion in section 3.D. Quite likely, **6.17** is the single most useful construction principle discovered so far in matroid theory.

Exercises VI.1

→ 1. Draw all combinatorial geometries with at most 5 points. (There are 17.)

2. Prove that for $n = 1, 2, 3$ there are exactly $2^n$ non-isomorphic matroids with $n$ points.

3. Complete the proof of **6.2**.

→ 4. Show that the subspace lattice of $\mathbf{AG}(n, K)$ is precisely the affine lattice $\mathscr{A}(n, K)$ of section I.2. (Hint: Show that $p$ is affinely dependent on $\{p_1, \ldots, p_k\}$ if and only if $p - p_1$ is linearly dependent on $\{p_2 - p_1, \ldots, p_k - p_1\}$.)

5. Let $\mathbf{M}(S)$ be a matroid, $A \subseteq C$ where $A$ is independent and $C$ spanning. Show the existence of a basis $B$ with $A \subseteq B \subseteq C$.

→ 6. Show that a matroid of rank $r$ contains at least $2^r$ flats.

7. Let $\mathbf{M}(S)$ be a matroid, $\mathfrak{I}$ the family of independent sets, and $A \subseteq S$. Define $\mathfrak{I}' \subseteq \mathfrak{I}$ by $X \in \mathfrak{I}' :\Leftrightarrow X \cap A = \varnothing$. Prove that $\mathfrak{I}'$ satisfies **6.9** and thus induces a matroid on $S$.

8. Prove in detail **6.10**.

→ 9. Strengthen **6.10**(ii): Let $B \neq B'$ be two bases of a matroid $\mathbf{M}(S)$, $p \in B$. Then there exists $q \in B'$ such that $(B - p) \cup q$ and $(B' - q) \cup p$ are both bases of $\mathbf{M}(S)$. (Cf. ex. II. 2.6.)

10.* Prove the following multiple exchange property: Let $B \neq B'$ be bases of a matroid $\mathbf{M}(S)$ and $A \subseteq B$. Then there exists $A' \subseteq B'$ such that $(B - A) \cup A'$ and $(B' - A') \cup A$ are both both bases of the matroid. (Greene)

→ 11. Let $C_1 \neq C_2$ be circuits of a matroid with $C_1 \cap C_2 \neq \varnothing$, $p \in C_1 - C_2$, $q \in C_2 - C_1$. Show that there is a circuit $C$ with $p, q \in C \subseteq C_1 \cup C_2$.

→ 12. Let $S$ be a finite set. Prove that $\mathfrak{H} \subseteq 2^S$ is the family of copoints of some matroid on $S$ if and only if

(i) $H \neq H' \in \mathfrak{H} \Rightarrow H \nsubseteq H', H' \nsubseteq H$
(ii) $H \neq H' \in \mathfrak{H}, p \notin H \cup H' \Rightarrow \exists K \in \mathfrak{H}$ with $(H \cap H') \cup p \subseteq K$.

→13. Let $G(V, E)$ be a finite graph and let $f\colon 2^E \to \mathbb{Z}$ be defined by $f(A) := |\bigcup_{\{i,j\}\in A} \{i, j\}| - 1$, i.e., $f(A) = \#\{\text{vertices incident with } A\} - 1$. Prove that $f$ is monotone and semimodular and describe the matroid $\mathbf{M}_f(E)$.

14. Reverse the rôle of vertices and edges in ex. 13 and describe the matroid $\mathbf{M}_f(V)$ induced by $f\colon 2^V \to \mathbb{N}_0$.

15.* Complete the proof of **6.17** by verifying the finite basis property for $\hat{f}$. Use the following steps:

(i) Let $A \subseteq S$ be infinite and $\mathscr{B}(A)$ the family of finite subsets of $A$. Prove that there exists $B_0 \in \mathscr{B}(A)$ with $\hat{f}(B_0) = \max_{B\in\mathscr{B}(A)} \hat{f}(B)$ and that $\hat{f}(B) = \hat{f}(B_0)$ for all $B_0 \subseteq B \in \mathscr{B}(A)$.

(ii) $C$ is called a *minimal set* of $B$ if $\hat{f}(B) = f(C) + |B - C|$. Show that the union and intersection of minimal sets is a minimal set.

(iii) Index the family $\{B \in \mathscr{B}(A)\colon B \supseteq B_0\}$ by $\{B_j\colon j \in J\}$. By (ii), there is a smallest minimal set in $B_j$ which shall be denoted by $C_j$. Show that

$$B_j \subseteq B_k \Rightarrow C_j \subseteq C_k,$$
$$C_l \cup C_m = C_t,$$

where $t$ is the index of $B_l \cup B_m$, and from this the existence of an index $j_0$ with

$$C_k \cap B_j = C_{j_0} \cap B_j \quad \text{for all } B_k \supseteq B_{j_0} \supseteq B_j.$$

(iv) Using the finite basis property of $f$, show that

$$f\left(\bigcup_{j\in J} C_j\right) = f(C) \quad \text{for some } C \subseteq \bigcup_j C_j,\ |C| < \infty$$
$$= f(C_k) \quad \text{for some } k.$$

(v) Complete the proof by showing that $B_j - (\bigcup_{k\in J} C_k)$ is constant for all $j \in J$ and hence that

$$\hat{f}(A) = \hat{f}(B_0) = f\left(\bigcup_{j\in J} C\right) + \left|A - \bigcup_{j\in J} C_j\right|.$$

## 2. Fundamental Examples

After having introduced the basic concepts, let us now study in more detail the most important classes of matroids.

### A. Linear Matroids and Function Spaces

When studying a class of algebraic objects it is a common approach to see how far an arbitrary member of the class can be represented in some known subclass. For matroids, a natural subclass is that of vector space matroids. Hence we give the following definitions.

**Definition.** Let $K$ be a field. A matroid $\mathbf{M}(S)$ is called *coordinatizable over* $K$ or $K$*-linear* if there exists a function $\phi: S \to V(n, K)$ such that for all $A \subseteq S$

$$A \text{ independent in } \mathbf{M}(S) \Leftrightarrow \{\phi(a): a \in A\} \text{ linearly independent in } V(n, K).$$

Notice that if $\mathbf{M}$ is $K$-linear we can choose $n = r(\mathbf{M})$.

Any such mapping $\phi$ is called a *coordinatization of* $\mathbf{M}(S)$ *over* $K$. A matroid is called *linear* if it is coordinatizable over some field and it is called *regular* if it is coordinatizable over every field. A matroid which can be coordinatized over $GF(q)$ is called $q$*-linear.*

Notice that a coordinatization $\phi$ need not be injective. For instance, all loops of $\mathbf{M}(S)$ must be mapped onto the single loop 0 in $V(n, K)$; similarly, parallel elements must be mapped onto parallel elements. It is, however, an easy consequence of the definition that the mapping $\phi_0$ on the underlying geometry $\mathbf{M}_0(S_0)$ in injective. Hence the lattice $L(S)$ is mapped by $\phi_0$ injectively into the vector space lattice $\mathscr{L}(n, K)$.

**Example.** The free matroid $\mathbf{FM}(S)$ on an $n$-set $S$ is regular since we just have to map $S$ onto a basis of $V(n, K)$ for any field $K$.

Throughout the book we shall mean by a field a commutative field. Most of the results can be extended to division rings (see the exercises). Linear matroids will occupy us again in the next chapter where we shall discuss their connections to geometry, graphs and matrices. In this section we concentrate on an important equivalent concept.

**Definition.** Let $S$ be a set, $K$ a field. Any finite-dimensional subspace $F$ of the $K$-vector space of all functions from $S$ to $K$ (with the usual addition and scalar multiplication) is called a *function space over* $K$.

For $A \subseteq S$, $U \subseteq F$ define

$$h(A) = \text{hull}(A) := \{f \in F: f(p) = 0 \text{ for all } p \in A\},$$
$$k(U) = \text{kernel}(U) := \{q \in S: g(q) = 0 \text{ for all } g \in U\}.$$

**6.19 Proposition.** *Let $F = F(S, K)$ be a function space over $K$. The pair $(h, k)$ is a Galois connection between the lattices $2^S$ and $2^F$, and hence $kh$ a closure on $S$, called the hull-kernel closure. We have, for $A \subseteq S$, $p \in S$*

$$p \in kh(A) \Leftrightarrow \underset{f \in F}{\forall} (f|_A = 0 \Rightarrow f(p) = 0).$$

*Proof.* Let $A \subseteq B \subseteq S$. Then $f(p) = 0$ for all $p \in B$ implies $f(p) = 0$ for $p \in A$, whence $h(A) \supseteq h(B)$. The other assertions are similarly verified. $\square$

**6.20 Theorem.** *Let $F = F(S, K)$ be a function space over $K$. $F$ together with the hull-kernel closure induces a matroid on $S$, called the* function space matroid $\mathbf{M}(F(S, K))$.

*Proof.* The finite basis property follows easily from **4.36**(iii) and the fact that $F$ has finite dimension. Let $p, q \in S$, $A \subseteq S$ with $p \notin kh(A)$, $p \in kh(A \cup q)$. We have to show that for all $f \in F, f|_{A \cup p} = 0 \Rightarrow f(q) = 0$. Assume otherwise and choose $g \in F$ with $g|_{A \cup p} = 0$, $g(q) \neq 0$. Since $p \notin kh(A)$ there must be $g_1 \in F$ with $g_1|_A = 0$, $g_1(p) \neq 0$. Now define $g_2 \in F$ by

$$g_2 := g_1(q)g - g(q)g_1 \in F.$$

Then $g_2|_A = 0$, $g_2(q) = 0$, but $g_2(p) \neq 0$, contradicting $p \in kh(A \cup q)$. □

As a corollary we obtain a very useful description of the bases and copoints of a function space matroid.

**6.21 Proposition.** *Let $\mathbf{M}(F(S, K))$ be a function space matroid. Then*:

(i) *If $g \in F$ vanishes on a basis of $\mathbf{M}(F(S, K))$ then $g = 0$ is the 0-function.*
(ii) *$H \subseteq S$ is a copoint of $\mathbf{M}(F(S, K))$ if and only if $H = \ker f$ for some $0 \neq f \in F$ and $H$ is maximal with respect to this property.*
(iii) *Let $B = \{b_1, \ldots, b_n\}$ be a basis of $\mathbf{M}(F(S, K))$. Then all sets $\{f_1, \ldots, f_n\} \subseteq F$ with $\ker f = kh(B - b_i)$ $(i = 1, \ldots, n)$ form a basis of the vector space $F(S, K)$. In particular, $r(\mathbf{M}(F(S, K))) = \dim F(S, K)$.*

*Proof.* (i) is an immediate consequence of **6.19**. Invoking **4.36**(ii), we see that the flats in $\mathbf{M}(F(S, K))$ are precisely the sets $k(U)$, where $U$ is a subspace of $F$. Hence it follows from the antiisomorphism in **4.36**(iii) that the copoints of $\mathbf{M}(F(S, K))$ are precisely the kernels of the one-dimensional subspaces of $F$, i.e., of single functions $f \neq 0$. Finally, let $f_i: S \to K$ be defined as in (iii). Since $f_i(b_j) \neq 0$ if and only if $i = j$, $\{f_1, \ldots, f_n\}$ is linearly independent in $F(S, K)$. Let $f \neq 0$ be any function in $F$ and $\lambda_i = f(b_i)/f_i(b_i)$, $i = 1, \ldots, n$. $f - \sum_{i=1}^n \lambda_i f_i$ vanishes on the basis $B$ and hence must be identically 0 by (i). □

We come to the proof of the basic result that the function space matroids over $K$ and the $K$-linear matroids comprise the same class of abstract matroids, a result which is, of course, nothing but the duality between a vector space and its dual space. This twofold aspect of linearity will prove very useful in the next chapter since for a given matroid it is often easier to construct a function space than to directly find a coordinatization.

**6.22 Theorem.** *Let $F(S, K)$ be a function space over $K$. We denote by $F^* = \operatorname{Hom}(F, K)$ the vector space dual of $F$. Then the mapping $\phi: S \to F^*$ defined by*

$$\phi(p) = L_p \qquad (p \in S)$$

*where* $L_p \in F^*$ *with* $L_p(f) = f(p)$ *for all* $f \in F$ *is a coordinatization of* $\mathbf{M}(F(S, K))$ *over* $K$.

*Conversely, Let* $\mathbf{M}(S)$ *be a K-linear matroid and* $\psi: S \to V(n, K)$ *a coordinatization. If* $V^*$ *is the vector space dual of* $V$ *then* $\mathbf{M}(S) \cong \mathbf{M}(V^*(S, K))$ *under the isomorphism* $p \to L_{\psi p}$.

*Proof.* To show that $\Phi$ preserves the dependency relations it clearly suffices to consider bases and circuits. Let $B = \{b_1, \ldots, b_n\}$ be a basis of $\mathbf{M}(F(S, K))$ and $f_1, \ldots, f_n \in F$ with $\ker f_i = kh(B - b_i)$, $i = 1, \ldots, n$. Since $L_{b_i}(f_j) \neq 0$ if and only if $i = j$, the set $\{L_{b_1}, \ldots, L_{b_n}\}$ is linearly independent in $F^*$. Now consider a circuit $C = \{b_0\ b_1, \ldots, b_k\}$ in $\mathbf{M}(F(S, K))$. We extend $\{b_1, \ldots, b_k\}$ to a basis $B = \{b_1, \ldots, b_n\}$ and define functions $f_i$, $i = 1, \ldots, n$, as before. We then have $b_0 \in kh(\{b_1, \ldots, b_k\}) \subseteq kh(B - b_i)$ for all $i > k$. Since according to **6.21**(iii) the $f_i$'s form a basis of $F$, the functional $L_{b_0} \in F^*$ is uniquely determined by its values on $f_1, \ldots, f_n$. Setting $\lambda_i = L_{b_0}(f_i)/L_{b_i}(f_i)$ we have $L_{b_0} = \sum_{i=1}^k \lambda_i L_{b_i}$, implying that $\phi C$ is linearly dependent in $F^*$. The converse construction $\mathbf{M} \to V^*$ is easily accomplished using suitable dual bases. □

**Example.** Consider the matroid $\mathbf{P}(K_5)$ induced by the complete graph $K_5$, a representation of which in Euclidean 3-space is shown in Figure 6.6. It consists of

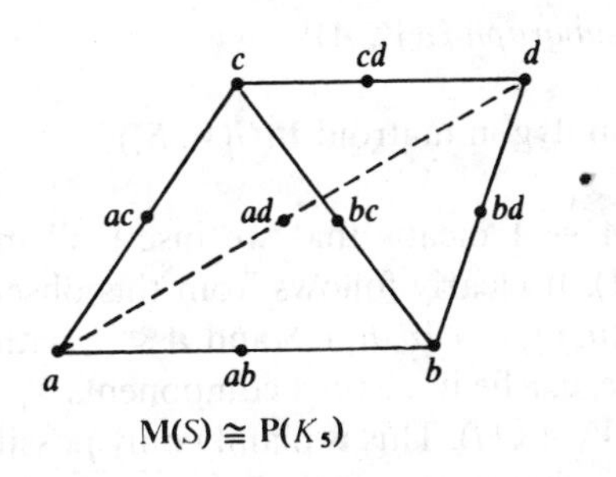


Figure 6.6

10 points (corresponding to the edges $S$ of $K_5$), 10 3-point lines (corresponding to the triangles) of which the curves $\{ab, ac, bc\}$, $\{ab, ad, bd\}$, $\{ac, ad, cd\}$ and $\{bc, bd, cd\}$ have been omitted in the figure for the sake of clarity. $\mathbf{P}(K_5)$ has the basis $B = \{a, b, c, d\}$ and the other 6 points are $ij$, where $ij$ is the unique third point on the line through $i$ and $j$. Let us define functions $f_i: S \to GF(2)$ with $\ker f_i = \overline{B - i}$, $i = a, b, c, d$. Then it is easy to see that $\mathbf{P}(K_5)$ is isomorphic to the function space matroid $\mathbf{M}(F(S, GF(2))$ where $F$ is the function space over $GF(2)$ spanned by $\{f_a, f_b, f_c, f_d\}$. Hence the matroid $\mathbf{P}(K_5)$ is $GF(2)$-linear. The most economical way

to represent $F$ is by a $4 \times 10$-matrix where we write $f_i(j)$ in position $(i,j)$, $i = a,\ldots,d$, $j \in S$:

| | $a$ | $b$ | $c$ | $d$ | $ab$ | $ac$ | $ad$ | $bc$ | $bd$ | $cd$ |
|---|---|---|---|---|---|---|---|---|---|---|
| $f_a$ | 1 | 0 | 0 | 0 | 1 | 1 | 1 | 0 | 0 | 0 |
| $f_b$ | 0 | 1 | 0 | 0 | 1 | 0 | 0 | 1 | 1 | 0 |
| $f_c$ | 0 | 0 | 1 | 0 | 0 | 1 | 0 | 1 | 0 | 1 |
| $f_d$ | 0 | 0 | 0 | 1 | 0 | 0 | 1 | 0 | 1 | 1 |

It is clear from the construction in **6.22** that the column vectors of this matrix represent a coordinatization of $\mathbf{P}(K_5)$ over $GF(2)$.

## B. Graphs

In this section we shall always mean by a graph a finite undirected graph and we shall usually denote a graph by $G(V, S)$ where $V$ is the vertex-set and $S$ the edge-set of the graph. The following result, which generalizes the construction of $\mathbf{P}(K_n)$ mentioned in section 1.A, was one of the starting points of matroid theory.

**6.23 Theorem** (Whitney). *Let $G(V, S)$ be a graph. The edge-set $S$ together with the closure $A \to \bar{A}$ where*

$$\bar{A} = \{e = \{u, v\} : u, v \in V \text{ are in the same connected component of the subgraph } G(V, A)\}$$

*is a matroid, called the* polygon matroid $\mathbf{P}(G(V, S))$.

*Proof.* The operator $A \to \bar{A}$ means that we insert all missing edges *within* the components of $G(V, A)$. It clearly follows from this observation that $A \to \bar{A}$ is a closure on $S$. Let $k = \{u, v\}$, $l = \{a, b\} \in S$ and $A \subseteq S$ with $k \notin \bar{A}$, $k \in \overline{A \cup l}$. By the definition of the closure, $u$, $v$ lie in distinct components $V_1$, $V_2$ of $G(V, A)$ but in the same component of $G(V, A \cup l)$. This is plainly only possible when $l$ also joins the components $V_1$ and $V_2$, whence we conclude $l \in \overline{A \cup k}$ (see Figure 6.7). □

**Definition.** A matroid $\mathbf{M}(S)$ is called *graphic* if it is isomorphic to the polygon matroid of some graph.

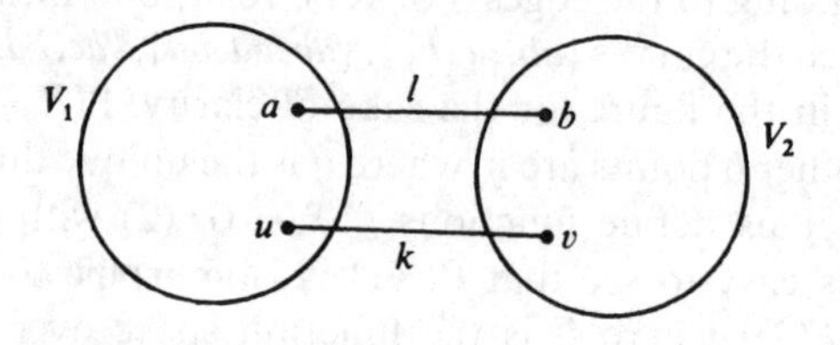


Figure 6.7

We shall discuss the characterization of graphic matroids in chapter VII. The smallest non-graphic matroid is the line consisting of 4 points. (Proof?) From the definition of the closure it is evident that the loops of a graph $G$ are precisely the loops of the polygon matroid $\mathbf{P}(G)$, and similarly that the parallel edges in $G$ are precisely the parallel elements in $\mathbf{P}(G)$.

**Example.** Let $G(V, S)$ be the graph in Figure 6.8. The closure $\bar{A}$ of $A \subseteq S$ is shown next to it.

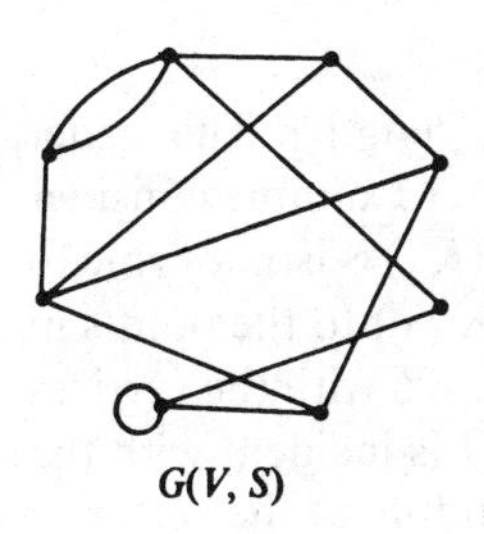


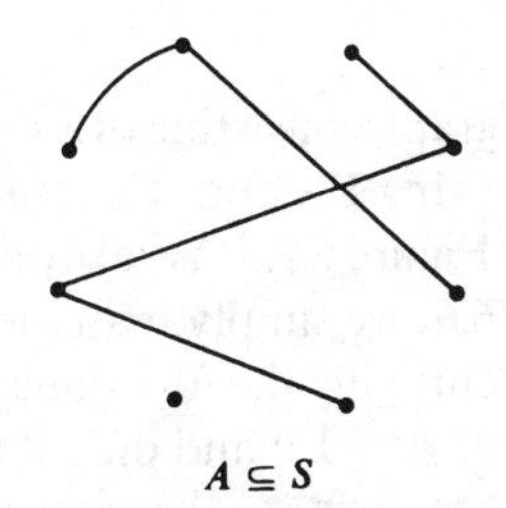


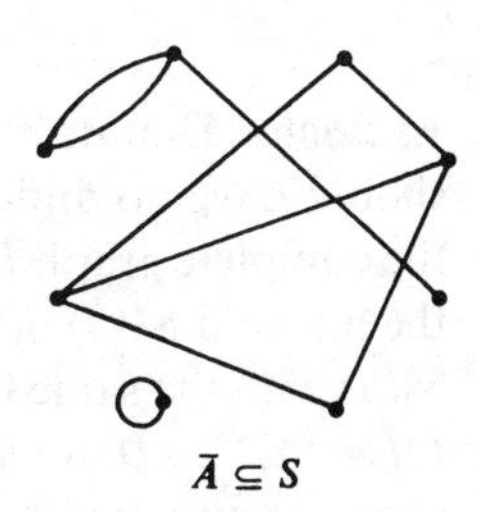


Figure 6.8

**6.24 Proposition.** *Let $\mathbf{P}(G(V, S))$ be the polygon matroid of $G(V, S)$. Then:*

(i) *$A \subseteq S$ is independent $\Leftrightarrow G(V, A)$ is a forest.*
(ii) *$B \subseteq S$ is a basis of $\mathbf{P}(G(V, S)) \Leftrightarrow G(V, B)$ is a spanning forest, i.e., the forest $G(V, B)$ has the same number of components as $G(V, S)$.*
(iii) *$C \subseteq S$ is a circuit $\Leftrightarrow C$ is a* polygon *in $G(V, S)$, i.e., $C$ is the edge-set of a circuit in $G(V, S)$.*
(iv) *$H \subseteq S$ is a copoint $\Leftrightarrow G(V, H)$ has precisely one component more than $G(V, S)$ and is maximal with respect to this property.*
(v) *$r(A) = |V| - k(A)$, where $k(A) = \#\{$components in $G(V, A)\}$.*

*Proof.* The edge-set $A$ of a forest is certainly independent since the deletion of any $e \in A$ splits the components containing $e$ into two parts implying $e \notin \overline{A - e}$. If, on the other hand, $G(V, A)$ contains a polygon $\{e_0, e_1, \ldots, e_t\}$, then by the definition of the closure $e_0 \in \overline{\{e_1, \ldots, e_t\}}$, whence $A$ is dependent. This proves (i), (ii), and (iii). Furthermore it says that $r(A)$ is the number of edges in a spanning forest of $G(V, A)$. Every tree has one edge less than the number of vertices. (See Harary [1, p. 33].) Hence if $V_1, \ldots, V_{k(A)}$ are the vertex-sets of the components of $G(V, S)$, then

$$r(A) = \sum_{i=1}^{k(A)} (|V_i| - 1) = |V| - k(A).$$

The characterization of the copoints is now an immediate corollary of this formula. □

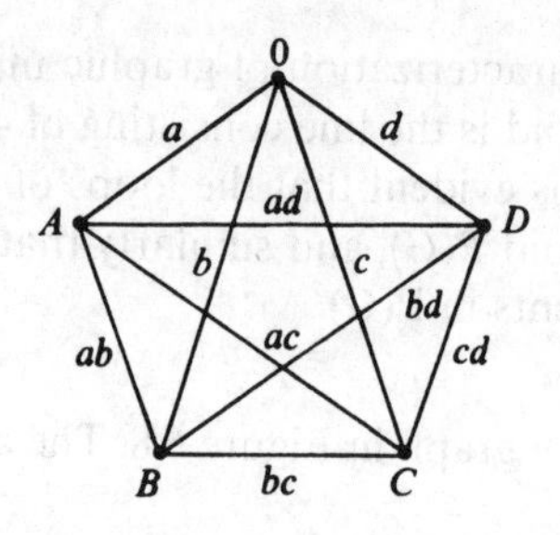


Figure 6.9

**Examples.** Denote by $C_n$ the graph consisting of a single cycle of length $n$. The reader should have no difficulty proving $\mathbf{P}(C_n) \cong \mathbf{U}_{n-1}(n)$. As another example consider the complete graph $K_5$ as in Figure 6.9. It is easily seen that $\mathbf{P}(K_5)$ is isomorphic to the matroid $\mathbf{M}(S)$ of Figure 6.6; we simply map the edges of $K_5$ onto the points in $\mathbf{M}(S)$ with the same label. Identifying the functions $f_i$ in Figure 6.6 with the vertices $I$, $I = A, \ldots, D$, we see that $f_i(j) = 1$ if and only if the vertex $I$ is incident with the edge $j$. Hence if we add another row $f_0$ to the matrix corresponding to the vertex 0:

| | $a$ | $b$ | $c$ | $d$ | $ab$ | $ac$ | $ad$ | $bc$ | $bd$ | $cd$ |
|---|---|---|---|---|---|---|---|---|---|---|
| $f_0$ | 1 | 1 | 1 | 1 | 0 | 0 | 0 | 0 | 0 | 0 |

and denote the new matrix by $I$, then $I$ is precisely the *incidence matrix* of $K_5$. Thus we have the result: The columns of the incidence matrix of $K_5$ are a coordinatization of the polygon matroid $\mathbf{P}(K_5)$ over $GF(2)$. We shall prove the same theorem for arbitrary graphs in section VII.3. One could not hope for a more elegant result. In particular, this will imply that all polygon matroids are $GF(2)$-linear; in fact, we shall see that they are all regular.

In addition to the drawing in Figure 6.6 we can represent $\mathbf{P}(K_5)$ in Euclidean 3-space as the 3-dimensional *Desarguesian configuration* with center $ce$, axis $\{ab, bd, ad\}$ and perspective triples $\{ae, be, de\}$ and $\{ac, bc, cd\}$ (see Figure 6.10). For this reason, $\mathbf{P}(K_5)$ is also called the *Desarguesian block* (see **7.14**).

## C. Transversal Matroids

Consider a binary relation $R \subseteq S \times I$ on the finite sets $S$ and $I$. We regard $R$ as "directed" from $S$ to $I$ and for $A \subseteq S$ set

$$R(A) := \bigcup_{a \in A} \{y \in I: (a, y) \in R\}.$$

For all $A, B \subseteq S$ we clearly have

$$A \subseteq B \Rightarrow R(A) \subseteq R(B),$$
$$R(A \cup B) = R(A) \cup R(B), \; R(A \cap B) \subseteq R(A) \cap R(B).$$

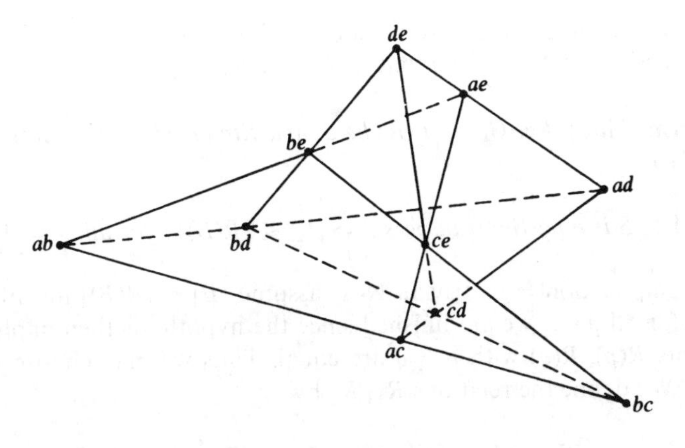


Figure 6.10

Hence it follows that $f: 2^S \to \mathbb{N}_0$ defined by

$$f(A) := |R(A)| \qquad (A \subseteq S)$$

is a monotone semimodular function with $f(\emptyset) = 0$. Hence $f$ induces by **6.17** a matroid on $S$, denoted by $\mathbf{T}(S, R, I)$.

In exactly the same way, we obtain a matroid $\mathbf{T}(I, R, S)$ by using the monotone semimodular function $g: 2^I \to \mathbb{N}_0$ given by $g(D) := |R(D)|$ where $R(D) = \bigcup_{d \in D} \{x \in S: (x, d) \in R\}$.

According to **6.17**, the independent sets of $\mathbf{T}(S, R, I)$ are characterized by the condition

$$(+) \qquad |B| \leq |R(B)| \quad \text{for all } B \subseteq A.$$

The following theorem of Hall characterizes sets satisfying $(+)$. It is one of the fundamental theorems of all combinatorics and the starting point for what is today known as transversal theory. First we need a few definitions. Any binary relation may be regarded as a bipartite graph with defining vertex-sets $S$ and $I$, or, to be more precise, as a directed bipartite graph $G(S \cup I, R)$ with all edges directed from $S$ to $I$, where $(a, b)$ is an edge of $G$ if and only if $(a, b) \in R$. We shall henceforth consider the binary relation $R \subseteq S \times I$ and the bipartite graph $G(S \cup I, R)$ to be one and the same and use them interchangeably. If $A \subseteq S$ then $R(A)$ is now the set of vertices in $I$ which are joined by at least one edge to $A$.

**Definition.** Let $G(S \cup I, R)$ be a bipartite graph. A *matching* $M$ is a set of edges no two of which have a common endpoint. We denote by $\text{match}_S(M)$ and $\text{match}_I(M)$ the set of endpoints of $M$ that lie in $S$ and $I$, respectively. We say that $A \subseteq S$ is a *partial transversal* in $S$ or that $A$ can be *matched into* $I$ if there is a matching $M$ with $A = \text{match}_S(M)$; similarly for $B \subseteq I$.

In other words, $A \subseteq S$ is a partial transversal in $G(S \cup I, R)$ if and only if there exists an injection $\phi: A \to I$ such that $(a, \phi a) \in R$ for all $a \in A$.

**6.25 Theorem** (Hall). *Let $G(S \cup I, R)$ be a bipartite graph on the finite vertex-sets $S$ and $I$. Then*

$$A \subseteq S \text{ is a partial transversal} \Leftrightarrow |B| \leq |R(B)| \quad \text{for all } B \subseteq A.$$

*Proof.* The implication $\Rightarrow$ is trivial. Now assume $|B| \leq |R(B)|$ for all $B \subseteq A$. If $|R(p)| = 1$ for all $p \in A$ we are finished since the hypothesis then implies that no two elements $R(p)$, $R(q)$ with $p \neq q$ are equal. Thus we may choose $p \in A$ with $|R(p)| \geq 2$. We define the relations $R_1, R_2$ by

$$R_1 := R - (p, q_1), \qquad R_2 := R - (p, q_2)$$

for $q_1 \neq q_2 \in R(p)$ and assert that at least one of the bipartite graphs $G(S \cup I, R_1)$ or $G(S \cup I, R_2)$ satisfies the condition of the theorem. If not, there exist sets $A_1, A_2 \subseteq A - p$ with

$$|R_1(p \cup A_1)| < |A_1| + 1, \qquad |R_2(p \cup A_2)| < |A_2| + 1.$$

Now $R_1(p \cup A_1) = (R(p) - q_1) \cup R(A_1)$, $R_2(p \cup A_2) = (R(p) - q_2) \cup R(A_2)$ and thus

$$R_1(p \cup A_1) \cup R_2(p \cup A_2) = R(p \cup A_1 \cup A_2)$$
$$R_1(p \cup A_1) \cap R_2(p \cup A_2) \supseteq R(A_1) \cap R(A_2) \supseteq R(A_1 \cap A_2).$$

By the hypothesis of the theorem this gives

$$|A_1| + |A_2| \geq |R_1(p \cup A_1)| + |R_2(p \cup A_2)| \geq |R(p \cup A_1 \cup A_2)| + |R(A_1 \cap A_2)|$$
$$\geq 1 + |A_1 \cup A_2| + |A_1 \cap A_2| = 1 + |A_1| + |A_2|,$$

which is absurd. Repeated application of this reduction step finally yields a subgraph with edge-set $R_t \subseteq R$ and $|R_t(p)| = 1$ for all $p \in A$ for which, as we have seen, a matching exists trivially. □

The analogous result holds, of course, for subsets $B$ of $I$.

The real importance of bipartite graphs and the matroids induced by them comes from the interpretation of a bipartite graph as a *set system.* Any family $\mathfrak{A} = \{A_i : i \in I\}$ of subsets of a set $S$ gives rise to a binary relation $R \subseteq S \times I$ by setting

$$(p, i) \in R :\Leftrightarrow p \in A_i,$$

and, clearly, any binary relation can be considered as a set system in this fashion. We can now translate all the concepts we have introduced for bipartite graphs into their counterparts for set systems. For instance, a subset $T \subseteq S$ is called a *transversal* or a *system of distinct representatives* of the set system $\mathfrak{A} = \{A_i : i \in I\}$ if there exists a bijection $\phi: T \to I$ with $p \in A_{\phi p}$ for all $p \in T$. This is, of course, the origin of the term transversal. $T \subseteq S$ is a *partial transversal* of $\mathfrak{A}$ if it is a transversal of some subfamily of $\mathfrak{A}$. We shall study set systems and their transversals in detail in chapter VIII, but for the present it is more convenient to continue considering bipartite graphs. The reader should have no difficulty translating the theorems on bipartite graphs into theorems on set systems.

To summarize our results in this section, we now know that the independent sets of the matroid $\mathbf{T}(S, R, I)$ induced by the function $f: A \to |R(A)|$ are precisely the partial transversals in $S$, with the analogous statement holding for $I$. Hence we have two interesting descriptions of the same matroid.

**6.26 Theorem** (Edmonds–Fulkerson). *Let $G(S \cup I, R)$ be a bipartite graph on the finite sets $S$ and $I$. The family of partial transversals in $S$ is the family of independent sets of a matroid on $S$, called the* transversal matroid $\mathbf{T}(S, R, I)$. *Similarly, the partial transversals in $I$ induce a transversal matroid on $I$, and the two matroids have the same rank (equal to the cardinality of the largest matching in $R$).* □

**Example.** Let $S = \{1, 2, 3, 4, 5\}$ and consider the set system $\{A_1, A_2, A_3, A_4\}$ with $A_1 = \{1, 2, 3\}$, $A_2 = \{1, 4\}$, $A_3 = \{2, 3, 5\}$, $A_4 = \{4\}$. The transversals of $\mathfrak{A}$ are $\{1, 2, 3, 4\}$, $\{1, 2, 4, 5\}$ and $\{1, 3, 4, 5\}$. The bipartite graph $G$ corresponding to $\mathfrak{A}$ and the two transversal matroids induced by $G$ are shown in Figure 6.11.

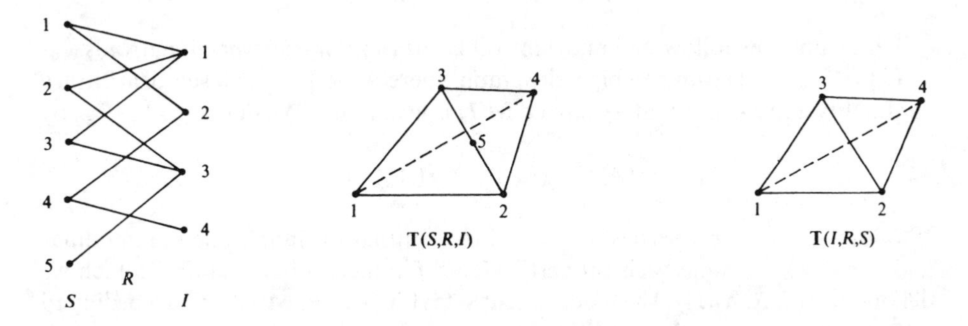


Figure 6.11

Notice that **6.26** implies that all maximal partial transversals in a bipartite graph have the same cardinality, a fact which is by no means obvious at first sight.

Any abstract matroid which is isomorphic to some matroid $\mathbf{T}(S, R, I)$ is called a *transversal matroid*. We shall discuss transversal matroids in some depth in chapter VII.4. The reader may convince himself that any matroid with at most 5 points is transversal but that the polygon matroid of the graph in Figure 6.12 is not transversal.

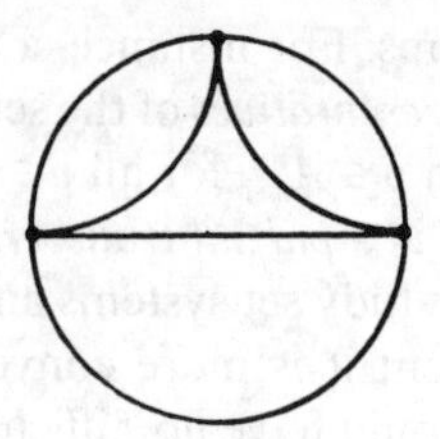

Figure 6.12

**6.27 Proposition.** *Let* $\mathbf{T}(S, R, I)$ *be the transversal matroid induced by the bipartite graph* $G(S \cup I, R)$. *Then*:

(i) $A \subseteq S$ *is independent* $\Leftrightarrow$ $A$ *is a partial transversal in* $S$.
(ii) $A \subseteq S$ *is a basis* $\Leftrightarrow$ $A$ *is a maximal partial transversal in* $S$.
(iii) $r(A) = \min_{B \subseteq A}(|R(B)| + |A - B|) = |A| - \max_{B \subseteq A} \delta(B)$ *where* $\delta(B) := |B| - |R(B)|$ *is the* defect *of* $B$.
(iv) $r(A) = \min |C|$ *over all sets* $C \subseteq S \cup I$ *with the property that any edge between* $A$ *and* $I$ *has at least one endpoint in* $C$. *Such sets* $C$ *are called* point covers *of* $A$.

*Proof.* We have already seen (i) and (ii). The rank formula (iii) is a consequence of **6.17** and Hall's theorem **6.25**. Any point cover $C = S_1 \cup I_1$ where $S_1 \subseteq A$ and $I_1 \subseteq I$ clearly satisfies $R(A - S_1) \subseteq I_1$. A minimal point cover must therefore be of the form $R(B) \cup (A - B)$, from which (iv) follows. $\square$

Let us note the following important generalization of transversal matroids.

Let $G(S \cup I, R)$ again be a bipartite graph where $S$ and $I$ are now sets of arbitrary cardinality and assume that a matroid $\mathbf{M}(I)$ is given on $I$. We define $f: 2^S \to \mathbb{N}_0$ by

$$f(A) := r(R(A)) \qquad (A \subseteq S).$$

It follows from the properties of a rank function that $f$ is monotone semimodular and possesses the finite basis property. Hence $f$ induces a matroid on $S$ which we denote by $\mathbf{T}(S, R, \mathbf{M}(I))$. The independent sets $A$ in $\mathbf{T}(S, R, \mathbf{M}(I))$ are, according to **6.17**, characterized by the condition

$$(++) \qquad |B| \leq r(R(B)) \quad \text{for all } B \subseteq A.$$

Of course, if $S$ and $I$ are finite and $\mathbf{M}(I)$ is the free matroid on $I$ then $\mathbf{T}(S, R, \mathbf{M}(I))$ is just the ordinary transversal matroid. It was observed by Rado that the sets $A \subseteq S$ satisfying $(++)$ can be characterized by means of matchings as in **6.25**.

**Definition.** Let $G(S \cup I, R)$ be a bipartite graph on the (not necessarily finite) sets $S$ and $I$ and $\mathbf{M}(I)$ a matroid on $I$. A subset $A \subseteq S$ is called an *independent partial transversal* in $S$ if it can be matched onto an independent set in $\mathbf{M}(I)$.

The following theorem is proved by the same reduction method as was used to prove **6.25** after noticing that $G(S \cup I, R)$ always contains a subgraph $G(S \cup I, R')$ with $|R'(p)| < \infty$ for all $p \in S$ which also satisfies (+,++).

**6.28 Theorem** (Rado). *Let $G(S \cup I, R)$ be a bipartite graph and $\mathbf{M}(I)$ a matroid on $I$. Then*:

$A \subseteq S$ *is an independent partial transversal* $\Leftrightarrow |B| \leq r(R(B))$ *for all* $B \subseteq A$. □

The extension of **6.26** now reads as follows

**6.29 Theorem** (Rado). *Let $G(S \cup I, R)$ be a bipartite graph and $\mathbf{M}(I)$ a matroid on $I$. The family of independent partial transversals in $S$ induces as family of independent sets a matroid on $S$, denoted by $\mathbf{T}(S, R, \mathbf{M}(I))$. The rank function of this matroid is given by*

$$r(A) = \min_{B \subseteq A} (r(R(B)) + |A - B|) \qquad (A \subseteq S). \quad \square$$

**6.28** and **6.29** open the door to a wide range of applications through the proper choice of matroids on $I$. We shall return to this topic in chapter VIII. For the moment, let us just note a particularly simple and useful example.

**6.30 Corollary.** *Let $\mathbf{M}$ be a matroid on $S$ and $f: S \to T$ a surjective mapping. If $\mathfrak{J}$ denotes the family of independent sets in $\mathbf{M}$ then $f(\mathfrak{J}) := \{f(A): A \in \mathfrak{J}\}$ is the family of independent sets of a matroid $f(\mathbf{M})$ on $T$.*

*Proof.* Consider the bipartite graph $G(T \cup S, f^{-1})$, i.e., $q \in T$ and $p \in S$ are joined by an edge if and only if $f(p) = q$. It is now easy to see that the independent partial transversals in $T$ are precisely the images of independent sets in $\mathbf{M}$. □

## D. Incidence Geometries

To illustrate the close relationship between matroids and incidence structures we end our survey of examples of matroids with a brief discussion of an especially interesting class of matroids arising in this way. The starting point and model for our axiom system are the properties in **2.32** defining synthetically a projective space.

Let $\mathfrak{P}$ be a set of *points* and $\mathfrak{K}, \mathfrak{F} \subseteq 2^{\mathfrak{P}}$ two non-empty families of subsets of $\mathfrak{P}$ whose members we call *curves* and *surfaces*, respectively. Let $n \in \mathbb{N}_0$. We posit the following axioms:

(i) Any $n + 1$ points lie on a unique curve and each curve contains at least $n + 1$ points.
(ii) Any $n + 2$ points not lying on a curve lie on a unique surface and each surface contains at least $n + 2$ points not lying on a curve.
(iii) A surface contains with every $n + 1$ points the whole curve determined by them.

$A \subseteq \mathfrak{P}$ is called a *subspace* if $A$ contains all curves and surfaces through any $n + 1$ or $n + 2$ points in $A$ determined by (i) or (ii). It is immediate that the intersection of subspaces is again a subspace. Hence we may define a closure $A \to \bar{A}$ where $\bar{A}$ is the smallest subspace containing $A$.

(iv) If two surfaces $F \neq F'$ lie in the closure of $n + 3$ points then $|F \cap F'| \neq n$.
(v) $A \to \bar{A}$ satisfies the finite basis axiom.

**Definition.** Any incidence structure $(\mathfrak{P}, \mathfrak{K}, \mathfrak{F})$ which satisfies (i) to (v) for some $n \in \mathbb{N}_0$ together with the closure defined there is called an *incidence geometry of grade n*, denoted by $\mathbf{G}(\mathfrak{P}, \mathfrak{K}, \mathfrak{F})$.

Before proving that any incidence geometry is indeed a combinatorial geometry let us see how some of our previous examples fit into this scheme.

**Examples.** If we take as curves the *points* and as surfaces the *lines* of a projective space $(\mathfrak{P}, \mathfrak{G})$ then the axioms are valid for $n = 0$. Hence any projective space is an incidence geometry of grade 0. In fact, any modular geometry is, by Birkhoff's theorem **2.55**, an incidence geometry of grade 0. Let $\mathbf{PG}(\mathfrak{P}, \mathfrak{G})$ be a projective space of $g$-dimension $n$ and let $H$ be an arbitrary copoint of $\mathbf{PG}(\mathfrak{P}, \mathfrak{G})$. The set $\mathfrak{P}' = \mathfrak{P} - H$ together with the system of subspaces $\{A' = A - H \neq \emptyset : A \text{ subspace in } \mathbf{PG}(\mathfrak{P}, \mathfrak{G})\} \cup \{\emptyset\}$ is called the *affine space* $\mathbf{AG}(\mathfrak{P}')$ *of g-dimension n*. It follows from the modularity of the subspace lattice of **PG** that all affine spaces obtained in this way (i.e., by deleting copoints) are isomorphic. In particular, we may unambiguously denote by $\mathbf{AG}(n, K)$ the affine space derived from $\mathbf{PG}(n, K)$. It is well-known and easily proved that $\mathbf{AG}(n, K)$ is indeed isomorphic to the affine geometry induced by the affine closure in section 1.A. If in $\mathbf{AG}(\mathfrak{P}')$ we take as curves the subspaces $A' = A - H \neq \emptyset$ of $g$-dimension 1 and as surfaces the subspaces $A'$ of $g$-dimension 2 then we can easily derive from the modular law in **PG** the axioms (i) to (v) above for $n = 1$. Hence affine spaces are incidence geometries of grade 1. For further classical examples, e.g., Möbius geometries, see Dembowski [1, ch. 6].

Suppose the incidence geometry $\mathbf{G}(\mathfrak{P}, \mathfrak{K}, \mathfrak{F})$ of grade $t - 1$ possesses only one surface, namely $\mathfrak{P}$. In this case, $\mathbf{G}(\mathfrak{P}, \mathfrak{K}, \mathfrak{F})$ has only three types of subspaces:

(a) all sets $A \subseteq \mathfrak{P}$ with $|A| \leq t - 1$;
(b) the curves;
(c) $\mathfrak{P}$.

Comparing this with **6.2** we see that $\mathbf{G}(\mathfrak{P}, \mathfrak{K}, \mathfrak{F})$ is precisely the covering matroid induced by the curves.

To describe the structure of arbitrary incidence geometries we need the following lemma whose proof is left to the reader.

**6.31 Lemma.** *Let $\mathbf{G}(\mathfrak{P}, \mathfrak{K}, \mathfrak{F})$ be an incidence geometry of grade n. Suppose $A \neq B$ are subspaces with $\{p_1, \ldots, p_n\} \subseteq A \cap B$. Then*

$$A \vee B = \bigcup_{a \in A, b \in B} \overline{\{p_1, \ldots, p_n, a, b\}}. \quad \square$$

Notice that for $n = 0$ this reduces to the inductive construction of subspaces in projective spaces.

**6.32 Theorem** (Wille). *Any incidence geometry* $\mathbf{G}(\mathfrak{P}, \mathfrak{K}, \mathfrak{F})$ *is a matroid. Furthermore, for all subspaces* $A$ *of rank equal to the grade of* $\mathbf{G}$ *we have in the subspace lattice*

$$[0, A] \text{ is distributive and } [A, 1] \text{ is modular.}$$

*Proof.* Let $\mathbf{G}(\mathfrak{P}, \mathfrak{K}, \mathfrak{F})$ be of grade $n$. We have to verify the exchange axiom. It is evident from the definition of an incidence geometry that all subsets $A$ with $|A| \leq n$ are closed. Let $p, q \in \mathfrak{P}, A \subseteq \mathfrak{P}$ with $p \notin \bar{A}, p \in \overline{A \cup q}$. If $A$ contains less than $n$ points then $A \cup q$ is closed and therefore $p = q$. Now assume $\{p_1, \ldots, p_n\} \subseteq A$, $B = \{p_1, \ldots, p_n, q\}$. By **6.31**,

$$\overline{A \cup q} = \overline{A \cup B} = \bigcup_{a \in \bar{A}} \overline{\{p_1, \ldots, p_n, a, q\}}$$

and hence

$$p \in \overline{\{p_1, \ldots, p_n, a, q\}} \quad \text{for some } a \in \bar{A}.$$

Since $q \notin \overline{\{p_1, \ldots, p_n, a\}}$ we infer that the points $p_1, \ldots, p_n, a, q$ determine a unique surface $F$ with $p \in F$. It follows from $p \notin \overline{\{p_1, \ldots, p_n, a\}}$, on the other hand, that the surface through the points $p_1, \ldots, p_n, a, p$ must also be $F$, whence we conclude

$$q \in \overline{\{p_1, \ldots, p_n, a, p\}} \subseteq \overline{A \cup p}.$$

Finally, let $A$ be a subspace of rank $n$. Then we already know that $|A| = n$ and thus $[0, A] \cong \mathscr{B}(n)$. To verify the modularity of $[A, 1]$ we have to show that $C \wedge H > A$ for any two subspaces $C, H \in [A, 1]$ with $r(C) = r(A) + 2$ and $H$ a copoint (see **2.43**). Let $A <\cdot B <\cdot C$ and assume $C \wedge H = A$. Then $C \leq B \vee H = \mathfrak{P}$. Set $A = \{p_1, \ldots, p_n\}$ and choose any $q \in C - B$. By **6.31**,

$$q \in \overline{\{p_1, \ldots, p_n, b, h\}}$$

for some $b \in B, h \in H$. Since $q \notin \overline{\{p_1, \ldots, p_n, b\}}$ we infer from the exchange axiom that $h \in \overline{\{p_1, \ldots, p_n, b, q\}} \subseteq C$. Hence $h \in C \cap H = A \subseteq B$, and thus $q \in B$, contradicting the assumption $q \notin B$. ☐

It can be shown that the converse to **6.32** also holds; hence **6.32** gives a lattice-theoretic characterization of incidence geometries (see the exercises).

EXERCISES VI.2

→ 1. Generalizing the definition of a coordinatization, we say that $\phi: S \to T$ is a *representation* of the matroid $\mathbf{M}(S)$ in the matroid $\mathbf{M}(T)$ if $A$ is independent in $\mathbf{M}(S)$ if and only if $\{\phi(a): a \in A\}$ is independent in $\mathbf{M}(T)$. Show that $\phi: S \to T$ is a representation if and only if $r(\phi(A)) = r(A)$ for all $A \subseteq S$.

→ 2. Generalize **6.20** to integral domains with unity. Extend **6.21** and **6.22** to this case.

→ 3. Prove that a simple matroid is free if and only if every coline is covered by precisely two copoints.

4. Prove that $\mathbf{U}_2(4)$ is the smallest non-graphic matroid. For what $k$ and $n$ is the uniform matroid $\mathbf{U}_k(n)$ graphic?

5. Show that the matroid constructed in ex. VI.1.13 is precisely the polygon matroid of the given graph.

→ 6. Show that the polygon matroid of the graph in Figure 6.12 is not transversal.

7. Verify that $\mathbf{P}(K_4)$ (and therefore $\mathbf{P}(K_n)$ for $n \geq 4$) is not transversal.

→ 8. Prove that every matroid with at most 5 elements is transversal and, similarly, every matroid $\mathbf{M}(S)$ with $r(S) \geq |S| - 2$. Are these the best possible bounds?

9.* Show that there are at least $2^n$ non-isomorphic transversal matroids. (Piff–Welsh)

→10. Complete the proof of **6.28**. (Hint: Let $G(S \cup I, R)$ be an arbitrary bipartite graph and $\mathbf{M}(I)$ a matroid on $I$. For every $C \subseteq S$ and $p \in C$ choose a basis $B_{p,C}$ of $R(C)$ with $B_{p,C} \cap R(p) \neq \varnothing$ (this is possible because $R(p) \neq \varnothing$). Now define $R' \subseteq R$ by $R'(p) = \bigcup_{q,C} B_{q,C} \cap R(p)$.)

→11. A covering matroid $\mathbf{M}(S, \mathfrak{H})$ corresponding to a $t$-partition $\mathfrak{H}$ is called a *t-design* (of index 1) if all sets in $\mathfrak{H}$ have the same cardinality. Set $v = |S|$, $b = |\mathfrak{H}|$, $k = |H|$ for all $H \in \mathfrak{H}$. Show that in a 2-design every point lies on the same number of sets $H \in \mathfrak{H}$, say $r$, and that

(i) $vr = bk$,
(ii) $v - 1 = r(k - 1)$,
(iii) $k = 3 \Rightarrow v \equiv 1$ or $3 \pmod 6$.

Construct 2-designs of index 1 with $k = 3$ for $v = 7$ and $v = 9$ and show that they are unique. Prove that, in fact, they are isomorphic to the Fano plane **F** and the affine plane **AG**(2, 3). Relate the notion of a 2-design to the strong JD-condition defined in ex. V.1.3.

12. A finite matroid is called *equicardinal* if all its hyperplanes have the same cardinality. Show that the following matroids are equicardinal:

(i) $\mathbf{FM}(n)$,
(ii) $\mathbf{PG}(n, q)$,
(iii) $\mathbf{AG}(n, q)$,
(iv) $\mathbf{U}_k(n)$,

and that they all satisfy the stronger condition that all flats of the same rank $k$ are equicardinal, for every $k$.

13.* Find all equicardinal graphic matroids. (Murty)

→14. Prove **6.31**. (Hint: It suffices to show that the set on the right-hand side is a flat. To this end, first prove that $A$ is a flat if and only if $\{p_1, \ldots, p_n, q, r\} \subseteq A$ for all pairs $q, r \in A$.)

15.* Prove the converse to **6.32**: A matroid $\mathbf{M}(S)$ is isomorphic to an incidence geometry of grade $n$ if and only if $\bar{\varnothing} = \varnothing, \bar{p} = p$ for all $p \in S$ (in case $n \geq 1$) and $[0, x]$ is distributive, $[x, 1]$ modular for all $x \in L(S)$ with $r(x) = n$. (Wille)

# 3. Construction of Matroids

After introducing a mathematical object one usually asks for construction methods, i.e., how to construct new objects from given ones. Two such constructions are well-known throughout mathematics, the subobject and quotient object construction. These two constructions called *restriction* and *contraction* here, are fundamental for all of the ensuing theory and will be discussed first. After that we study the *product* and *sum* of matroids, and finally two less well-known constructions, the *one-point extension* and the *truncation* of matroids.

To appreciate the geometric background we shall always try to picture a construction geometrically on the lattice of flats. In many cases this interpretation will make the construction at hand more transparent, suggesting further concepts as well as some simplifications.

## A. Restriction and Contraction

We shall pursue in this and the next sections the following course: First we define the construction for arbitrary matroids, then study the relationship of independent sets, bases, etc., of the old matroid with those of the new matroid and finally apply the results to the principal examples of section 2.

**Definition.** Let $\mathbf{M}(S)$ be a matroid with closure $J$, and $A \subseteq S$.

(i) The set $A$ together with the operator $J_A: B \to J(B) \cap A$ for $B \subseteq A$ is called the *restriction of* $\mathbf{M}(S)$ *to* $A$.
(ii) The set $S - A$ together with the operator $J_{S/A}: B \to J(B \cup A) - A$ for $B \subseteq S - A$ is called the *contraction of* $\mathbf{M}(S)$ *through* $A$ or the *contraction to* $S - A$.

**6.33 Proposition.** *Let* $\mathbf{M}(S)$ *be a matroid with closure* $J$, $A \subseteq S$.

(i) *The restriction of* $\mathbf{M}(S)$ *to* $A$ *is a matroid, denoted by* $\mathbf{M}(S) \cdot A$. $\mathbf{M}(S) \cdot A$ *is also called the* submatroid generated by $A$.
(ii) *Let* $L(A)$ *be the lattice of flats of* $\mathbf{M}(S) \cdot A$. *The mapping* $\phi: L(A) \to [0, J(A)] \subseteq L(S)$ *defined by* $\phi B = J(B)$ *is an order-isomorphism which preserves suprema (but in general not infima). If, in particular,* $A$ *is a flat of* $\mathbf{M}(S)$ *then* $L(A) \cong [0, A] \subseteq L(S)$.

*Proof.* $J_A$ is surely a closure on $A$ which satisfies the finite basis condition. Let $p, q \in A, B \subseteq A$ with $p \notin J_A(B), p \in J_A(B \cup q)$. Then $p \notin J(B) \cap A$, i.e., $p \notin J(B)$ and $p \in J(B \cup q) \cap A$. The exchange axiom for $J$ implies $q \in J(B \cup p) \cap A = J_A(B \cup p)$. To prove (ii), consider the mappings $\phi: L(A) \to [0, J(A)]$ and $\psi: [0, J(A)] \to L(A)$ given by

$$\phi(B) = J(B) \qquad (B \in L(A))$$

$$\psi(C) = C \cap A \qquad (C \in [0, J(A)]).$$

$\phi$ and $\psi$ are monotone with $\psi\phi = id$ and we have, for $B, C \in L(A)$,

$$\begin{aligned}\phi(B \vee C) &= \phi(J_A(B \cup C)) = J(J_A(B \cup C)) \\ &= J(B \cup C) = J(B) \vee J(C) = \phi(B) \vee \phi(C).\end{aligned}$$

The last assertion is clear since in this case we also have $\phi\psi = id$. □

If there is no danger of confusion we write $\mathbf{M}(A)$ for the submatroid generated by $A$. The following proposition is an immediate consequence of **6.33**.

**6.34 Proposition.** *Let $\mathbf{M}(A)$ be the submatroid of $\mathbf{M}(S)$ generated by $A$, and $r$ and $r_A$ the rank functions of $\mathbf{M}(S)$ and $\mathbf{M}(A)$, respectively. Then for all $B \subseteq A$:*

(i) *$B$ independent in $\mathbf{M}(A) \Leftrightarrow B$ independent in $\mathbf{M}(S)$.*
(ii) *$B$ basis of $\mathbf{M}(A) \Leftrightarrow B$ basis of $A$ in $\mathbf{M}(S)$.*
(iii) *$B$ circuit in $\mathbf{M}(A) \Leftrightarrow B$ circuit in $\mathbf{M}(S)$.*
(iv) *$r_A(B) = r(B)$.* □

**6.35 Proposition.** *Let $\mathbf{M}(S)$ be a matroid with closure $J$, $A \subseteq S$.*

(i) *The contraction of $\mathbf{M}(S)$ through $A$ is a matroid, denoted by $\mathbf{M}(S)/A$.*
(ii) *Let $L(S/A)$ be the lattice of flats of $\mathbf{M}(S)/A$. Then $L(S/A) \cong [J(A), 1] \subseteq L(S)$ by means of the lattice isomorphism $\phi B = J(B \cup A)$, $B \in L(S/A)$.*

*Proof.* Let us just verify the isomorphism $\phi$. Define $\psi: [J(A), 1] \to L(S/A)$ by $\psi C = C - A$ for $C \in [J(A), 1]$. Then

$$J_{S/A}(C - A) = J((C - A) \cup A) - A = C - A \quad \text{for all } C \in [J(A), 1],$$

and thus

$$\psi\phi B = \psi(J(B \cup A)) = J(B \cup A) - A = J_{S/A}(B) = B \quad \text{for all } B \in L(S/A),$$

$$\phi\psi C = \phi(C - A) = J(C) = C \quad \text{for all } C \in [J(A), 1]. \quad \square$$

**6.36 Proposition.** *Let* $\mathbf{M}(S)/A$ *be the contraction of* $\mathbf{M}(S)$ *through* $A$, *and* $r$, $r_{S/A}$ *the rank functions of* $\mathbf{M}(S)$ *and* $\mathbf{M}(S)/A$, *respectively. Then for all* $B \subseteq S - A$:

(i) $B$ *independent in* $\mathbf{M}(S)/A \Leftrightarrow B \cup C$ *independent in* $\mathbf{M}(S)$ *for all independent sets* $C \subseteq A$.
(ii) $B$ *basis of* $\mathbf{M}(S)/A \Leftrightarrow B \cup C$ *basis of* $\mathbf{M}(S)$ *for all bases* $C$ *of* $A$.
(iii) $B$ *circuit in* $\mathbf{M}(S)/A \Leftrightarrow B = C - A \neq \emptyset$, *where* $C$ *is a circuit in* $\mathbf{M}(S)$ *and* $B$ *is minimal with this property.*
(iv) $r_{S/A}(B) = r(B \cup A) - r(A)$.

*Proof.* The isomorphism $L(S/A) \cong [J(A), 1]$ implies (iv) which in turn implies (i) and (ii). Let $B$ be a circuit in $\mathbf{M}(S)/A$, then by (i) $B$ is either a circuit in $\mathbf{M}(S)$ or B is independent and there exists an independent set $D \subseteq A$ in $\mathbf{M}(S)$ such that $B \cup D$ is dependent in $\mathbf{M}(S)$. If we choose $D_0 \subseteq A$ to be minimal among all these sets $D$, then $C_0 = B \cup D_0$ is a circuit in $\mathbf{M}(S)$ with $B = C_0 - A$. Conversely, any set $C - A \neq \emptyset$ with $C$ a circuit in $\mathbf{M}(S)$ is dependent in $\mathbf{M}(S)/A$ and hence contains a circuit of $\mathbf{M}(S)/\mathrm{A}$. □

Before passing to the examples let us note some useful formulas which are easily proved using **6.34** and **6.36**.

**6.37 Proposition.** *Let* $\mathbf{M}(S)$ *be a matroid*, $B \subseteq A \subseteq S$. *Then*:

(i) $(\mathbf{M}(S) \cdot A) \cdot B = \mathbf{M}(S) \cdot B$.
(ii) $(\mathbf{M}(S)/B)/(A - B) = \mathbf{M}(S)/A$.
(iii) $(\mathbf{M}(S) \cdot A)/B = (\mathbf{M}(S)/B) \cdot (A - B)$. □

**Definition.** A contraction of a restriction is called a *minor* of the matroid. By **6.37**(iii), minors can also be defined as restrictions of contractions.

**6.37** implies that a minor of a minor is again a minor. The lattice $L(A/B)$ of the minor $(\mathbf{M}(S) \cdot A)/B$ is, according to our results, embedded in the interval $[J(B), J(A)]$ of $L(S)$. If, in particular, $A$ is a flat, then $L(A/B) \cong [J(B), J(A)]$. Hence any interval of $L(S)$ is the lattice of flats of some minor of $\mathbf{M}(S)$.

Now consider a function space $F(S, K)$. For $A \subseteq S$ we define two new function spaces over $K$:

$$F(S, K) \cdot A := \{f|_A : f \in F\},$$

$F(S, K) \cdot A$ is called the function space *restricted to* $A$.

$$F(S, K)/A := \{f|_{S-A} : f \in F \text{ with } f|_A \equiv 0\},$$

$F(S, K)/A$ is the function space *contracted through* $A$. Clearly, $F(S, K) \cdot A$ and $F(S, K)/A$ are again function spaces.

**6.38 Proposition.** *Let $F(S, K)$ be a function space. Then*

(i) $\mathbf{M}(F(S, K)) \cdot A \cong \mathbf{M}(F(S, K) \cdot A)$,
(ii) $\mathbf{M}(F(S, K))/A \cong \mathbf{M}(F(S, K)/A)$.

*Proof.* We prove (ii). Let $B \subseteq S - A$ and $J$, $J'$, $J''$ the closure operators of $\mathbf{M}(F(S, K))$, $\mathbf{M}(F(S, K))/A$ and $\mathbf{M}(F(S, K)/A)$, respectively. Then

$$\begin{aligned} J'(B) = J(B \cup A) - A &= \left\{p \in S - A \colon \underset{f \in F}{\forall} (f|_{B \cup A} \equiv 0 \Rightarrow f(p) = 0\right\} \\ &= \left\{p \in S - A \colon \underset{f \in F}{\forall} f|_A \equiv 0 \Rightarrow (f|_B \equiv 0 \Rightarrow f(p) = 0)\right\} \\ &= \left\{p \in S - A \colon \underset{f' \in F/A}{\forall} (f'|_B \equiv 0 \Rightarrow f'(p) = 0\right\} = J''(B). \quad \square \end{aligned}$$

**6.38** says that the restriction of a function space matroid is precisely the matroid of the restricted space; similarly for the contraction. This yields the following important corollary.

**6.39 Corollary** (Tutte). *Any minor of a K-linear matroid is K-linear. In particular, any minor of a regular matroid is regular.* $\square$

**Example.** Consider the matroid $\mathbf{M}(S)$ as in Figure 6.6 and the coordinatization there. Let $A = \{a, b, c, ab, bd\}$, $B = \{a, b, ab\}$. The restriction $\mathbf{M}(S) \cdot A$ is then the function space matroid generated by the functions $f' \colon A \to GF(2)$ whereas $\mathbf{M}(S)/B$ is generated by $f''_c, f''_d \colon S - B \to GF(2)$:

| | $a$ | $b$ | $c$ | $ab$ | $bd$ |
|---|---|---|---|---|---|
| $f'_a$ | 1 | 0 | 0 | 1 | 0 |
| $f'_b$ | 0 | 1 | 0 | 1 | 1 |
| $f'_c$ | 0 | 0 | 1 | 0 | 0 |
| $f'_d$ | 0 | 0 | 0 | 0 | 1 |

| | $c$ | $d$ | $ac$ | $ad$ | $bc$ | $bd$ | $cd$ |
|---|---|---|---|---|---|---|---|
| $f''_c$ | 1 | 0 | 1 | 0 | 1 | 0 | 1 |
| $f''_d$ | 0 | 1 | 0 | 1 | 0 | 1 | 1 |

We proceed analogously for graphs. We define restrictions and contractions within the graph itself and prove then that the polygon matroid of the restricted (contracted) graph corresponds precisely to the matroid restriction (contraction).

**Definition.** Let $G(V, S)$ be a finite graph, $A \subseteq S$. The *restriction* $G(V, S) \cdot A$ is the graph obtained by deleting the edges in $S - A$; in other words $G(V, S) \cdot A$ is the subgraph $G(V, A)$. The *contraction* $G(V, S)/A$ is the following graph: The vertices of $G(V, S)/A$ are the connected components of $G(V, A)$. The edge-set is $S - A$ where the endpoints of $e \in S - A$ are those components of $G(V, A)$ which contained the original endpoints of $e$ in $G(V, S)$.

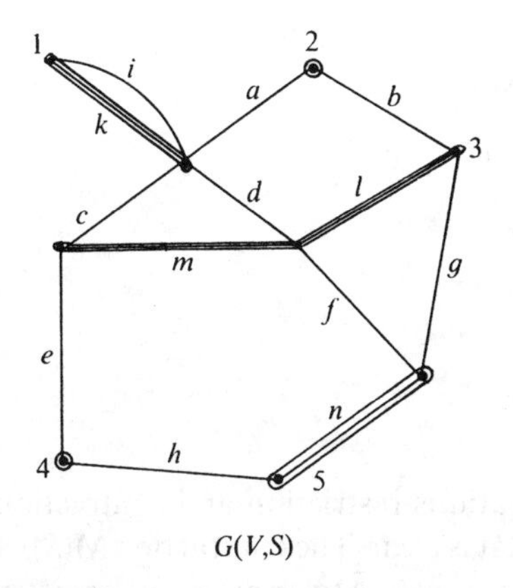

$G(V,S)$

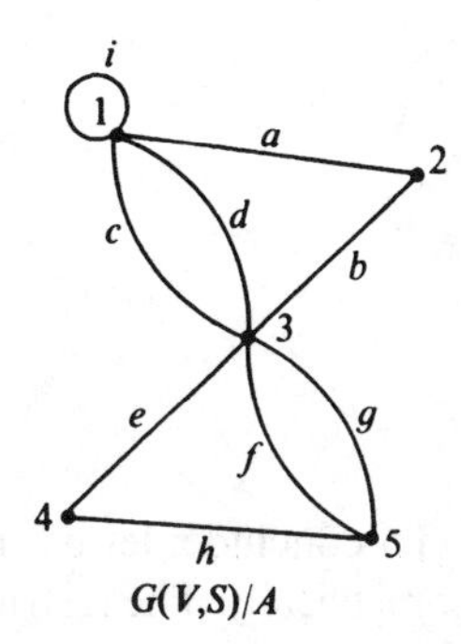

$G(V,S)/A$

Figure 6.13

**Example.** Figure 6.13 shows a graph $G(V, S)$ and its contraction through $A = \{k, l, m, n\}$. Notice that we obtain $G(V, S)/A$ by identifying, in turn, the endpoints of the edges in $A$, i.e., by contracting these edges to a single vertex; hence the name contraction.

**6.40 Proposition.** *Let $G(V, S)$ be a graph. Then*:

(i) $\mathbf{P}(G(V, S)) \cdot A \cong \mathbf{P}(G(V, S) \cdot A)$.
(ii) $\mathbf{P}(G(V, S))/A \cong \mathbf{P}(G(V, S)/A)$.

*Proof.* Left to the reader. □

**6.41 Corollary** (Tutte). *Any minor of a graphic matroid is graphic.* □

The situation is different for transversal matroids. If we restrict a binary relation $R \subseteq S \times I$ to a subset $A \subseteq S$ by defining $R_A := R \cap (A \times I)$ then we clearly have (e.g., by **6.34**(i))

$$\mathbf{T}(A, R_A, I) \cong \mathbf{T}(S, R, I) \cdot A.$$

and hence:

**6.42 Proposition.** *Any restriction of a transversal matroid is transversal.* □

The contraction of a transversal matroid, however, need not be transversal any more. As an example, consider the graph of Figure 6.14. It is easy to see that the polygon matroid of this graph is transversal. If, however, we contract through the edge $e$ we obtain, by **6.40**(ii), the polygon matroid of the graph in Fig. 6.12 which is not transversal. The class of transversal matroids is therefore not closed with respect to taking minors. We shall study the smallest "minor-closed" class containing the transversal matroids in section VII.4.

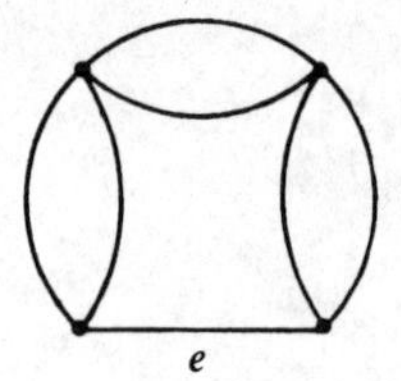


Figure 6.14

To conclude, let us try to visualize the operations restriction and contraction geometrically. The restriction as reduction to a flat is clear. The contraction $\mathbf{M}(S)/A$ is best understood as a *projection* of $\mathbf{M}(S)$ from the center $\bar{A}$ to an external geometry of rank $r(S) - r(A)$; see Figure 6.15 for an example. Since any contraction is, by **6.37**, a sequence of point contractions we may assume that all projections have a point as center.

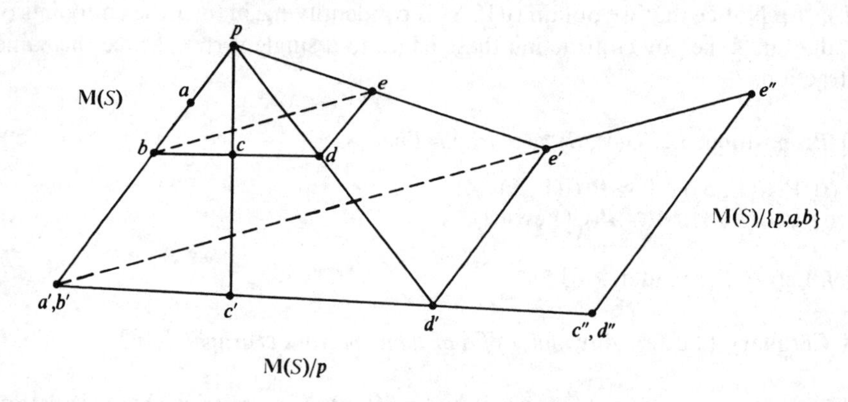


Figure 6.15

## B. Product and Sum

The direct product of matroids is defined as one would expect. Since this is the only product appearing in this book we shall omit the word "direct."

**Definition.** Let $\mathbf{M}_i(S_i)$ be matroids on the pairwise disjoint sets $S_i$ with closure operators $J_i$, $i = 1, \ldots, t$. The *product* $\prod_{i=1}^{t} \mathbf{M}_i$ is the matroid on $\bigcup_{i=1}^{t} S_i$ induced by the closure

$$J\colon \bigcup_{i=1}^{t} A_i \to \bigcup_{i=1}^{t} J_i(A_i) \qquad (A_i \subseteq S_i).$$

The $\mathbf{M}_i$'s are called the *factors* of $\prod_{i=1}^{t} \mathbf{M}_i$.

The following proposition is immediate.

**6.43 Proposition.** *The product* $\prod_{i=1} \mathbf{M}_i(S_i)$ *is a matroid on* $\bigcup_{i=1}^t S_i$ *with its lattice of flats* $L(S) \cong \prod_{i=1}^t L_i(S_i)$. *Furthermore, we have, for* $A = \bigcup_{i=1}^t A_i$, $A_i \subseteq S_i$:

(i) $A$ *independent in* $\prod \mathbf{M}_i \Leftrightarrow A_i$ *independent in* $\mathbf{M}_i$ *for all* $i$.
(ii) $A$ *basis of* $\prod \mathbf{M}_i \Leftrightarrow A_i$ *basis of* $\mathbf{M}_i$ *for all* $i$.
(iii) $A$ *circuit in* $\prod \mathbf{M}_i \Leftrightarrow A = A_i$ *circuit in* $\mathbf{M}_i$ *for some* $i$ *and* $A_j = \varnothing$ *for* $j \neq i$.
(iv) $r(A) = \sum_{i=1}^t r_i(A_i)$. □

A matroid which cannot be decomposed into a product of smaller matroids is called *connected*. Notice that any loop is a factor. Hence a connected matroid is either a single loop or has no loops. We shall study the connectivity of matroids in more detail in section 4.C.

We now consider a construction method for matroids which are defined on the same set.

**Definition.** Let $\mathbf{M}_1(S), \ldots, \mathbf{M}_t(S)$ be matroids on $S$ with rank functions $r_i$. The *sum* $\sum_{i=1}^t \mathbf{M}_i(S)$ is the matroid on $S$ induced by the monotone semimodular function $\sum_{i=1}^t r_i$.

**6.44 Theorem** (Nash-Williams). *Let* $\sum_{i=1}^t \mathbf{M}_i(S)$ *be the sum of the matroids* $\mathbf{M}_i(S)$ *with rank functions* $r_i$. *The following conditions are equivalent for* $A \subseteq S$:

(i) $A$ *is independent in* $\sum_{i=1}^t \mathbf{M}_i(S)$.
(ii) $A = \bigcup_{i=1}^t A_i$, *where* $A_i$ *is independent in* $\mathbf{M}_i(S)$, *for all* $i$.
(iii) $A = \dot{\bigcup}_{i=1}^t A_i$ *is a disjoint union, where* $A_i$ *is independent in* $\mathbf{M}_i(S)$, *for all* $i$.

*The rank function* $r$ *of* $\sum_{i=1}^t \mathbf{M}_i(S)$ *is given by*

$$r(A) = \min_{B \subseteq A} \left( \sum_{i=1}^t r_i(B) + |A - B| \right).$$

*Proof.* The last assertion is the rank formula in **6.17**. We trivially have (iii) ⇒ (ii). To see (ii) ⇒ (i) we use **6.17**. Let $B = \bigcup_{i=1}^t B_i$, $B_i \subseteq A_i$ for all $i$, be any subset of $A$. Assuming (ii), we have $|B| \leq \sum_{i=1}^t |B_i| = \sum_{i=1}^t r_i(B_i) \leq \sum_{i=1}^t r_i(B)$ which, by **6.17**, is just the definition for independence in $\sum_{i=1}^t \mathbf{M}_i(S)$. It remains to be shown that (i) ⇒ (iii). Choose $t$ pairwise disjoint sets $S_i$ of the same cardinality as $S$ and $t$ bijections $\phi_i: S \to S_i$. Defining $\phi_i(B)$ to be independent in $S_i$ if and only if $B$ is independent in $\mathbf{M}_i(S)$, we obtain matroids $\mathbf{N}_i(S_i) \cong \mathbf{M}_i(S)$ for all $i$. Denote by $r_i'$ the rank function of $\mathbf{N}_i(S_i)$, i.e., $r_i'(\phi_i(A)) = r_i(A)$ for all $A \subseteq S$, and by $r'$ the rank function of $\prod_{i=1}^t \mathbf{N}_i(S_i)$. Now consider the binary relation $R \subseteq S \times \bigcup_{i=1}^t S_i$ given by

$$R = \{(p, \phi_i(p)) : p \in S, i = 1, \ldots, t\}$$

The matroid $\mathbf{T}(S, R, \prod_{i=1}^{t} \mathbf{N}_i(S_i))$ is, according to our discussion of transversal matroids, induced by the monotone semimodular function $f(A) = r'(R(A))$. Since

$$r'(R(A)) = r'\left(\bigcup_{i=1}^{t} \Phi_i(A)\right) = \sum_{i=1}^{t} r_i'(\Phi_i(A)) = \sum_{i=1}^{t} r_i(A) \quad \text{for all } A \subseteq S$$

it follows that the functions $f$ and $\sum_{i=1}^{t} r_i$ are, in fact, the same, or in other words, that $\sum_{i=1}^{t} \mathbf{M}_i(S) = \mathbf{T}(S, R, \prod_{i=1}^{t} \mathbf{N}_i(S_i))$. Using the description of independent sets as independent partial transversals we obtain the following chain of equivalences:

$A$ independent in $\sum_{i=1}^{t} \mathbf{M}_i \Leftrightarrow \exists$ injection $\phi: A \to \bigcup_{i=1}^{t} S_i$ such that $\phi(A)$ is independent in $\prod_{i=1}^{t} \mathbf{N}_i(S_i) \Leftrightarrow \exists$ injection $\phi: A \to \bigcup_{i=1}^{t} S_i$ such that $\phi(A) = \bigcup_{i=1}^{t} \phi_i(A_i)$ with $\phi_i(A_i)$ independent in $\mathbf{N}_i(S_i)$ for all $i \Leftrightarrow A = \bigcup_{i=1}^{t} A_i$ is a disjoint union with $A_i$ independent in $\mathbf{M}_i(S)$ for all $i$. □

Notice that $\sum_{i=1}^{t} \mathbf{M}_i(S)$ is also the matroid $f(\prod_{i=1}^{t} \mathbf{N}_i(S_i))$ of **6.30** induced by the surjective function $f: \bigcup_{i=1}^{t} S_i \to S$ where $f(p) = \phi_i^{-1}(p)$ for $p \in S_i$.

In the case where we take the sum of $t$ copies of the same matroid $\mathbf{M}(S)$, **6.44** yields very interesting packing and covering theorems for matroids. In the first instance we seek to pack as many disjoint "big" (=spanning) sets into $S$ as possible. In the second, we want to cover $S$ by as few "small" (=independent) sets as possible.

**6.45 Proposition.** *Let $\mathbf{M}(S)$ be a matroid. $S$ is the union of $t$ disjoint spanning sets if and only if $S$ contains $t$ disjoint bases if and only if*

$$|S - B| \geq t(r(S) - r(B)) \quad \textit{for all } B \subseteq S.$$

*Proof.* The first equivalence is obvious. $S$ contains $t$ disjoint bases if and only if the rank of the $t$-fold sum $\mathbf{M} + \cdots + \mathbf{M}$ is at least $t \cdot r(S)$. Now apply the rank formula of **6.44**. □

As a corollary we obtain an expression for the *packing number* pack(**M**) of a finite matroid **M** where pack(**M**) denotes the maximal number of disjoint bases in **M**.

**6.46 Proposition.** *Let $\mathbf{M}(S)$ be a finite matroid. Then*

$$\operatorname{pack}(\mathbf{M}) = \min_{B \subseteq S,\, r(B) \neq r(S)} \left\lfloor \frac{|S - B|}{r(S) - r(B)} \right\rfloor = \min_{S \neq B \in L(S)} \left\lfloor \frac{|S - B|}{r(S) - r(B)} \right\rfloor. \quad \square$$

**Example.** Consider the vector space $V(n, q)$. We have

$$\operatorname{pack}(\mathbf{M}(V(n, q)) = \min_{0 \leq k < n} \left\lfloor \frac{q^n - q^k}{n - k} \right\rfloor = \left\lfloor \frac{q^n - 1}{n} \right\rfloor.$$

Since there are $q^n - 1$ non-zero vectors in $V(n, q)$ and since any basis contains $n$ vectors, $V(n, q)$ can be optimally packed.

For graphs $G(V, S)$, pack($\mathbf{P}(G)$) is the maximal number of edge-disjoint forests contained in $G$. For the complete graph $K_n$, one easily obtains

$$\text{pack}(\mathbf{P}(K_n)) = \left\lfloor \frac{n}{2} \right\rfloor = \begin{cases} \dfrac{n}{2} & \text{if } n \text{ is even} \\ \dfrac{n-1}{2} & \text{if } n \text{ is odd.} \end{cases}$$

Again, since any tree contains $n - 1$ edges and since $\binom{n}{2}$ is the total number of edges, $\mathbf{P}(K_n)$ can be optimally packed.

Let us turn to the covering problem. By **6.44**, $\mathbf{M}(S)$ contains $t$ independent sets whose union is $S$ if and only if $S$ is independent in the sum $\mathbf{M} + \cdots + \mathbf{M}$ ($t$ times). It is clear that $S$ is the union of $t$ independent sets if and only if it is the disjoint union of such sets. Application of the rank formula in **6.44** yields the following result.

**6.47 Proposition.** *Let* $\mathbf{M}(S)$ *be a matroid. $S$ is then the union of $t$ disjoint independent sets if and only if*

$$|B| \leq t \cdot r(B) \quad \textit{for all } B \subseteq S. \quad \square$$

Analogously, we define the *covering number* $\text{cov}(M)$ of a matroid $\mathbf{M}$ as the minimal number of independent sets whose union is $S$. Since in the presence of loops this is impossible, we have to assume that $\mathbf{M}(S)$ has no loops. By **6.47**, $\text{cov}(\mathbf{M}(S)) = \max_{\varnothing \neq B \subseteq S} \lceil |B|/r(B) \rceil = \max_{0 \neq B \in L(S)} \lceil |B|/r(B) \rceil$. Anticipating the result of the next section, that any matroid $\mathbf{M}$ decomposes into a product $\mathbf{M} = \prod \mathbf{M}_i$ of connected submatroids $\mathbf{M}_i$, we deduce from $|B|/r(B) = (|B_1| + \cdots + |B_s|)/(r(B_1) + \cdots + r(B_s))$, $B_i$ connected in $B$, that $|B|/r(B) \leq \max_{1 \leq i \leq s}(|B_i|/r(B_i))$, and hence the following formula.

**6.48 Proposition.** *Let* $\mathbf{M}(S)$ *be a finite matroid without loops. Then*

$$\text{cov}(\mathbf{M}) = \max_{0 \neq B \in L(S)} \left\lceil \frac{|B|}{r(B)} \right\rceil \quad \textit{over all connected flats } B \neq 0. \quad \square$$

**Example.** For graphs $G(V, S)$ the covering number gives the minimum number of disjoint forests into which $G(V, S)$ can be decomposed. In graph theory this number is called the *arboricity* of the graph. Let $G(V, S)$ have the connected graph components $G(V_1, S_1), \ldots, G(V_l, S_l)$. It is almost immediate (use, e.g., **6.43**(i)) that $\mathbf{P}(G(V, S)) \cong \prod_{k=1}^{l} \mathbf{P}(G(V_k, S_k))$. In computing the arboricity we may, by **6.48**, thus confine ourselves to connected subgraphs of $G(V, S)$. Using **6.24**(v) we therefore obtain the following formula (Nash–Williams):

*Let $G(V, S)$ be a loopless graph. The arboricity* $\operatorname{arb}(G)$ *is then given by*

$$\operatorname{arb}(G) = \max_{k \geq 2} \left\lceil \frac{e_k(G)}{k-1} \right\rceil$$

*where $e_k(G)$ denotes the maximal number of edges among all connected subgraphs of $G$ on $k$ vertices.*

For the complete graph $K_n$ we plainly have $e_k(K_n) = \binom{k}{2}$, and thus

$$\left\lceil \frac{e_k(K_n)}{(k-1)} \right\rceil = \left\lceil \frac{k}{2} \right\rceil$$

implying

$$\operatorname{arb}(K_n) = \left\lceil \frac{n}{2} \right\rceil = \begin{cases} \frac{n}{2} & \text{if } n \text{ is even} \\ \frac{n+1}{2} & \text{if } n \text{ is odd.} \end{cases}$$

Hence $K_n$ can be optimally covered by trees. Figure 6.16 shows a decomposition of $K_7$ into 4 trees. The matroid sum with its applications to packing and covering problems is one of the most successful concepts in matroid theory. It not only makes transparent the underlying principles common to all these results but it also makes possible much easier proofs than were previously known.

As a final application we note a pleasing characterization of transversal matroids.

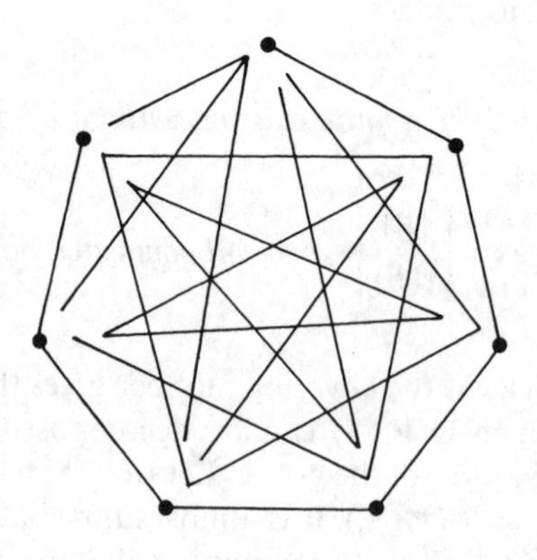

Figure 6.16

**6.49 Proposition.** *A finite matroid is a transversal matroid if and only if* $\mathbf{M}(S) = \sum \mathbf{M}_i(S)$ *is the sum of matroids* $\mathbf{M}_i(S)$ *all of rank* 1.

*Proof.* Let $\mathbf{M}(S) = \mathbf{T}(S, R, I)$ and set $B_j := R^{-1}(j), j \in I$. We define the matroid $\mathbf{M}_j(S)$ by $\overline{\varnothing} = S - B_j$ and $\bar{p} = S$ for all $p \in B_j$. By **6.44**, $A \subseteq S$ is independent in $\sum_{i \in I} \mathbf{M}_i(S)$ if and only if for some $J \subseteq I$, $A = \{p_j : j \in J\}$ with $p_j \in B_j$ which means that $A$ is a partial transversal. The converse is clear. □

**6.50 Corollary.** *The sum of transversal matroids is a transversal matroid.* □

By analogy to the decomposition of matroids into connected matroids we may call $\mathbf{M}$ *irreducible* if $\mathbf{M}$ is not the sum of two non-trivial matroids. Very little is known about general properties of irreducible matroids. As an example, the reader may verify that the Fano matroid $\mathbf{F}$ and $\mathbf{P}(K_4)$ are both irreducible.

## C. Extension of Matroids

A natural problem for any axiomatic structure is that of classifying the possible extensions. Let $\mathbf{M}(S)$ be a matroid. Any matroid $\mathbf{M}(S \cup p)$ which contains $\mathbf{M}(S)$ as a submatroid is called a *one-element extension* of $\mathbf{M}(S)$. Our goal is to determine the possible extensions *within* the given matroid $\mathbf{M}(S)$.

**Definition.** Two sets $A, B \subseteq S$ are said to form a *modular pair* in the matroid $\mathbf{M}(S)$ if

$$r(A \cap B) + r(A \cup B) = r(A) + r(B).$$

A modular pair of flats is called a *modular pair* of the lattice $L(S)$.

The following observation is immediate.

**6.51.** *If* $A, B$ *are a modular pair in* $\mathbf{M}(S)$ *then* $\bar{A}, \bar{B}$ *are a modular pair in* $L(S)$. *The converse is, in general, false.* □

**Definition.** A *modular filter* $\mathfrak{M}$ of the matroid $\mathbf{M}(S)$ is a family $\mathfrak{M} \subseteq 2^S$ such that

(i) $\mathfrak{M}$ is a filter in $2^S$;
(ii) $A, B \in \mathfrak{M}$, $A, B$ modular pair $\Rightarrow A \cap B \in \mathfrak{M}$.

A modular filter in $L(S)$ is defined analogously.

Every modular filter $\mathfrak{M}$ of $\mathbf{M}(S)$ induces a modular filter $M$ in $L(S)$, namely $M = \{\bar{A} : A \in \mathfrak{M}\}$. Conversely, if $N$ is a modular filter in $L(S)$, then $\mathfrak{N} = \{A : \bar{A} \in N\}$ is a modular filter in $\mathbf{M}(S)$.

**6.52 Proposition.** *Let* $\mathbf{M}(S \cup p)$ *be a matroid with closure* $J$ *and rank function* $r$, $p \notin S$. *The set* $\mathfrak{M} = \{A \subseteq S : p \in J(A)\}$ *is then a modular filter of the submatroid* $\mathbf{M}(S)$.

*Proof.* $\mathfrak{M}$ is certainly a filter. By **6.8**(iii), $A \in \mathfrak{M}$ is equivalent to $r(A \cup p) = r(A)$. Hence if $A, B \in \mathfrak{M}$ are a modular pair in $\mathbf{M}(S)$, then

$$\begin{aligned} r((A \cap B) \cup p) &\leq r(A \cup p) + r(B \cup p) - r(A \cup B \cup p) = r(A) + r(B) - r(A \cup B) \\ &= r(A \cap B), \end{aligned}$$

and thus $A \cap B \in \mathfrak{M}$. □

The main result of this section asserts the converse to **6.52**: Every modular filter in $\mathbf{M}(S)$ determines a unique one-element extension of $\mathbf{M}(S)$. If $\mathfrak{M}$ is a modular filter in $\mathbf{M}(S)$ induced by $A \in \mathfrak{M} \Leftrightarrow p \in J(A)$ then we also have $A \in \mathfrak{M} \Leftrightarrow J_S(A) \in \mathfrak{M}$. This means different modular filters in $\mathbf{M}(S)$ may generate the same one-point extension, but to every such extension there belongs a unique modular filter of the lattice $L(S)$. The correspondence "modular filter in $L(S)$" $\leftrightarrow$ "one-element extension" is bijective and hence there are as many extensions as there are modular filters in $L(S)$.

**6.53 Theorem** (Crapo). *Let $\mathfrak{M}$ be a modular filter of the matroid $\mathbf{M}(S)$. Then there is a unique one-element extension $\mathbf{M}(S \cup p)$ such that $\mathfrak{M} = \{A \subseteq S : p \in J_{S \cup p}(A)\}$.*

*Proof.* We denote by $A \to \bar{A}$ and $r$ the closure and rank function in $\mathbf{M}(S)$. Let $\bar{r} : S \cup p \to \mathbb{N}_0$ be defined by

(i) $\bar{r}(A) = r(A)$, for $A \subseteq S$;
(ii) $\bar{r}(A \cup p) = r(A) + 1$, for $A \subseteq S$, $\bar{A} \notin \mathfrak{M}$;
(iii) $\bar{r}(A \cup p) = r(A)$, for $A \subseteq S$, $\bar{A} \in \mathfrak{M}$.

We want to verify the rank axioms **6.14** for $\bar{r}$. This will also prove the uniqueness of the extension since the rank function of $\mathbf{M}(S \cup p)$ must be defined in this way. The function $\bar{r}$ is clearly monotone and satisfies the finite basis axiom. To verify the semimodularity we distinguish two cases:

(a) pairs $A \cup p$, $B$ with $A, B \subseteq S$;
(b) pairs $A \cup p$, $B \cup p$ with $A, B \subseteq S$.

Now for all $A, B \subseteq S$

$$\bar{r}(A \cup B \cup p) - \bar{r}(A \cup B) \leq \bar{r}(A \cup p) - \bar{r}(A).$$

To see this notice that the left-hand side is always $\leq 1$, and equal to 1 if $\overline{A \cup B} \notin \mathfrak{M}$ and thus $\bar{A} \notin \mathfrak{M}$ in which case the right-hand side is also 1. It follows that

$$\begin{aligned} \bar{r}(A \cup B \cup p) - \bar{r}(A \cup p) &\leq \bar{r}(A \cup B) - \bar{r}(A) = r(A \cup B) - r(A) \\ &\leq r(B) - r(A \cap B) = \bar{r}(B) - \bar{r}(A \cap B), \end{aligned}$$

proving case (a). In case (b) we have to show

$$\bar{r}((A \cap B) \cup p) + \bar{r}(A \cup B \cup p) \leq \bar{r}(A \cup p) + \bar{r}(B \cup p).$$

If $\overline{\bar{A} \cup \bar{B}} \notin \mathfrak{M}$ then also $\bar{A}$, $\bar{B}$, $\overline{\bar{A} \cap \bar{B}} \notin \mathfrak{M}$ whence the semimodularity reduces to a valid semimodular inequality in $\mathbf{M}(S)$. If, on the other hand, $\overline{\bar{A} \cup \bar{B}} \in \mathfrak{M}$ then the inequality can be violated only if $\bar{A} \in \mathfrak{M}$, $\bar{B} \in \mathfrak{M}$ and $A$, $B$ are a modular pair in $\mathbf{M}(S)$. In this case, $\bar{A}$, $\bar{B}$ are also a modular pair which implies $\overline{\bar{A} \cap \bar{B}} = \bar{A} \cap \bar{B} \in \mathfrak{M}$. □

**6.53** enables us in principle to construct step by step all *finite matroids*. The reader ought to carry through these one-element extensions up to 5 elements.

The following description of the flats of the extension $\mathbf{M}(S \cup p)$ is easily seen from **6.53**.

**6.54 Proposition.** *Let the extension* $\mathbf{M}(S \cup p)$ *be determined by the modular filter* $M$ *of* $L(S)$. *The flats of* $\mathbf{M}(S \cup p)$ *are of three types*:

(a) *all flats* $A$ *of* $\mathbf{M}(S)$ *with* $A \notin M$;
(b) *all sets* $A \cup p$ *where* $A$ *is a flat of* $\mathbf{M}(S)$ *with* $A \in M$;
(c) *all sets* $A \cup p$ *where* $A$ *is a flat of* $\mathbf{M}(S)$ *which is not in* $M$ *and is not covered in* $L(S)$ *by a flat of* $M$. □

To get a clearer picture we may identify the flats under (a) and (b) with the flats of the original matroid $\mathbf{M}(S)$. Those of (c) are the new flats. If $L_e \subseteq L(S)$ is the poset of these new flats in $L(S \cup p)$, then

$$L(S \cup p) = L(S) \cup \{A \cup p : A \in L_e\}$$

with the additional covering relations

(i) $A \lessdot A \cup p$ for $A \in L_e$;
(ii) $A \cup p \lessdot B \cup p$ for $A \in L_e$, $B \in M$, $A \subseteq B$ with $r(B) = r(A) + 2$.

Notice that any flat $C \notin M \cup L_e$ is covered by exactly one flat in $M$ because of the modularity of the filter $M$.

**Example.** Consider the matroids $\mathbf{M}(S)$ and $\mathbf{M}(S \cup p)$ as given in their Euclidean representations:

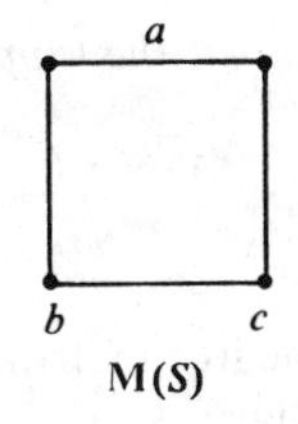

$\mathbf{M}(S)$

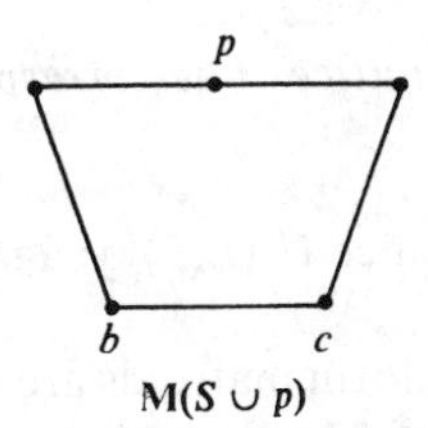

$\mathbf{M}(S \cup p)$

We have $M = \{a, 1\}$ and $L_e = \{0, b, c\}$ with the corresponding lattices shown in Figure 6.17.

The two extremal cases when the modular filter $M = \emptyset$ or $M = L(S)$ correspond to the extensions by a loop, or by an element $p$ which is independent of all of $\mathbf{M}(S)$, respectively. In the latter case $\mathbf{M}(S \cup p) = \mathbf{M}(S) \times \mathbf{M}(p)$; $p$ is then called a *coloop*.

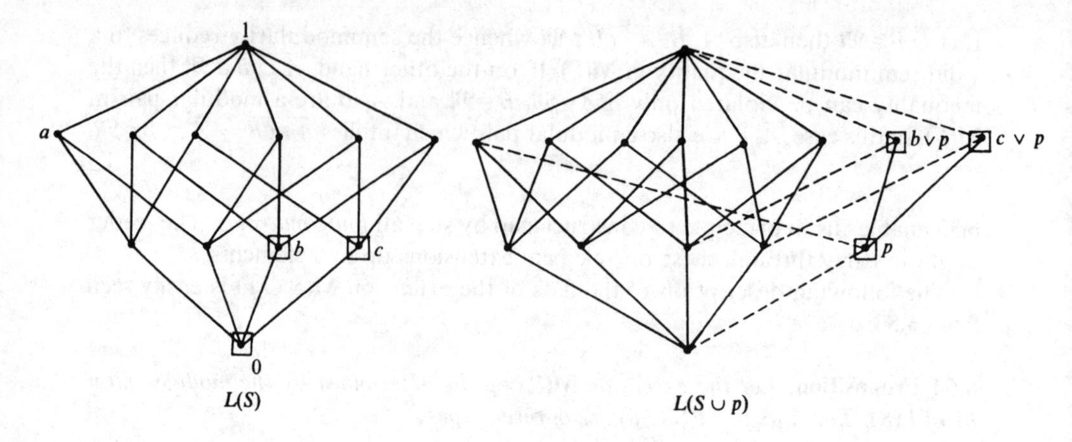


Figure 6.17

## D. Truncation

The last two constructions we want to discuss are induced by two natural operations on the lattice $L$ of flats. We cut off (truncate) the top part (as with the uniform matroids) or the bottom part, respectively, of $L$ and consider the matroids corresponding to the new lattices. Whereas any upper truncation of a geometric lattice is again a geometric lattice and hence induces a matroid, for the lower truncation this is no longer true. In this case, however, there exists a natural completion of the truncated lattice to a geometric lattice, called the *Dilworth completion.*

**6.55 Proposition.** *Let* $\mathbf{M}(S)$ *be a matroid of rank* $n \geq 2$ *and* $L(S)$ *its lattice of flats,* $1 \leq k \leq n - 1$. *The lattice* $U_k(L(S))$ *obtained from* $L(S)$ *by deleting all flats of rank* $\geq k$ *and replacing them by a new maximal element, i.e.,*

$$U_k(L(S)) = \{x \in L(S): r(x) < k\} \cup \{1\},$$

*is a geometric lattice. The corresponding matroid is called the (upper)* truncation $U_k(\mathbf{M}(S))$.

*Proof.* Trivial since $U_k(L(S))$ is again semimodular. □

Thus the uniform matroids are the truncations of the free matroids. Despite the obviousness of **6.55** we can deduce a very useful corollary.

**6.56 Proposition.** *Let* $L$ *be a finite geometric lattice of rank* $n \geq 3$ *and denote by* $W_k$, $0 \leq k \leq n$, *its level numbers. Then*

(i) $W_1 < W_k$ *for* $2 \leq k \leq n - 2$;
(ii) $W_1 \leq W_{n-1}$ *and* $W_1 = W_{n-1}$ *if and only if* $L$ *is modular.*

*Proof.* We have already seen (ii) in **4.54**. Since $U_k(L)$ is a geometric lattice for all $k \geq 1$ we infer $W_1 \leq W_k$ for all $k \geq 1$. Suppose $W_1 = W_k$ for some $k$ between 2 and $n - 2$. By **4.54**, $U_{k+1}(L)$ must then be modular. Choose elements $a < x < b$ of rank $k - 2$, $k$ and $k + 2$, respectively. By **2.41**, any minimal $x$-complement $y$ in $[a, b]$ has rank $k$ in $L$, thus $x$ and $y$ are copoints in $U_{k+1}(L)$ with $x \wedge y$ not a coline, violating the modularity of $U_{k+1}(L)$. □

The upper truncation $U_k(\mathbf{M}(S))$ may be interpreted geometrically as a projection from a center outside $\mathbf{M}(S)$ onto a geometry of rank $k$ in general position relative to $\mathbf{M}(S)$; see Figure 6.18.

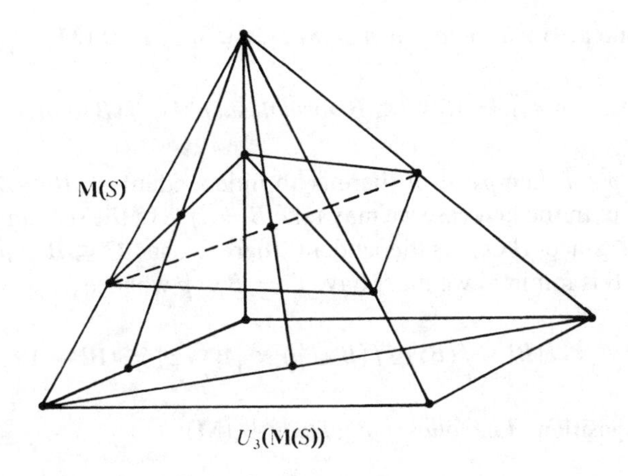


Figure 6.18

As already mentioned, the lower truncation $L_k(L)$ of a geometric lattice does not give a geometric lattice except in the trivial cases $k = n - 2$ or $k = n - 1$. If, for instance, we truncate $\mathscr{B}(4)$ at the point level we obtain a lattice with $\binom{4}{2} = 6$ points and 4 copoints which, by **6.56**(ii), is not geometric. Our goal is to embed $L_k(L)$ into as small a geometric lattice as possible by adding further flats.

To avoid trivial cases we assume $L$ has rank $n \geq 4$ and $1 \leq k \leq n - 3$. Let $\rho$ denote the rank function in $L$. The points of $L_k(L)$ are thus the flats $x \in L$ with $\rho(x) = k + 1$; we denote this set by $S_{k+1}$. The function $f: 2^{S_{k+1}} \to \mathbb{Z}$ defined by

$$f(\varnothing) = -k$$
$$f(A) = \rho(\sup_L A) - k \qquad (\varnothing \neq A \subseteq S_{k+1})$$

is clearly monotone, semimodular and possesses the finite basis property. Furthermore we have $f(p) = 1$ for all $p \in S_{k+1}$. Therefore, by **6.18**, $f$ induces a matroid on $S_{k+1}$, called the *Dilworth completion* $D_k(\mathbf{M})$, whose corresponding lattice is called the Dilworth completion $D_k(L)$. (Notice that we need the general version **6.18** since $f$ would not be semimodular if we require $f(\varnothing) \geq 0$.) To justify the name completion we want to show that every flat $x \in L$ with $\rho(x) \geq k + 1$ is—regarded as the point set $A_x = \{p \in S_{k+1} : p \leq x\}$—again a flat in $D_k(L)$. To avoid confusion we denote by $\rho$ and $r$ the rank functions in $L$ and $D_k(L)$, respectively. The lattice operations in $L$ are $\wedge_L$ and $\vee_L$, those in $D_k(L)$ are $\wedge$ and $\vee$. $A \to \bar{A}$ is the closure in $D_k(\mathbf{M})$. Notice that $x = \sup_L A_x$ for all $x \in L$ with $\rho(x) \geq k+1$.

We know from **6.18** that for $A \subseteq S_{k+1}$:

$$A \text{ independent in } D_k(\mathbf{M}) \Leftrightarrow |B| \leq f(B) \text{ for all } \varnothing \neq B \subseteq A.$$

We call an independent set $A \subseteq S_{k+1}$ *normal* if $|A| = f(A)$. For instance, all points $p \in S_{k+1}$ are normal.

**6.57 Lemma.** *We have for all $A \subseteq S_{k+1}$, $p \in S_{k+1}$ in $D_k(\mathbf{M})$*

$$p \in \bar{A} \Leftrightarrow \exists B \subseteq A,\ B \text{ normal, such that } f(B \cup p) = f(B).$$

*Proof.* If $p \in \bar{A}$ then $p \in A$ or there is an independent set $B \subseteq A$ such that $B \cup p$ is dependent. In the first case we may take $B = \{p\}$. In the second case choose $B$ to be minimal. Since $B \cup p$ is dependent there exists $C \subseteq B \cup p$ with $f(C) < |C|$. Because $B$ is minimal we must have $C = B \cup p$ whence

$$|B| \leq f(B) \leq f(B \cup p) < |B \cup p| = |B| + 1. \quad \square$$

**6.58 Proposition.** *The following holds in $D_k(\mathbf{M})$:*

(i) $A_x = \{p \in S_{k+1} : p \leq x\}$ *is a flat of $D_k(\mathbf{M})$ for all $x \in L_k(L)$. The $A_x$'s are henceforth called $L$-*flats.
(ii) *Every basis of $A_x$ is normal and if, conversely, a flat $A$ of $D_k(\mathbf{M})$ has a normal basis then $A$ is an $L$-flat.*
(iii) $r(A_x) = \rho(x) - k = f(A_x)$ *for all $L$-flats $A_x$. In particular, $x \lessdot_L y$ in $L$ implies $A_x \lessdot A_y$ in $D_k(L)$.*
(iv) $A_x \wedge A_y = A_{x \wedge_L y}$ *for all $L$-flats $A_x$, $A_y$.*

*Proof.* Consider $A_x$ and $p \in \overline{A_x}$, $p \in S_{k+1}$. By **6.57**, there is a normal set $B \subseteq A_x$ with $f(B \cup p) = f(B)$. This is equivalent to

$$\rho(\sup_L(B \cup p)) = \rho(\sup_L B \vee_L p) = \rho(\sup_L B)$$

whence

$$p \leq \sup_L B \leq_L \sup_L A_x = x,$$

i.e., $p \in A_x$, proving (i). For a basis $B$ of $A_x$, we have

$$r(A_x) = |B| \leq f(B) = \rho(\sup_L B) - k \leq \rho(\sup_L A_x) - k = \rho(x) - k.$$

On the other hand, consider any maximal chain $p_1 <\cdot_L p_2 <\cdot_L \cdots <\cdot_L p_{\rho(x)-k} = x$ in $L$ from some $p_1 \in S_{k+1}$ to $x$. Since all $A_{p_i}$'s are flats of $D_k(\mathbf{M})$ we must have

$$r(A_x) \geq \rho(x) - k$$

which implies $|B| = f(B)$ for any basis $B$ of $A_x$, and also the rank formula (iii). Conversely, let $A$ be a flat of $D_k(\mathbf{M})$ with a normal basis $B$. Setting $x = \sup_L B$ we have for $A \subseteq A_x$

$$r(A) = |B| = f(B) = \rho(\sup_L B) - k = \rho(x) - k = r(A_x),$$

i.e., $A = A_x$.

The last assertion follows from (i) and $A_x \cap A_y \supseteq A_{x \wedge_L y} \supseteq A_x \cap A_y$. $\square$

Our main result describes all flats of the Dilworth completion $D_k(\mathbf{M})$.

**6.59 Theorem.** *Let $\mathbf{M}(S)$ be a matroid with lattice $L$ and rank function $\rho$, $1 \leq k \leq n-1$. The set $A \subseteq S_{k+1}$ is a flat of the Dilworth completion $D_k(\mathbf{M})$ if and only if*

$$A = A_{x_1} \cup A_{x_2} \cup \cdots \cup A_{x_t}$$

*with*

$$\rho\left(\bigvee_{j=1}^{s}{}_L\, x_{i_j}\right) > \sum_{j=1}^{s} \rho(x_{i_j}) - (s-1)k \quad \textit{for all } \{x_{i_1}, \ldots, x_{i_s}\},\ 2 \leq s \leq t.$$

*Notice that the $A_{x_i}$'s are pairwise disjoint since $\rho(x_i \vee_L x_j) > \rho(x_i) + \rho(x_j) - k$ implies $\rho(x_i \wedge_L x_j) < k$.*

*Proof.* Let $A$ be a flat of $D_k(\mathbf{M})$ and $A_{x_1}, \ldots, A_{x_t}$ the maximal $L$-flats contained in $A$; clearly $A = \bigcup_{i=1}^{t} A_{x_i}$. We want to prove that the conditions of the theorem hold for the $A_{x_i}$'s. For $t = 1$ there is nothing to show, so assume $t \geq 2$. First we prove that any two $A_{x_i} \neq A_{x_j}$ are disjoint. Suppose on the contrary that $A_{x_i} \cap A_{x_j} = A_{x_i \wedge_L x_j} \neq \varnothing$ and choose $w \in L$ with $x_i \wedge_L x_j <\cdot_L w \leq x_j$. Then $x_i <\cdot_L x_i \vee_L w$, hence $A_{x_i} <\cdot A_{x_i \vee_L w} = A_{x_i} \vee A_w \leq A$, contradicting the maximality of $A_{x_i}$. Next we show that $B_i \cup B_j$ is independent in $D_k(\mathbf{M})$ for any two bases $B_i$ of $A_{x_i}$ and $B_j$ of $A_{x_j}$. If not, there is in $B_j = \{p_1, p_2, \ldots\}$ a minimal $l$ such that $B_i \cup p_1 \cup \cdots \cup p_l$ is independent but $p_{l+1} \in \overline{B_i \cup p_1 \cup \cdots \cup p_l}$. By **6.57**, there exists a normal set $C \subseteq B_i \cup p_1 \cup \cdots \cup p_l$ with $f(C \cup p_{l+1}) = f(C)$. This implies that $B_i \cap C \neq \varnothing$, that $B_i \cup C$ is independent and hence, by the normality of $B_i$ and $C$, that

$$\begin{aligned} |B_i \cup C| &\leq f(B_i \cup C) \leq f(B_i) + f(C) - f(B_i \cap C) \\ &\leq |B_i| + |C| - |B_i \cap C| = |B_i \cup C|. \end{aligned}$$

Thus $B_i \cup C$ is a normal set and therefore spans an $L$-flat, containing $A_{x_i}$ and $p_{l+1}$, in contradiction to the maximality of $A_{x_i}$. We conclude by induction that $B_{i_1} \dot\cup B_{i_2} \dot\cup \cdots \dot\cup B_{i_s}$ is independent for any set of bases of $A_{x_{i_1}}, \ldots, A_{x_{i_s}}$. The maximality of the $A_{x_i}$'s implies that $\dot\bigcup_{j=1}^{s} B_{i_j}$ is *not* normal for $s \geq 2$ whence

$$\rho\left(\bigvee_{j=1}^{s}{}_L\, x_{i_j}\right) = f\left(\bigcup_{j=1}^{s} A_{x_{i_j}}\right) + k \geq f\left(\bigcup_{j=1}^{s} B_{i_j}\right) + k > \left|\bigcup_{j=1}^{s} B_{i_j}\right| + k$$

$$= \sum_{j=1}^{s} |B_{ij}| + k = \sum_{j=1}^{s} r(A_{x_{i_j}}) + k = \sum_{j=1}^{s} \rho(x_{i_j}) - (s-1)k.$$

That, conversely, precisely those disjoint unions $A_{x_1} \cup \cdots \cup A_{x_t}$ which satisfy the conditions of the theorem give a flat of $D_k(\mathbf{M})$ follows without difficulty by induction on $t$. □

As a corollary we obtain an interesting rank formula for the sets in $D_k(\mathbf{M})$ different from the one in **6.18**.

**6.60 Corollary.** *Let $D_k(\mathbf{M})$ be the Dilworth completion of the matroid* $\mathbf{M}$.

(i) *If $\varnothing \neq A \subseteq S_{k+1}$ then $r(A) = \min_\pi \sum_i f(A_i)$ over all partitions $\pi = A_1|A_2|\cdots$ of $A$.*
(ii) *If $A \neq \varnothing$ is a flat then $r(A) = \sum_{i=1}^{t} r(A_{x_i}) = \sum_{i=1}^{t} \rho(x_i) - kt$ where $A_{x_1}, \ldots, A_{x_t}$ are the maximal L-flats contained in $A$.*

*Proof.* (ii) was shown in **6.59**. For any partition $A = A_1|A_2|\cdots$ we have, by **6.18**, $r(A) \leq \sum r(A_i) \leq \sum f(A_i)$. If $\bar{A} = A_{x_1} \cup \cdots \cup A_{x_t}$ is the partition of $\bar{A}$ into its maximal $L$-flats then $A_{x_1} \cap A|\cdots|A_{x_t} \cap A$ is a partition of $A$ and we infer, from $f(A_{x_i} \cap A) \leq f(A_{x_i})$ for all $i$ and **6.58**(iii), that

$$r(A) = r(\bar{A}) = \sum_{i=1}^{t} r(A_{x_i}) = \sum_{i=1}^{t} f(A_{x_i}) \geq \sum_{i=1}^{t} f(A_{x_i} \cap A). \quad \square$$

**Examples.** Consider $\mathbf{D}_1(n) := D_1(\mathbf{FM}(n))$. The points of $\mathbf{D}_1(n)$ are all *pairs* $\{i, j\}$ of an $n$-set $S$. Condition **6.59** says in this case that the flats of $\mathbf{D}_1(n)$ are all disjoint unions $A = A_1 \dot\cup \cdots \dot\cup A_t$. Identifying $A$ with the partition $A_1|A_2|\cdots|A_t|\cdot|\cdot|\cdots$, adding the elements of $S - A$ as singleton blocks, we see that the lattice of $\mathbf{D}_1(n)$ is just the partition lattice $\mathscr{P}(S)$, i.e., $\mathbf{D}_1(n) \cong \mathbf{P}(K_n)$. One can similarly describe the matroids $\mathbf{D}_k(n)$ for $1 < k \leq n - 3$. The situation is particularly simple for $k = n - 3$. In this case the points are all $(n-2)$-subsets of $S$. Apart from the $(n-1)$-subsets of $S$ (the $L$-lines) the only other lines are the trivial lines $A_1 \cup A_2$ with $|A_1| = |A_2| = n - 2$, $|A_1 \cap A_2| \leq n - 4$, i.e., $|A_1 \cap A_2| = n - 4$. $\mathbf{D}_{n-3}(n)$ contains therefore $\binom{n}{2}$ points, $n$ non-trivial lines and $3\binom{n}{4}$ trivial lines. Figure 6.19 shows the Euclidean representation of $\mathbf{D}_2(5)$.

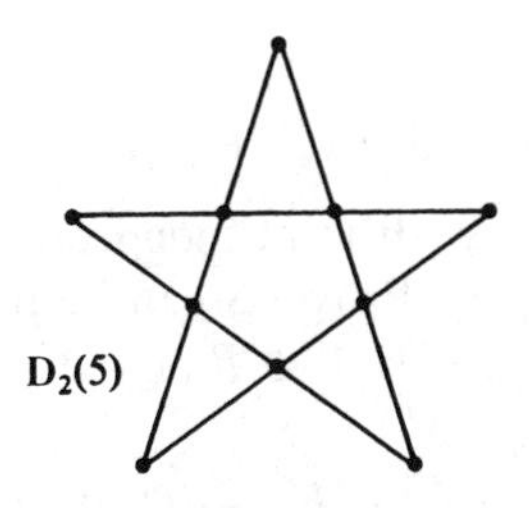


Figure 6.19

Geometrically, the completion $D_k(\mathbf{M})$ can again be regarded as a projection of $\mathbf{M}$ onto an external geometry of rank $n - k$. The Desarguesian block in Figure 6.10 shows the projection of $\mathbf{FM}(4)$ spanned by $\{ae, be, ce, de\}$ onto $\mathbf{P}(K_4)$ generated by $\{ab, ac, ad, bc, bd, cd\}$.

## EXERCISES VI.3

→ 1. Consider the set $\mathfrak{G}$ of finite matroids (say, on the baseset $\mathbb{N}$) and a commutative ring $R$. A function $f: \mathfrak{G} \to \mathrm{R}$ is called a *chromatic invariant* in $R$ if for all $\mathbf{M} = \mathbf{M}(S) \in \mathfrak{G}$:

(i) $\mathbf{M} \cong \mathbf{N} \Rightarrow f(\mathbf{M}) = f(\mathbf{N})$;
(ii) $\mathbf{M} = \mathbf{M}_1 \times \mathbf{M}_2 \Rightarrow f(\mathbf{M}) = f(\mathbf{M}_1) \cdot f(\mathbf{M}_2)$;
(iii) $f(\mathbf{M}) = f(\mathbf{M} \backslash q) + f(\mathbf{M}/q)$ for all points $q$ which are not a loop or a coloop of $\mathbf{M}$ where $\mathbf{M} \backslash q$ and $\mathbf{M}/q$ denote the restriction and the contraction onto $S - q$, respectively. Prove:

(i) The *characteristic polynomial* $\chi(\mathbf{M}; \lambda) := \sum_{T \subseteq S} (-1)^{|T|} \lambda^{r(S) - r(T)}$ is a chromatic invariant in $\mathbb{Z}[\lambda]$.
(ii) The *rank generating function*

$$\sigma(\mathbf{M}; u, v) := \sum_{i,j} a_{ij} u^i v^j,$$

where $a_{ij} := \#\{T \subseteq S : r(S) - r(T) = i, |T| - r(T) = j\}$ is a chromatic invariant in $\mathbb{Z}[u, v]$.

→ 2. Let $\mathbf{M}$ be a finite loopless matroid and let $\mu$ be the Möbius function of its lattice of flats. Show that

$$\chi(\mathbf{M}; \lambda) = \sum_{a \in L} \mu(0, a) \lambda^{r(1) - r(a)},$$

i.e., $\chi(\mathbf{M}; \lambda)$ is precisely the characteristic polynomial of the lattice $L$ as in **4.20**. Let $\mu(\mathbf{M}) := \mu_L(0, 1)$ and deduce that $(-1)^{r(\mathbf{M})} \mu(\mathbf{M})$ is a chromatic invariant in $\mathbb{Z}$ for any loopless matroid $\mathbf{M}$.

3. Prove **6.40**.

4. Compute pack($\mathbf{P}(K_{m,n})$) and cov($\mathbf{P}(K_{m,n})$).

→ 5. Let $B_1, \ldots, B_t$ be pairwise disjoint independent sets in the matroid $\mathbf{M}(S)$ and set $S' := S - \bigcup_{i=1}^{t} B_i$. Prove: $S$ can be partitioned into pairwise disjoint independent sets $A_i$ with $A_i \supseteq B_i$, $i = 1, \ldots, t$, if and only if

$$|A| \leq \sum_{i=1}^{t} (r(A \cup B_i) - r(B_i)) \quad \text{for all } A \subseteq S'.$$

(Edmonds–Fulkerson)

→ 6. Let $\mathbf{M}_1$ and $\mathbf{M}_2$ be matroids on a set $S$ and $f: S \to T$ surjective. Prove:
(i) $f(\mathbf{M}_1 + \mathbf{M}_2) = f(\mathbf{M}_1) + f(\mathbf{M}_2)$;
(ii) $\mathbf{M}$ transversal $\Rightarrow f(\mathbf{M})$ transversal.

→ 7. Construct all matroids on up to 5 points by one-element extension starting from the two matroids on one point.

8.* Use **6.53** to show that are at least $2^n$ non-isomorphic matroids on $n$ points.

9. A subset $A$ of the copoints of a matroid $\mathbf{M}(S)$ is said to form a *linear class* if $c, d \in A, c \wedge d <\cdot c, d \Rightarrow z \in A$ for every copoint $z \cdot> c \wedge d$. Show:
(i) If $A$ is a linear class, then $M := \{w: w \leq z <\cdot 1 \Rightarrow z \in A\}$ is a modular filter.
(ii) If $M$ is a modular filter, then $A := \{z \in M: z <\cdot 1\}$ is a linear class.

→10. Use **6.53** to show that for every $n \geq 2$ there exist geometric lattices $L(S)$ and $L(S \cup p)$, $|S| = n$, which have isomorphic posets above the point level. (Hint: Construct a modular filter in $L(S)$ which has $p$ as the only flat of type (iii) in **6.54**.)

11. A *strong map* between geometric lattices $P$ and $L$ is a function $f: P \to L$ such that

(i) $f(\sup A) = \sup(f(a): a \in A)$ for all $A \subseteq P$;
(ii) $x <\cdot y \Rightarrow f(x) = f(y)$ or $f(x) <\cdot f(y)$;
(iii) $f(0) = 0$.

Let $\mathbf{M}(S)$ be a matroid, $A \subseteq S$. Show that the injection $\phi: L(A) \to L(S)$ with $\phi B = \bar{B}$ as in **6.33** and the contraction $\psi: L(S) \to L(S/A)$, $\psi B = \overline{B \cup A} - A$ in **6.35** are strong maps.

→12. Let $f: L \to L$ be a strong map of a geometric lattice into itself and denote by $Q = f(L)$ the image lattice. $f$ is called *elementary* if $r(Q) = r(L) - 1$. Establish the following correspondence between elementary strong maps on $L$ and modular filters of $L$:

(i) Let $f$ be an elementary strong map on $L$, $Q = f(L)$. Then

$$M_f := \{f(x): r_Q(f(x)) < r(x), x \in L\}$$

is a modular filter $\neq \emptyset, L$ in $L$.

(ii) Let $M \neq \emptyset$, $L$ be a modular filter in $L$ and $L_e$ the set of flats of type **6.54**(iii). Then $f_M: L \to L$ given by

$$f_M(x) := \begin{cases} x & \text{if } x \in M \cup L_e \\ y & \text{if } x \notin M \cup L_e \text{ and } x \lessdot y \end{cases}$$

is an elementary strong map.

(iii) The correspondence in (i) and (ii) is bijective.

13.* According to ex. 12, an elementary map $f$ on a geometric lattice $L(S)$ can be described as follows: We first embed $L(S)$ in the extension $L(S \cup p)$ induced by the modular filter $M_f$ and then contract $L(S \cup p)$ through the point $p$. Generalize this to the following statement: Every strong map on $L$ can be factored into an injection followed by a contraction.

→ 14. An elementary strong map $f$ on $L$ is called *principal* if the corresponding filter $M_f$ is a principal filter in $L$, i.e., if $M_f = M(a) := \{x \in L : x \geq a\}, a \neq 0$. A matroid $\mathbf{M}(S)$, $|S| = n < \infty$, is called *principal* if it can be derived from $\mathbf{FM}(S)$ by a sequence of $n - r(S)$ principal maps

$$\mathbf{FM}(S) \xrightarrow{f_1} \mathbf{M}_1 \xrightarrow{f_2} \mathbf{M}_2 \xrightarrow{f_3} \cdots \xrightarrow{f_{n-r(S)}} \mathbf{M}(S).$$

Since a principal map $f$ with filter $M_f = M(a)$ is uniquely determined by every set $A$ with $\bar{A} = a$, we may also describe a principal matroid by

$$\mathbf{FM}(S) \xrightarrow{A_1} \mathbf{M}_1 \xrightarrow{A_2} \mathbf{M}_2 \xrightarrow{A_3} \cdots \xrightarrow{A_{n-r(S)}} \mathbf{M}(S)$$

where $M_i = M_{f_i} = M(J_{i-1}(A_i))$, $J_{i-1}$ closure in $\mathbf{M}_{i-1}$. It is convenient to allow $f_i = id$ corresponding to $A_i \subseteq J_{i-1}(\emptyset)$. Such $A_i$'s are called *trivial*. Let $\mathbf{M}(S)$ be the principal matroid defined by the chain

$$\mathbf{FM}(S) \xrightarrow{A_1} \mathbf{M}_1 \xrightarrow{A_2} \cdots \xrightarrow{A_t} \mathbf{M}(S),$$

and $J_i$ the closure in $\mathbf{M}_i$, $i = 1, \ldots, t$, and $J$ the closure in $\mathbf{M}(S)$. Prove:

(i) The $A_i$'s can be interchanged so that all non-trivial sets appear first without altering the matroid $\mathbf{M}(S)$.

(ii) $B \subseteq J_1(B) \subseteq J_2(B) \subseteq \cdots \subseteq J(B)$ for all $B \subseteq S$.

(iii) $r(\mathbf{M}) = |S| - \#\{i : A_i \text{ non-trivial}\}$.

(iv) $r(B) = \min_{C \supseteq B}(|C| - h(C)) = |J(B)| - h(J(B))$ where

$$h(B) := \#\{i : 1 \leq i \leq t, B \supseteq A_i, A_i \text{ non-trivial}\}.$$

(v) $B$ independent in $\mathbf{M}(S) \Leftrightarrow h(C) \leq |C - B|$ for all $C \supseteq B$.

(Dowling–Kelly)

15.* Let $\mathbf{M}(S)$ be a matroid of rank $n$. We say that the matroid $\mathbf{N}(S)$ is an *erection* of $\mathbf{M}(S)$ if $r(\mathbf{N}) = n + 1$ and $\mathbf{M} \cong U_n(\mathbf{N})$. For instance, the erections of the free matroid $\mathbf{FM}(n)$ are clearly the covering matroids induced by $n$-partitions. Call $A \subseteq S$ *k-closed* if and only if $\bigcup_{B \subseteq A, |B| \leq k} \bar{B} \subseteq A$. Let $\mathbf{M}(S)$ be a simple matroid of rank $n$. Prove: A family $\mathfrak{H} \subseteq 2^S$ is the family of copoints of an erection of $\mathbf{M}$ if and only if

(i) each $A \in \mathfrak{H}$ spans $\mathbf{M}$;
(ii) each $A \in \mathfrak{H}$ is $(n-1)$-closed;
(iii) every basis of $\mathbf{M}$ is contained in a unique member of $\mathfrak{H}$.

(Crapo)

## 4. Duality and Connectivity

A basis of a matroid is both a maximal independent set and a minimal spanning set. This relationship "spanning" $\leftrightarrow$ "independent" gives rise to the first concept in matroid theory which was neither suggested by related notions in linear algebra nor was carried over from lattice theory—the concept of the *dual matroid* $\mathbf{M}^{\perp}(S)$. The notion of duality is one of the most powerful tools in matroid theory with important applications to many branches of combinatorics, in particular to graph theory. The notation $\mathbf{M}^{\perp}$ was chosen to point out the connection between duality and the concept of orthogonality in the geometric sense as exemplified by the function space matroid. A thorough study of matroid duality forms the first part of this section. We close this introductory chapter with a discussion of connected matroids already mentioned in the previous section.

### A. Duality

We shall assume throughout most of this section that $S$ is finite. A theory of duality in arbitrary matroids has been proposed by several authors (see, e.g., Higgs [3]) but the technical difficulties contained in all of these attempts are for our purposes unnecessarily distracting from the main theme.

The following theorem is the basic result.

**6.61 Theorem** (Whitney). *Let* $\mathbf{M}(S)$ *be a finite matroid and* $\mathfrak{B}$ *the family of its bases. The family* $\mathfrak{B}^{\perp} := \{C \subseteq S : C = S - B$ *for some* $B \in \mathfrak{B}\}$ *of bases complements satisfies the basis axioms* **6.10** *and hence defines a matroid* $\mathbf{M}^{\perp}(S)$, *called the* dual matroid *of* $\mathbf{M}(S)$.

*Proof.* Only the exchange axiom **6.10**(ii) needs to be verified. Let $C \neq D \in \mathfrak{B}^{\perp}$, $c \in C - D$. We have to find $d \in D$ with $(C - c) \cup d \in \mathfrak{B}^{\perp}$. Since $c \notin S - C \in \mathfrak{B}$, $c \in C - D$ there exists, by **6.12**, a unique circuit $K$ of $\mathbf{M}(S)$ such that $c \in K \subseteq (S - C) \cup c$. $K$ cannot lie entirely in the basis $S - D$. Choosing $d \in K - (S - D)$ we have $d \in D \cap (S - C)$, $c \notin \overline{(S - C) - d}$ (because of $d \in K$), and thus $((S - C) - d) \cup c \in \mathfrak{B}$, i.e., $(C - c) \cup d \in \mathfrak{B}^{\perp}$. □

**6.62 Corollary.** *For every finite matroid* $\mathbf{M}(S)$

$$\mathbf{M}^{\perp\perp}(S) = \mathbf{M}(S). \quad \square$$

It should be emphasized that matroid duality is related neither to the usual duality in vector spaces nor to the duality notion in lattices. For this reason (and its connection to orthogonality) the term "orthogonal matroid" appears preferable but we have followed the most widely used terminology to avoid further confusion. It is useful to introduce the following notions: A set $A \subseteq S$ which is a circuit of the dual matroid $\mathbf{M}^{\perp}(S)$ is called a *cocircuit* of $\mathbf{M}(S)$. Similarly, the bases of $\mathbf{M}^{\perp}(S)$, i.e., the bases complements of $\mathbf{M}(S)$, are called the *cobases* of $\mathbf{M}(S)$, etc. One exception to this rule: The names copoint, coline, coplane and corank keep their usual meaning. It follows from **6.61** that every proposition about a matroid $\mathbf{M}(S)$ induces a "co-proposition." This is called the *duality principle* for matroids. For example, the dual proposition of **6.12** reads: If $B$ is a basis complement and $p \notin B$ then there is a unique cocircuit $C$ with $p \in C \subseteq B \cup p$.

**6.63 Proposition.** *Let* $\mathbf{M}(S)$ *be a finite matroid,* $A \subseteq S$. *Then*:

(i) *$A$ independent in $\mathbf{M}(S) \Leftrightarrow S - A$ spans $\mathbf{M}^{\perp}(S)$.*
(ii) *$A$ basis of $\mathbf{M}(S) \Leftrightarrow S - A$ basis of $\mathbf{M}^{\perp}(S)$.*
(iii) *$A$ cocircuit in $\mathbf{M}(S) \Leftrightarrow A = S - H$, $H$ copoint in $\mathbf{M}(S)$.*
(iv) *$r^{\perp}(A) = |A| - r(S) + r(S - A)$ where $r^{\perp}$ is the rank function of $\mathbf{M}^{\perp}(S)$.*
(v) *$r^{\perp}(S) = |S| - r(S)$.*

*Proof.* (i) and (ii) are already contained in **6.61**. To prove (iii), we have: $A$ cocircuit in $\mathbf{M}(S) \Leftrightarrow A$ minimally dependent in $\mathbf{M}^{\perp}(S) \Leftrightarrow S - A$ maximally non-spanning in $\mathbf{M}(S) \Leftrightarrow S - A$ copoint in $\mathbf{M}(S)$. It is easy to see that $r^{\perp}$ as defined in (iv) satisfies the rank axioms **6.14**. Since $r^{\perp}(A) = |A| \Leftrightarrow r(S) = r(S - A) \Leftrightarrow S - A$ spans $\mathbf{M}(S) \overset{(i)}{\Leftrightarrow} A$ independent in $\mathbf{M}^{\perp}(S)$, $r^{\perp}$ is indeed the rank function of $\mathbf{M}^{\perp}(S)$. $\square$

According to **6.63**(iii) we may define the cocircuits of an arbitrary matroid simply as the complements of the copoints.

**Examples.** The dual matroid of the free matroid $\mathbf{FM}(n)$ is of rank 0 and hence consists of $n$ loops. The coloops of an arbitrary matroid are characterized by the condition $0 = r^{\perp}(p) = 1 - r(S) + r(S - p)$, i.e., $r(S) > r(S - p)$. Hence $p$ is a coloop if and only if $p$ is contained in every basis of $\mathbf{M}(S)$ or equivalently if and only if $\mathbf{M}(S) \cong \mathbf{M}(S - p) \times \mathbf{M}(p)$, in agreement with the definition in section 3.C.

The polygon matroid $\mathbf{P}(K_4)$ is an example of a self-dual matroid $\mathbf{P}(K_4) \cong \mathbf{P}(K_4)^{\perp}$, with the isomorphism interchanging the points $\{1, 3\}$ and $\{2, 4\}$ in Figure 6.4 and keeping the others fixed. A necessary and sufficient condition for $id: \mathbf{M}(S) \to \mathbf{M}^{\perp}(S)$ to be an isomorphism is that every cobasis of $\mathbf{M}(S)$ is a basis, and conversely. An example of such an identically-self-dual matroid is the affine geometry $\mathbf{AG}(3, 2)$ in Figure 6.5.

**6.63** yields an easy proof of a theorem which belongs to the packing and covering results of section 3.B.

**6.64 Proposition** (Edmonds). *Let $\mathbf{M}_1$ and $\mathbf{M}_2$ be matroids on the finite set $S$ with rank functions $r_1$ and $r_2$. There exists a $t$-set $B \subseteq S$ which is independent in both $\mathbf{M}_1$ and $\mathbf{M}_2$ if and only if*

$$r_1(A) + r_2(S - A) \geq t \quad \textit{for all } A \subseteq S.$$

*In particular,*

$$\max_{\substack{B \text{ indep.} \\ \text{in } \mathbf{M}_1 \text{ and } \mathbf{M}_2}} |B| = \min_{A \subseteq S} (r_1(A) + r_2(S - A)).$$

*Proof.* If the $t$-set $B$ is independent in both $\mathbf{M}_1$ and $\mathbf{M}_2$ then $S - B$ spans the matroid $\mathbf{M}_2^\perp$ and therefore contains a basis of $\mathbf{M}_2^\perp$, whence by **6.44**

$$(+) \qquad r(\mathbf{M}_1 + \mathbf{M}_2^\perp) \geq t + r_2^\perp(S).$$

Conversely, assume $(+)$. Any basis $A$ of $\mathbf{M}_2^\perp(S)$ is independent in $\mathbf{M}_1 + \mathbf{M}_2^\perp$. Hence, by **6.44**(iii), there exists a disjoint union $D = D_1 \dot\cup D_2$ with $D \supseteq A$ and $|D| = t + r_2^\perp(S)$, such that $D_1$ is independent in $\mathbf{M}_1$ and $D_2$ is independent in $\mathbf{M}_2^\perp(S)$. Assume $|D_2| < r_2^\perp(S)$. Then, by **6.9**(ii), there is an element $p \in D_1 \cap A$ such that $D = (D_1 - p) \dot\cup (D_2 \cup p)$ is again such a union. Continuing in this way we arrive at a disjoint pair $(B, C)$, $D = B \dot\cup C$, where $B$ is independent in $\mathbf{M}_1$, $C$ is independent in $\mathbf{M}_2^\perp$, and $|B| = t$, $|C| = r_2^\perp(S)$. This means that $S - C$ is independent in $\mathbf{M}_2(S)$ and contains the independent $t$-set $B$. Hence the existence of a $t$-set $B$ as specified in the theorem is equivalent to $(+)$, which, in turn, is equivalent to the condition of the theorem by the rank formula in **6.44** and **6.63**(iv). $\square$

As a corollary we can now extend the rank formulas in **6.27** and **6.29**. Let $G(S \cup I, R)$ be a bipartite graph and suppose we are given matroids $\mathbf{M}_1(S)$ and $\mathbf{M}_2(I)$ on $S$ and $I$, respectively. A matching $M$ is now a set of edges such that $\text{match}_S(M)$ is independent in $\mathbf{M}_1(S)$ and $\text{match}_I(M)$ is independent in $\mathbf{M}_2(I)$.

**6.65 Proposition.** *Let $G(S \cup I, R)$ be a finite bipartite graph and let $\mathbf{M}_1(S)$ and $\mathbf{M}_2(I)$ be matroids on $S$ and $I$, respectively, with rank functions $r_1$ and $r_2$. Then*

$$\max_{M \text{ matching}} |M| = \min_{A \subseteq S} (r_1(S - A) + r_2(R(A))).$$

*Proof.* The inequality max $\leq$ min is easily seen. To prove the converse we define the following matroids $\mathbf{L}_1(R)$ and $\mathbf{L}_2(R)$ on the edge-set $R$. The independent sets in $\mathbf{L}_1(R)$ are $\varnothing$ and all sets $\{\{p_1, q_1\}, \ldots, \{p_k, q_k\}\} \subseteq R$ for which $\{p_1, \ldots, p_k\}$ is independent in $\mathbf{M}_1(S)$. It is straightforward to verify the axioms **6.9**. In an analogous way, we define the matroid $\mathbf{L}_2(R)$. Now we plainly have:

$$M \subseteq R \text{ is a matching} \Leftrightarrow M \text{ is independent in } \mathbf{L}_1(R) \text{ and } \mathbf{L}_2(R).$$

Applying **6.64** this gives

$$\max_{M \text{ matching}} |M| = \min_{B \subseteq R}(r'_1(B) + r'_2(R - B)),$$

where $r'_1$ and $r'_2$ are the rank functions in $\mathbf{L}_1$ and $\mathbf{L}_2$, respectively. Let $B$ be a subset of $R$ which attains the minimum and let $A = \{p \in S: p \text{ is incident with no edge of } B\}$. Then by the definition of $\mathbf{L}_1$, $r_1(S - A) = r'_1(B)$. For the same reason, we have $r_2(R(A)) \leq r'_2(R - B)$, and hence altogether

$$\max_{M \text{ matching}} |M| \geq \min_{A \subseteq S}(r_1(S - A) + r_2(R(A))). \quad \square$$

**6.65** is valid for arbitrary bipartite graphs but requires a different argument (see the exercises for more details).

To illustrate the duality principle let us collect a few results which follow immediately from **6.63**. The dual statement is marked with a prime. Notice that **6.66** is valid for any matroid.

**6.66 Proposition.** *We have in a matroid* $\mathbf{M}(S)$:

(i) $B$ *is a basis* $\Leftrightarrow$ $B$ *is a minimal set which intersects every cocircuit nontrivially.*
(i') $B'$ *is a cobasis* $\Leftrightarrow$ $B'$ *is a minimal set which intersects every circuit nontrivially.*
(ii) $C$ *is a cocircuit* $\Leftrightarrow$ $C$ *is a minimal set which intersects every basis nontrivially.*
(ii') $C'$ *is a circuit* $\Leftrightarrow$ $C'$ *is a minimal set which intersects every cobasis nontrivially.* $\square$

A very useful result (especially for the coordinatization theory of the next chapter) is the following proposition (which may be dualized as above).

**6.67 Proposition.** *Let* $\mathfrak{K}$ *and* $\mathfrak{C}$ *be the families of circuits and cocircuits of a matroid* $\mathbf{M}(S)$, *respectively. Then*

(i) $K \in \mathfrak{K} \Leftrightarrow |K \cap C| \neq 1$ *for all* $C \in \mathfrak{C}$ *and* $K$ *is a minimal set with this property.*
(ii) $A$ *is a union of circuits* $\Leftrightarrow |A \cap C| \neq 1$ *for all* $C \in \mathfrak{C} \Leftrightarrow |A \cap B| \neq 1$ *for all* $B$ *which are unions of cocircuits* $\Leftrightarrow p \in \overline{A - p}$ *for all* $p \in A$.

*Proof.* As (i) is a consequence of (ii) we prove (ii). Clearly, $A$ is a union of circuits if and only if $p \in \overline{A - p}$ for all $p \in A$. Since cocircuits are the complements of copoints it follows that the unions of cocircuits are precisely the complements $B = S - T$ of flats $T \in L(S)$. Now if $A \cap B = \{p\}$, then $A - p \subseteq S - B \in L(S)$; hence $\overline{A - p} \subseteq S - B$, i.e. $p \notin \overline{A - p}$. On the other hand if $p \notin \overline{A - p}$ then, by **2.40**(ii), there exists a copoint $H$ with $\overline{A - p} \subseteq H$, $p \notin H$, implying $A \cap (S - H) = \{p\}$. $\square$

Finally, let us study the effect of duality on the operations restriction, contraction and product.

**6.68 Proposition.** *Let* $\mathbf{M}(S)$ *be a finite matroid. We have for* $A \subseteq S$:

(i) $(\mathbf{M}(S) \cdot A)^{\perp} = \mathbf{M}^{\perp}(S)/(S - A)$.
(i′) $(\mathbf{M}(S)/A)^{\perp} = \mathbf{M}^{\perp}(S) \cdot (S - A)$.
(ii) *If* $\mathbf{M}(S) \cong \mathbf{M}_1(S_1) \times \mathbf{M}_2(S_2)$ *then* $\mathbf{M}^{\perp}(S) \cong \mathbf{M}_1^{\perp}(S_1) \times \mathbf{M}_2^{\perp}(S_2)$.

*Proof.* (ii) is immediately clear. (i) and (i′) are dual and follow readily from the formulas for the various rank functions. □

B. Examples

Consider a function space matroid $\mathbf{M}(F(S, K))$. Just as in our discussion of the restriction and contraction, we seek a new function space $F^{\perp}(S, K)$ such that $\mathbf{M}(F)^{\perp} \cong \mathbf{M}(F^{\perp})$.

**Definition.** Two functions $f, g$ from the finite set $S$ into the field $K$ are called *orthogonal*, denoted by $f \perp g$, if

$$\sum_{p \in S} f(p)g(p) = 0.$$

Let $F(S, K)$ be a finite function space. The set

$$F^{\perp}(S, K) := \{g: S \to K \text{ with } g \perp f \text{ for all } f \in F\}$$

is clearly again a function space over $K$, and we have $F \subseteq F^{\perp\perp}$.

**6.69 Theorem.** *Let* $F(S, K)$ *be a finite function space. Then*

$$\mathbf{M}(F(S, K))^{\perp} \cong \mathbf{M}(F^{\perp}(S, K)).$$

*Proof.* Let $B$ be a basis of $\mathbf{M}(F(S, K))$. We want to show that $S - B$ is a basis of $\mathbf{M}(F^{\perp}(S, K))$. Let us reformulate **6.21**(ii) in terms of cocircuits. We define the *support* $\|f\|$ of a function $f$ by $\|f\| := \{p \in S: f(p) \neq 0\}$, i.e., $\|f\| = S - \ker(f)$. **6.21**(ii) now says that the cocircuits of $\mathbf{M}(F(S, K))$ are precisely the minimal non-empty supports. We define, for each basis $B$, functions $f_b$, $b \in B$, by

$$f_b(p) := \begin{cases} 0 & \text{if } p \in \overline{B - b} \\ \neq 0 & \text{if } p \notin \overline{B - b}. \end{cases}$$

Since $K$ is a field, we may assume $f_b(b) = 1$ for all $b \in B$. Let $g_c: S \to K$ for $c \in S - B$ be defined by

$$g_c(q) := \begin{cases} 0 & \text{if } q \in (S - B) - c \\ 1 & \text{if } q = c \\ -f_b(c) & \text{if } q = b \in B. \end{cases}$$

We then have for all $c \in S - B, b \in B$,

$$\sum_{p \in S} g_c(p) f_b(p) = g_c(b) f_b(b) + g_c(c) f_b(c) = 0,$$

hence $g_c \in F^{\perp}(S, K)$ since the $f_b$'s form a basis of the vector space $F(S, K)$. Furthermore, the $g_c$'s are obviously linearly independent. If a cocircuit $C$ of $\mathbf{M}(F^{\perp}(S, K))$ were entirely contained in $B$ and $\|h\| = C$ with $0 \neq h \in F^{\perp}$, then $h \perp f_b$ for all $b \in B$ and thus

$$0 = \sum_{p \in S} h(p) f_b(p) = \sum_{p \in B} h(p) f_b(p) = h(b),$$

i.e., $h = 0$, contrary to the hypothesis. Hence $S - B$ has a non-empty intersection with every cocircuit of $\mathbf{M}(F^{\perp}(S, K))$ and (considering the family $\{g_c\}$) is a minimal set with this property. Invoking **6.66**(i) we infer that $S - B$ is a basis of $\mathbf{M}(F^{\perp}(S, K))$. This now implies

$$\dim F^{\perp} = r(\mathbf{M}(F^{\perp}(S, K))) = |S| - r(\mathbf{M}(F(S, K))) = |S| - \dim F$$

and

$$\dim F^{\perp\perp} = |S| - \dim F^{\perp} = \dim F;$$

hence $F^{\perp\perp} = F$. If, conversely, $S - B$ is a basis of $\mathbf{M}(F^{\perp}(S, K))$ then, by the same argument as above, $B$ is a basis of $\mathbf{M}(F^{\perp\perp}(S, K)) = \mathbf{M}(F(S, K))$. □

**6.70 Corollary.** *If a finite matroid is K-linear then so is the dual matroid. In particular, the dual matroid of a regular matroid is regular.* □

**Example.** The Fano matroid **F** and its dual $\mathbf{F}^{\perp}$ are $GF(2)$-linear (see Figure 20). We shall prove in the next chapter that **F**, and therefore $\mathbf{F}^{\perp}$, can be coordinatized only over fields of characteristic 2. All lines in $\mathbf{F}^{\perp}$ contain 2 points (hence $\mathbf{F}^{\perp}$ is a covering matroid); there are 7 planes of cardinality 4 (=complements of the lines in **F**) and 7 planes of cardinality 3 (=lines of **F**).

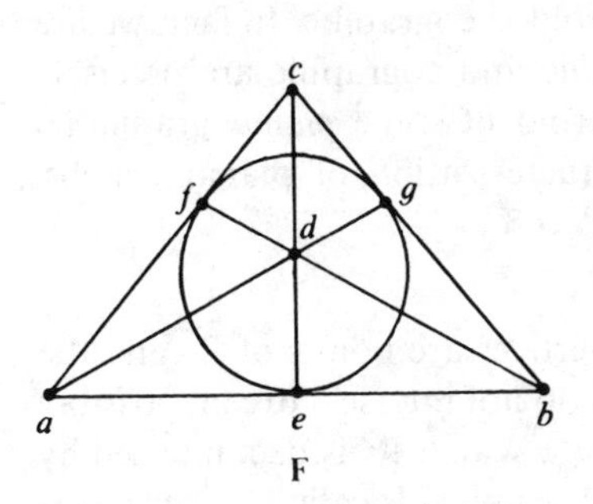


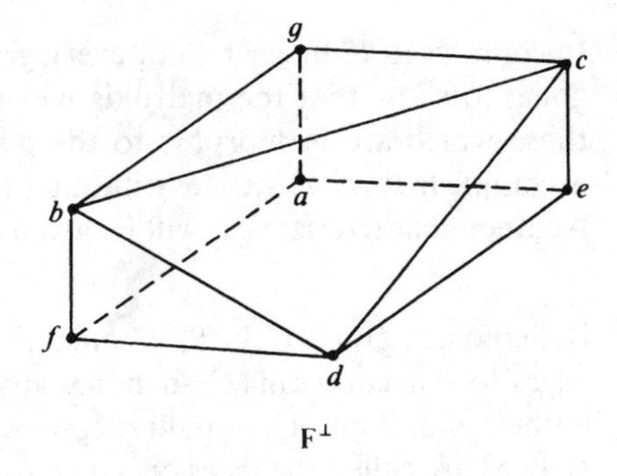


Figure 6.20

We turn to graphs. We can give a very useful graphical definition of a cocircuit in $\mathbf{P}(G(V, S))$. Recall from **6.24**(iv) that the cocircuits in $\mathbf{P}(G(V, S))$ are precisely the minimal edge-sets connecting two components of the corresponding copoint. Read differently, this says that the cocircuits are the minimal edge-sets whose removal increases the number of components in the graph. In graph theory these sets are called *minimal edge-cutsets* or *bonds*. We shall always use the short term *bonds*. Notice that the set $\mathrm{St}(v)$ of edges which are not loops and which are incident with a vertex $v$ contains a bond whenever $\mathrm{St}(v) \neq \emptyset$. $\mathrm{St}(v)$ is called the *star* centered at $v$.

In the graph of Figure 6.21, the sets $\{a, b, c\}$, $\{c, d, e\}$ are bonds. The one-element bonds are those edges whose deletion disconnects the component in which the edge is contained. These edges are the coloops of $\mathbf{P}(G(V, S))$. In graph theory, coloops are called *bridges* of the graph. In the graph of Figure 6.21, $f$ is the only bridge.

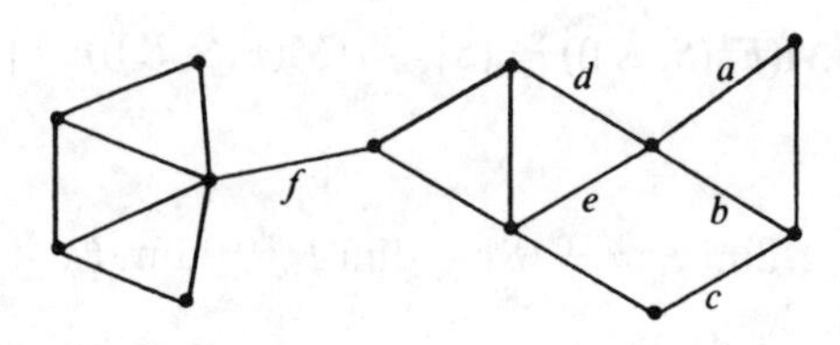


Figure 6.21

**Definition.** Let $G(V, S)$ be a graph. The dual matroid of the polygon matroid is called the *bond matroid* of $G(V, S)$, denoted by $\mathbf{B}(G(V, S))$. A matroid is called *cographic* if it is isomorphic to the bond matroid of some graph.

By **6.41** and **6.68** we have:

**6.71 Proposition.** *Any minor of a cographic matroid is cographic.* $\square$

In contrast to $K$-linearity, not every graphic matroid is cographic. In fact, we are going to show that the matroids which are graphic and cographic are precisely those which are isomorphic to the polygon matroid of some *planar* graph. To accomplish this we quote without proof a characterization of planar graphs. Another characterization will be given in section VII.3.

**Definition.** A graph $G(V, S)$ is called *plane* if the vertices are points of $\mathbb{R}^2$ and the edges Jordan curves in $\mathbb{R}^2$, such that any two edges do not intersect except possibly at their endpoints. The smallest connected parts into which $\mathbb{R}^2$ is decomposed by $G(V, S)$ are called the *faces* or *regions* of $G(V, S)$. A graph $G$ is called *planar* if $G$ is isomorphic to some plane graph $H$; $H$ is then called a *plane representation* of $G$.

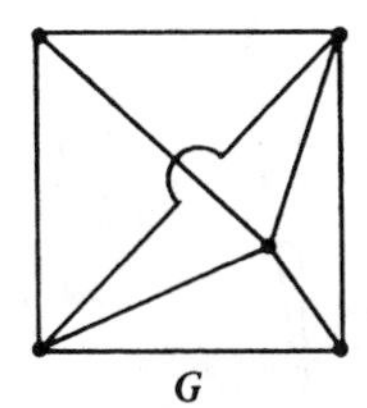


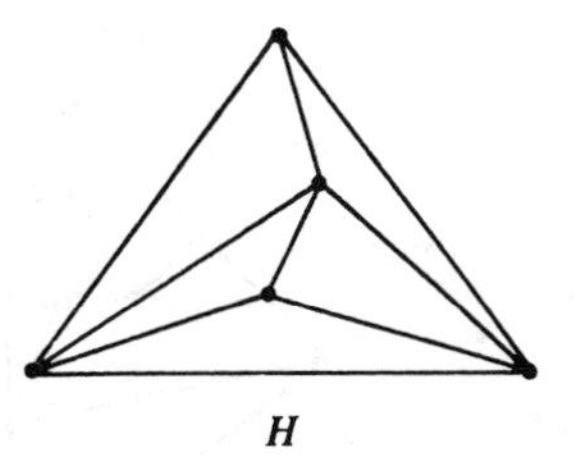


Figure 6.22

The distinction between planar and plane graphs seems artificial at first sight but we shall see that a planar graph possesses, in general, several different embeddings into the plane.

**Example.** Figure 6.22 shows a planar graph $G$ and a plane representation $H$. We quote the following well-known theorem on planar graphs, a proof of which can be found in Harary [1, p. 109]. An alternative treatment is given in section VII.3.B.

**6.72 Theorem** (Kuratowski). *A graph $G(V, S)$ is planar if and only if it does not contain a subgraph which can be contracted to $K_5$ or $K_{3,3}$. In matroid terminology: $G(V, S)$ is planar if and only if $\mathbf{P}(G(V, S))$ has no minor $\mathbf{M} \cong \mathbf{P}(K_5)$ or $\mathbf{M} \cong \mathbf{P}(K_{3,3})$.*[1] □

**6.73 Proposition.** *$\mathbf{P}(K_5)$ and $\mathbf{P}(K_{3,3})$ are not cographic.*

*Proof.* Suppose $\mathbf{P}(K_5)$ is cographic with $\mathbf{P}(K_5) \cong \mathbf{B}(G(V, S))$ for some graph $G$, where we can assume that $G$ has no isolated vertices. Then we have $r(\mathbf{P}(G(V, S)) = 10 - r(\mathbf{P}(K_5)) = 6$ and hence $|V| \geq 7$ according to **6.24**(v). Furthermore, all polygons $C$ in $K_5$, and hence all bonds $C$ in $G(V, S)$, satisfy $|C| \geq 3$. This means that $|\mathrm{St}(v)| \geq 3$ and thus $\gamma(v) \geq 3$ for every vertex $v$, implying

$$20 = 2|S| = \sum_{v \in V} \gamma(v) \geq 7.3 = 21,$$

which is a contradiction. The assertion for $K_{3,3}$ is proved similarly. □

Combining the last three results, we conclude that the polygon matroid of a non-planar graph cannot be cographic. To prove the converse, for every planar graph $G$ we have to find another graph $G^*$ such that $\mathbf{P}(G) \cong \mathbf{B}(G^*)$.

Let $G(V, S)$ be a plane graph. We construct the *dual graph* $G^*$ as follows. We choose a vertex $v_i^*$ in the interior of each face $F_i$ of $G$ ; these are the vertices of $G^*$. Then, for each edge $e \in S$, we draw a new edge $e^*$ which crosses $e$ (but no other edge

[1] The knowledgeable reader will notice that this is not quite what is usually known as Kuratowski's theorem. The two versions are, however, equivalent (see Harary [1, p. 113]).

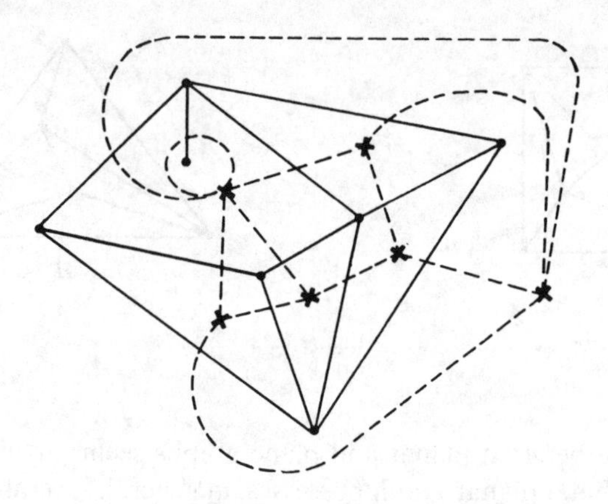

Figure 6.23. Plane graph and its dual.

of $G$) and joins the vertices $v_i^*$, $v_j^*$ corresponding to the faces $F_i$, $F_j$ adjacent to $e$; these are the edges of $G^*$ (see Figure 6.23). If $G$ is a planar graph then different plane representations may give rise to non-isomorphic dual graphs (see the exercises for an example). The main result below, however, asserts that the polygon matroids of all dual graphs are isomorphic.

**6.74 Proposition.** *Let $G(V, S)$ be a plane graph and $G^*(V^*, S^*)$ the dual graph. The mapping $\phi: S \to S^*$, $\phi e = e^*$ is an isomorphism,*

$$\mathbf{P}(G(V, S)) \cong \mathbf{B}(G^*(V^*, S^*)).$$

*Proof.* A polygon $K$ of $G(V, S)$ contains one or more faces $F_i$ of $G(V, S)$ in its interior. The removal of the edge-set $K^* = \{e^* : e \in K\}$ increases the number of components in $G^*$ by at least one since the vertices corresponding to the faces inside $K$ can no longer be joined to vertices corresponding to faces outside $K$. $K^*$ is therefore an edge-cutset in $G^*$ and, since it is clearly minimal, a bond of $G^*$. One proves similarly "$C^*$ bond in $G^* \Rightarrow C$ polygon in $G$." □

An immediate consequence is the well-known Euler formula for plane graphs.

**6.75 Proposition.** *Let $G(V, S)$ be a connected plane graph and $F$ the set of faces. Then*

$$|V| - |S| + |F| = 2.$$

*Proof.* Upon noticing that the dual graph of a connected plane graph is connected, the theorem follows from $r(\mathbf{P}) + r(\mathbf{P}^\perp) = |S|$. □

**6.72** and **6.74** together yield the result announced earlier.

**6.76 Theorem** (Whitney). *A matroid* $\mathbf{M}(S)$ *is graphic and cographic if and only if* $\mathbf{M}(S) \cong \mathbf{P}(G(V, S))$ *for some planar graph* $G(V, S)$. *A graph is planar if and only if its polygon matroid is cographic, or equivalently, if and only if its bond matroid is graphic.* □

Motivated by **6.76** we define a matroid to be *planar* if it is both graphic and cographic.

**6.77 Corollary.** *Any minor of a planar matroid is planar; the dual of a planar matroid is planar.* □

There are also interesting descriptions of the dual matroids of our third fundamental class, the transversal matroids. We shall discuss one of them in the next chapter; see also the exercises. Notice that the dual of a transversal matroid is, in general, not transversal. As an example, consider the planar matroid $\mathbf{P}(K_{2,3})$. It is easily seen that $\mathbf{P}(K_{2,3})$ is transversal. The dual matroid of $\mathbf{P}(K_{2,3})$ is, by **6.74**, the polygon matroid of the graph $K_{2,3}^*$ shown in fig. 6.12, which, as was noted there, is not transversal.

C. Connectivity

To conclude this introductory chapter, we study the decomposition of matroids into their smallest indecomposable parts, analogous to the decomposition of geometric lattices discussed in chapter II.

**Definition.** A *separator* of a matroid $\mathbf{M}(S)$ is a subset $T \subseteq S$ such that

$$\mathbf{M}(S) = \mathbf{M}(S) \cdot T \times \mathbf{M}(S) \cdot (S - T).$$

Hence $T$ is a separator if and only if $\mathbf{M}(S)$ factors into the product of submatroids generated by $T$ and $S - T$. The following result is the companion theorem to **2.45**.

**6.78 Theorem.** *In a matroid* $\mathbf{M}(S)$ *the following properties are equivalent for* $T \subseteq S$:

(i) *$T$ is a separator.*
(ii) *For each circuit $C$ either $C \subseteq T$ or $C \subseteq S - T$.*
(iii) $\mathbf{M}(S) \cdot T = \mathbf{M}(S)/(S - T)$.
(iv) $r(T) + r(S - T) = r(S)$.
(v) $T \cup \overline{\varnothing} = \overline{T}$ *and* $\overline{T}$ *is a separator of* $L(S)$ *in the sense of* **2.45**.

*Proof.* (i) ⇒ (ii). Immediate from **6.43**. (ii) ⇒ (iii). The circuits of $\mathbf{M}(S) \cdot T$ are precisely the circuits $C$ of $\mathbf{M}(S)$ with $C \subseteq T$. Using **6.36**(iii) and (ii), it follows that these are also the circuits of $\mathbf{M}(S)/(S - T)$. (iii) ⇒ (iv). This is implied by the rank formulae for restriction and contraction. (iv) $\Rightarrow$ (v). Suppose there exists $p \in S$ with $p \notin T$, $p \notin \overline{\varnothing}$ and $p \in \overline{T}$. Then $p \in S - T$ and hence

$$r(S) - r(S - T) = r(T) = r(T \cup p) \geq r(S) + r(p) - r(S - T)$$
$$= r(S) - r(S - T) + 1,$$

which is impossible. We now verify **2.45**(iii) for $x = \overline{T}$, $x' = \overline{S - T}$. We have $\overline{T} \cap \overline{S - T} = \overline{\varnothing}$, $r(\overline{T}) + r(\overline{S - T}) = r(S)$ and trivially $\bar{p} \leq \overline{T}$ or $\bar{p} \leq \overline{S - T}$ for all $p \in S$. (v) $\Rightarrow$ (i). By the definition of a lattice separator, $L(S) = [0, \overline{T}] \times [0, \overline{S - T}]$, $\overline{T} = T \cup \overline{\varnothing}$, $\overline{S - T} = (S - T) \cup \overline{\varnothing}$, which implies $r(A) = r_T(A \cap T) + r_{S-T}(A \cap (S - T))$ for all $A \subseteq S$. □

**6.79 Corollary.** *Let* $\mathbf{M}(S)$ *be a matroid. The union and intersection of separators is again a separator as is the set complement of a separator. Hence the separators form a distributive sublattice of* $2^S$. *In particular, every point* $p \in S$ *is contained in a unique smallest separator* $\neq \varnothing$ *and the family of these minimal separators forms a partition of* $S$. □

**Definition.** $\varnothing$ and $S$ are always separators. A matroid which only possesses these trivial separators is called *connected.*

Notice that, by **6.78**(ii), any loop is a minimal separator $\neq \varnothing$. Hence a connected matroid has no loops unless it consists of just a single loop. **6.78**(v) then implies that a matroid with at least two elements is connected if and only if it has no loops and its lattice of flats is indecomposable.

By analogy to our discussion of geometric lattices, we now describe the finest decomposition of a matroid. The following lemma is an immediate consequence of **6.78**(iv) and the various rank formulas.

**6.80 Lemma.** *Let* $T$ *be a separator of* $\mathbf{M}(S)$ *and* $U \subseteq T$. $U$ *is then a separator of* $\mathbf{M}(S) \cdot T$ *if and only if* $U$ *is a separator of* $\mathbf{M}(S)$. □

**6.81 Theorem.** *Let* $\mathbf{M}(S)$ *be a matroid and* $\{T_i : i \in I\}$ *the minimal separators* $\neq \varnothing$ *of* $\mathbf{M}(S)$. *Then*

$$\mathbf{M}(S) \cong \prod_{i \in I} \mathbf{M}(T_i)$$

*is the unique decomposition of* $\mathbf{M}(S)$ *into connected submatroids* $\neq \varnothing$. *The product is well-defined since* $\mathbf{M}(T_i)$ *is either a single loop* ($r(T_i) = 0$) *or loopless* ($r(T_i) > 0$). *The submatroids* $\mathbf{M}(T_i)$ *are called the* components *of* $\mathbf{M}(S)$. □

By translating **2.51** we could give a description of connected matroids in terms of a "perspectivity relation." Since, however, we would have to take special care of loops and parallel elements, the following characterization, suggested by **6.78**(ii), has proved more useful.

**6.82 Proposition.** *Let* $\mathbf{M}(S)$ *be a matroid. The relation* $p \sim q$ *if and only if* $p = q$ *or there exists a circuit containing both* $p$ *and* $q$ *is an equivalence relation on the point set* $S$ *whose equivalence classes are precisely the minimal separators* $\neq \varnothing$.

*Proof.* Only the transitivity needs verification. Let $\mathfrak{K}$ be the family of circuits. Suppose the assertion is not true. Then there must be triples $p, q, r \in S$ with $p \sim q$, $q \sim r$, i.e., $\{p, q\} \subseteq C_1 \in \mathfrak{K}$, $\{q, r\} \subseteq C_2 \in \mathfrak{K}$, but $p \nsim r$. Among all these triples choose $p, q, r$ such that $|C_1 \cup C_2|$ is minimal. Applying the strong circuit exchange axiom **6.13**(ii'), we deduce the existence of $C_3, C_4 \in \mathfrak{K}$ with

$$p \in C_3 \subseteq (C_1 \cup C_2) - q,$$
$$r \in C_4 \subseteq (C_1 \cup C_2) - q.$$

Clearly, $C_3 \cap (C_2 - C_1) \neq \varnothing$ and $C_4 \cap (C_1 - C_2) \neq \varnothing$. Using $p \in C_1$, $r \in C_4$ and the minimality of $|C_1 \cup C_2|$, we infer from $C_4 \cap (C_1 - C_2) \neq \varnothing$ that $C_1 \cup C_4 = C_1 \cup C_2$, and thus $C_4 - C_1 = C_2 - C_1$. Again by the minimality of $|C_1 \cup C_2|$, $p \in C_3$, $r \in C_4$ and $|C_3 \cup C_4| < |C_1 \cup C_2|$, we have $C_3 \cap C_4 = \varnothing$. Together this gives $\varnothing \neq C_3 \cap (C_2 - C_1) = C_3 \cap (C_4 - C_1) \subseteq C_3 \cap C_4 = \varnothing$, which is impossible.

To complete the proof, let $T$ be a minimal separator and $p \in T$. It is obvious from **6.78**(ii) that the $\sim$-equivalence class $[p]$ containing $p$ lies fully in $T$. On the other hand, since $\sim$ is transitive, there is no circuit containing points from both $[p]$ and its complement. Hence $[p]$ is itself a separator and therefore $[p] = T$. □

**6.83 Corollary.** *A matroid is connected if and only if any two elements are contained in a circuit.* □

The following corollary of **6.83** is useful. (Exercises)

**6.84 Proposition.** *Let* $\mathbf{M}(S)$ *be a connected matroid,* $p \in S$. *Then either the restriction* $\mathbf{M}(S) \cdot (S - p)$ *or the contraction* $\mathbf{M}(S)/\{p\}$ *is connected.* □

Before turning to examples, let us consider the separators of the dual matroid.

**6.85 Proposition.** *Let* $\mathbf{M}(S)$ *be any matroid,* $p, q \in S$. *Then there is a circuit containing both* $p$ *and* $q$ *if and only if there is a cocircuit containing* $p$ *and* $q$. *Hence* $T$ *is a separator of* $\mathbf{M}(S)$ *if and only if* $C \subseteq T$ *or* $C \subseteq S - T$ *for every cocircuit* $C$.

*Proof.* We may assume $r(S) \geq 2$. If $p \neq q$ are contained in a circuit then either $p$ and $q$ are parallel, in which case there is a copoint not containing $p$ and $q$ (**2.40**(i)), or $p$ and $q$ are contained in a minimal separator $\bar{T}$ of the lattice $L(S)$ and hence are perspective in the sense of **2.51**, which, by definition, means that they are contained in a cocircuit. The whole argument is reversible. □

Connectivity is thus a self-dual property, implying the following corollary in which the last statement is easily verified using **6.68**.

**6.86 Theorem.** *A finite matroid* $\mathbf{M}(S)$ *and its dual matroid* $\mathbf{M}^{\perp}(S)$ *have the same separators.* $\mathbf{M}(S)$ *is connected if and only if* $\mathbf{M}^{\perp}(S)$ *is connected. The components of* $\mathbf{M}^{\perp}(S)$ *are the dual matroids of the components of* $\mathbf{M}(S)$. □

Let us illustrate our results for graphs. There is an interesting (and hopefully not confusing) relationship between matroid connectivity and graph connectivity.

**Definition.** Let $G(V, S)$ be a graph. A subset $A \subseteq V$ is called a *separating vertex-set* if the removal of $A$ and its incident edges results in a disconnected graph. $G(V, S)$ is called *k-connected* if $G$ contains no separating set with less than $k$ vertices.

Hence a graph is 1-connected if and only if it is connected in the usual sense. A vertex whose removal disconnects a connected graph is called a *cut point*. Thus cut points are the vertex analogues of bridges. Since loops have no influence on the connectivity of a graph we shall mostly consider graphs without loops.

**6.87 Proposition.** *Let* $G(V, S)$ *be a connected graph without loops and with at least* 3 *vertices. Then the following conditions are equivalent:*

(i) $\mathbf{P}(G(V, S))$ *is connected.*
(ii) *Any two edges lie on a common polygon.*
(iii) *Any two edges lie on a common bond.*
(iv) $G(V, S)$ *is 2-connected.*
(v) $\mathrm{St}(v)$ *is a bond for all* $v \in V$.

*Proof.* We already know the equivalence of (i), (ii) and (iii). (ii) ⇒ (iv). If $G$ has a cut point $v$ then the subgraph induced by $V - v$ decomposes into at least two connected components with, say, vertex sets $V_1, \ldots, V_t$. It follows that any edge joining $v$ to a vertex in $V_1$ cannot lie on a common polygon with any edge joining $v$ to $V_2$. (iv) ⇒ (v). If St(v) properly contains a bond, then $v$ is clearly a cut point of $G(V, S)$. (v) ⇒ (iii). Let $e, f \in S$ and let $v_0, v_1, \ldots, v_k$ be a path in $G(V, S)$ with $e = \{v_0, v_1\}$ and $f = \{v_{k-1}, v_k\}$. Any two successive edges $\{v_{i-1}, v_i\}, \{v_i, v_{i+1}\}$ lie on the common bond $\mathrm{St}(v_i)$. Since, by **6.85**, the relation "on a common bond" is transitive, the result follows. □

In the graph of Figure 6.24, $v$ is a cut point. The set $\{e, f\}$ is a bond properly contained in $\mathrm{St}(v)$.

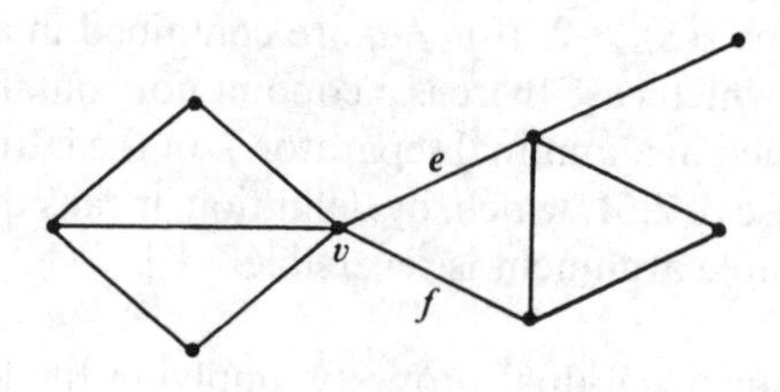


Figure 6.24

**6.88 Corollary.** *Let $G(V, S)$ be a 2-connected planar graph without loops. Then the dual graph of any plane representation of $G$ is also 2-connected without loops.* □

In graph theory, the maximal 2-connected subgraphs and the loops are called the *blocks* of the graph. Hence the blocks are just the minimal separators of the polygon matroid and are therefore edge-disjoint. A useful consequence is that we can realize any graphic matroid by a *connected* graph since we may identify the blocks at a common vertex and delete all isolated vertices without altering the matroid.

EXERCISES VI.4

1.* Extend **6.65** to arbitrary bipartite graphs by the following approach. Let $G(S \cup I, R)$ be a bipartite graph and $\mathbf{M}_1(S)$ and $\mathbf{M}_2(I)$ matroids on $S$ and $I$, respectively. Let $M$ be a matching in the sense of **6.65** with $\text{match}_S\, M = A$, $\text{match}_I\, M = B$. An *augmenting chain* relative to $M$ is a sequence $\{a'_0, b'_1\}$, $\{b_1, a_1\}$, $\{a'_1, b'_2\}, \ldots, \{b_n, a_n\}, \{a'_n, b'_{n+1}\}$ of $2n + 1$ distinct pairs ($n \geq 0$) such that

   (i) $\{a_i, b_i\} \in M, (i = 1, \ldots, n)$,
   (ii) $\{a'_i, b'_{i+1}\} \in R - M, (i = 0, \ldots, n), a'_0 \in S - \bar{A}, b'_{n+1} \in I - \bar{B}$,
   (iii) $a_i \in \bar{A}, a_i \notin \overline{A - \{a_1, \ldots, a_i\} \cup \{a'_1, \ldots, a'_{i+1}\}}, (i = 1, \ldots, n)$ and $b'_i \in \bar{B}$, $b'_i \notin \overline{B - \{b_1, \ldots, b_i\} \cup \{b'_1, \ldots, b'_{i+1}\}}, (i = 1, \ldots, n)$.

   Prove that a matching $M$ has maximal cardinality if and only if there is no augmenting chain with respect to $M$ and deduce from this fact **6.65**. (Aigner–Dowling)

2. Verify **6.66**.

→ 3. Dualize **6.34**, **6.36**, **6.45**, and **6.47**.

4. Let $\mathfrak{S}_1$ and $\mathfrak{S}_2$ denote the spanning sets of the finite matroids $\mathbf{M}_1$ and $\mathbf{M}_2$, respectively. Prove that $\mathfrak{S}_1 \wedge \mathfrak{S}_2 := \{A_1 \cap A_2 : A_1 \in \mathfrak{S}_1, A_2 \in \mathfrak{S}_2\}$ is the family of spanning sets of a matroid $\mathbf{M}_1 \wedge \mathbf{M}_2$ and show further that $(\mathbf{M}_1 \wedge \mathbf{M}_2)^\perp = \mathbf{M}_1^\perp + \mathbf{M}_2^\perp$.

→ 5. Let $G$ and $H$ be the plane graphs as shown in the figure.

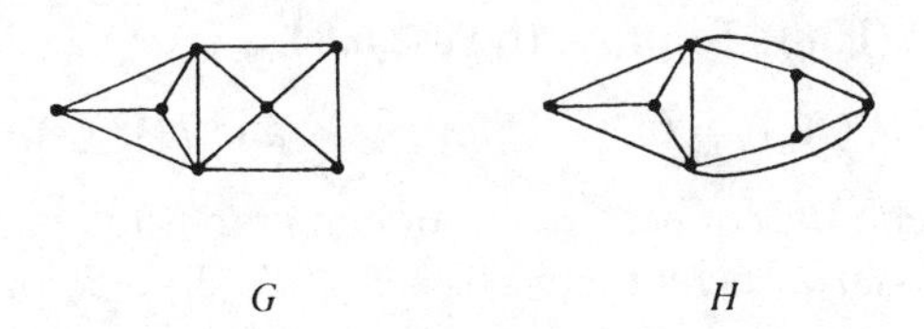


   Prove: $G \cong H$ but $G^* \ncong H^*$. Is 7 the minimal possible number of vertices for this situation, assuming that $G$ and $H$ are simple?

6. Show that $\mathbf{P}(K_{3,3})$ is not cographic.

→ 7.* A finite matroid is called *cotransversal* if it is the dual of a transversal matroid. Let $S$ be an $n$-set and $\mathfrak{A} = \{A_1, \ldots, A_t\} \subseteq 2^S$. Prove that the transversal matroid $\mathbf{T}(S; \mathfrak{A})$ (i.e., the matroid whose independent sets are the partial transversals in $\mathfrak{A}$) is dual to the principal matroid $\mathbf{M}(S)$ defined by the sequence

$$\mathbf{FM}(S) \xrightarrow{A_1} \mathbf{M}_1 \xrightarrow{A_2} \mathbf{M}_2 \to \cdots \xrightarrow{A_t} \mathbf{M}(S)$$

in the sense of ex. VI.3.14. Hence a matroid is cotransversal if and only if it is principal. (Hint: We may assume that the first $s$ mappings $\mathbf{M}_{i-1} \xrightarrow{A_i} \mathbf{M}_i$ are non-trivial and the remaining $t - s$ are trivial, whence $r(\mathbf{M}) = n - s$. Show that $B \subseteq S$ spans $S$ in $\mathbf{T}(S; \mathfrak{A})$ if and only if $S - B$ is independent in $\mathbf{M}(S)$ by using ex. VI.3.14 and the rank formula in $\mathbf{T}(S; \mathfrak{A})$.) (Brown–Dowling–Kelly)

8. We know from ex. VI.2.7 that $\mathbf{P}(K_4)$ is not transversal. Prove this again by using the previous exercise.

→ 9. Prove **6.80**.

10. Let $\mathbf{M}(S)/A$ and $\mathbf{M}(S)/B$ be connected with $A \cup B \neq S$. Show that $\mathbf{M}(S)/A \cap B$ is also connected. Can we dispense with the assumption $A \cup B \neq S$?

→ 11. Prove **6.84**. (Hint: Use the description of circuits of minors given in **6.34** and **6.36**.)

→ 12. Let $\mathfrak{G}$ be the set of finite matroids as in ex. VI.3.1 and define $\beta\colon \mathfrak{G} \to \mathbb{Z}$ by

$$\beta(\mathbf{M}(S)) := (-1)^{r(S)} \sum_{T \subseteq S} (-1)^{|T|} r(T).$$

Prove:

(i) $\beta(\mathbf{M}) = \beta(\mathbf{M}\backslash q) + \beta(\mathbf{M}/q)$ for all $q$ not a loop or coloop of $\mathbf{M}$.
(ii) $\beta(\text{coloop}) = 1$, $\beta(\text{loop}) = 0$.
(iii) For $|S| \geq 2$, $\beta(\mathbf{M}) = \beta(\mathbf{M}^\perp)$.
(iv) $\beta(\mathbf{M}) \geq 0$ for all $\mathbf{M} \in \mathfrak{G}$, and $\beta(\mathbf{M}) = 0$ if and only if $\mathbf{M}$ is a single loop or $\mathbf{M}$ is not connected.

(Crapo) (Hint: To prove (iv) use **6.84**.)

13. Verify $\beta(\mathbf{P}(K_n)) = (n - 2)!$ for $n \geq 2$ and $\beta(\mathbf{U}_k(n)) = \binom{n-2}{k-1}$.

14. We extend the connectivity notion in graphs to matroids. A matroid $\mathbf{M}(S)$ is called *k-separated* if there exists $T \subseteq S$, $|T| \geq k$, $|S - T| \geq k$ such that

$$r(T) + r(S - T) = r(S) + k - 1.$$

Thus $\mathbf{M}$ is 1-separated if and only if it is disconnected. For a finite matroid $\mathbf{M}$, prove that $\mathbf{M}$ is $k$-separated if and only if $\mathbf{M}^\perp$ is $k$-separated. (Tutte)

→15. Call a graph $G(V, E)$ *k-separated* if there exist $k$ vertices after whose deletion $k$ edges are separated from $k$ other edges (i.e., there exists $U \subseteq E$ with $|U| \geq k, |E - U| \geq k$ such that $U$ and $E - U$ have $k$ vertices in common). $G$ is 0-separated if it is not connected. We say $G$ is $n - T$*-connected* if $G$ is not $k$-separated for any $k < n$. Prove:

(i) $G$ is 1-$T$-connected $\Leftrightarrow$ $G$ is connected.
(ii) $G$ is 2-$T$-connected $\Leftrightarrow$ $\mathbf{P}(G)$ is connected.
(iii) $G$ connected and $k$-separated $\Rightarrow$ $\mathbf{P}(G)$ $k$-separated.

Note: The converse to (iii) also holds. (Tutte)

## Notes

The theory of matroids originated with the papers of Whitney [7], Birkhoff [5], and Mac Lane [1], and the book of Van der Waerden [1] where he formulated the abstract properties of dependence in vector spaces and field extensions; see Crapo–Rota [1] and Welsh [1] for historical comments. Since Rado's papers [1, 2, 3], Dilworth's [1, 2, 5], and, in particular, Tutte's work [5, 6, 10], the field has seen extensive research branching out into representation problems, transversal theory, algorithms and other areas. In this chapter the point of view was taken that construction of matroids and a discussion of the basic examples provide the most natural introduction to the theory. Section 1 is based on Whitney [7] and Edmonds–Fulkerson [1]; see also Pym–Perfect [1] and Mirsky–Perfect [2]. The examples of section 2 again appear in Whitney [7] and Edmonds–Fulkerson [1]; for incidence geometries, see Wille [2] and the comprehensive book by Dembowski [1]. Section 3 follows the treatment by Tutte [10], Nash–Williams [2], Crapo [1], and Crapo–Rota [1]. The notion of duality, perhaps the most important new concept, was expounded first by Whitney [7], Minty [2], and Lehman [1], and studied further by Higgs [3] and many others. The theory of chromatic invariants was invented by Tutte [2, 3]; see also Brylawski [2], Crapo [6], and Aigner [1, vol. II, ch. VII].

Chapter VII

# Matroids: Further Theory

After our discussion of the basics of matroid theory in the previous chapter, we are now going to study in detail the most important classes of matroids: Linear matroids, binary and regular matroids, graphic and transversal matroids. The emphasis lies here on the characterization of these matroids and on applications to concrete combinatorial problems.

## 1. Linear Matroids

We have already observed in section VI.2.A that a coordinatization $\phi$ of a matroid over a field is, in general, not an injective mapping, but that $\phi$ induces an injective mapping on the simple matroid underlying $\mathbf{M}$. The reader is reminded that a field always means commutative field for us here.

**7.1 Proposition.** *A matroid* $\mathbf{M}(S)$ *is K-linear if and only if its underlying geometry* $\mathbf{M}_0(S_0)$ *is K-linear.*

*Proof.* Let $\phi: S \to V(n, K)$ be a coordinatization of $\mathbf{M}$. We choose, for every $\bar{p} \nsubseteq \overline{\varnothing}$, $p \in S$, a fixed element $0 \neq a_p \in \overline{\phi p}$ (this is possible since $\bar{p} = \bar{q} \Leftrightarrow \overline{\phi p} = \overline{\phi q}$) and define $\phi_0: S_0 \to V(n, K)$ by

$$\phi_0(\bar{p}) := a_p \qquad (\bar{p} \in S_0).$$

$\phi_0$ is then a coordinatization of $\mathbf{M}_0(S_0)$. The converse is clear. □

According to this proposition, we may confine ourselves to geometries when studying the coordinatizability of a matroid and shall on occasion do so. Geometrically, **7.1** says that a coordinatization of a geometry $\mathbf{M}$ is an embedding of $\mathbf{M}$ into a projective space $\mathbf{PG}(n, K)$. It follows, in particular, that configurations such as those of Desargues or Pappos must be present in a linear geometry. This observation will enable us to construct small non-linear matroids.

## A. Coordinatization Theorems

We begin with some simple algebraic facts.

**7.2 Proposition.** *A $K$-linear matroid is also $K'$-linear for every extension $K'$ of $K$.* □

For some linear matroids this is the most we can say.

**7.3 Proposition.** *The projective geometry* $\mathbf{PG}(n, K)$ *with $n \geq 2$ is linear only over extensions of $K$.* □

This follows immediately from the fact that a Desarguesian projective space of $g$-dimension at least 2 uniquely determines the field $K$ (see **2.34** and **2.35**). For a proof see, for instance, Baer [1, ch. 7].

**Example.** Let $\mathbf{F}$ be the Fano plane of Figure 6.20. By **7.3**, $\mathbf{F}$ can only be coordinatized over fields of characteristic 2. We can give a direct proof of this easily. Let $\phi\colon \mathbf{F} \to V(3, K)$ be a coordinatization over some field $K$. Without loss of generality we may assume $\phi(a) = (1, 0, 0)$, $\phi(b) = (0, 1, 0)$, $\phi(c) = (0, 0, 1)$, and $\phi(d) = (1, 1, 1)$. Since $e$ depends on $\{a, b\}$ and $\{c, d\}$, we must have $\phi(e) = \lambda(1, 1, 0)$ and similarly $\phi(f) = \mu(1, 0, 1)$, $\phi(g) = \nu(0, 1, 1)$ with $\lambda, \mu, \nu \neq 0$. The vectors $\phi(e), \phi(f), \phi(g)$ are linearly dependent, whence $\det(\phi(e), \phi(f), \phi(g)) = \lambda\mu\nu(-2) = 0$, which is only possible in characteristic 2.

If in $\mathbf{F}$ we replace the line $\{e, f, g\}$ by the 3 trivial lines $\{e, f\}$, $\{e, g\}$, and $\{f, g\}$, then we obtain a new matroid $\mathbf{F}_1$, which, by the same argument as for $\mathbf{F}$, is coordinatizable only over fields of characteristic $\neq 2$. It follows that the product $\mathbf{M} = \mathbf{F} \times \mathbf{F}_1$ cannot be coordinatized over any field $K$ since otherwise the restrictions to $\mathbf{F}$ and $\mathbf{F}_1$ would be $K$-linear **(6.39)**, implying char $K = 2$ and char $K \neq 2$. Hence $\mathbf{F} \times \mathbf{F}_1$ is a non-linear matroid on 14 points.

**Definition.** The *characteristic set* $\text{ch}(\mathbf{M})$ of a matroid $\mathbf{M}$ is the set of possible characteristics of coordinatizing fields, i.e., $\text{ch}(\mathbf{M}) := \{\text{char } K : \mathbf{M} \text{ is } K\text{-linear}\}$.

Let $\mathbb{P}$ be the set of primes together with 0. An interesting problem is to determine which subsets of $\mathbb{P}$ are characteristic sets of some matroid. So far, we have seen that the following characteristic sets occur:

| | |
|---|---|
| $\varnothing$ | non-linear matroids |
| $\mathbb{P}$ | Boolean algebras, regular matroids |
| $\{p\}$ | $\mathbf{PG}(n, K)$, $n \geq 2$, char $K = p$ |
| $\mathbb{P} - \{2\}$ | $\mathbf{F}_1$. |

It is known that $|\text{ch}(\mathbf{M})| = \infty$ implies $0 \in \text{ch}(\mathbf{M})$. The converse is true for finite matroids, but not for infinite matroids since $\text{ch}(\mathbf{PG}(n, \mathbb{Q})) = \{0\}$ for $n \geq 2$ (Rado,

Vamos). Furthermore, it follows from results of Dowling [3] that any cofinite set of primes (together with 0) is the characteristic set of some matroid.

We may ask next which operations on matroids preserve $K$-linearity. Restriction, contraction, and dualization where shown to preserve $K$-linearity in chapter VI. The following propositions concern the product and sum.

**7.4 Proposition.** *The product of matroids is $K$-linear if and only if all factors are. In particular,* $\mathrm{ch}(\prod_i \mathbf{M}_i) = \bigcap_i \mathrm{ch}(\mathbf{M}_i)$.

*Proof.* We already know that the $K$-linearity of $\prod \mathbf{M}_i$ implies the $K$-linearity of each $\mathbf{M}_i$. For the converse it suffices to consider two factors $\mathbf{M}_1(S_1)$ and $\mathbf{M}_2(S_2)$. Let $\Phi: S_1 \to V(n_1, K)$ and $\Phi_2: S_2 \to V(n_2, K)$ be coordinatizations of $\mathbf{M}_1$ and $\mathbf{M}_2$, respectively. Then $\Phi: S_1 \cup S_2 \to V(n_1 + n_2, K)$ given by

$$\Phi(p) := \begin{cases} (\Phi_1(p), \underbrace{0, \ldots, 0}_{n_2}) & (p \in S_1) \\ (\underbrace{0, \ldots, 0}_{n_1}, \Phi_2(p)) & (p \in S_2) \end{cases}$$

is a coordinatization of $\mathbf{M}_1(S_1) \times \mathbf{M}_2(S_2)$ over $K$. The last statement follows from the fact that two fields of the same characteristic have a common extension. □

Before discussing the sum let us make one observation. If a matroid $\mathbf{M}(S)$ is coordinatizable over a finite field $K$, then the underlying geometry $\mathbf{M}_0(S_0)$ must be finite since $V(n, K)$ is finite. Because of this and **7.1**, we may confine our discussion to finite matroids as, for example, in the following propositions.

**7.5 Proposition** (Piff–Welsh). *Let $\mathbf{M}(S)$ be a finite matroid. Then there exists an integer $n$ such that if $\mathbf{M}(S)$ is $K$-linear and $|K| \geq n$, then the matroid $f(\mathbf{M})$ on $T$ is $K$-linear for all surjections $f: S \to T$.*

*Proof.* It suffices to prove the theorem for sets $S$ and $T$ with $|T| = |S| - 1$. Let $S = \{p_0, p_1, \ldots, p_s\}$, $T = \{q_1, \ldots, q_s\}$ and $f(p_0) = f(p_1) = q_1$, $f(p_i) = q_i$ for $i = 2, \ldots, s$. Suppose $\Phi: S \to V(n, K)$ is a coordinatization of $\mathbf{M}(S)$. We define $\Psi: T - q_1 \to V(n, K)$ by

$$\Psi(q_i) := \Phi(p_i) \qquad (i = 2, \ldots, s)$$

and show that we can find $\lambda_0, \lambda_1 \in K$ such that $\Psi: T \to V(n, K)$ with $\Psi(q_1) = \lambda_0 \Phi(p_0) + \lambda_1 \Phi(p_1)$ is a coordinatization of $f(\mathbf{M})$ over $K$, if $|K|$ is large enough. By the definition of $f(\mathbf{M})$, we have for $B \subseteq T - q_1$:

$$\begin{aligned} B \text{ independent in } f(\mathbf{M}) \Leftrightarrow f^{-1}(B) \text{ independent in } \mathbf{M} \Leftrightarrow \Phi(f^{-1}(B)) \\ = \Psi(B) \text{ is linearly independent in } V(n, K). \end{aligned}$$

Hence we may confine ourselves to sets $B = A \cup q_1 \subseteq T$. If $p_0$ and $p_1$ are parallel in $\mathbf{M}$ then $\Psi(q_1) = \Phi(p_0)$ clearly yields a coordinatization. Assume $p_0$ and $p_1$ are not parallel. For an independent set $A \cup q_1 \subseteq T$ we define the subspace $W(A) \subseteq V(n, K)$ by $W(A) := \overline{\Psi(A)} \cap \overline{\{\Phi(p_0), \Phi(p_1)\}}$. Since $A \cup q_1$ is independent, at least one of the elements $p_0$ or $p_1$ is not in the closure of $f^{-1}(A)$ in $\mathbf{M}$ and hence $\Phi(p_0) \notin \overline{\Psi(A)}$ or $\Phi(p_1) \notin \overline{\Psi(A)}$. $W(A)$ is therefore of rank 0 or 1 in $\mathbf{M}(V(n, K))$. It follows that $W(A_i) \subsetneqq \overline{\{\Phi(p_0), \Phi(p_1)\}}$ for any independent set $B_i = q_1 \cup A_i$ of $f(\mathbf{M})$ containing $q_1$, and therefore $\bigcup_i W(A_i) \subsetneqq \overline{\{\Phi(p_0), \Phi(p_1)\}}$ if $|K|$ is large enough. We conclude that some linear combination $\lambda_0 \Phi(p_0) + \lambda_1 \Phi(p_1)$ is not in $\bigcup_i W(A_i)$ and hence independent of *all* $\Psi(A_i)$'s. Defining $\Psi(q_1) := \lambda_0 \Phi(p_0) + \lambda_1 \Phi(p_1)$, we have that $\Psi(A \cup q_1)$ is linearly independent whenever $A \cup q_1$ is independent in $f(\mathbf{M})$. If, on the other hand, $C \cup q_1$ is a circuit in $f(\mathbf{M})$, then both sets $f^{-1}(C) \cup p_0$ and $f^{-1}(C) \cup p_1$ are dependent in $\mathbf{M}$, whence $\Phi(p_0) \in \overline{\Psi(C)}$, $\Phi(p_1) \in \overline{\Psi(C)}$, and thus $\Psi(q_1) \in \overline{\Psi(C)}$. □

The exercises contain an example showing that the cardinality condition in **7.5** cannot be weakened in general. By the remark after **6.44** and **6.49** we now have the following corollaries.

**7.6 Corollary.** *Let $\mathbf{M}_1, \ldots, \mathbf{M}_t$ be matroids on the finite set S. Then there exists an integer n such that if $\mathbf{M}_1, \ldots, \mathbf{M}_t$ are K-linear and $|K| \geq n$ then $\sum_{i=1}^t \mathbf{M}_i$ is K-linear.* □

**7.7 Corollary.** *Let $\mathbf{M}$ be a transversal or cotransversal matroid. Then there exists $n \in \mathbb{N}$ such that $\mathbf{M}$ is K-linear whenever $|K| \geq n$. In particular,* $\mathrm{ch}(\mathbf{M}) = \mathbb{P}$. □

Notice that **7.7** supplies another proof that the Fano plane is neither transversal nor cotransversal.

We come to the main result of this section, a characterization of $K$-linear matroids or, equivalently, of $K$-function space matroids.

**7.8 Theorem** (Tutte). *Let $\mathbf{M}(S)$ be a matroid and $\mathfrak{H}$ the family of copoints. $\mathbf{M}(S)$ is K-linear if and only if for every $H \in \mathfrak{H}$ there exists a function $f_H: S \to K$ such that*

(i) $\ker f_H = H$,
(ii) *if three distinct copoints $H_1, H_2, H_3$ contain a common coline then there exist $\lambda_1, \lambda_2, \lambda_3 \in K - \{0\}$ with*

$$\lambda_1 f_{H_1} + \lambda_2 f_{H_2} + \lambda_3 f_{H_3} = 0.$$

*Proof.* Let $r(\mathbf{M}) = n$. If $\mathbf{M}(S) = \mathbf{M}(F(S, K))$, then, by **6.21**(ii), there exist functions $f_H \in F(S, K)$ with $\ker f_H = H$ for all $H \in \mathfrak{H}$. Let $W = H_1 \wedge H_2 \wedge H_3$ be a coline and $B = \{b_3, \ldots, b_n\}$ a basis of $W$. We extend $B$ to a basis $B \cup b_1$ of $H_1$ and to a basis $B \cup b_2$ of $H_2$. $B \cup b_1 \cup b_2$ is then a basis of $\mathbf{M}$. The system of linear equations

$$\begin{aligned} \lambda_2 f_{H_2}(b_1) + \lambda_3 f_{H_3}(b_1) &= 0 \\ \lambda_1 f_{H_1}(b_2) \qquad\qquad + \lambda_3 f_{H_3}(b_2) &= 0 \end{aligned}$$

possesses a non-trivial solution $\lambda_1, \lambda_2, \lambda_3$; in fact $\lambda_1, \lambda_2, \lambda_3 \in K - \{0\}$, and we obtain $\sum_{i=1}^{3} \lambda_i f_{H_i}(b) = 0$ for all $b \in B$, and thus $\sum_{i=1}^{3} \lambda_i f_{H_i} = 0$ by **6.21**(i).

Now suppose there are functions $f_H$, $H \in \mathfrak{H}$, satisfying the conditions of the theorem. Let $F(S, K)$ be the function space spanned by the $f_H$'s, $B = \{b_1, \ldots, b_n\}$ a basis of $\mathbf{M}(S)$, and $H_i = \overline{B - b_i}$ for $i = 1, \ldots, n$. The functions $f_{H_i}$ corresponding to the copoints $H_i$ are obviously independent. To show that $f_{H_1}, \ldots, f_{H_n}$ span $F(S, K)$ we use downward induction on $|H \cap B|$ for $H \in \mathfrak{H}$. If $|H \cap B| = n - 1$ then $H = H_i$ for some $i$, and the assertion is trivial. Let us say $H$ has *type k* if $|H \cap B| = k$. Assume that all $f_H$ with $H$ of type at least $k + 1$ are linear combinations of the $f_{H_i}$'s, and let $H$ be of type $k$, say $|H \cap B| = \{b_1, \ldots, b_k\}$. We extend $H \cap B$ to a basis $\{b_1, \ldots, b_k, c_{k+1}, \ldots, c_{n-1}\}$ of $H$. $W = \overline{\{b_1, \ldots, b_k, c_{k+1}, \ldots, c_{n-2}\}}$ is a coline in $\mathbf{M}$ with $W < \cdot H$. Since $B \not\subseteq W$ we can find $b' \in B - W$. The copoint $H' = \overline{\{b_1, \ldots, b_k, c_{k+1}, \ldots, c_{n-2}, b'\}}$ is of type $\geq k + 1$, hence different from $H$, and we also have $W <\cdot H'$. Again, since $B \not\subseteq H'$ there exists $b'' \in B - H'$. $H'' = \overline{\{b_1, \ldots, b_k, c_{k+1}, \ldots, c_{n-2}, b''\}}$ is then another copoint of type $\geq k + 1$, different from $H$ and $H'$, which also covers $W$. By condition (ii), we have $f_H \in \overline{\{f_{H'}, f_{H''}\}}$ and thus $f_H \in \overline{\{f_{H_1}, \ldots, f_{H_n}\}}$ by induction.

To conclude the proof we show that $\mathbf{M}(S)$ and $\mathbf{M}(F(S, K))$ possess the same set of bases. Let $B$ be a basis of $\mathbf{M}(S)$ and $f_{H_i}$ the functions as above. Any $f \in F(S, K)$ is, by our previous argument, a linear combination, $f = \sum_{i=1}^{n} \lambda_i f_{H_i}$. Hence if $f$ vanishes on $B$ then $0 = f(b_i) = \lambda_i f_{H_i}(b_i)$, i.e., $\lambda_i = 0$ for all $i$, which implies that $B$ is a basis of $\mathbf{M}(F(S, K))$ by **6.21**. On the other hand, if $B$ is an $n$-set, but not a basis of $\mathbf{M}(S)$, then $B$ is contained in a copoint $H$. Hence there exists $f_H \in F(S, K)$ with $f_H(b) = 0$ for all $b \in B$, and $B$ is not a basis of $\mathbf{M}(F(S, K))$ by **6.21**. □

Matrices are convenient for describing finite linear matroids. Let $\mathbf{M}(S)$ be $K$-linear of rank $n$ with $|S| = s$, and $\Phi: S \to V(n, K)$ a coordinatization. The $n \times s$-matrix $R$ over $K$ whose columns are the vectors $\Phi(p)$, $p \in S$, is called a *coordinatization matrix of* $\mathbf{M}$ over $K$. $\mathbf{M}(S)$ is thus isomorphic to the submatroid $\mathbf{M}(R)$ of $\mathbf{M}(V(n, K))$ generated by the columns of $R$. On the other hand, if $R$ is any $n \times s$-matrix over $K$ then the columns of $R$ induce a submatroid of $M(V(n, K))$, sometimes called the *matrix matroid* $\mathbf{M}(R)$. Thus finite $K$-linear matroids and matrix matroids over $K$ comprise the same class of matroids, and it will often be useful to study a $K$-linear matroid in matrix form. Needless to say, the word matroid stems from this relation.

The following proposition is clear from linear algebra.

**7.9 Proposition.** *Let $R$ be an $n \times s$-matrix over $K$.*

(i) *Then $\mathbf{M}(R') \cong \mathbf{M}(R)$ for any $K$-matrix $R'$ which is derived from $R$ by a sequence of the following operations: (a) deletion of a row consisting of 0's; (b) permutation of rows or columns; (c) multiplication of rows or columns by $0 \neq \lambda \in K$.*

(ii) *If $N$ is a non-singular $n \times n$-matrix over $K$, then $\mathbf{M}(NR) \cong \mathbf{M}(R)$.* □

Let $\mathbf{M}(S)$ be a finite $K$-linear matrix and $B$ a basis of $\mathbf{M}(S)$. It is a consequence of **7.9** that $\mathbf{M}(S)$ has a coordinatization matrix $R_B$ of the form

$$R_B = [I_n, A_{n,s-n}]$$

where the columns of the identity matrix $I_n$ correspond to the elements of $B$. We say that $R_B$ is a *standard matrix* relative to $B$ or that $R_B$ is in *standard form*. The rows of $R_B$ correspond precisely to the functions $f_i$, with $\ker f_i = \overline{B - b_i}$, $f(b_i) = 1$ and $f_i(p)$ in position $(i, p)$, as discussed in the example after **6.22**.

**7.10 Proposition.** *Let $\mathbf{M}(S)$ be a finite $K$-linear matroid and $R_B = [I_n, A]$ a standard matrix relative to the basis $B$. Then $R_{S-B} = [-A^T, I_{s-n}]$ is a standard matrix of $\mathbf{M}^{\perp}(S)$ relative to the basis $S - B$ where $A^T$ is the transpose of $A$.*

*Proof.* Consider the bases $\{f_i\}$ and $\{g_j\}$ as in the proof of **6.69**. □

**Example.** Let $\mathbf{M}(S)$ be the $\mathbb{Q}$-linear matroid with standard matrix $R_B$,

$$R_B = \begin{bmatrix} \boxed{1} & 0 & \boxed{0} & \boxed{2 \quad 4 \quad -7 \quad 1} \\ 0 & 1 & 0 & 0 \quad 3 \quad 2 \quad 0 \\ \boxed{0} & 0 & \boxed{1} & \boxed{5 \quad -2 \quad 3 \quad -1} \end{bmatrix}.$$

Then

$$R_{S-B} = \begin{bmatrix} -2 & 0 & -5 & 1 & 0 & 0 & 0 \\ -4 & -3 & 2 & 0 & 1 & 0 & 0 \\ 7 & -2 & -3 & 0 & 0 & 1 & 0 \\ -1 & 0 & 1 & 0 & 0 & 0 & 1 \end{bmatrix}$$

is a standard matrix of $\mathbf{M}^{\perp}(S)$.

Let $R_B = [I_n, A]$ be a standard matrix of $\mathbf{M}(S)$ relative to the basis $B$ and let $A'$ be a submatrix of $A$. Suppose $Z_{i_1}, \ldots, Z_{i_k}$ and $S' \subseteq S - B$ are the rows and columns of $A'$. We add to $A'$ the submatrix $I_k$ of $I_n$ consisting of all rows and columns with indices $i_1, \ldots, i_k$. As an example consider the submatrix consisting of the boxed elements in the example above. **6.38** immediately yields the following useful result.

**7.11 Proposition.** *If $R = [I_n, A]$ is a standard matrix of $\mathbf{M}(S)$ relative to the basis $B$, then $R' = [I_k, A']$ as defined above is a standard matrix of the minor $(\mathbf{M}(S)/B - B') \cdot (S' \cup B')$ relative to the basis $B' = \{b_{i_1}, \ldots, b_{i_k}\} \subseteq B$.* □

The following application of **7.11** is the key result for our study of regular matroids in section 2.

**7.12 Lemma.** *Let $R_1 = [I_n, A_1]$ and $R_2 = [I_n, A_2]$ be two $n \times s$-matrices over fields $K_1$ and $K_2$, respectively. The mapping $\phi$ which carries each column of $R_1$ onto the corresponding column of $R_2$ is an isomorphism between the matroids $\mathbf{M}(R_1)$ and $\mathbf{M}(R_2)$ if and only if $\det_{K_1} N_1 = 0 \Leftrightarrow \det_{K_2} N_2 = 0$ for any square submatrix $N_1 \subseteq A_1$ and its corresponding submatrix $N_2 \subseteq A_2$. In particular, if $\mathbf{M}(R_1) \overset{\phi}{\cong} \mathbf{M}(R_2)$, then $A_1$ has a 0-entry if and only if $A_2$ has a 0-entry in the corresponding position.*

*Proof.* If $\phi$ is a matroid isomorphism then corresponding minors must be isomorphic. Since the columns of $N_1$ are, by **7.11**, the coordinatizing vectors of some minor of $\mathbf{M}(R_1)$, they must be dependent (i.e., $\det_{K_1} N_1 = 0$) if and only if the same holds for $N_2$. Conversely, if the condition of the theorem is satisfied we prove that $B$ is a basis in $\mathbf{M}(R_1)$ if and only if the corresponding columns are a basis of $\mathbf{M}(R_2)$. In other words, we have to show that an $n \times n$-submatrix $B \subseteq R_1$ is non-singular over $K_1$ if and only if the corresponding matrix is non-singular over $K_2$. For $B \subseteq A_1$ this follows directly from the hypothesis. If $B = B' \cup B''$ where $B' \subseteq I_n$, $B'' \subseteq A_1$, then by expanding $\det B$ with respect to the columns in $B'$ we have $\det_{K_1} B = \pm \det_{K_1} N$, where $N$ is a square submatrix of $A_1$, whence we may apply the hypothesis. □

## B. Geometric Configurations

We know from **2.34** that the projective spaces of $g$-dimension $n \geq 3$ are precisely the geometries $\mathbf{PG}(n, K)$ over division rings $K$. Hence from $g$-dimension 3 on (i.e., from rank 4 on) the coordinatization problem reduces to the question whether the given geometry can be embedded in a projective space. For rank 3, i.e., for planes, the situation is different (ranks 1 and 2 are trivial). We shall show that *any* rank 3 geometry can be embedded in a projective plane $\mathbf{E}$. Therefore, the question now is to decide whether $\mathbf{E} \cong \mathbf{PG}(2, K)$ for some field $K$.

**7.13 Proposition.** *Any geometry $\mathbf{M}(S)$ of rank 3 can be embedded as a submatroid in a projective plane $\mathbf{E}$.*

*Proof.* Any two points in $\mathbf{M}(S)$ lie on a single line, but it may happen that two lines do not intersect. Hence, in order to construct $\mathbf{E}$, we must add new intersection points, then new connecting lines, and so on. Formally, we proceed by induction. We set $\mathbf{E}_0 = \mathbf{M}(S)$. Let $\mathbf{E}_k$ be already constructed. Then, for any two lines $l_i$, $l_j$ of $\mathbf{E}_k$ which do not intersect, we add a new point $P_{ij}$, and similarly, for any two points $P_r$, $P_s$ of $\mathbf{E}_k$ which do not lie on a common line, we add a new line $l_{rs}$. We then define $\mathbf{E}_{k+1} := \mathbf{E}_k \cup \bigcup \{P_{ij}\} \cup \bigcup \{l_{rs}\}$ with the additional incidences $P_{ij} \in l_i \cap l_j$, $\{P_r, P_s\} \subseteq l_{rs}$. $\mathbf{E} = \bigcup_{k \geq 0} \mathbf{E}_k$ is now easily seen to be a projective plane or a product of a line with a point in which case, by **7.4**, $\mathbf{E}$ can be embedded in a projective plane. Finally, it is clear that $\mathbf{M}(S)$ is a submatroid of $\mathbf{E}$. □

It is obvious that the preceding construction embeds any finite rank 3 geometry which is not already a projective plane in a *countable* projective plane. The question of when a finite rank 3 geometry can be embedded in a *finite* projective plane is one of the most interesting open problems in projective geometry.

Among the most beautiful results in projective geometry are those theorems which characterize the coordinatizing domain by the existence or non-existence of certain configurations in the plane. We quote as examples the following two classical theorems.

**7.14.** In a projective space $\mathbf{PG}(n, K)$ over a division ring $K$ and with $n \geq 2$ the *Theorem of Desargues* holds. This means the following: Let $g$, $h$ and $k$ be three lines through the point 0 (called the center of perspectivity) and take 3 pairs of points $a, A \in g$; $b, B \in h$; $c, C \in k$. If we denote by $X$, $Y$, and $Z$ the (distinct) points of intersection of the lines $\{\overline{a, b}\}$, $\{\overline{A, B}\}$; $\{\overline{a, c}\}$, $\{\overline{A, C}\}$; and $\{\overline{b, c}\}$, $\{\overline{B, C}\}$, respectively, then $X$, $Y$ and $Z$ are collinear (axis of perspectivity).

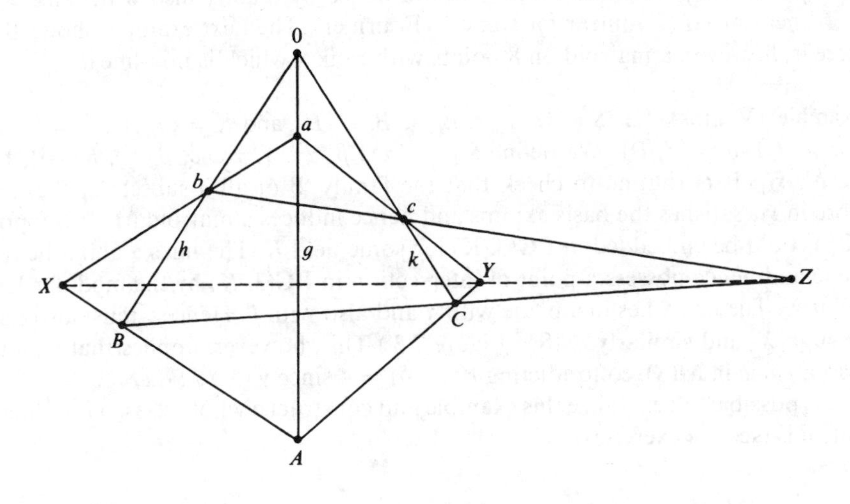


Figure 7.1

Hence, in Figure 7.1, if we replace the line $\{X, Y, Z\}$ by three trivial lines then the new matroid is non-linear. (In fact, it cannot be coordinatized over any division ring.)

**7.15.** In a projective plane $\mathbf{PG}(n, K)$ over a field $K$ and with $n \geq 2$ the *Theorem of Pappos* holds. This means the following: Let $g$ and $h$ be two lines and $a, b, c \in g$, $A, B, C \in h$ six distinct points. If we denote by $X$, $Y$ and $Z$ the (distinct) points of intersection of the lines $\{\overline{a, B}\}$, $\{\overline{b, A}\}$; $\{\overline{a, C}\}$, $\{\overline{c, A}\}$; $\{\overline{b, C}\}$, $\{\overline{c, B}\}$, then $X, Y, Z$ are collinear.

Hence in Figure 7.2, if we replace the line $\{X, Y, Z\}$ by three trivial lines then the matroid we obtain is non-linear.

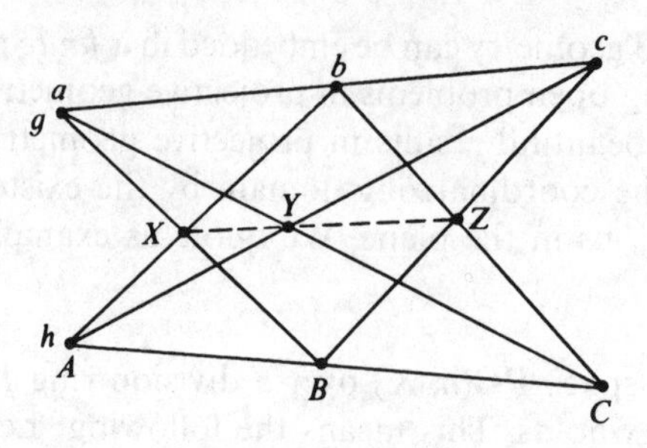


Figure 7.2

It is known that every matroid with at most 7 points is $\mathbb{Q}$-linear except for **F** and $\mathbf{F}^\perp$ for which $\mathrm{ch}(\mathbf{F}) = \mathrm{ch}(\mathbf{F}^\perp) = \{2\}$. Furthermore it is known that every matroid with 8 points and rank $\leq 3$ (and hence by duality also with rank $\geq 5$) is $\mathbb{Q}$-linear or $GF(2^k)$-linear for some $k$ (Fournier). The next example shows that there is, however, a matroid on 8 points with rank 4 which is non-linear.

**Example** (Vamos). Let $S = \{a, b, c, d, A, B, C, D\}$ and $\alpha = \{a, A\}$, $\beta = \{b, B\}$ $\gamma = \{c, C\}$, $\delta = \{d, D\}$. We define $\mathfrak{H}_1 := \{\alpha \cup \beta, \alpha \cup \gamma, \alpha \cup \delta, \beta \cup \gamma, \beta \cup \delta\}$, but $\gamma \cup \delta \notin \mathfrak{H}_1$. It is routine to check that the family $\mathfrak{B}$ of all 4-subsets of $S$ except those in $\mathfrak{H}_1$ satisfies the basis axioms and hence induces a matroid $\mathbf{M}(S)$. Suppose $\mathbf{M}(S)$ could be embedded in $\mathbf{PG}(3, K)$ for some field $K$. The lines $\alpha$ and $\beta$ lie in a plane and hence possess a point of intersection in $\mathbf{PG}(3, K)$ by the modular law; call it $X$. The line $\gamma$ lies in a plane with $\alpha$ and also with $\beta$. Hence $\gamma$ must also pass through $X$, and similarly $\delta$. (See Figure 7.3.) This, however, implies that $\gamma$ and $\delta$ span a *plane* in $\mathbf{M}(S)$, contradicting $r(\gamma \cup \delta) = 4$ since $\gamma \cup \delta \notin \mathfrak{H}_1$.

It is possible to generalize this example and construct a whole class of non-linear matroids (see the exercises).

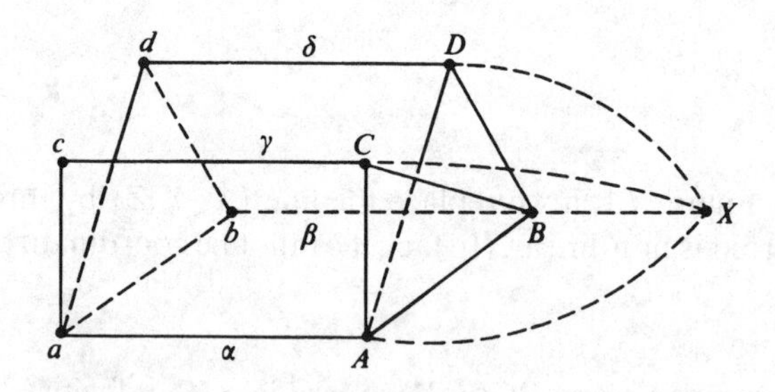


Figure 7.3

## C. The Critical Problem

To conclude our general remarks on linear matroids let us study an important extremal problem.

**Definition.** Let $S \subseteq V(n, K)$ be a set of vectors not containing the 0-vector. The *critical exponent* crit($S$) of $S$ is defined as the minimal number $c$ of copoints $H_1, \ldots, H_c$ in $\mathbf{M}(V(n, K))$ such that

$$S \cap \bigcap_{i=1}^{c} H_i = \varnothing.$$

Clearly, $1 \le \operatorname{crit}(S) \le n$ for all $S \neq \varnothing$ with $0 \notin S$.

**7.16 Proposition.** *Let $S \subseteq V(n, K)$, $0 \notin S$. If $k(S)$ denotes the maximal rank of a subspace $U$ in $\mathbf{M}(V(n, K))$ such that $S \cap U = \varnothing$, then*

$$k(S) = n - \operatorname{crit}(S).$$

*Proof.* If $U$ is such a subspace of maximal rank $k(S)$ then there are $n - k(S)$ copoints $H_i$ with $\bigcap H_i = U$; hence $n - k(S) \ge \operatorname{crit}(S)$. If, on the other hand, $H_1, \ldots, H_c$ is a minimal set of copoints with $\bigcap H_i \cap S = \varnothing$, then, by the modularity of the rank function in $\mathbf{M}(V(n, K))$, we have $r(\bigcap_{i=1}^{c} H_i) = n - c$, and thus $n - \operatorname{crit}(S) \le k(S)$. □

The computation of crit($S$) and $k(S)$ are thus dual extremal problems in the sense that the solution of either determines the solution of the other. We shall illustrate this duality with an example later on.

Suppose $\mathbf{M}(S)$ is $K$-linear without loops and $\Phi: S \to V(n, K)$ is a coordinatization. It is plausible at first sight that different coordinatizations $\Phi$ yield different critical exponents crit($\Phi S$). That this is not so is the main result of this section, where the critical exponent is shown to be an invariant of the lattice of flats. We confine ourselves to finite matroids and finite fields $GF(q)$. Recall the definition of the characteristic polynomial $\chi(L; x)$ of a finite poset with rank function $r$ (see **4.20**):

$$\chi(L; x) = \sum_{a \in L} \mu(0, a) x^{r(1) - r(a)}$$

where $\mu$ is the Möbius function of $L$. It is convenient to consider linear functionals instead of copoints in the following theorem. Each copoint is the kernel of some functional, and two linear functionals have the same kernel if and only if they are multiples of each other. Furthermore, since each linear functional on $V(n, q)$ is uniquely determined by its values on a basis there are precisely $q^n$ linear functionals.

**7.17 Theorem** (Crapo–Rota). *Let $\varnothing \neq S \subseteq V(n, q)$, $0 \notin S$, and $L = L(S)$ be the lattice of flats of the submatroid generated by $S$. Then the number of ordered sequences $(f_1, \ldots, f_k)$ of $k$ linear functionals such that*

$$S \cap \bigcap_{i=1}^{k} \ker f_i = \varnothing$$

*equals $\chi(L; q^k)$.*[1]

[1] By an ordered sequence we mean "ordered, with repetitions".

*Proof.* For $z \in L$, let $g(z)$ be the number of sequences $(f_1, \ldots, f_k)$ with

$$z \leq \bigcap_{i=1}^{k} \ker f_i.$$

Clearly, $g(z) = q^{k(r(1)-r(z))}$. Let $f(z)$ be the number of sequences $(f_1, \ldots, f_k)$ such that

$$z \leq \bigcap_{i=1}^{k} \ker f_i \quad \text{and} \quad z < y \in L \Rightarrow y \nleq \bigcap_{i=1}^{k} \ker f_i.$$

Since any sequence $(f_1, \ldots, f_k)$ determines a unique maximal flat $w \in L$ with $w \leq \bigcap_{i=1}^{k} \ker f_i$ we have

$$g(z) = \sum_{y \geq z} f(y),$$

and thus by Möbius inversion

$$f(z) = \sum_{y \geq z} \mu_L(z, y) g(y).$$

By the definition of $f$, $f(0)$ counts the number of sequences $(f_1, \ldots, f_k)$ with $S \cap \bigcap_{i=1}^{k} \ker f_i = \varnothing$. Hence $f(0)$ is the number we seek and we obtain

$$f(0) = \sum_{y \in L} \mu_L(0, y) q^{k(r(1)-r(y))} = \chi(L; q^k). \quad \square$$

**7.18 Corollary.** *Let $\varnothing \neq S \subseteq V(n, q)$, $0 \notin S$, and $L(S)$ be the lattice of flats induced by S. Then*

(i) $\chi(L; q^k) = 0$ *for* $k = 0, 1, \ldots, \operatorname{crit}(S) - 1$
(ii) $\chi(L; q^k) > 0$ *for* $k \geq \operatorname{crit}(S)$. $\square$

By **7.17** we may speak unambiguously of *the critical exponent of a finite q-linear matroid* $\mathbf{M}(S)$ without loops. **7.18** shows that $\operatorname{crit}(\mathbf{M}) = \min_{k \in \mathbb{N}} k$ such that $\chi(L(S); q^k) > 0$.

An obvious corollary of **7.16** is the following characterization of matroids with critical exponent 1.

**7.19 Corollary.** *A finite q-linear geometry has critical exponent* 1 *if and only if it is isomorphic to a submatroid of the affine geometry* $\mathbf{AG}(n, q)$. $\square$

**Example.** We describe two important combinatorial extremal problems which are instances of the duality pointed out in **7.16**.

*Coding Problem*. The *weight* $w(\alpha)$ of a vector $\alpha \in V(n, q)$ is defined to be the number of coordinates $\neq 0$ in $\alpha$. The function $d(\alpha, \beta) := w(\beta - \alpha)$ is a metric on $V(n, q)$. In coding theory one is interested in the *minimal dimension* $d(U) := \min_{\alpha \neq \beta \in U} d(\alpha, \beta)$ of a subspace $U \subseteq V(n, q)$. By linearity we have

$$d(U) := \min_{0 \neq \alpha \in U} w(\alpha).$$

The coding problem can now be stated as follows. Given $n$, $q$ and $d$, determine the maximal rank $C(n, q, d)$ among all subspaces $U \subseteq V(n, q)$ with $d(U) \geq d$.

Since subspaces correspond to linear codes and the distance from 0 to the number of errors that can be corrected by this code, the coding problem asks for as large a code as possible which corrects a given number of errors. (See, for instance, Berlekamp [1, ch. 1].)

*Packing Problem*. We call a set $S \subseteq V(r, q)$ *d-independent* if any $d$ vectors of $S$ are linearly independent. Problem: Given $r$, $q$ and $d$, what is the largest cardinality $P(r, q, d)$ of a $d$-independent set in $V(r, q)$?

Notice that the concept of $d$-independence has an obvious analogue in an arbitrary matroid. The connection between the coding and packing problem is given by the following proposition.

**7.20 Proposition.** *Let $n$, $r$ and $d$ be positive integers, and $q$ a prime power. Then*

$$C(n, q, d + 1) \geq n - r \Leftrightarrow P(r, q, d) \geq n.$$

*Proof*. Let $R$ be any $r \times n$-matrix over $GF(q)$. Suppose the columns of $R$ form a $d$-independent set in $V(r, q)$. Any non-zero vector $\alpha \in V(n, q)$ which is orthogonal to all row vectors of $R$ has $w(\alpha) \geq d + 1$, since otherwise the columns corresponding to the non-zero entries in $\alpha$ would not be linearly independent. Hence the nullspace $U$ of $R$ satisfies $d(U) \geq d + 1$ and has rank $r(U) \geq n - r$. Conversely, if the nullspace of $R$ has this property then the columns of $R$ are a $d$-independent set. Since any subspace of $V(n, q)$ of rank $\geq n - r$ is the nullspace of some $r \times n$-matrix, the theorem follows. □

To see the connection to the critical problem, consider the set $S_{n,q,d} := \{0 \neq \alpha \in V(n, q) : w(\alpha) \leq d\}$ for all $d \geq 1$. Then by **7.16**

$$C(n, q, d + 1) = n - \operatorname{crit}(S_{n,q,d}).$$

Hence, the complete solution to any one of the problems of computing $C$, $P$ or *crit* yields the solution to the other two problems. Let us summarize the various relations implied by **7.18** and **7.20**.

**7.21 Proposition.** *We have*:

(i) $C(n, q, d + 1) = \max_{0 \le k \le n} k$ *such that* $\chi(L(S_{n,q,d}); q^{n-k}) > 0$.
(ii) $P(r, z, d) = \max_{0 \le n \le z^r - 1}$ *such that* $\chi(L(S_{n,q,d}); q^r) > 0$. □

The coding and packing problems therefore lead to the study of the roots of the polynomial $\chi(L(S_{n,q,d}); x)$. For $d = 1$ we have, trivially, $L(S_{n,q,1}) \cong \mathscr{B}(n)$ for all $q$, and thus by **4.20**(ii)

$$\chi(L(S_{n,q,1}); x) = (x - 1)^n,$$

leading to the obvious formulas $C(n, q, 2) = n - 1$, $P(r, q, 1) = q^r - 1$. For $q = 2$ and $d = 2$, the set $S_{n,2,2} = \{0 \neq \alpha \in V(n, 2): w(\alpha) \le 2\}$ consists of all vectors with one or two coordinates equal to 1 and the rest equal to 0. It follows that $|S_{n,2,2}| = \binom{n}{2} + \binom{n}{1} = \binom{n+1}{2}$. We number the positions of the coordinates 1 to $n$ and consider the complete graph $K_{n+1}$ on the vertex-set $V = \{0, 1, \ldots, n\}$. Define $\Phi: S_{n,2,2} \to V^{(2)}$ by

$$\Phi\alpha = \begin{cases} \{i, j\} & \text{if } w(\alpha) = 2 \text{ and } i, j \text{ are the 1-coordinates} \\ \{0, i\} & \text{if } w(\alpha) = 1 \text{ and } i \text{ is the 1-coordinate.} \end{cases}$$

It is easy to see (and will be proved in **7.50**) that $\phi$ is a matroid isomorphism between $\mathbf{M}(S_{n,2,2})$ and $\mathbf{P}(K_{n+1})$. Using **4.20**(iv) we therefore have

$$\chi(L(S_{n,2,2}); x) = (x - 1)(x - 2) \cdots (x - n),$$

and thus

$$\operatorname{crit}(S_{n,2,2}) = \lceil \log_2(n + 1) \rceil, \qquad C(n, 2, 3) = n - \lceil \log_2(n + 1) \rceil,$$
$$P(r, q, 2) = 2^r - 1.$$

The most interesting example of a critical problem arises in connection with the coloring of graphs. It will be discussed in section 3.C.

EXERCISES VII.1

→ 1. Let $V(p + 1, p)$ be a $(p + 1)$-dimensional vector space over $GF(p)$, $p$ prime, and let $\{v_1, \ldots, v_{p+1}\}$ be a basis, $v = \sum_{i=1}^{p+1} v_i$. Prove that the submatroid $\mathbf{M}(S) \subseteq \mathbf{M}(V(p + 1, GF(p)))$ generated by

$$S = \{v_1, \ldots, v_{p+1}, v, v - v_1, \ldots, v - v_{p+1}\}$$

has characteristic set $\operatorname{ch}(\mathbf{M}) = \{p\}$. (Lazarson)

2. Use the construction of ex. 1 on the vector space $V(n + 1, \mathbb{Q})$ and show that the resulting matroid $\mathbf{M}$ has $\operatorname{ch}(\mathbf{M}) = \mathbb{P} - \{p \text{ prime}: p \le n\}$.

3.* Show that for a finite matroid $\mathbf{M}$, $0 \in \operatorname{ch}(\mathbf{M})$ if and only if $|\operatorname{ch}(\mathbf{M})| = \infty$. (Rado–Vamos)

→ 4. Consider the vector space $V(4, 2)$ and a basis $\{v_1, v_2, v_3, v_4\}$, and let $\mathbf{M}(S)$ be the submatroid generated by

$$S = \{v_1, v_2, v_3, v_4, v_3 + v_4, v_2 + v_3 + v_4, v_1 + v_2 + v_3 + v_4\}.$$

Let $T = \{a_1, a_2, a_3, a_4, a_5, a_6\}$ be any set with 6 elements. Show that the matroid $f(\mathbf{M})$ on $T$ induced by the surjection $f: S \to T$, $f(v_i) = a_i$, $i = 1, \ldots, 4$, $f(v_3 + v_4) = f(v_2 + v_3 + v_4) = a_5$, and $f(v_1 + v_2 + v_3 + v_4) = a_6$ is not coordinatizable over $GF(2)$.

5.* If $\mathbf{M}(S)$ is coordinatizable over an infinite field $K$, prove that any truncation $U_k(\mathbf{M})$ is also coordinatizable over $K$. Is this true for finite fields as well?

6. Verify **7.9** and **7.10**.

→ 7. Consider a 3-design $(S, \mathfrak{H})$ with $|S| = 22$ and $|H| = 6$ for all $H \in \mathfrak{H}$. (Such designs exist.) Prove that the corresponding covering matroid $\mathbf{M}(S, \mathfrak{H})$ is non-linear and show that $\mathbf{M}(S, \mathfrak{H})$ cannot even be embedded in a modular geometry. (Hint: Using **6.56**, show that every interval $[p, 1]$, $p \in S$, is modular and conclude from this that any two sets $H \neq H' \in \mathfrak{H}$ are either disjoint or intersect in exactly 2 points. Now use a similar argument as in the Vamos example.) This example is best possible since it is known that any geometry all of whose upper intervals of length 4 are modular can be embedded in a modular geometry (Wille).

8. Verify that Desargues' Theorem holds in $\mathbf{PG}(n, K)$, $n \geq 2$, over any division ring $K$, and that Pappus' theorem holds in $\mathbf{PG}(n, K)$, $n \geq 2$, over any field $K$.

→ 9. Consider the geometry $\mathbf{M}$ with Euclidean representation as shown in the figure; $\mathbf{M}$ has 8 points and 8 3-point lines. Show that $\mathbf{M}$ is $K$-linear if and only if the equation $x^2 - x + 1 = 0$ is solvable in $K$. Conclude that $\mathbf{M}$ is, in particular, $GF(3)$-linear and determine a coordinatization. (Mac Lane)

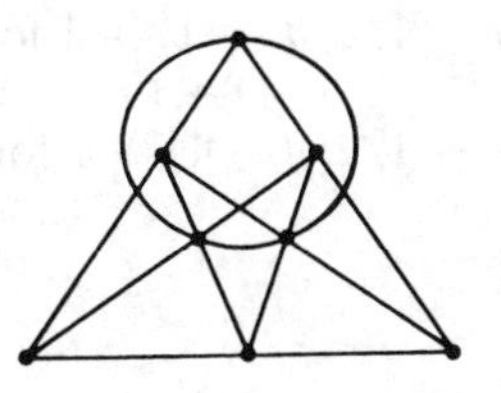

10. Generalize the Vamos example: Let $S = \dot{\bigcup}_{i=1}^{4} A_i, |A_i| = s_i \geq 2, i = 1, \ldots, 4.$ The bases of $S$ are all $(\sum_{i=1}^{4} s_i - 4)$-sets $B$ such that

(i) $A_i \cup A_j \not\subseteq B$ for all $\{i, j\} \neq \{3, 4\}$, $1 \leq i < j \leq 4$,
(ii) $B = A_3 \cup A_4 \cup B_1 \cup B_2$ with $B_1 \subseteq A_1$, $B_2 \subseteq A_2$ and $|A_1 - B_1| = |A_2 - B_2| = 2$.

Show that (i) and (ii) defines a geometry $\mathbf{M}(S)$ and that $\mathbf{M}(S)$ is non-linear.

→11. Let $\mathbf{M}(S)$ be a linear geometry with rank function $r$. Show: For all 4-tuples $A_1, A_2, A_3, A_4$ of subsets of $S$

$$r(A_1) + r(A_2) + r(A_3 \cup A_4) + \sum_{i=3}^{4} r(A_1 \cup A_2 \cup A_i) \leq \sum_{\substack{1 \leq i < j \leq 4 \\ \{i, j\} \neq \{3, 4\}}} r(A_i \cup A_j).$$

(Ingleton) (Hint: Show that the $A_i$'s can be assumed to be flats and use then the modular equality.)

12. Let $\varnothing \neq S \subseteq V(n, 2), 0 \notin S$. Show that the submatroid $\mathbf{M}(S)$ has critical exponent 1 if and only if all circuits in $\mathbf{M}(S)$ have even length.

13. Consider the set $S_{n,q,d}$ as defined in the packing problem. Prove that

$$\chi(L(S_{n,q,2}); x) = \prod_{i=0}^{n-1} (x - (q-1)i - 1).$$

Study the lattice $L(S_{n,q,2})$ and prove for this "$q$-partition lattice" properties analogous to $\mathscr{P}(n+1)$. (Dowling)

→14. Show:

(i) $\chi(L(S_{3,2,3}); x) = (x-1)(x-2)(x-4)$
(ii) $\chi(L(S_{4,2,3}); x) = (x-1)(x-2)(x-4)(x-7)$
(iii) $\chi(L(S_{5,2,3}); x) = (x-1)(x-2)(x-4)(x-8)(x-10)$
(iv) $\chi(L(S_{6,2,3}); x) = (x-1)(x-2)(x-4)(x-8)(x^2 - 26x + 175)$.

15. Show:

(i) $P(r, q, d) \geq P(r-1, q, d) + 1$
(ii) $P(r, q, d) \leq P(r-1, q, d-1) + 1$ for $d \geq 3$
(iii) $P(r, q, d) \leq q^{r-d+2}/(q-1) + (d-2)$ for $d \geq 2$
(iv) $P(r, 2, d) = P(r-1, 2, d-1) + 1$ for $d \geq 3$, $d$ odd
(v) $P(r, 2, 3) = 2^{r-1}$.

## 2. Binary Matroids

The basic problem of finding necessary and sufficient conditions for a matroid to be $K$-linear has been studied and solved most satisfactorily when $K = GF(2)$. In this section we describe several characterizations of such matroids. Each of these characterizations will produce interesting results for graphs and we shall see, conversely, that many theorems on graphs have natural extensions to $GF(2)$-linear matroids. Section B deals with regular matroids and their characterizations.

## A. Characterization of Binary Matroids

**Definition.** A matroid is called *binary* if it is coordinatizable over $GF(2)$.

Notice that a simple binary matroid of rank $n$ has at most $2^n - 1$ points. In view of **7.1** we shall often consider only finite binary matroids.

Let $\mathfrak{M}$ be a family of matroids (or more precisely of isomorphism classes of matroids) such that the following holds: If $\mathbf{M}$ is in $\mathfrak{M}$ then any minor of $\mathbf{M}$ is in $\mathfrak{M}$. We then call $\mathfrak{M}$ a *minor closed class.* Examples of minor closed classes are free matroids, $K$-linear, linear, regular, graphic, and cographic matroids. Every matroid $\mathbf{G} \notin \mathfrak{M}$ yields a necessary condition for the description of $\mathfrak{M}$ since, by definition, $\mathbf{M} \in \mathfrak{M}$ implies that $\mathbf{M}$ has no minor isomorphic to $\mathbf{G}$. Since taking minors is a transitive operation we are particularly interested in the *minimal* matroids $\mathbf{G}$ (minimal with respect to minors) which do not lie in $\mathfrak{M}$. These minimal matroids are called *obstructions* or *forbidden minors* of the family $\mathfrak{M}$. In general, a family $\mathfrak{M}$ need not have any such minimal matroids not in $\mathfrak{M}$, but it is an easy exercise to prove that all of the families listed above possess a complete list of obstructions.

The problem, then, is to determine for a minor closed class $\mathfrak{M}$ the list of obstructions. For instance, we remarked in section 1 that the Fano matroid $\mathbf{F}$ and its dual $\mathbf{F}^{\perp}$ are not $\mathbb{Q}$-linear, but that any proper minor is. Hence $\mathbf{F}$ and $\mathbf{F}^{\perp}$ are obstructions for $\mathbb{Q}$-linearity. The family of obstructions need not be finite. It is known, for example, that there exist infinitely many minimal non-$\mathbb{Q}$-linear matroids (Vamos). For binary matroids, however, the answer is very simple, the 4-point line $\mathbf{U}_2(4)$ being the only obstruction.

**7.22 Theorem.** *For a matroid* $\mathbf{M}(S)$ *the following conditions are equivalent*:

(i) $\mathbf{M}(S)$ *is binary.*
(ii) $\mathbf{M}(S)$ *does not have a minor isomorphic to* $\mathbf{U}_2(4)$.
(iii) *For every interval* $[a, b] \subseteq L(S)$ *of length* 2 *we have* $|[a, b]| \leq 5$.
(iv) *Every coline of* $\mathbf{M}(S)$ *is contained in at most* 3 *copoints.*

*Proof.* Since in $\mathbf{PG}(n, 2)$ every line has 3 points, the 4-point line $\mathbf{U}_2(4)$ cannot be binary. Together with **6.39** this gives the implication (i) $\Rightarrow$ (ii) and by **6.37** (ii) $\Rightarrow$ (iii). (iii) $\Rightarrow$ (iv) is trivial. To prove (iv) $\Rightarrow$ (i) we use the coordinatization theorem **7.8**. For each copoint $H$ of $\mathbf{M}(S)$ we define $f_H: S \to GF(2)$ by

$$f_H(p) = \begin{cases} 0 & \text{if } p \in H \\ 1 & \text{if } p \notin H. \end{cases}$$

We have to verify condition (ii) in **7.8**. Let $H_1$, $H_2$ and $H_3$ be distinct copoints with $H_1 \wedge H_2 \wedge H_3 = W$ where $W$ is a coline. We have to show $(f_{H_1} + f_{H_2} + f_{H_3})(p) = 0$ for all $p \in S$. This holds trivially for $p \in W$. By the hypothesis, $S - W = (H_1 - W) \cup (H_2 - W) \cup (H_3 - W)$. Hence $p \in S - W$ is a root of precisely two of the functions $f_{H_i}$ implying $(f_{H_1} + f_{H_2} + f_{H_3})(p) = 0$. $\square$

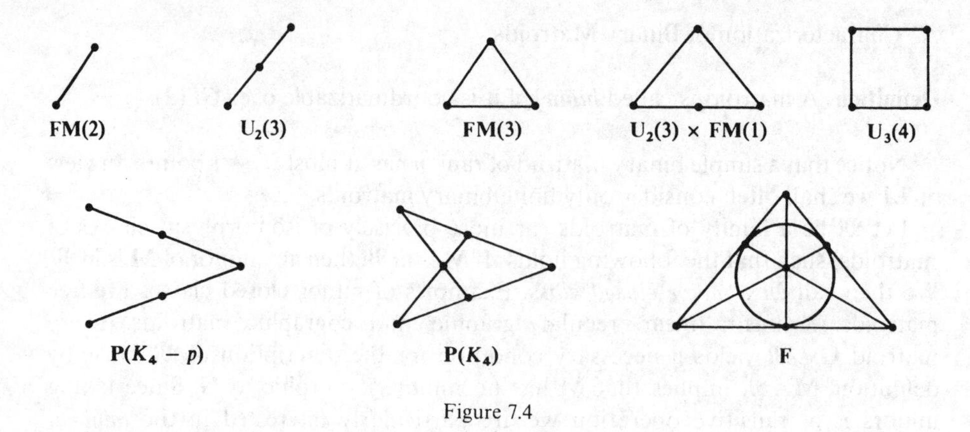


Figure 7.4

**Example.** Figure 7.4 depicts all binary geometries of rank 2 and 3.

**7.23 Corollary** (Tutte–Whitney). *Any graphic or cographic matroid is binary.*

*Proof.* We know from **6.24**(v) that any coline of a polygon matroid $\mathbf{P}(G(V, S))$, regarded as a subgraph, has precisely two graph components more than $G(V, S)$ and every copoint has one graph component more. This immediately implies that any coline is contained in 2 or 3 copoints. □

**7.22** together with **6.22** yields a very simple coordinatization of binary matroids $\mathbf{M}(S)$. By the remark at the beginning we may assume that $S$ is finite. We index the columns of a matrix $R$ by $S$ and the rows by the family $\mathfrak{H}$ of copoints. If we write the 0,1-functions $f_H$, $\ker f_H = H$, as rows of $R$, then the columns will give us a coordinatization of $\mathbf{M}(S)$. The functions $f_H$ may also be regarded as the *characteristic* functions of the cocircuits $S - H$ with $f_H + f_{H'}$ being the "addition" of the corresponding cocircuits. This observation leads to the second characterization of binary matroids.

**Definition.** Let $A_1, \ldots, A_t$ be subsets of $S$. We define the *sum of the $A_i$'s modulo* 2 as the set

$$A_1 + \cdots + A_t := \{p \in S : p \text{ is contained in an odd number of the } A_i\text{'s}\}.$$

In particular, $A_1 + A_2 = (A_1 - A_2) \cup (A_2 - A_1)$ is the *symmetric difference* of $A_1$ and $A_2$. A *group of sets* $\mathfrak{S} \subseteq 2^S$ is a non-empty family of *finite subsets* of $S$ which is closed with respect to addition modulo 2.

**7.24 Proposition.** *Let $\mathfrak{S}$ be a group of sets on $S$. Then*:

(i) *The minimal non-empty sets in $\mathfrak{S}$ satisfy the elimination axiom for circuits* **6.13**(ii).
(ii) *Each $A \in \mathfrak{S}$ is a disjoint union of minimal non-empty sets in $\mathfrak{S}$.*

*Proof.* Let $C \neq D$ be minimal non-empty sets in $\mathfrak{S}$ and $p \in C \cap D$. Then $p \notin C + D \in \mathfrak{S}$, i.e., $C + D \subseteq (C \cup D) - p$, which implies that $(C \cup D) - p$ contains a minimal non-empty set of $\mathfrak{S}$. For $A \in \mathfrak{S}$ choose $A_0 \in \mathfrak{S}$ to be a maximal subset of $A$ which is a disjoint union of minimal sets in $\mathfrak{S}$. If $A_0 \subsetneqq A$ then $\varnothing \neq A - A_0 = A_0 + A \in \mathfrak{S}$. $A - A_0$ contains therefore a minimal non-empty set $C \in \mathfrak{S}$, whence $C \cup A_0 = C + A_0 \in \mathfrak{S}$, $A_0 \subsetneqq C \cup A_0 \subseteq A$. This contradicts the maximality of $A_0$. □

**7.24**(i) shows that the minimal non-empty sets in a group of sets $\mathfrak{S}$ induce a unique matroid $\mathbf{M}(S, \mathfrak{S})$ if the finite basis axiom **6.13**(iii) is also satisfied. Furthermore, **7.24**(ii) immediately implies the following result.

**7.25 Corollary.** *Let* $\mathbf{M}(S)$ *be a matroid. If there is a group of sets* $\mathfrak{S}$ *on* $S$ *with* $\mathbf{M}(S) = \mathbf{M}(S, \mathfrak{S})$ *then* $\mathfrak{S}$ *is uniquely determined, namely*

$$\mathfrak{S} = \{A \subseteq S : A = \varnothing \text{ or } A \text{ is the union of disjoint circuits of } \mathbf{M}(S)\}. \quad \square$$

This suggests the following definitions.

**Definition.** Let $\mathbf{M}(S)$ be a matroid. The *cycles* of $\mathbf{M}(S)$ are the empty set and all disjoint unions of finitely many circuits in $\mathbf{M}(S)$. The *cocycles* of $\mathbf{M}(S)$ are the empty set and all disjoint unions of finitely many cocircuits of $\mathbf{M}(S)$. The group of sets generated by the cycles is called the *cycle group* of $\mathbf{M}(S)$. The *cocycle group* is similarly defined in the case that all cocircuits are finite.

**7.26 Theorem.** *For a matroid* $\mathbf{M}(S)$ *the following conditions are equivalent*:

(i) $\mathbf{M}(S)$ *is binary.*
(ii) *Any non-empty modulo 2 sum* $C_1 + \cdots + C_k$ *of circuits in* $\mathbf{M}(S)$ *is the union of disjoint circuits.*
(iii) $\mathbf{M}(S) = \mathbf{M}(S, \mathfrak{S})$ *for some group of sets* $\mathfrak{S}$.
(iv) *The symmetric difference of any two distinct circuits in* $\mathbf{M}(S)$ *is the union of disjoint circuits.*

*Proof.* (i) ⇒ (ii). Suppose $\Phi: S \to V(n, 2)$ is a coordinatization of $\mathbf{M}(S)$. Regarding a finite $s$-set $\Phi(C)$, $C \subseteq S$, of vectors as columns of an $n \times s$-matrix $R$ over $GF(2)$ we note that every row of $R$ must contain an even number of 1's if $C$ is a circuit. Hence, if $C_1, \ldots, C_k$ are circuits in $\mathbf{M}(S)$ then $\Phi(C_1 + \cdots + C_k)$ is a matrix $R'$ whose rows also contain an even number of 1's, which implies that $\Phi(C_1 + \cdots + C_k)$ and thus $C_1 + \cdots + C_k$ is dependent. Removing a circuit $C$ from $C_1 + \cdots + C_k$ we again obtain a matrix $\Phi(C_1 + \cdots + C_k - C)$ with this property. We may continue this process until $C_1 + \cdots + C_k$ is exhausted.

(ii) ⇒ (iii). Let $\mathfrak{S}$ be the family of cycles of $\mathbf{M}(S)$. The hypothesis assures that $\mathfrak{S}$ is a group of sets with $\mathbf{M}(S) = \mathbf{M}(S, \mathfrak{S})$.

(iii) ⇒ (iv). Immediate from **7.24**(ii).

(iv) $\Rightarrow$ (i). Let $\mathfrak{K}$ be the family of circuits in $\mathbf{M}(S)$ and $F(S, GF(2)) = \{f: S \to GF(2)$ such that $|\|f\| \cap A|$ is even for all $A \in \mathfrak{K}\}$. Hence if we identify $f$ with its support $\|f\|$, then $F(S, GF((2))$ consists of all subsets of $S$ which intersect every circuit in an even number of points. $F(S, GF(2))$ is clearly a function space; we want to show $\mathbf{M}(S) = \mathbf{M}(F(S, GF(2)))$. By the dual statement of **6.67**(ii), any cocircuit of $\mathbf{M}(F(S, GF(2)))$ contains a cocircuit of $\mathbf{M}(S)$. It remains to be shown that every cocircuit of $\mathbf{M}(S)$ cuts every circuit in an even number of points. Suppose otherwise and choose a circuit $A$ and a cocircuit $C$ such that $|C \cap A|$ is odd and minimal. From **6.67**(i) we infer that $|C \cap A| \geq 3$. Let $p \neq q \in C \cap A$. Since $S - C$ is a copoint, $q \in (S - C) \cup p$. Hence there is a circuit $B$ with $q \in B \subseteq (S - C) \cup \{p, q\}$ and thus $B \cap C = \{p, q\}$. Taking the symmetric difference $A + B$ we have $|C \cap (A + B)| = |C \cap A| - 2$. Since, by hypothesis, $A + B = K_1 \dot{\cup} \cdots \dot{\cup} K_t$ with $K_i \in \mathfrak{K}$, there must be a circuit $K_i$ with $|C \cap K_i| < |C \cap A|$ and $|C \cap K_i|$ odd, contradicting the minimality of $|C \cap A|$. $\square$

Condition **7.26**(iv) may be thought of as a stronger form of the circuit elimination axiom characterizing binary matroids. If $S$ is finite we know that $\mathbf{M}^{\perp}(S)$ is binary whenever $\mathbf{M}(S)$ is binary. Hence, in **7.26**, we may replace circuits by cocircuits, i.e.,

$$\begin{aligned} \mathbf{M}(S) &= \mathbf{M}(S, \mathfrak{S}), \quad \text{where } \mathfrak{S} \text{ is the cycle group} \\ \mathbf{M}^{\perp}(S) &= \mathbf{M}(S, \mathfrak{S}^{\perp}), \quad \text{where } \mathfrak{S}^{\perp} \text{ is the cocycle group.} \end{aligned}$$

The proof (iv) $\Rightarrow$ (i) in **7.26** shows that $\mathfrak{S}$ and $\mathfrak{S}^{\perp}$ with modulo 2 addition are vector spaces over $GF(2)$ of dimension $|S| - r(\mathbf{M})$ and $r(\mathbf{M})$, respectively. Let us summarize this in the following proposition.

**7.27 Proposition.** *Let $\mathbf{M}(S)$ be a finite binary matroid with cycle group $\mathfrak{S}$ and cocycle group $\mathfrak{S}^{\perp}$. Then:*

(i) *$\mathfrak{S}$ and $\mathfrak{S}^{\perp}$ with modulo 2 addition are vector spaces over $GF(2)$ with* $\dim \mathfrak{S} = |S| - r(\mathbf{M})$ *and* $\dim \mathfrak{S}^{\perp} = r(\mathbf{M})$.
(ii) $\mathfrak{S}^{\perp} = \{C \subseteq S : |C \cap A|$ *even for all* $A \in \mathfrak{S}\}$. $\square$

Since the circuits generate the whole cycle group $\mathfrak{S}$, we also have $\mathfrak{S}^{\perp} = \{C \subseteq S : |C \cap A|$ even for all circuits $A\}$. This leads to our last characterization of binary matroids.

**7.28 Theorem.** *Let $\mathbf{M}(S)$ be a matroid and $\mathfrak{K}, \mathfrak{C}$ the families of circuits and cocircuits, respectively. $\mathbf{M}(S)$ is binary if and only if $|K \cap C|$ is even for any $K \in \mathfrak{K}$ and $C \in \mathfrak{C}$.*

*Proof.* We have seen the necessity in the proof of (iv) $\Rightarrow$ (i) in **7.26**. For the converse we use **7.22**(iv). Suppose $W$ is a coline which is contained in 4 distinct copoints $H_1, \ldots, H_4$. Choose a basis $B'$ of $W$ and 4 points $p_i \in H_i - W$, $i = 1, \ldots, 4$. $B = B' \cup p_1 \cup p_2$ is then a basis of $\mathbf{M}(S)$. If $K$ is the unique circuit with $p_3 \in K \subseteq B \cup p_3$

(see **6.12**), then $K$ must contain $p_1$ and $p_2$ since otherwise $p_3$ would be in $H_1$ or $H_2$. Hence $\{p_1, p_2, p_3\} \subseteq K$, and thus $|K \cap (S - H_4)| = 3$, in violation of the hypothesis. □

We already know that every polygon matroid $\mathbf{P}(G(V, S))$ is binary. There are interesting graph-theoretic descriptions of the cycles and cocycles of the matroid $\mathbf{P}(G(V, S))$. Call a graph $G(V, S)$ *Eulerian* if all degrees $\gamma(v)$ are even. First we need an easy lemma whose proof is left to the reader.

**7.29 Lemma.** *Let $G(V, E)$ be a finite graph with $E \neq \varnothing$ and $\gamma(v) \neq 1$ for all $v \in V$. Then $G$ contains a non-trivial circuit.* □

**7.30 Proposition.** *The cycles of the matroid $\mathbf{P}(G(V, S))$ are precisely the edge-sets of Eulerian subgraphs.*

*Proof.* The edge-sets of Eulerian subgraphs are clearly closed with respect to addition modulo 2; hence they generate a group $\mathfrak{S}'$. To prove $\mathfrak{S} = \mathfrak{S}'$ it suffices to show that the minimal non-empty Eulerian edge-sets are precisely the polygons. A polygon is certainly a minimal Eulerian edge-set. Conversely, if $G(V, C)$ is a non-empty Eulerian subgraph then no vertex has degree 1, whence, by the lemma, $C$ contains a polygon $C'$ and therefore must be equal to $C'$. □

It follows from **7.25** and **7.30** that a graph is Eulerian if and only if its edge-set can be partitioned into polygons. Figure 7.5 gives an Eulerian graph and a decomposition into polygons. We turn to the cocycles. Call a set $C \subseteq S$ of edges a *bipartition* of the graph $G(V, S)$ if $C = \varnothing$ or if there exists a partition $V = V_1 \dot\cup V_2$ such that $C = \{e \in S\colon e$ has one endpoint in $V_1$ and the other in $V_2\}$. For example, the bold edges of the graph in Figure 7.6 constitute a bipartition of this graph.

**7.31 Proposition.** *The cocycles of the matroid $\mathbf{P}(G(V, S))$ are precisely the bipartitions of $G(V, S)$.*

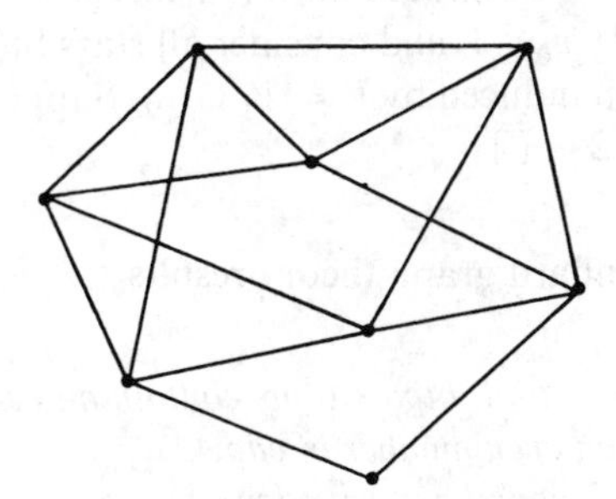
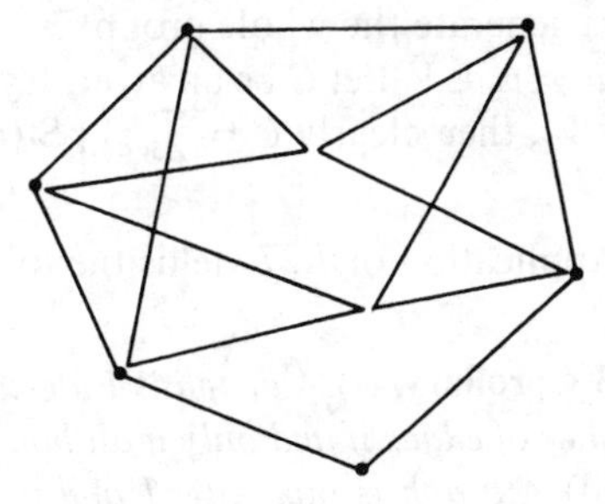

Figure 7.5

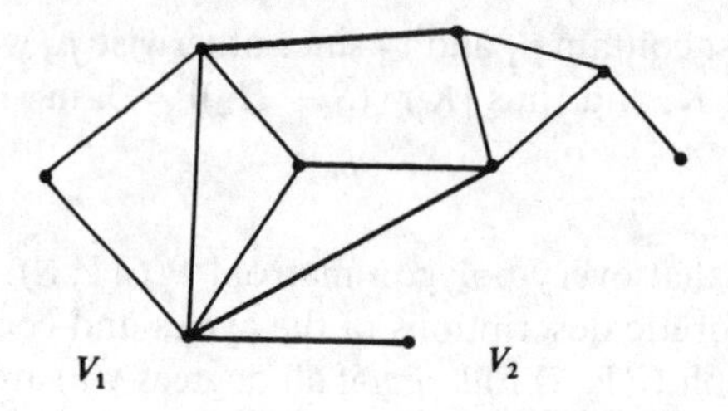


Figure 7.6. Bipartition.

*Proof.* If $C$ is a bipartition then it is clear that $|C \cap K|$ is even for every polygon $K$ (we have to switch back and forth between $V_1$ and $V_2$). Hence $C \in \mathfrak{S}^\perp$. Now suppose $C \in \mathfrak{S}^\perp$ where, without loss of generality, $G(V, C)$ is a connected graph. We define the relation $\approx$ on $V$ by

$$v \approx w :\Leftrightarrow \exists v, w\text{-path } G(V, W) \text{ such that } |W \cap C| \text{ is even.}$$

It is readily seen that $\approx$ is an equivalence relation with at most two equivalence classes; call them $V_1$ and $V_2$. We want to show that $C$ is precisely the bipartition induced by $V_1$ and $V_2$. Every edge $e = \{u, v\}$ with $u \in V_1, v \in V_2$ is in $C$ since otherwise the edge $e$ would be a $u, v$-path $W$ with $|W \cap C| = 0$. If, on the other hand, the edge $e = \{u, v\} \in C$ has both endpoints in, say, $V_1$ then we would have a $u, v$-path $W$ with $|W \cap C| =$ even and hence a polygon $K = W \cup e$ with $|K \cap C| =$ odd, contradicting **7.27**(ii). □

We know from **6.87** that the stars $\mathrm{St}(v_i)$ are bonds and hence cocycles if the graph $G(V, S)$ is 2-connected. The following result is useful.

**7.32 Proposition.** *Let $G(V, S)$ be a 2-connected graph, $|V| \geq 2$. Any set of $|V| - 1$ stars $\mathrm{St}(v_i)$ is then a basis of the cocycle group $\mathfrak{S}^\perp$ of $\mathbf{P}(G(V, S))$.*

*Proof.* Since $\dim \mathfrak{S}^\perp = r(\mathbf{P}(G, (V, S))) = |V| - 1$, we have to show that any $|V| - 1$ stars generate the whole group $\mathfrak{S}^\perp$. Choose any $v_0 \in V$ and consider all stars $\mathrm{St}(v)$ for $v \neq v_0 \in V$. Let $C$ be an arbitrary bipartition induced by $V = V_1 \dot\cup V_2$. Suppose $v_0 \in V_2$, then clearly $C = \sum_{v \in V_1} \mathrm{St}(v)$ modulo 2. □

Application of **7.27** yields the following standard graph theory results.

**7.33 Corollary.** (i) *A graph is Eulerian if and only if all bipartitions contain an even number of edges if and only if all bonds contain an even number of edges.*

(ii) *A graph is bipartite if and only if all polygons have even length if and only if the edge-set can be partitioned into disjoint bonds.* □

On the other hand, many results on graphs permit straightforward generalizations to binary matroids; see the exercises for some examples.

### B. Regular Matroids

Perhaps the most famous result to date in all of matroid theory is Tutte's characterization of regular matroids by means of forbidden minors. A proof of this theorem is too long to be presented here; we shall, however, deduce a few equivalent descriptions of regular matroids which will lead in a natural way to the characterization theorem. An interesting feature of regular matroids is their close relationship with an important class of matrices, the unimodular matrices. Let us therefore note a few facts about these matrices.

**Definition.** An $n \times s$-matrix $R$ over $\mathbb{Z}$ is called *unimodular* if $\det_{\mathbb{Q}} A = 0$ or $\pm 1$ for every square submatrix $A$ of $R$. We call a finite matroid $\mathbf{M}$ *unimodular* if $\mathbf{M}$ possesses a unimodular coordinatization matrix.

Note that the entries of a unimodular matrix are all 0 or $\pm 1$.

**7.34 Lemma.** *The following conditions are equivalent for a finite matroid* $\mathbf{M}$ *of rank* $n$:

(i) $\mathbf{M}$ *is unimodular.*
(ii) $\mathbf{M}$ *has a unimodular coordinatization matrix* $R = [I_n, A]$ *in standard form.*

*Proof.* We have to show (i) $\Rightarrow$ (ii). Let $R'$ be a unimodular coordinatization matrix with $n$ rows, where we may assume that the submatrix $B'$ of the first $n$ columns is non-singular. Let $N = [n_{ij}] = B'^{-1}$ (over $\mathbb{Q}$). Then $\det N = \pm 1$ and by a well-known formula in linear algebra we have $n_{ij} = \pm \det B'_{j,i}/\det B' = 0$ or $\pm 1$, where $B'_{ij}$ is the submatrix obtained from $B'$ by deleting its $i$-th row and $j$-th column. Therefore the matrix $R = NR' = [I_n, A]$ has all its entries in $\mathbb{Z}$. If $C$ is any $n \times n$-submatrix of $R$ and $C'$ the corresponding submatrix of $R'$ then $C = NC'$, and hence $\det C = \det N \cdot \det C' = 0$ or $\pm 1$ by the unimodularity of $R'$. Now let $B$ be a $k \times k$-submatrix of $R$. If $B \subseteq I_n$ then certainly $\det B = 0$ or $\pm 1$. Otherwise, suppose $i_1, \ldots, i_k$ are the rows of $B$ and $j_1, \ldots, j_k$ the columns, of which the columns $j_{l+1}, \ldots, j_k$ belong to $A$. By expanding $\det B$ with respect to the columns $j_1, \ldots, j_l$ we obtain $\det B = \pm \det D$ where $D$ is a square submatrix of $A$. Hence we may assume that the $k \times k$-matrix $B$ is contained in $A$. We enlarge $B$ to an $n \times n$-submatrix $C$ by adding the remaining elements in the columns $j_1, \ldots, j_k$ and the columns $\{1, \ldots, n\} - \{i_1, \ldots, i_k\}$. By expanding $\det C$ with respect to the columns in $I_n$ we obtain $\det B = \pm \det C = 0$ or $\pm 1$ by our previous argument. □

We come to the main theorem. By **7.1**, a matroid is regular if and only if the underlying geometry (which is finite) is regular. Hence we may confine ourselves to finite matroids.

**7.35 Theorem.** *Let* $\mathbf{M}(S)$ *be a finite matroid. Then the following conditions are equivalent*:

(i) $\mathbf{M}(S)$ *is regular.*
(ii) $\mathbf{M}(S)$ *is binary and ternary* ($= GF(3)$-*linear*).
(iii) $\mathbf{M}(S)$ *is binary and* $K$-*linear for some field* $K$ *with* char $K \neq 2$.
(iv) $\mathbf{M}(S)$ *is unimodular.*
(v) $\mathbf{M}(S) = \mathbf{M}(F(S, \mathbb{Q}))$ *and for every copoint* $H$ *there exists* $f_H \in F(S, \mathbb{Q})$ *with* $\ker f_H = H$ *and* $f_H(p) = \pm 1$ *for all* $p \notin H$.

*Proof.* The implications (i) $\Rightarrow$ (ii) $\Rightarrow$ (iii) are trivial.

(iii) $\Rightarrow$ (iv). Let $R = [I_n, A]$ be a coordinatization matrix of $\mathbf{M}$ over $K$ where char $K \neq 2$. We want to show that it is possible to transform $R$ into a unimodular matrix $R'$ by suitably multiplying rows and columns by non-zero scalars. This will prove (iv) by **7.9**(i). The following fact is the crux of the proof. If $R' = [I_n, A']$ is a *binary* coordinatization matrix of $\mathbf{M}(S)$, then by **7.12** we must have $a_{ij} \neq 0$ in $R$ if and only if $a'_{ij} = 1$ in $R'$. In other words, the substitution $a_{ij} \neq 0 \to 1$, $a_{ij} = 0 \to 0$ transforms $R$ into a binary coordinatization matrix $R'$, whence, by **7.12** again, we have $\det_K C = 0 \Leftrightarrow \det_{GF(2)} C' = 0$ for all square submatrices $C \subseteq R$ and their counterparts $C' \subseteq R'$. In particular, if $B$ is a $2 \times 2$-submatrix of $A$ all of whose elements are $\neq 0$ then $\det_K B = 0$, since $B' = [\begin{smallmatrix}1 & 1\\ 1 & 1\end{smallmatrix}]$ has $\det_{GF(2)} B' = 0$. Let us denote this fact by (+). Let $Z_1, \ldots, Z_n$ be the rows of $R$. Suppose the elements $\neq 0$ in $Z_1$ are $a_{11} = 1, a_{1,i_2}, \ldots, a_{1,i_s}$. We divide column $i_j$ by $a_{1,i_j}$, $j = 2, \ldots, s$, thus making all entries in $Z_1$ either 0 or 1. Assume that all elements of $Z_1, \ldots, Z_{k-1}$ have been made either 0 or $\pm 1$ by suitable multiplications of rows and columns, and let $a_{k,j_1}, \ldots, a_{k,j_t}$ be the non-zero elements of $Z_k$. We set

$$C_{j_l} := \{i: 1 \leq i \leq k-1, a_{i,j_l} \neq 0\}.$$

If $i \in C_{j_l} \cap C_{j_m}$ then we have a submatrix of the form

$$\begin{bmatrix} \pm 1 & \pm 1 \\ a_{k,j_l} & a_{k,j_m} \end{bmatrix},$$

whence, by (+), $a_{k,j_l} = \pm a_{k,j_m}$. Hence, if we multiply the columns $j_l$ by $a_{k,j_l}^{-1}$ then $Z_k$ contains only 0's and 1's whereas $Z_i$, for $i = 1, \ldots, k-1$, contains $0, \pm 1$, and $\pm b_i \neq 0$ in the columns $j_l$ for which $i \in C_{j_l}$. We now multiply each row $Z_i$, $i = 1, \ldots, k-1$, containing at least one $b_i$ (i.e., which does not have 0's in all the columns $j_1, \ldots, j_t$) by $b_i^{-1}$. The new matrix is of the form shown below.

$$\begin{array}{c} h \\ i \\ j \\ k \end{array}
\begin{bmatrix}
\pm 1 & & & & \pm b_h^{-1} & & \\
0 & \pm 1 & & & & \pm b_i^{-1} & \\
0 & 0 & \cdots & 0 & \pm 1 & \cdots & \pm 1 \\
1 & 1 & \cdots & 1 & 0 & \cdots & 0
\end{bmatrix}$$

$$\begin{array}{ccc} j_1 & \qquad\qquad & j_t \end{array}$$

The columns $j_1, \ldots, j_t$ are already as required. Hence our last step will be to multiply each column $s \notin \{j_1, \ldots, j_t\}$ by some $b_h$. Thus we have to make sure that if $b_h^{-1}, b_i^{-1}$ are in column $s$ then $b_h = \pm b_i$. If $h, i \in C_{j_l}$ for some $l$ then we have a submatrix

$$\begin{array}{c} h \\ i \end{array} \begin{bmatrix} \pm 1 & \pm b_h^{-1} \\ \pm 1 & \pm b_i^{-1} \end{bmatrix}, \\ \quad j_l \qquad s$$

whence $b_h = \pm b_i$ by (+). In the other case we obtain a submatrix of the form shown below which is also singular over $GF(2)$, again implying $b_h = \pm b_i$.

$$\begin{array}{c} h \\ i \\ k \end{array} \begin{bmatrix} \pm 1 & 0 & \pm b_h^{-1} \\ 0 & \pm 1 & \pm b_i^{-1} \\ 1 & 1 & 0 \end{bmatrix} \\ \qquad\qquad s$$

The final difficulty is that column $s$ may also contain some entries $\pm 1$ from a row, say $Z_j$, which has only 0's in the columns $j_1, \ldots, j_t$. In this case, first we have to multiply $Z_j$ by $b_h^{-1}$ if $\pm b_h^{-1}$ is the common $b$-value in column $s$. Hence it remains to be shown that if $Z_j$ has $\pm 1$ in columns $s, s' \notin \{j_1, \ldots, j_t\}$, then $b_h = \pm b_i$ for the $b$-value $b_h$ in column $s$, and $b_i$ in column $s'$, respectively. This is certainly the case by our previous argument if $b_h^{-1}, b_i^{-1}$ are both in column $s$ or both in column $s'$. Hence we assume that this is not so. If $h, i \in C_{j_l}$, then we have a submatrix

$$\begin{array}{c} h \\ i \\ j \end{array} \begin{bmatrix} \pm 1 & b_h^{-1} & 0 \\ \pm 1 & 0 & \pm b_i^{-1} \\ 0 & \pm 1 & \pm 1 \end{bmatrix} \\ \quad j_l \qquad s \qquad s'$$

which is singular over $GF(2)$ so that $b_h = \pm b_i$. In the other case, we obtain a submatrix

$$\begin{array}{c} h \\ i \\ j \\ k \end{array} \begin{bmatrix} \pm 1 & 0 & \pm b_h^{-1} & 0 \\ 0 & \pm 1 & 0 & \pm b_i^{-1} \\ 0 & 0 & \pm 1 & \pm 1 \\ 1 & 1 & 0 & 0 \end{bmatrix}.$$

This matrix is again singular over $GF(2)$, whence $b_h = \pm b_i$. We finally show that $R' = [I_n, A']$, regarded as a matrix over $\mathbb{Z}$, is unimodular. (+) implies that $R'$ contains no $2 \times 2$-submatrix with all elements $\neq 0$ and precisely one element $= 1$

or precisely one element $= -1$. Let us denote this fact by $(++)$. Suppose $M = [m_{ij}]$ is a $k \times k$-submatrix with $\det_{\mathbb{Q}} M \neq 0$, and, without loss of generality, $m_{11} \neq 0$. We make the column below $m_{11}$ to 0 by suitable additions, thus obtaining

$$M' = \begin{bmatrix} m_{11} & N \\ 0 & M_1 \end{bmatrix}$$

with $\det_{\mathbb{Q}} M_1 = \pm\det_{\mathbb{Q}} M$. $(++)$ implies that all entries in $M_1$ are 0, $\pm 1$ since otherwise $M$ has a submatrix

$$\begin{bmatrix} m_{11} & 1 \\ -m_{11} & 1 \end{bmatrix}$$

or

$$\begin{bmatrix} m_{11} & -1 \\ -m_{11} & -1 \end{bmatrix}.$$

$M_1$ also satisfies $(++)$, because, if

$$B_1 = \begin{bmatrix} b_{ij} & b_{ih} \\ b_{lj} & b_{lh} \end{bmatrix}$$

is a submatrix of $M_1$ excluded by $(++)$, then for the submatrix

$$B = \begin{bmatrix} m_{11} & m_{1j} & m_{1h} \\ m_{i1} & m_{ij} & m_{ih} \\ m_{l1} & m_{lj} & m_{lh} \end{bmatrix}$$

of $M$ we would have $\det_{\mathbb{Q}} B = m_{11} \cdot \det_{\mathbb{Q}} B_1 = \pm 2$, i.e., $\det_K B \neq 0$, but $\det_{GF(2)} B = 0$, contradicting **7.12**. Now we choose an element $\neq 0$ in the first column of $M_1$ and make the rest of the column 0. The matrix $M_2$ again has all entries equal to 0 or $\pm 1$ and satisfies $(++)$. After $k$ steps this yields $\det_{\mathbb{Q}} M = \pm 1$.

(iv) $\Rightarrow$ (v). Let $R = [I_n, A]$ be a unimodular coordinatization matrix with respect to the basis $B = \{b_1, \ldots, b_n\}$ (such a matrix exists by **7.34**). We define functions $f_i: S \to \mathbb{Q}$ corresponding to the rows of $R = [r_{ip}]$, $i = 1, \ldots, n$, $p \in S$, i.e., $f_i(p) = r_{ip}$, and denote by $F(S, \mathbb{Q})$ the function space generated by the $f_i$'s. We have to show that for an arbitrary copoint $H$ there exists a function $f_H \in F$ with $\ker f_H = H$ and $f_H(p) = \pm 1$ for all $p \notin H$. Let $|H \cap B| = n - 1 - k$, $0 \leq k \leq n - 1$. For $k = 0$, $H = H_i$ for some $i$, whence we may set $f_H = f_{H_i}$. Suppose then $k > 0$ and $H \cap B = \{b_{k+2}, \ldots, b_n\}$. We expand $H \cap B$ to a basis $B' = (H \cap B) \cup \{c_1, \ldots, c_k\}$ of $H$ and choose among all functions $f \in F(S, \mathbb{Q})$ with $\ker f = H$ a function $f_H$ for which $f_H(b_{k+1}) = 1$. From $f_H = \sum_{i=1}^{n} \lambda_i f_{H_i}$ we infer

$\lambda_i = f_H(b_i)$ and thus $f_H = \sum_{i=1}^{k+1} f_H(b_i) f_i$. Let $p \notin H$ and consider the system of linear equations:

$$\begin{aligned} f_H(b_1)f_1(c_1) + f_H(b_2)f_2(c_1) + \cdots + f_{k+1}(c_1) &= 0 \\ &\vdots \\ f_H(b_1)f_1(c_k) + f_H(b_2)f_2(c_k) + \cdots + f_{k+1}(c_k) &= 0 \\ f_H(b_1)f_1(p) + f_H(b_2)f_2(p) + \cdots + f_{k+1}(p) &= f_H(p). \end{aligned}$$

By Cramer's rule,

$$f_H(p) = \pm \frac{\det[f_i(c_j), f_i(p)]_{i=1,\ldots,k+1,\, j=1,\ldots,k}}{\det[f_i(c_j)]_{i,j=1,\ldots,k}} = \pm 1.$$

(v) $\Rightarrow$ (i). $\mathbf{M}(S)$ is $\mathbb{Q}$-linear; hence, by **7.2**, it suffices to show that $\mathbf{M}(S)$ is linear over every prime field $GF(p)$. Let $R = [I_n, A]$ be a $\mathbb{Q}$-coordinatization matrix with respect to the basis $B = \{b_1, \ldots, b_n\}$ and $f_i: S \to \mathbb{Q}$ the 0, $\pm 1$-functions corresponding to the rows of $R$ as specified in (v). We regard $R$ as a matrix $R_p$ over $GF(p) = \mathbb{Z}/p\mathbb{Z}$ and prove that $R_p$ is a coordinatization matrix of $\mathbf{M}(S)$ over $GF(p)$. To this end it suffices to show that any $n \times n$-submatrix $C$ of $R$ has $\det_{\mathbb{Q}} C = 0$ if and only if $\det_{GF(p)} C = 0$, since this implies that the matrix matroids induced by $R$ and $R_p$ have the same set of bases. If $\det_{\mathbb{Q}} C = 0$ then certainly $\det_{GF(p)} C = 0$. In the other case, the elements $b_{j_1}, \ldots, b_{j_n}$ corresponding to the columns of $C$ are a basis $B'$ of $\mathbf{M}(S)$. Hence by the hypothesis there exist 0, $\pm 1$-functions $g_i: S \to \mathbb{Q}$ such that $\ker g_i = \overline{B' - b_{j_i}}$, $g_i(b_{j_i}) = 1$. Let $R'$ be the matrix whose rows $Z_i'$ correspond to the functions $g_i$. The $Z_i'$'s are linear combinations of the rows of $R$ with coefficients 0, $\pm 1$. Hence $R' = NR$ for a $\mathbb{Z}$-matrix $N$, and thus $1 = \det_{\mathbb{Q}} N \cdot \det_{\mathbb{Q}} C$, i.e., $\det_{\mathbb{Q}} C = \pm 1 = \det_{GF(p)} C$. $\square$

Notice that step (iii) $\Rightarrow$ (iv) shows, in particular, that any standard coordinatization matrix $R = [I_n, A]$ of a regular matroid over some field $K$ with char $K \neq 2$ which contains only 0 and $\pm 1$ is already unimodular.

**7.36 Proposition.** *Let* $\mathbf{M}$ *be a finite regular matroid of rank* $n$ *and* $R$ *an* $n$*-rowed coordinatization matrix of* $\mathbf{M}$ *over* $\mathbb{Q}$ *all of whose* $n \times n$*-submatrices have determinant equal to* 0 *or* $\pm 1$. *(In particular, this is the case when* $R = [I_n, A]$ *is in standard form and all elements of* $A$ *are equal to* 0 *or* $\pm 1$.*) Then*

$$\det RR^T = \text{number of bases of } \mathbf{M}.$$

*Proof.* By the Binet–Cauchy theorem on determinants we have $\det RR^T = \sum_B (\det B)^2$ where the sum extends over all $n \times n$-submatrices $B$ of $R$. By our hypothesis, $(\det B)^2 = 0$ or 1, and $(\det B)^2 = 1$ if and only if the columns of $B$ correspond to a basis of $\mathbf{M}$. $\square$

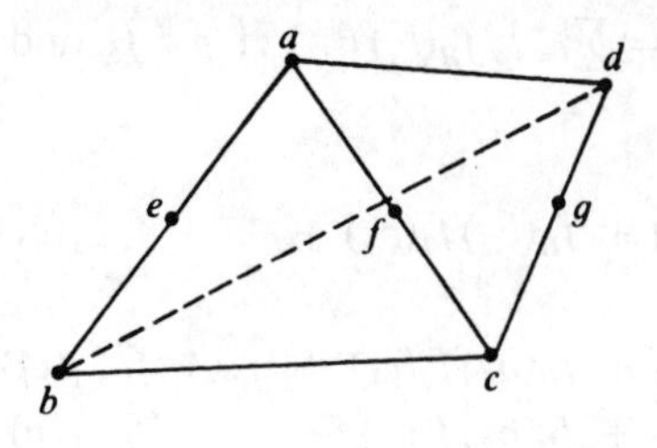


Figure 7.7

**Example.** Let **M** be the matroid with Euclidean representation as in Figure 7.7. It is easy to see that the coordinatization matrix

$$R = \begin{bmatrix} 1 & 0 & 0 & 0 & -1 & -1 & 0 \\ 0 & 1 & 0 & 0 & -1 & 0 & 0 \\ 0 & 0 & 1 & 0 & 0 & -1 & -1 \\ 0 & 0 & 0 & 1 & 0 & 0 & -1 \end{bmatrix}$$

is regular. From

$$RR^T = \begin{bmatrix} 3 & 1 & 1 & 0 \\ 1 & 2 & 0 & 0 \\ 1 & 0 & 3 & 1 \\ 0 & 0 & 1 & 2 \end{bmatrix}$$

we conclude that the number of bases is $\det RR^T = 21$.

Condition (iv) in **7.35** says that the matroids which are regular are precisely those which can be coordinatized by unimodular matrices. Regarding the $+1$ and $-1$ entries as *orientation* of the elements of **M** relative to the cocycles (corresponding to the rows), we are led to a characterization of regular matroids which is the precise generalization of the coordinatization of oriented graphs to be studied in the next section.

Let $\mathbf{M}(S)$ be a finite matroid. We define the *circuit matrix* $\mathscr{K} = [k_{ij}]$ over $\mathbb{Z}$ as the incidence matrix of circuits and elements. That is, the rows of $\mathscr{K}$ correspond to the circuits, the columns correspond to the elements of $S$ and we write $k_{ij} = 1$ or 0 depending on whether $j \in S$ is in the $i$-th circuit or not. Similarly, we define the *cocircuit matrix* $\mathscr{C} = [c_{ij}]$, where we use the same indexing of the columns as in $\mathscr{K}$. $\mathbf{M}(S)$ is called *orientable* if there exists an assignment of $\pm 1$ to the 1's in $\mathscr{K}$ and $\mathscr{C}$ such that $\tilde{\mathscr{C}}\tilde{\mathscr{K}}^T = 0$ for the "oriented" matrices $\tilde{\mathscr{C}}$ and $\tilde{\mathscr{K}}$.

**7.37 Theorem** (Minty). *A finite matroid is regular if and only if it is orientable.*

*Proof.* If $\mathbf{M}(S)$ is regular then we assign to the cocircuit matrix $\mathscr{C}$ the values $\pm 1$ according to **7.35**(v). Since $\tilde{\mathscr{C}}$ is also a coordinatization matrix over $GF(3)$ there exist $\lambda_{i_k} = \pm 1$, $k = 1, \ldots, t$, such that $\sum_{k=1}^{t} \lambda_{i_k} S_{i_k} = 0$ whenever the columns $S_{i_1}, \ldots, S_{i_t}$ of $\tilde{\mathscr{C}}$ correspond to a circuit $K$ of $\mathbf{M}(S)$. Hence if we assign to the row of $\mathscr{K}$ corresponding to $K$ the values $\lambda_{i_k}$ and do this for every circuit, then $\tilde{\mathscr{C}}\tilde{\mathscr{K}}^T = 0$. Conversely, let $\mathbf{M}(S)$ be orientable and $\tilde{\mathscr{C}}$, $\tilde{\mathscr{K}}$ the oriented circuit and cocircuit matrices, respectively, with $\tilde{\mathscr{C}}\tilde{\mathscr{K}}^T = 0$. By **7.35**(v) it suffices to show that $\tilde{\mathscr{C}}$ is a coordinatization matrix over $\mathbb{Q}$. If $K = \{i_1, \ldots, i_t\}$ is a circuit of $\mathbf{M}(S)$ then the columns $S_{i_1}, \ldots, S_{i_t}$ of $\tilde{\mathscr{C}}$ are linearly dependent over $\mathbb{Q}$ because $\tilde{\mathscr{C}}\tilde{\mathscr{K}}^T = 0$. If, on the other hand, $B = \{j_1, \ldots, j_n\}$ is a basis of $\mathbf{M}(S)$ then there are cocircuits $C_i = S - \overline{B - j_i}$ with $j_i \in C_i$ and $j_k \notin C_i$ for $k \neq i$. This immediately implies that the columns $S_{j_1}, \ldots, S_{j_n}$ of $\tilde{\mathscr{C}}$ are linearly independent over $\mathbb{Q}$, and the proof is complete. □

Another consequence of **7.35** is that a characterization of ternary matroids together with any of the known descriptions of binary matroids implies a characterization of regular matroids. In particular, a complete list of obstructions for $GF(3)$-linearity would yield the list of obstructions for regularity. We can immediately give 4 minimal non-ternary matroids: the 5-point line $\mathbf{U}_2(5)$ (any line in $\mathbf{PG}(n, 3)$ has 4 points!), the Fano plane $\mathbf{F}$, and their duals (Figure 7.8):

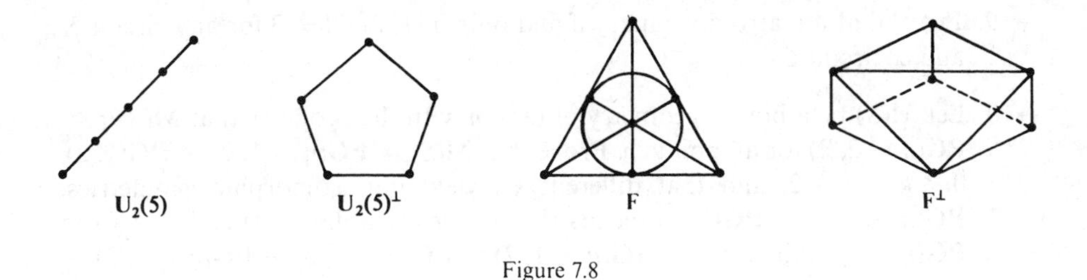


Figure 7.8

It is known (Bixby, Seymour) that these 4 matroids indeed constitute the complete list of obstructions for $GF(3)$-linearity. Since $\mathbf{U}_2(5)$ and its dual contain the 4-point line as a minor we conclude that the 4-point line $\mathbf{U}_2(4)$, $\mathbf{F}$ and $\mathbf{F}^\perp$ are the only obstructions for regularity. This is the famous coordinatization theorem of Tutte proved by him by other means.

**7.38 Theorem** (Tutte). *A matroid is regular if and only if it is binary and contains no minor isomorphic to* $\mathbf{F}$ *or* $\mathbf{F}^\perp$. □

EXERCISES VII.2

→ 1. Show that the following matroid classes can all be characterized by obstructions: free matroids, linear matroids, $K$-linear matroids, regular matroids, graphic matroids, cographic matroids.

2. Show: If the finite matroid $\mathbf{M}$ is an obstruction for $K$-linearity, then so is $\mathbf{M}^{\perp}$.

→ 3. Determine the obstructions for the class of free matroids.

4.* Prove that the class of $\mathbb{Q}$-linear matroids possesses infinitely many obstructions. (Vamos) (Hint: Use ex. VII.1.2 and ex. VII.1.3.)

→ 5. Prove that a matroid is binary if and only if for all bases $B \neq B'$ the following holds: For every $p \in B$ there exists an even number of elements $q \in B'$ such that $(B - p) \cup q$ and $(B' - q) \cup p$ are both bases.

6. Prove **7.29**.

7. Show: A connected graph $G(V, S)$ is Eulerian if and only if there is a closed sequence of edges which contains every edge of $G$ exactly once. How can this be generalized to arbitrary binary matroids?

8. The following conjecture generalizes a result on the existence of a Hamiltonian circuit in graphs. Let $\mathbf{M}$ be a finite connected binary matroid of rank $n$ such that every cocircuit $C$ has $|C| \geq (n + 1)/2$. Then there exists a circuit $K$ with $|K| = n + 1$. Prove this is true for all binary matroids up to 8 elements.

→ 9. Prove that a matroid is binary if and only if $|K \cap C| \neq 3$ for any circuit $K$ and cocircuit $C$.

→ 10. Let $\mathbf{M}(S)$ be a binary geometry of rank $n$ with the property that $\mathbf{M}(S)/p \cong \mathbf{PG}(n - 2, 2)$ for all points $p$. Prove that $\mathbf{M}(S) \cong \mathbf{PG}(n - 1, 2) - \mathbf{PG}(k, 2)$, $0 \leq k \leq n - 2$, and that different $k$'s yield non-isomorphic geometries. $\mathbf{PG}(n - 1, 2) - \mathbf{PG}(k, 2)$ means that we delete a flat of rank $k + 1$ from $\mathbf{PG}(n - 1, 2)$; hence, e.g., $\mathbf{AG}(n - 1, 2) \cong \mathbf{PG}(n - 1, 2) - \mathbf{PG}(n - 2, 2)$.

11. Prove: If $\mathbf{M}(S)$ is an incidence geometry of grade $n \geq 3$ in which every coline is contained in exactly 3 copoints, then $\mathbf{M}(S) \cong \mathbf{U}_{n-1}(n)$.

→ 12.* Let $\mathbf{M}(S)$ be a geometry in which every coline is contained in exactly 3 copoints. Show:

(i) $\mathbf{M}(S)$ graphic $\Rightarrow \mathbf{M}(S) \cong \mathbf{U}_{n-1}(n)$ or $\cong \mathbf{P}(K_n)$ for some $n \in \mathbb{N}$.
(ii) $\mathbf{M}(S)$ cographic $\Rightarrow \mathbf{M}(S) \cong \mathbf{U}_{n-1}(n)$ for some $n \in \mathbb{N}$ or $\cong \mathbf{P}(K_4)$ or $\cong \mathbf{B}(K_{3,3})$.

(Aigner) (Hint: Suppose $\mathbf{M}(S) \cong \mathbf{P}(G(V, S))$ with $G \neq K_n$; reduce $G$ step by step to a polygon.)

→ 13. Let $R$ be a coordinatization matrix of the finite matroid $\mathbf{M}(S)$ over $K$. Show:

(i) $p$ loop $\Leftrightarrow S_p = 0$ (column corresponding to $p$).
(ii) $p$ coloop $\Leftrightarrow$?
(iii) $R - S_p$ is a coordinatization matrix of $\mathbf{M}(S).(S - p)$.

(iv) If $p \notin \overline{\emptyset}$, then $\mathbf{M}(S)/p$ is coordinatized by the matrix $R'$ obtained as follows: Apply suitable elementary operations to $R$ until $S_p$ contains only one coordinate different from zero. Then delete the corresponding row and $S_p$ itself.

14. Prove the regularity of graphic matroids using **7.37**.
15. Prove that the smallest regular matroids which are not graphic are $\mathbf{B}(K_5)$ and $\mathbf{B}(K_{3,3})$.

# 3. Graphic Matroids

We have remarked in several places that graph theory stood at the very beginning of matroid theory and significantly influenced its development. In this section we shall demonstrate, conversely, that several concepts and methods in matroid theory can be successfully applied to graphs, providing a better understanding of such diverse topics as embedding and connectivity, networks and the coloring of graphs.

## A. Connectivity and Embeddings

To what extent is a graph determined by its polygon matroid? Obviously, if two graphs are isomorphic then so are their polygon matroids. On the other hand, $\mathbf{P}(G) \cong \mathbf{P}(G')$ need not imply that the graphs $G$ and $G'$ are isomorphic if one of them is not connected (see the remark at the end of chapter VI). But even if $G$ and $G'$ are connected they need not be isomorphic since, for instance, any two trees on $n$ vertices have isomorphic polygon matroids ($\cong \mathbf{FM}(n-1)$). Not even 2-connection is enough as the graphs in Figure 7.9 show. It is readily seen that $\mathbf{P}(G) \cong \mathbf{P}(G')$, whereas we have $G \ncong G'$.

**7.39 Lemma.** *Let $G(V, S)$ be a 3-connected graph without loops and at least 3 vertices, and $H$ a copoint of $\mathbf{P}(G(V, S))$. Then $\mathbf{P}(G(V, H))$ is connected if and only if $H = S - \mathrm{St}(v)$ for some $v \in V$.*

*Proof.* Since, for $H = S - \mathrm{St}(v)$, $G(V, H)$ is a 2-connected graph, $\mathbf{P}(G(V, H))$ is connected by **6.87**. Any other copoint $H$ has as a subgraph of $G$ at least two non-trivial graph components, hence $\mathbf{P}(G(V, H))$ is not connected. □

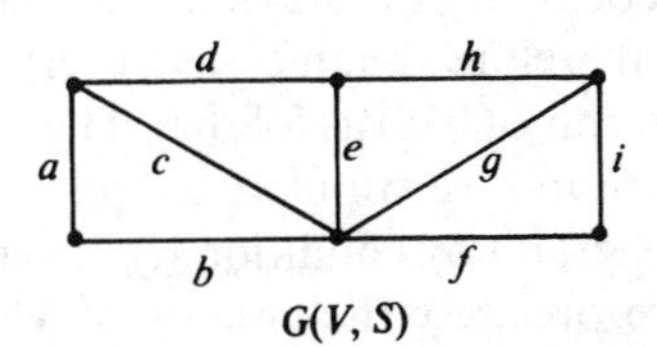


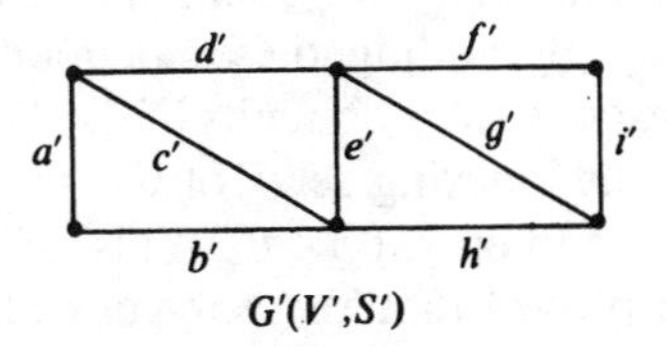


Figure 7.9

**7.40 Theorem** (Whitney). *Let $G(V, S)$ be a 3-connected graph without loops and at least 3 vertices and $G'(V', S')$ a graph without isolated vertices. Then $\mathbf{P}(G(V, S)) \cong \mathbf{P}(G'(V', S'))$ if and only if $G(V, S) \cong G'(V', S')$.*

*Proof.* By **6.87**, $G'(V', S')$ must be 2-connected which, by **6.24**(v), implies that $|S| = |S'|$ and $|V| = |V'|$. Let $\phi: S \to S'$ be an isomorphism from $\mathbf{P}(G(V, S))$ onto $\mathbf{P}(G'(V', S'))$. $\phi$ maps any (matroid-) connected copoint $H = S - \mathrm{St}(v)$ onto a (matroid-) connected copoint $H' = S' - \mathrm{St}(v')$. The map $\varphi: V \to V'$ defined by $\varphi(v) := v'$ is injective, and thus a bijection because of $|V| = |V'|$. It remains to be shown that $\varphi$ preserves vertex-edge incidences. We have, for $v \in V$ and $e \in S$,

$$v \notin e \Leftrightarrow e \notin \mathrm{St}(v) \Leftrightarrow e \in S - \mathrm{St}(v) \Leftrightarrow \phi e \in S' - \mathrm{St}(\varphi v) \Leftrightarrow \varphi(v) \notin \phi e. \quad \square$$

Exercise VI.4.5 contains a planar 2-connected graph which has two non-isomorphic dual graphs. As a corollary to **7.40** we can now assert by **6.74** that a planar 3-connected graph has a unique embedding (up to isomorphism) since it is easy to see that any dual graph of a 3-connected plane graph is 3-connected.

**7.41 Corollary.** *A 3-connected planar graph without loops can be embedded in only one way into the plane.* $\square$

The relationship between the vertices of a graph and its connected copoints suggests the following characterization of graphic matroids.

**7.42 Theorem.** *Let $\mathbf{M}(S)$ be a finite binary matroid. Then $\mathbf{M}(S)$ is graphic if and only if there exists a family $\mathfrak{C}_0$ of cocircuits in $\mathbf{M}(S)$ such that*

(i) *every $p \in S$ is contained in at most two cocircuits of $\mathfrak{C}_0$,*
(ii) *$\mathfrak{C}_0$ spans the cocycle group $\mathfrak{S}^\perp$.*

*Proof.* By considering the connected components of $\mathbf{M}(S)$ separately, we may assume that $\mathbf{M}(S)$ is connected and has rank at least 2. If $\mathbf{M}(S) \cong \mathbf{P}(G(V, S))$, then we set $\mathfrak{C}_0 = \{\mathrm{St}(v): v \in V\}$ and (i) and (ii) follow from **7.32**. Assume, conversely, the existence of a family $\mathfrak{C}_0 = \{C_\alpha: \alpha \in V\}$ satisfying (i) and (ii). Since by **2.40**(i) every element is in at least one cocircuit, condition (ii) insures that every $p \in S$ is contained in at least one member of $\mathfrak{C}_0$. The cocycle $C = \sum_{\alpha \in V} C_\alpha$ modulo 2 is the set of all elements of $S$ which are in precisely one cocircuit of $\mathfrak{C}_0$. $C$ is either the empty set or the union of disjoint cocircuits by the dual statement of **7.26**(ii). By adding these cocircuits we may therefore assume that every $p \in S$ is in exactly two cocircuits of $\mathfrak{C}_0$. We define a graph $G(V, S)$ as follows. The vertex-set is the indexing set $V$ of $\mathfrak{C}_0$; the edge-set is $S$ and we stipulate $\alpha \in p$ for $\alpha \in V$, $p \in S$ if and only if $p \in C_\alpha$. This gives a loopless graph by condition (i). It remains to be proved that the polygons of $G(V, S)$ are precisely the circuits of $\mathbf{M}(S)$. If $K$ is a circuit of $\mathbf{M}(S)$ then, by **6.67**(i), no cocircuit $C_\alpha$ has exactly one element in common with $K$. For the graph $G(V, S)$ this means that no vertex $\alpha \in V$ has degree

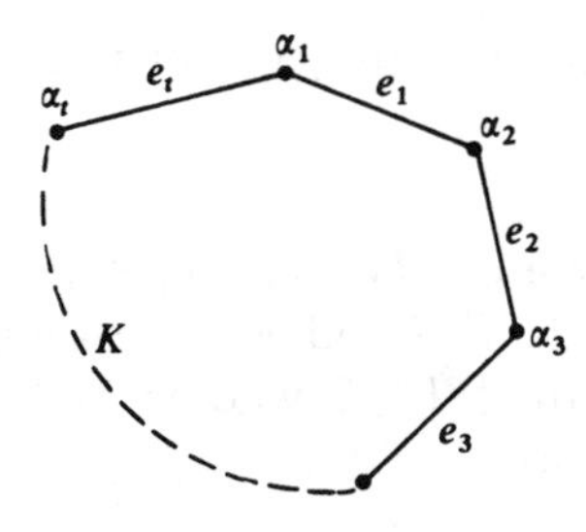


Figure 7.10

1 in the subgraph $G(V, K)$. Hence $K$ contains a polygon of $G(V, S)$ by **7.29**. Now let $K$ be a polygon of $G(V, S)$. We are going to show that $K$ is a union of circuits of $\mathbf{M}(S)$, thus proving the theorem. Suppose otherwise. Then, by **6.67**(ii), there is a cocircuit $C$ of $\mathbf{M}(S)$ with $|K \cap C| = 1$. We label the vertices and edges of $K$ with $\alpha_1, \ldots, \alpha_t$ and $e_1, \ldots, e_t$ as in Figure 7.10 and assume that $K \cap C = \{e_1\}$. Since $\mathfrak{C}_0$ spans $\mathfrak{S}^\perp$ we have $C = \sum_{\beta \in B} C_\beta$ modulo 2 for some $B \subseteq V$. By (i) and the definition of $G(V, S)$, the edge $e_1$ is contained in precisely $C_{\alpha_1}$ and $C_{\alpha_2}$ of $\mathfrak{C}_0$, whence we may assume $\alpha_1 \notin B$, $\alpha_2 \in B$. Again by (i), $e_2$ is contained in precisely $C_{\alpha_2}$ and $C_{\alpha_3}$, whence we deduce $\alpha_3 \in B$ because $e_2 \notin C$. Similarly, we deduce $\alpha_4 \in B, \ldots, \alpha_t \in B$ and finally $\alpha_1 \in B$; this contradicts the assumption $\alpha_1 \notin B$. □

As an application of our results we can give a new characterization of planar graphs due to Mac Lane. Since by **6.76** a graph is planar if and only if its bond matroid is graphic, we deduce that the dual statement of **7.42** characterizes planar graphs: A graph is planar if and only if there exists a family of polygons $\mathfrak{K}_0$ such that: (i) each edge lies in at most two polygons of $\mathfrak{K}_0$, (ii) $\mathfrak{K}_0$ spans the whole cycle group $\mathfrak{S}$. Of course, these polygons (when they exist) are just the boundaries of the faces of a plane representation (see Figure 7.11).

It is quite instructive, however, not to use **6.76** but to deduce Mac Lane's characterization directly from a general theory of embeddings. We shall confine

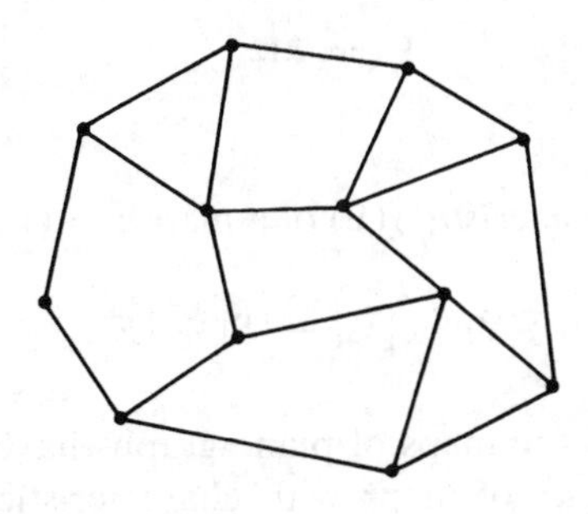

Figure 7.11

ourselves to 2-connected graphs; the general result will follow by looking at each block separately.

**Definition.** A *map* $\mathfrak{L}$ is a triple $\mathfrak{L} = (V, S, F)$ together with two incidence relations $I_1 \subseteq V \times S$ and $I_2 \subseteq S \times F$. $V$, $S$ and $F$ are assumed to be finite, non-empty and mutually disjoint sets with $|S| \geq 2$, whose members are called *vertices, edges*, and *faces*, respectively. We set

$$\partial_1 e := \{v \in V: vI_1e\} \qquad (e \in S)$$
$$\partial_2 f := \{e \in S: eI_2 f\} \qquad (f \in F)$$

and require

(i) $|\partial_1 e| = 2$ for all $e \in S$.

Hence the incidence $I_1$ induces a graph $G(V, S)$, called the *skeleton* $G(\mathfrak{L})$ of $\mathfrak{L}$. $G(\mathfrak{L})$ is required to be connected.

(ii) $\partial_2 f$ is a polygon in $G(\mathfrak{L})$ for all $f \in F$.

For $v \in V$ we define the stars $\mathrm{St}_1(v) := \{e \in S: vI_1e\}$, $\mathrm{St}_2(v) := \{f \in F: \mathrm{St}_1(v) \cap \partial_2 f \neq \emptyset\}$.

(iii) The graph $G(\mathrm{St}_1(v), \mathrm{St}_2(v))$ induced by $I_2$ is a polygon for all $v \in V$.

The axioms (ii) and (iii) say that every face is bounded by a polygon and that the faces adjacent to a vertex $v$ are cyclically arranged around $v$ (see Figure 7.12). If there is no danger of confusion we shall just write $\partial$ for $\partial_2$.

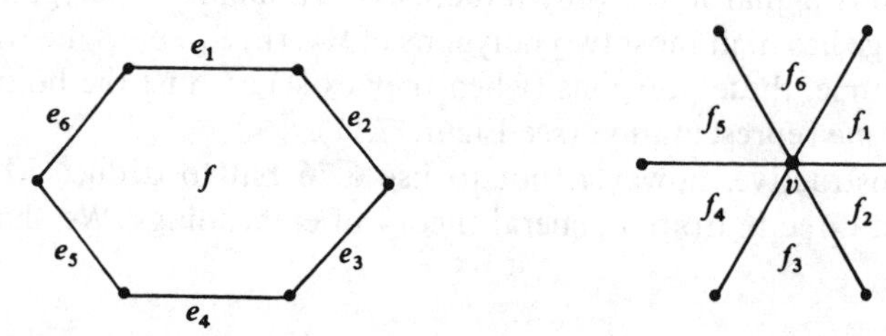


Figure 7.12

**Definition.** The *Euler characteristic* $\chi(\mathfrak{L})$ of a map $\mathfrak{L} = (V, S, F)$ is given by

$$\chi(\mathfrak{L}) := |V| - |S| + |F|.$$

We already know that the maps of plane graphs have characteristic 2 (**6.75**). The following are examples of maps with characteristic 1 and 0. We give the skeleton and the faces.

**Example.**

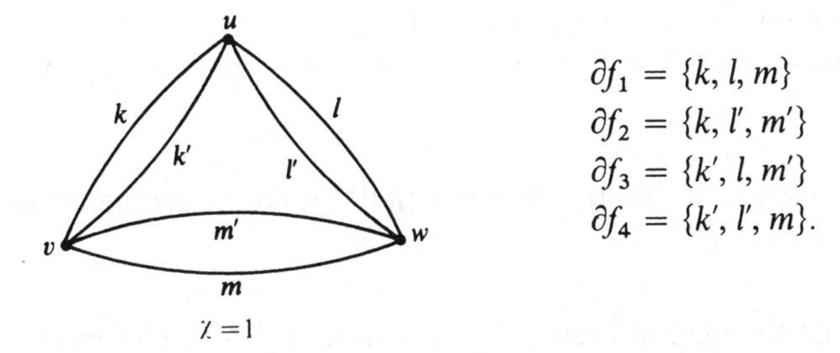


$$\partial f_1 = \{k, l, m\}$$
$$\partial f_2 = \{k, l', m'\}$$
$$\partial f_3 = \{k', l, m'\}$$
$$\partial f_4 = \{k', l', m\}.$$

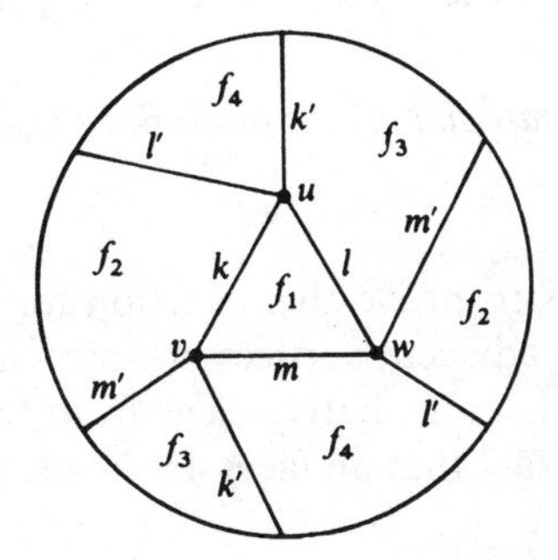


Figure 7.13

A realization of this map is possible on the real projective plane (see Figure 7.13). We draw a circular disk and identify opposite points of the boundary.

**Example.**

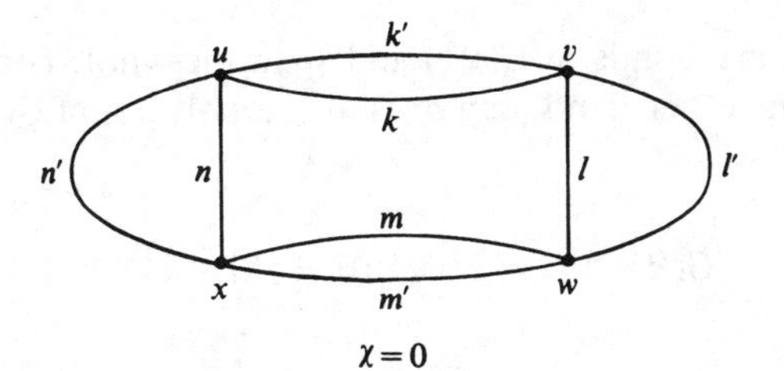


$$\partial f_1 = \{k, l, m, n\}$$
$$\partial f_2 = \{k, l', m, n'\}$$
$$\partial f_3 = \{k', l, m', n\}$$
$$\partial f_4 = \{k', l', m', n'\}.$$

This map may be realized on the torus, as seen in Figure 7.14.

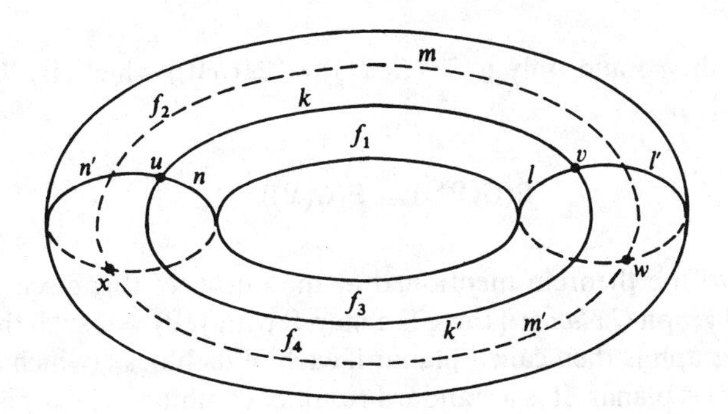


Figure 7.14

**7.43 Proposition.** *Let $\mathfrak{L} = (V, S, F), I_1, I_2$ be a map. Then $\mathfrak{L}^* = (F, S, V)$ with the inverse incidences $I_2^* \subseteq F \times S$ and $I_1^* \subseteq S \times V$ is also a map. $\mathfrak{L}^*$ is called the* dual map *of $\mathfrak{L}$.*

*Proof.* Only the proof that the skeleton $G(\mathfrak{L}^*)$ is connected needs some care; it is left to the reader. □

Notice that the map of Figure 7.14 is self-dual. The reader ought to determine the dual map of Figure 7.13. Clearly $\mathfrak{L}^{**} = \mathfrak{L}$ for any map $\mathfrak{L}$.

**7.44 Lemma.** *Let $\mathfrak{L}$ be a map and $\mathfrak{L}^*$ its dual. Both skeletons $G(\mathfrak{L})$ and $G(\mathfrak{L}^*)$ are 2-connected graphs.*

*Proof.* It obviously suffices to prove the assertion for $G(\mathfrak{L})$. For every $v \in V$ we have to show that any two adjacent vertices of $v$ are joined by a path not passing through $v$. This, however, is immediately clear from the cyclical arrangement of the faces around $v$ and the fact that all faces are bordered by polygons. □

The following proposition singles out planar maps.

**7.45 Proposition.** *Let $\chi(\mathfrak{L})$ be the characteristic of the map $\mathfrak{L}$. Then $\chi(\mathfrak{L}) \leq 2$ and $\chi(\mathfrak{L}) = 2$ if and only if $\mathbf{P}(G(\mathfrak{L})) \overset{id}{\cong} \mathbf{B}(G(\mathfrak{L}^*))$.*

*Proof.* The sets $\partial f$, $f \in F$, are bonds in $G(\mathfrak{L}^*)$ and span the whole cocycle group $\mathfrak{S}^\perp(G(\mathfrak{L}^*))$ by **7.32**. On the other hand, any $\partial f$ is also a polygon of $G(\mathfrak{L})$ whence, by **7.27**(i),

$$|F| - 1 = r(\mathfrak{S}^\perp(G(\mathfrak{L}^*))) \leq r(\mathfrak{S}(G(\mathfrak{L}))) = |S| - |V| + 1,$$

and thus

$$\chi(\mathfrak{L}) \leq 2.$$

We have equality if and only if $\mathfrak{S}^\perp(G(\mathfrak{L}^*)) = \mathfrak{S}(G(\mathfrak{L}))$ which, by **7.26**(iii), is the case if and only if

$$\mathbf{B}(G(\mathfrak{L}^*)) \overset{id}{\cong} \mathbf{P}(G(\mathfrak{L})). \quad \square$$

We come to the theorem mentioned at the outset of this discussion. We call a 2-connected graph $G$ *planar* if there is a map $\mathfrak{L}$ with $\chi(\mathfrak{L}) = 2$ such that $G \cong G(\mathfrak{L})$. An arbitrary graph is then called planar if each of its blocks (which are not just a loop or bridge) is planar. It is a standard result of combinatorial topology that this definition is equivalent to the one given in section VI.4.B.

**7.46 Theorem** (Mac Lane). *A graph $G(V, S)$ is planar if and only if there exists a family $\mathfrak{K}_0 = \{K_\alpha : \alpha \in A\}$ of polygons such that*

(i) *each edge lies on at most two polygons of $\mathfrak{K}_0$,*
(ii) *$\mathfrak{K}_0$ spans the cycle group $\mathfrak{S}$.*

*Proof.* By looking at each 2-connected block separately we may assume that $G(V, S)$ is 2-connected and has at least 2 edges. If $G(V, S)$ is planar, $G = G(\mathfrak{L})$, where $\mathfrak{L} = (V, S, F)$ is a map with $\chi(\mathfrak{L}) = 2$, then we choose $\mathfrak{K}_0 = \{\partial f : f \in F\}$. Condition (i) is immediately clear whereas (ii) follows from **7.45**. Conversely, let $\mathfrak{K}_0$ satisfy (i) and (ii). Just as in the proof of **7.42**, we may assume that every edge lies on precisely two polygons of $\mathfrak{K}_0$. We define the map $\mathfrak{L} = (V, S, \mathfrak{K}_0)$ by setting

$$(v, e) \in I_1 :\Leftrightarrow v \in e \text{ in } G(V, S) \quad \text{and} \quad (e, K_\alpha) \in I_2 :\Leftrightarrow e \in K_\alpha$$

and show that $\mathfrak{L}$ is a map with $\chi(\mathfrak{L}) = 2$. The defining conditions (i) and (ii) for a map are trivially true. By the hypothesis, the graph $G(\mathrm{St}_1(v), \mathrm{St}_2(v))$ is regular of degree 2. It remains to be shown that it is also connected. Let $e \neq e'$ in $\mathrm{St}_1(v)$. Since $G(V, S)$ is 2-connected, by **6.83**, there exists a polygon $K$ of $G(V, S)$ with $\mathrm{St}_1(v) \cap K = \{e, e'\}$. By condition (ii), $K = \sum_{\beta \in B} K_\beta$ modulo 2 for some $B \subseteq A$. If $e \in K_{\alpha_1}$ with $\alpha_1 \in B$ then $K_{\alpha_1} \in \mathrm{St}_2(v)$. We have $K_{\alpha_1} \cap \mathrm{St}_1(v) = \{e, e_1\}$. Hence $e_1 = e'$ or there is $\alpha_2 \in B$ with $e_1 \in K_{\alpha_2} \in \mathrm{St}_2(v)$, $K_{\alpha_2} \neq K_{\alpha_1}$. Continuing in this fashion we obtain a sequence $K_{\alpha_1}, \ldots, K_{\alpha_t} \in \mathrm{St}_2(v)$ with $\alpha_1, \ldots, \alpha_t \in B$ and $K_{\alpha_{i-1}} \cap K_{\alpha_i} \neq \varnothing$ for $i = 2, \ldots, t$ and $e' \in K_{\alpha_t}$. Hence $e'$ can be reached from $e$ in the graph $G(\mathrm{St}_1(v), \mathrm{St}_2(v))$, proving that $\mathfrak{L} = (V, S, \mathfrak{K}_0)$ is indeed a map. Since $\{\partial f : f \in \mathfrak{K}_0\}$ spans the cocycle group $\mathfrak{S}^\perp(G(\mathfrak{L}^*))$ in any map and since, by (ii), $\mathfrak{K}_0$ in this case also spans the cycle group $\mathfrak{S}(G(\mathfrak{L}))$, we infer that $\mathfrak{S}^\perp(G(\mathfrak{L}^*)) = \mathfrak{S}(G(\mathfrak{L}))$ and hence, by **7.27**(i),

$$|\mathfrak{K}_0| - 1 = |S| - |V| + 1,$$

i.e., $\chi(L) = 2$. □

## B. Homology and Networks

We pursue two goals in this section. First, we want to show that graphic matroids are regular, and secondly, we want to establish the connection between graphic matroids and electrical networks via Kirchhoff's laws. This connection was historically one of the starting points of matroid theory and can be used for an axiomatization of graphic matroids (see Minty [2]). We consider a finite undirected graph $G(V, S)$ and orient the edges in some fashion. The directed graph is then denoted by $\vec{G}(V, S)$. To avoid confusion, we agree that the words polygon, cycle, etc., always refer to polygon, cycle, etc., of the undirected graph. In $\vec{G}(V, S)$ we speak explicitly of a directed polygon, directed cycle, and so on. If $e = (u, v)$ is in $\vec{G}(V, S)$ directed from $u$ to $v$, then we set $u = e^-$ and $v = e^+$.

Let $W$ be an integral domain, for instance $\mathbb{Z}$ or a field.

**Definition.** A 0-*chain* on $\vec{G}(V, S)$ over $W$ is a mapping $g: V \to W$; a 1-*chain* is a mapping $f: S \to W$. $\mathfrak{K}_0(\vec{G}, W)$ and $\mathfrak{K}_1(\vec{G}, W)$ denote the $W$-modules of all 0-chains and 1-chains, respectively, with the usual addition and scalar multiplication.

We now define, for $v \in V$ and $e \in S$,

$$\eta(v, e) := \begin{cases} 1 & \text{if } v = e^+, v \neq e^- \\ -1 & \text{if } v = e^-, v \neq e^+ \\ 0 & \text{if } v = e^+ = e^- \text{ or } v \notin \{e^-, e^+\}. \end{cases}$$

The *boundary operator* $\partial$ is the mapping $\partial: \mathfrak{K}_1 \to \mathfrak{K}_0$ given by

$$(\partial f)(v) := \sum_{e \in S} \eta(v, e) f(e) = \sum_{e \in S,\, e^+ = v} f(e) - \sum_{e \in S,\, e^- = v} f(e) \qquad (f \in \mathfrak{K}_1).$$

The *coboundary operator* $\delta: \mathfrak{K}_0 \to \mathfrak{K}_1$ is defined by

$$(\delta g)(e) := \sum_{v \in V} \eta(v, e) g(v) = g(e^+) - g(e^-) \qquad (g \in \mathfrak{K}_0).$$

Consider the 1-chain $f$ over $\mathbb{Z}$ of the graph in Figure 7.15. The values of the boundary $\partial f$ are encircled. Check the obvious equality $\sum_{v \in V} (\partial f)(v) = 0$.

We now define two important submodules of $\mathfrak{K}_1(\vec{G}, W)$.

**Definition.** Let $\vec{G}(V, S)$ be a directed graph. A 1-chain $f$ is called a *cycle* on $\vec{G}$ over $W$ if $\partial f = 0$. The cycles clearly form a submodule of $\mathfrak{K}_1$, called the *cycle module* $\mathfrak{Z}(\vec{G}, W)$. A 1-chain $h$ is called a *coboundary* if $h = \delta g$ for some $g \in \mathfrak{K}_0$. The coboundaries also form a submodule of $\mathfrak{K}_1$, called the *coboundary module* $\mathfrak{C}(\vec{G}, W)$. A cycle (coboundary) $f$ is called *elementary* if $f \neq 0$ and $f$ has minimal support $\| f \|$ among all non-zero cycles (coboundaries).

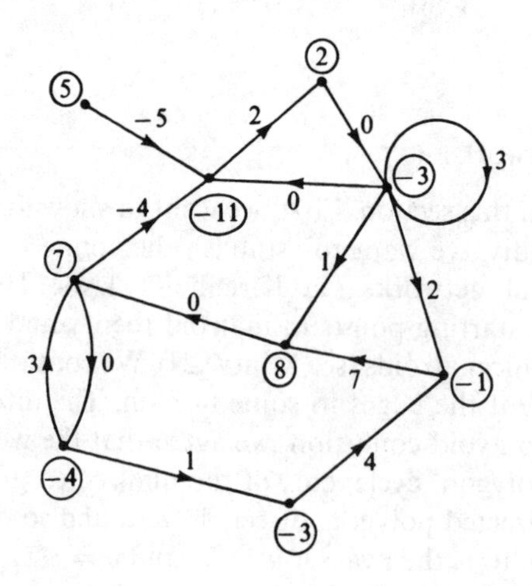


Figure 7.15

If $W$ is a field, then we often speak of the *cycle space* $\mathfrak{Z}(\vec{G}, W)$ and *coboundary space* $\mathfrak{C}(\vec{G}, W)$.

It is instructive to examine these definitions when $W = GF(2)$. In this case the orientation of $\vec{G}(V, S)$ is irrelevant because $1 = -1$ and $\eta(v, e)$ is just the vertex-edge incidence relation on $G(V, S)$ (excluding loops). Hence $f$ is a binary cycle if and only if $(\partial f)(v) = \sum_{e \in \mathrm{St}(v)} f(e) = 0$ over $GF(2)$, i.e., if and only if $\|f\|$, regarded as edge-set, has even degree at each vertex. Hence $f$ is a binary cycle if and only if $\|f\|$ is a cycle (=Eulerian) in $G(V, S)$ as defined in the previous section. Similarly, $h$ is a binary coboundary if and only if $\|h\|$ is a cocycle (=bipartition) in $G(V, S)$ in the sense of the previous section. In other words, $\mathfrak{C}(\vec{G}, GF(2))$ and $\mathfrak{Z}(\vec{G}, GF(2))$ yield the usual binary coordinatizations of the polygon matroid $\mathbf{P}(G(V, S))$ and the bond matroid $\mathbf{B}(G(V, S))$, respectively. It is this result which we now want to generalize to arbitrary fields.

**7.47 Theorem** (Tutte). *Let $K$ be a field, $G(V, S)$ a finite undirected graph and $\vec{G}$ an arbitrary orientation of $G$. Then we have the matroid isomorphisms*

(i) $\mathbf{P}(G(V, S)) \cong \mathbf{M}(\mathfrak{C}(\vec{G}, K))$
(ii) $\mathbf{B}(G(V, S)) \cong \mathbf{M}(\mathfrak{Z}(\vec{G}, K))$.

*Proof.* To prove (i) we have to show that the bonds of $G(V, S)$ coincide with the minimal non-empty supports $\|h\|$ with $h \in \mathfrak{C}(\vec{G}, K)$. Let $C$ be a bond with defining vertex-sets $V_1$ and $V_2$ (in the graph component which $C$ disconnects; see Figure 7.16). We define $g\colon V \to K$ by

$$g(v) := \begin{cases} 1 & \text{if } v \in V_1 \\ 0 & \text{otherwise.} \end{cases}$$

The coboundary $h = \delta g$ then clearly satisfies

$$h(e) = \begin{cases} \pm 1 & \text{if } e \in C \\ 0 & \text{otherwise.} \end{cases}$$

Hence $C = \|h\|$ and therefore $C$ contains a cocircuit of $\mathbf{M}(\mathfrak{C}(\vec{G}, K))$. Conversely, let $h = \delta g \in \mathfrak{C}(\vec{G}, K)$ have minimal non-empty support $\|h\| = D$. Then $(\delta g)(e) = 0$

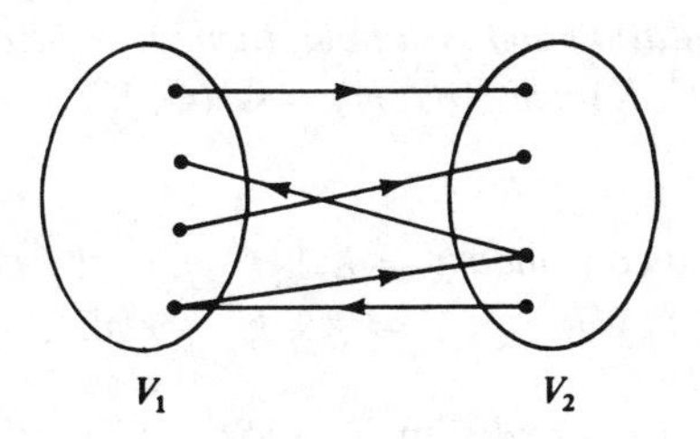


Figure 7.16

for all $e \in H = S - D$ so that the 0-chain $g$ is constant on the graph components of $G(V, H)$. It follows that $H$ is a flat in $\mathbf{P}(G(V, S))$ and hence that $D = S - H$ contains a bond of $G(V, S)$.

The proof of (ii) is just as easy. A polygon $C$ of $G(V, S)$ is the support of $f \in \mathfrak{Z}(\vec{G}, K)$ given by

$$f(e) := \begin{cases} 1 & \text{if } e \in C \text{ and } e \text{ is oriented clockwise} \\ -1 & \text{if } e \in C \text{ and } e \text{ is oriented counterclockwise} \\ 0 & \text{otherwise.} \end{cases}$$

Hence $C$ contains a cocircuit of $\mathbf{M}(\mathfrak{Z}(\vec{G}, K))$ (cf. Figure 7.17). Conversely, a cocircuit $D$ of $\mathbf{M}(\mathfrak{Z}(\vec{G}, K))$, regarded as subgraph of $G(V, S)$, has no vertex of degree 1 and hence contains a polygon. □

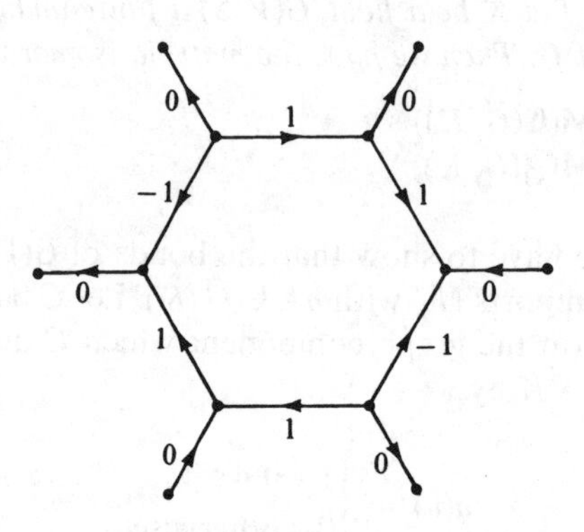


Figure 7.17

The importance of **7.47** rests on the fact that *any* orientation $\vec{G}$ yields a coordinatization of $G(V, S)$. We shall successfully apply this observation to coloring problems in section C. The following statements are immediate corollaries of **7.47**.

**7.48 Corollary.** *We have, for a graph $G(V, S)$:*

(i) $\mathbf{P}(G(V, S))$ *and* $\mathbf{B}(G(V, S))$ *are regular.*
*If $\vec{G}$ is any orientation and $K$ a field, then we have as function spaces*

(ii) $\mathfrak{C}(\vec{G}, K) = \mathfrak{Z}^{\perp}(\vec{G}, K)$ *and* $\mathfrak{Z}(\vec{G}, K) = \mathfrak{C}^{\perp}(\vec{G}, K)$.
*Hence*

$$\begin{aligned} &h \text{ is a coboundary} \Leftrightarrow h \perp f \quad \text{for all } f \in \mathfrak{Z}(\vec{G}, K), \\ &f \text{ is a cycle} \qquad\quad\ \Leftrightarrow f \perp h \quad \text{for all } h \in \mathfrak{C}(\vec{G}, K). \end{aligned}$$

(iii) $\dim \mathfrak{C} = |V| - k(G)$ *and* $\dim \mathfrak{Z} = |S| - |V| + k(G)$, *where $k(G)$ is the number of connected graph components of $G$.*

(iv) $f \in \mathfrak{Z}$ *is elementary* $\Leftrightarrow \|f\|$ *is a polygon in G, and every elementary cycle is a multiple of a* 0, $\pm 1$*-cycle*; $h \in \mathfrak{C}$ *is elementary* $\Leftrightarrow \|h\|$ *is a bond in G, and every elementary coboundary is a multiple of a* 0, $\pm 1$*-coboundary.* □

In the proof of **7.47**, for every bond and polygon $C$ we constructed a coboundary and cycle $f$, respectively, such that $\ker f = S - C$ and $f(p) = \pm 1$ for $p \in C$. This is precisely the orientation of the cocircuit and circuit matrices which make $\mathbf{P}(G(V, S))$ and $\mathbf{B}(G(V, S))$ into oriented matroids. Hence we could have proved the regularity of graphic matroids by directly appealing to **7.37**. A closer analysis of coboundaries and cycles is, however, useful for the applications to networks and colorings.

Next, we quote another famous result of Tutte's, characterizing graphic and cographic matroids within the class of regular matroids.

**7.49 Theorem** (Tutte). *Let* $\mathbf{M}(S)$ *be a finite matroid. Then*:

(i) $\mathbf{M}(S)$ *is graphic if and only if it is regular and contains no minor isomorphic to* $\mathbf{B}(K_5)$ *or* $\mathbf{B}(K_{3,3})$.
(ii) $\mathbf{M}(S)$ *is cographic if and only if it is regular and contains no minor isomorphic to* $\mathbf{P}(K_5)$ *or* $\mathbf{P}(K_{3,3})$. □

**7.49** together with **7.45** yields a proof of Kuratowski's characterization **6.72** of planar graphs. In addition, **7.49** gives the complete list of obstructions for graphic matroids as $\{\mathbf{U}_2(4), \mathbf{F}, \mathbf{F}^{\perp}, \mathbf{B}(K_5), \mathbf{B}(K_{3,3})\}$, dually for cographic matroids.

Let $\vec{G}$ be an orientation of the graph $G(V, S)$ and $K$ a field. For every $v \in V$ we define the 0-chain $\underline{v}: V \to K$ by

$$\underline{v}(x) := \begin{cases} 1 & \text{if } x = v \\ 0 & \text{if } x \neq v. \end{cases}$$

The 0-chains $\underline{v}$ clearly span $\mathfrak{K}_0$, whence it follows from the linearity of $\delta$ that $\{\delta\underline{v}: v \in V\}$ spans the coboundary space $\mathfrak{C}(\vec{G}, K)$. By the definition of $\delta$ we clearly have $\|\delta\underline{v}\| = \mathrm{St}(v)$; this yields a new proof of **7.32**. If we write the 1-chains $\delta\underline{v}$ as rows of a matrix $R = [a_{v,e}]$, then the columns of $R$ give us a coordinatization of $\mathbf{P}(G(V, S))$ over any field. Since

$$a_{v,e} := \begin{cases} 1 & \text{if } v = e^+, v \neq e^- \\ -1 & \text{if } v = e^-, v \neq e^+ \\ 0 & \text{otherwise,} \end{cases}$$

we obtain the following result, already announced in section VI.2.B (see Figure 6.9), where by the same argument as in the proof (v) ⇒ (i) of **7.35**, $R$ is easily seen to be unimodular.

**7.50 Proposition.** *Let $G(V, S)$ be a graph and $I = [c_{v,e}]$ its vertex-edge incidence matrix, i.e.,*

$$c_{v,e} := \begin{cases} 1 & \text{if } v \in e, e \text{ not a loop} \\ 0 & \text{otherwise.} \end{cases}$$

*In any column of $I$ not corresponding to a loop (where all entries are 0) we change arbitrarily one of the 1's to $-1$. The resulting matrix $R$ is then a coordinatization matrix of $\mathbf{P}(G(V, S))$ over any field $K$.* □

Now let $G(V, S)$ be a connected graph with $|V| = n + 1$, and hence $r(\mathbf{P}(G(V, S))) = n$. If we delete a row from the matrix $R$ above, we obtain a unimodular $n$-rowed coordinatization matrix $R'$ of $\mathbf{P}(G(V, S))$. Application of **7.36** to $R'$ yields the following result.

**7.51 Corollary.** *Let $G(V, S)$ be a connected graph without loops, $V = \{v_0, v_1, \ldots, v_n\}$. If $M = [m_{ij}]$, $1 \leq i, j \leq n$, denotes the symmetric $n \times n$-matrix defined by*

$$m_{ij} := \begin{cases} \gamma(v_i) & \text{if } v_i = v_j \\ (-1) . \#\{\text{edges between } v_i, v_j\} & \text{if } v_i \neq v_j, \end{cases}$$

*then*

$$\det M = \#\{\text{spanning trees in } G(V, S)\}. \quad \square$$

**Examples.** For the graph in Figure 7.18 we have

$$M = \begin{bmatrix} 2 & 0 & -1 & 0 & 0 & 0 \\ 0 & 2 & 0 & -1 & 0 & 0 \\ -1 & 0 & 5 & -1 & -2 & 0 \\ 0 & -1 & -1 & 4 & 0 & -1 \\ 0 & 0 & -2 & 0 & 3 & -1 \\ 0 & 0 & 0 & -1 & -1 & 2 \end{bmatrix}$$

$$\det M = 129.$$

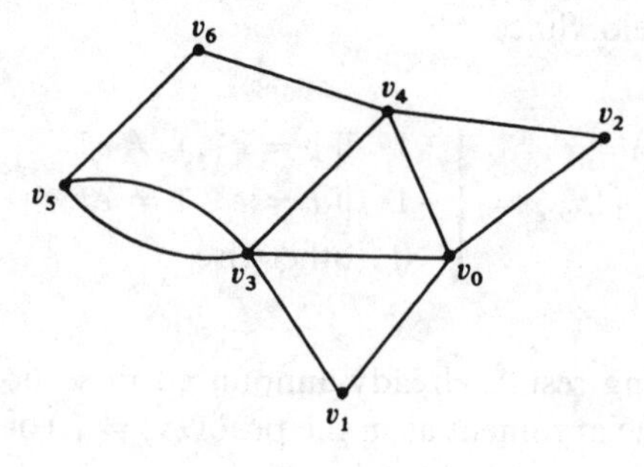


Figure 7.18

For the complete graph $K_n$, we obtain the $(n-1) \times (n-1)$-matrix

$$M = \begin{bmatrix} n-1 & -1 & & -1 \\ -1 & n-1 & & -1 \\ \vdots & -1 & \cdots & \vdots \\ -1 & -1 & & n-1 \end{bmatrix}.$$

By adding rows $2, \ldots, n-1$ to the first row and then subtracting the first column from all other columns we easily compute $\det M = n^{n-2}$ which is our old formula **5.12**.

Let us summarize our results.

(i) A *cycle* $f$ of a directed graph $\vec{G}(V, S)$ over $W$ is a mapping $f: S \to W$ such that for every $v \in V$ the total *flow* $\sum_{e^+ = v} f(e)$ *into* the vertex $v$ equals the total *flow* $\sum_{e^- = v} f(e)$ *out* from $v$ (see Figure 7.19). In the language of electrical networks, $f$ obeys *Kirchhoff's first law*.

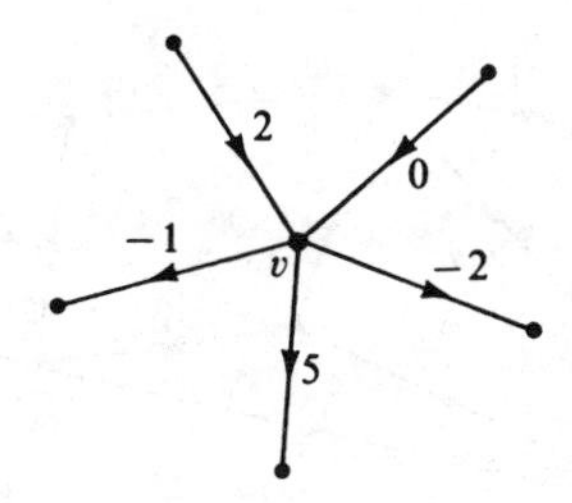


Figure 7.19

(ii) The *coboundaries* $h$ are, by **7.48**(ii, iv), precisely those mappings $h: S \to W$ which are orthogonal to all elementary cycles taking only the values $0, \pm 1$. Hence $h$ is a coboundary if and only if, for every directed polygon $C$, we have $\sum_{e \in C_1} h(e) = \sum_{e \in C_2} h(e)$, where $C = C_1 \dot\cup C_2$ is the partition of $C$ into clockwise and counterclockwise oriented edges (see Figure 7.20). In the language of electrical networks, $h$ obeys *Kirchhoff's second law*.

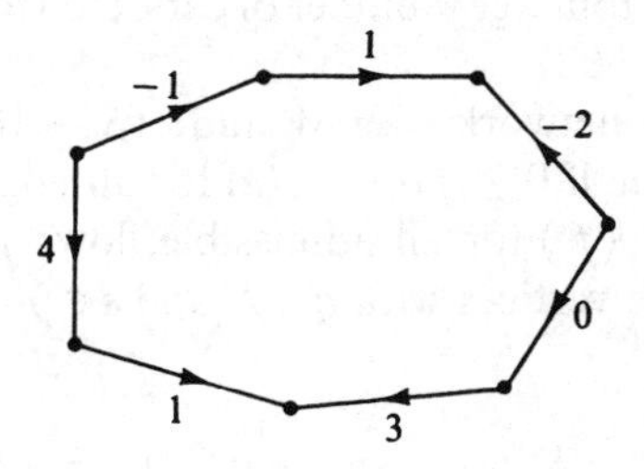


Figure 7.20

Applications of these ideas to the theory of electrical networks are described in the literature. (See, e.g., Duffin [3, 4], Minty [1, 2], or Ford–Fulkerson [1].) Here we shall discuss briefly only one important aspect—capacitated flows in networks.

**Definition.** Let $W$ be an ordered integral domain (e.g., $\mathbb{Z}$ or $\mathbb{R}$) and $W^+ := \{a \in W : a \geq 0\}$. A finite directed graph $\vec{G}(V, S)$ together with two specified vertices $q \neq s$ is called a *network* $\vec{G}(q, s)$ with *source* $q$ and *sink* $s$. A *flow in* $\vec{G}(q, s)$ *from* $q$ *to* $s$ is a 1-chain $f: S \to W$ such that

$$(\partial f)(v) = 0 \quad \text{for all } v \neq q, s.$$

The *value* $w(f)$ of the flow $f$ is given by $w(f) := (\partial f)(s)$. $f$ is called an *elementary flow* if the support $\|f\|$ is a path from $q$ to $s$ in $G(V, S)$. (Note: Not all edges in $\|f\|$ have to be directed from $q$ to $s$.)

For example the flow of the network in Figure 7.21 has value 3.

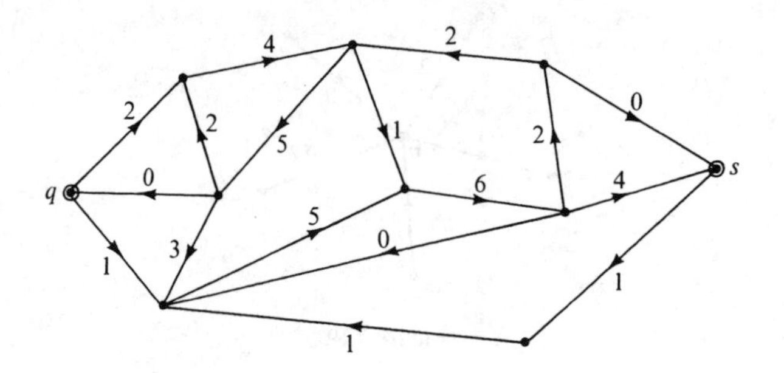


Figure 7.21

The main problem is the following. We are given a capacity function $c$ on the edges of the network. Determine a flow $f$ which has the maximal possible value among all flows $f$ with $0 \leq f(e) \leq c(e)$ for all edges $e$.

Think, for instance, of the network as a system of one-way streets and of the capacity as the maximum possible traffic on each street. We then seek an optimal traffic flow (e.g., in terms of tonnage, number of cars, etc.) for the total road system.

**Definition.** Let $\vec{G}(q, s)$ be a network over $W$ and $c: S \to W^+$ a capacity function. A flow $f$ is called *admissible* if $0 \leq f(e) \leq c(e)$ for all edges $e$. $f$ is *optimal* if $f$ is admissible and if $w(f) \geq w(f')$ for all admissible flows $f'$. A *cut* of $\vec{G}(q, s)$ is a partition $V = X \,\dot{\cup}\, Y$ of the vertices with $q \in X$ and $s \in Y$; the *capacity of the cut* $c(X, Y)$ is defined as

$$c(X, Y) := \sum_{e \in S} c(e) \quad \text{over all } e \in S \text{ with } e^- \in X,\ e^+ \in Y.$$

**7.52 Lemma.** *For each admissible flow $f$ and cut $(X, Y)$ we have*

$$w(f) \leq c(X, Y).$$

*Proof.* Set $S(A, B) := \{e \in S : e^- \in A, e^+ \in B\}$ for any partition $V = A \dot{\cup} B$. Since $(\partial f)(v) = 0$ for all $v \in V - \{q, s\}$ we infer that, for any admissible flow $f$ and any cut $(X, Y)$,

$$\begin{aligned} w(f) &= (\partial f)(s) = \sum_{y \in Y} (\partial f)(y) \\ &= \sum_{e \in S(X, Y)} f(e) - \sum_{e \in S(Y, X)} f(e) \leq \sum_{e \in S(X, Y)} f(e) \leq \sum_{e \in S(X, Y)} c(e) = c(X, Y). \quad \square \end{aligned}$$

According to **7.52** the values of admissible flows are bounded above by the capacities of the cuts. The *existence of* an optimal flow over $\mathbb{R}$ can be proved by standard continuity arguments. If $W = \mathbb{Z}$ (and we shall need only this case), there are only finitely many admissible flows, whence the existence of an optimal flow is trivially assured. The following theorem is the basic maximum-minimum theorem on flows and the starting point for our last topic—transversal theory.

How can we enlarge the value of an admissible flow $f$? We have to increase the outflow from $q$, keeping the net flow through each intermediate vertex equal to 0. This suggests the following definition. A path $A = \{q, v_1, v_2, \ldots, v_t = x\}$ is called an *augmenting path* from $q$ to $x$ if $f(e_i) < c(e_i)$ for each "forward" edge $e_i: v_{i-1} \to v_i$ and $0 < f(e_j)$ for each "backward" edge $e_j: v_j \to v_{j-1}$. If $f$ is any flow then we define the partition $V = X_f \dot{\cup} Y_f$ by

$$\begin{aligned} X_f &:= \{v \in V : v = q \text{ or there exists an augmenting path from } q \text{ to } v\}, \\ Y_f &:= V - X_f. \end{aligned}$$

**7.53 Theorem** (Ford–Fulkerson). *Let $\vec{G}(q, s)$ be a network with capacity $c: S \to W^+$. Then the following conditions are equivalent for an admissible flow $f$:*

(i) *$f$ is optimal.*
(ii) *There is no augmenting path from $q$ to $s$.*
(iii) *$(X_f, Y_f)$ is a cut.*

*If $f$ is optimal then $w(f) = c(X_f, Y_f)$ whence we have*

$$\max_{f \text{ admissible}} w(f) = \min_{(X, Y) \text{ cut}} c(X, Y).$$

*Proof.* (i) $\Rightarrow$ (ii). If $A$ is an augmenting path from $q$ to $s$, then we define an elementary flow $f_A: S \to W$ by

$$f_A(e) := \begin{cases} 1 & \text{if } e \in A \text{ and } e \text{ is a forward edge} \\ -1 & \text{if } e \in A \text{ and } e \text{ is a backward edge} \\ 0 & \text{if } e \notin A. \end{cases}$$

We set $\lambda_1 = \min(c(e) - f(e): e \in A$ forward edge), $\lambda_2 = \min(f(e): e \in A$ backward edge) and $\lambda = \min(\lambda_1, \lambda_2)$. Then $\lambda > 0$, and $f' = f + \lambda f_A$ is again an admissible flow with $w(f') = w(f) + \lambda > w(f)$.

(ii) $\Rightarrow$ (iii). Trivial.

(iii) $\Rightarrow$ (i). By the definition of $(X_f, Y_f)$ we have

$$\begin{aligned} f(e) &= c(e) \quad \text{for } e \in S(X_f, Y_f), \\ f(e) &= 0 \qquad \text{for } e \in S(Y_f, X_f), \end{aligned}$$

where $S(X_f, Y_f)$ and $S(Y_f, X_f)$ are defined as in the proof of **7.52**. Hence

$$w(f) = \sum_{e \in S(X_f, Y_f)} f(e) - \sum_{e \in S(Y_f, X_f)} f(e) = \sum_{e \in S(X_f, Y_f)} c(e) = c(X_f, Y_f).$$

$f$ is thus optimal by **7.52**, which at the same time proves the maximum-minimum formula. $\square$

Notice that in the case $W = \mathbb{Z}$, **7.53** provides an *algorithm* for finding an optimal flow. We start with the 0-flow and add elementary flows as long as augmenting paths exist. It is not true, however, that this algorithm must terminate after a finite number of steps over $\mathbb{R}$ although optimal flows over $\mathbb{R}$ always exist. Let $\vec{G}(q, s)$ be the capacitated network over $\mathbb{Z}$ shown in Figure 7.22. We start with the 0-flow and construct an optimal flow of value 6.

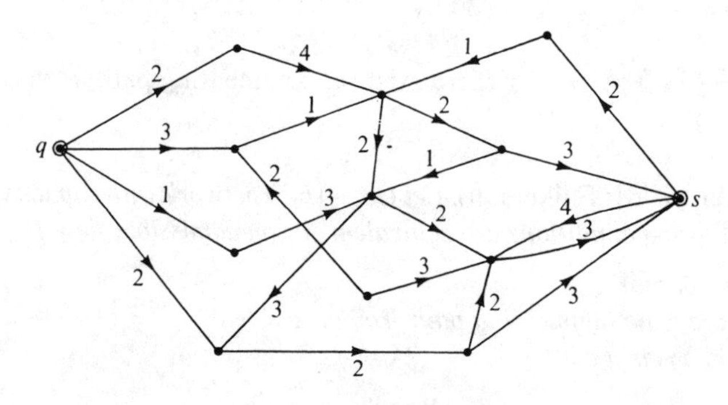


Figure 7.22

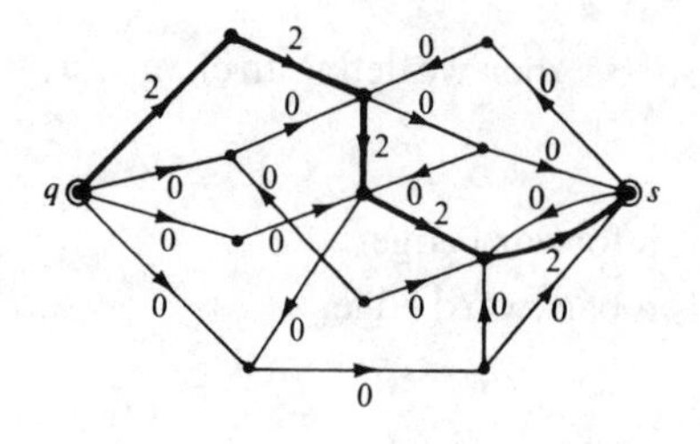


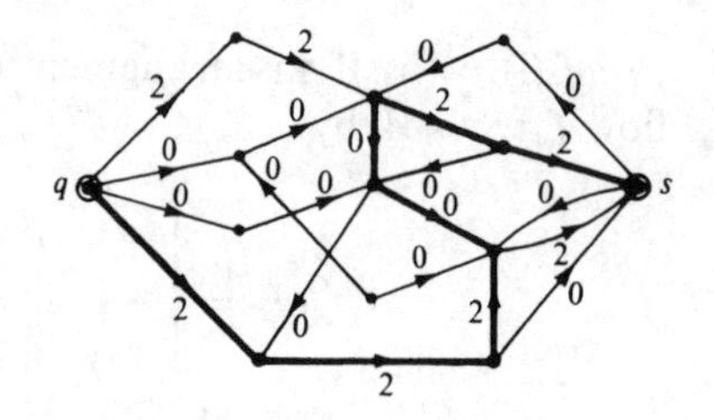

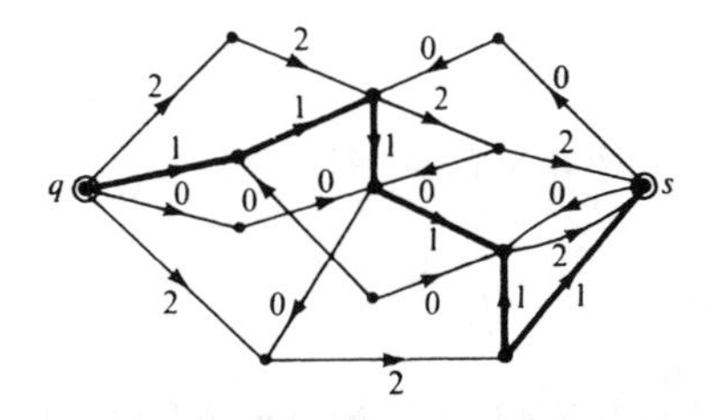


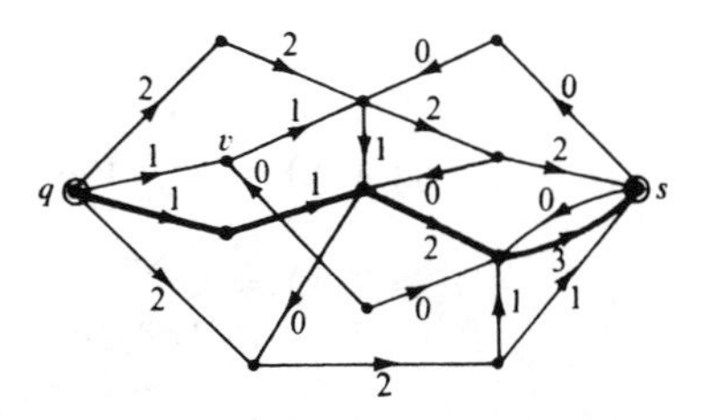


Figure 7.23

The elementary flows are marked heavily at each step of Figure 7.23. The last drawing shows the cut $\{q, v\} \dot{\cup} S - \{q, v\}$ corresponding to $f$.

## C. Colorings

We recall from **3.67** that a *coloring* of a graph $G(V, S)$ is a function $c: V \to F$ into a set of colors $F$ such that $c(u) \neq c(v)$ whenever $\{u, v\} \in S$. The *chromatic polynomial* $c(G; x)$ counts at $x = r \in \mathbb{N}$ the number of $r$-colorings, i.e., of colorings with $r$ or less colors. A coloring of the *plane map* $\mathfrak{L}$ is a coloring of the faces of $\mathfrak{L}$ such that any two adjacent faces receive different colors. In other words, the map colorings of $\mathfrak{L}$ correspond to the graph colorings of the dual graph $G^*$ where $G = G(\mathfrak{L})$. The coloring problem for plane maps is therefore equivalent to the coloring problem for planar graphs.

We want to study in this section some coloring results and conjectures from an algebraic point of view. We know that graphic and cographic matroids are regular and can thus be embedded in the vector space matroid $\mathbf{M}(V(n, q))$. The following two results give the connection between the chromatic polynomial and the chromatic number of a graph with invariants of the corresponding polygon matroid.

**7.54 Proposition.** *Let $G(V, S)$ be a graph with $k(G)$ graph components and chromatic polynomial $c(G; x)$, and let $L_G$ be the lattice of flats of $\mathbf{P}(G(V, S))$ with characteristic polynomial $\chi(L_G; x)$. If $G(V, S)$ contains loops, then $c(G; x) = 0$; otherwise*

$$c(G; x) = x^{k(G)}\chi(L_G; x).$$

*Proof.* If $G(V, S)$ has loops then no colorings are possible. Assume therefore $G$ to be loopless. To any mapping $f: V \to \{1, \ldots, x\}$, we assign the set $S(f) \subseteq S$ defined by

$$S(f) := \{\{u, v\} \in S: f(u) = f(v)\}.$$

It is immediate that $S(f)$ is a flat of $\mathbf{P}(G(V, S))$ for every $f$. Furthermore, we have

$$f \text{ is a coloring} \Leftrightarrow S(f) = \varnothing.$$

We define, for all $A, B \in L_G$,

$$f(A, x) := |\{f\colon V \to \{1, 2, \ldots, x\} \text{ such that } S(f) = A\}|,$$

$$g(B, x) := \sum_{A \geq B \in L_G} f(A, x).$$

The number $g(B, x)$ thus counts all mappings $f$ with $S(f) \geq B$. By the definition of $S(f)$, $f$ satisfies $S(f) \geq B$ if and only if $f$ is constant on each graph component of the subgraph $G(V, B)$. Hence

$$g(B, x) = x^{k(B)} = x^{|V| - r(B)},$$

and thus by Möbius inversion

$$f(B, x) = \sum_{A \geq B \in L_G} \mu_{L_G}(B, A) x^{|V| - r(A)}.$$

We conclude that

$$c(G; x) = f(\varnothing, x) = \sum_{A \in L_G} \mu_{L_G}(\varnothing, A) x^{|V| - r(A)} = x^{k(G)} \chi(L_G; x). \quad \square$$

**Example.** For a tree $G$ we have $L_G \cong \mathscr{B}(n - 1)$; hence $c(G; x) = x\chi(\mathscr{B}(n - 1); x) = x(x - 1)^{n-1}$.

**7.55 Theorem.** *Let $G(V, S)$ be a graph and $\vec{G}$ an arbitrary orientation of its edges. Let $q$ be a prime power and $l \in \mathbb{N}$. The following conditions are then equivalent:*

(i) *$G(V, S)$ is $q^l$-colorable.*
(ii) *There exists an $l$-tuple of coboundaries $(\delta c_1, \ldots, \delta c_l)$ in $\mathfrak{C}(\vec{G}, GF(q))$ such that*

$$((\delta c_1)(e), \ldots, (\delta c_l)(e)) \neq (0, \ldots, 0) \quad \text{for all } e \in S.$$

(iii) *The critical exponent* $\operatorname{crit}(\mathbf{P}(G(V, S)))$ *relative to coordinatization of* $\mathbf{P}(G(V, S))$ *over $GF(q)$ is at most $l$.*

*Proof.* (i) $\Leftrightarrow$ (ii). We choose as color set $F$ all $q^l$ $l$-tuples $(c_1, \ldots, c_l)$, $c_i \in GF(q)$. Any mapping $c\colon V \to F$ can thus be represented by the $l$-tuple of 0-chains $c_i\colon V \to GF(q)$ be setting

$$c(v) := (c_1(v), \ldots, c_l(v)) \qquad (v \in V).$$

Now

$$\begin{aligned} c \text{ is a coloring of } G(V, S) &\Leftrightarrow c(v) \neq c(w) \text{ for all } \{v, w\} \in S \\ &\Leftrightarrow \text{for all } e \in S \text{ there exists } i \text{ with } (\delta c_i)(e) \neq 0 \\ &\Leftrightarrow ((\delta c_1)(e), \ldots, (\delta c_l)(e)) \neq (0, \ldots, 0) \text{ for all } e \in S. \end{aligned}$$

(i) $\Leftrightarrow$ (iii). By **7.54** and **7.18**, we have: $G(V, S)$ is $q^l$-colorable $\Leftrightarrow$ $G(V, S)$ has no loops and $\chi(L_G; q^l) > 0 \Leftrightarrow \text{crit}(\mathbf{P}(G(V, S))) \leq l$. $\square$

**7.56 Corollary.** *Let c be the critical exponent of* $\mathbf{P}(G(V, S))$ *over* $GF(q)$. *Then*

$$q^{c-1} < \text{chrom } G \leq q^c,$$

*where* chrom $G$ *is the chromatic number of* $G$. $\square$

The equivalence of (i) and (ii) in **7.55** suggests the following definition.

**Definition.** Let $\mathbf{M}(S) = \mathbf{M}(F(S, GF(q)))$ be a $q$-linear matroid. $\mathbf{M}(S)$ is called *l-colorable over* $GF(q)$ if and only if the critical exponent of $\mathbf{M}(S)$ with respect to $GF(q)$ is at most $l$, i.e., if and only if there exist $c_1, \ldots, c_l \in F(S, GF(q))$ such that

$$(c_1(p), \ldots, c_l(p)) \neq (0, \ldots, 0) \quad \text{for all } p \in S.$$

By **7.17**, the existence of an $l$-coloring is independent of the function space used to coordinatize $\mathbf{M}(S)$. **7.55** opens the door to a wealth of coloring theorems. Let us look at the simplest case $q = 2$ and $l = 1$. Here the results are just restatements of **7.33**.

**7.57 Proposition.** (i) $\mathbf{P}(G(V, S))$ *is 1-colorable over* $GF(2) \Leftrightarrow G(V, S)$ *is 2-colorable* $\Leftrightarrow$ $G(V, S)$ *is bipartite* $\Leftrightarrow$ *all polygons of* $G(V, S)$ *have even length.*

(ii) $\mathbf{B}(G(V, S))$ *is 1-colorable over* $GF(2) \Leftrightarrow G(V, S)$ *is an Eulerian graph.*

(iii) *A plane map* $\mathfrak{L} = (V, S, F)$ *is 2-colorable* $\Leftrightarrow$ *the skeleton* $G(\mathfrak{L})$ *is an Eulerian graph.* $\square$

When $q = p$ is a prime, $p \neq 2$, we have to first orient $\vec{G}(V, S)$ to obtain results. By **7.55**, the existence of a coboundary $h$ with $h(e) \neq 0$ for all $e$ with respect to some orientation $\vec{G}$ implies the existence of such a coboundary relative to any orientation. Hence, for a given coloring problem, we may choose the most suitable orientation. As an exercise, the reader should establish conditions for $\mathbf{P}(G(V, S))$ to be 1-colorable over $GF(3)$.

One of the most famous problems in all of mathematics concerns the coloring of planar graphs. It had been conjectured since about 1840 that any loopless planar graph is 4-colorable. The planar graph $K_4$ shows that less than 4 colors will not suffice in general, and it is a well-known and easy theorem of graph theory that 5 colors will always suffice. It appears that after many unsuccessful attempts the 4-color conjecture has finally been settled in the affirmative by Appel and Haken [1]. (See Journal Graph Th. 1, No. 3 (1977).) In this context we have, by **7.55**, the following proposition ($q = 2, l = 2$).

**7.58 Proposition.** *The following conditions are equivalent for a graph $G(V, S)$.*

(i) *$G(V, S)$ is 4-colorable.*
(ii) *$\mathbf{P}(G(V, S))$ is 2-colorable over $GF(2)$.*
(iii) *There exists a pair of bipartite subgraphs $G(V, S_1)$, $G(V, S_2)$ with $S = S_1 \cup S_2$.* □

The following statements are therefore all equivalent formulations of the 4-color theorem.

**7.59 Theorem** (Appel–Haken).

(i) *Every loopless planar graph is 4-colorable.*
(ii) *Every bridgeless planar map is 4-colorable.*
(iii) *Every loopless planar matroid is 2-colorable over $GF(2)$.*
(iv) *Every loopless planar graph is the union of two bipartite graphs.*
(v) *Every bridgeless planar graph is the union of two Eulerian graphs.* □

The existence of parallel edges has plainly no influence on the colorability of a graph. Also, it is clear that every simple plane graph can be embedded in a *maximal plane* graph in which every face is bordered by exactly 3 edges, by adding new edges. Hence, if these maximal plane graphs can all be 4-colored, then so can all planar graphs. The dual of a maximal plane graph is regular of degree 3. Thus to prove the 4-color theorem it suffices to verify it for all *bridgeless 3-regular planar* graphs. This was historically one of the approaches to proving the 4-color conjecture. Let us therefore take a closer look at the class of 3-regular graphs in general.

**Definition.** An *edge-coloring* of the graph $G(V, S)$ is a mapping $\bar{c}: S \to F$ (set of colors), such that, for all $e \neq e' \in S$,

$$e \cap e' \neq \varnothing \Rightarrow \bar{c}(e) \neq \bar{c}(e').$$

If $|F| = s$ then $\bar{c}$ is an *s-coloring* of the edges.

An edge-coloring thus assigns different colors to edges with a common endpoint. Define the *edge-graph* $\bar{G}(\bar{V}, \bar{S})$ by taking as vertex-set $\bar{V} = S$ the set of edges in $G$, joining two edges in $\bar{G}$ if and only if they have a common endpoint in $G$. The edge-colorings of $G$ are then precisely the usual vertex-colorings of $\bar{G}$. (See Figure 7.24 for an example.)

**7.60 Proposition.** *Let $G(V, S)$ be a 3-regular graph (loops are counted twice). Then the bond matroid $\mathbf{B}(G(V, S))$ is 2-colorable over $GF(2)$ if and only if the edges of $G(V, S)$ are 3-colorable.*

*Proof.* If $G(V, S)$ possesses a loop then it also has a bridge and no colorings exist. Now suppose $\mathbf{B}(G(V, S))$ is 2-colorable over $GF(2)$. By **7.55**, there are

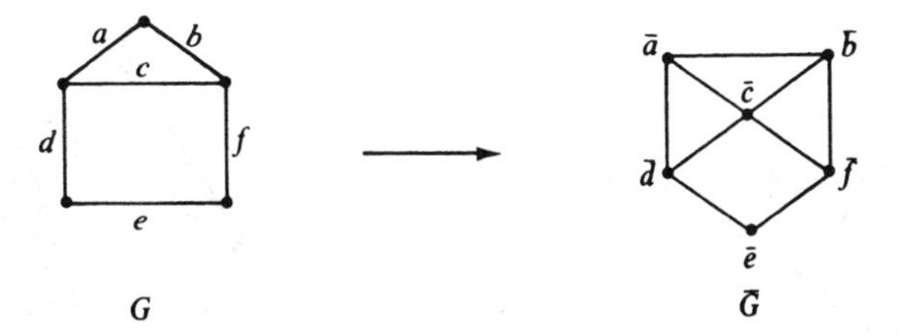


Figure 7.24

two Eulerian subgraphs $G(V, S_1)$ and $G(V, S_2)$ with $S = S_1 \cup S_2$. We define $\bar{c}: S \to \{(1, 0), (0, 1), (1, 1)\}$ by

$$\bar{c}(e) = \begin{cases} (1, 0) & \text{if } e \in S_1 - S_2 \\ (0, 1) & \text{if } e \in S_2 - S_1 \\ (1, 1) & \text{if } e \in S_1 \cap S_2. \end{cases}$$

Suppose $\bar{c}(e) = \bar{c}(e')$ for two adjacent edges $e = \{u, v\}$, $e' = \{u, w\}$. Since $S_1$ and $S_2$ are Eulerian edge-sets, they both contain either none or precisely two of the edges incident with $u$. It follows that the third edge beside $e$ and $e'$ could be neither in $S_1$ nor in $S_2$. The whole argument is reversible. □

As a corollary we obtain two well-known coloring theorems for plane graphs.

**7.61 Theorem** (Heawood–Tait). *Let $G(V, S)$ be a plane 3-regular graph. Then the following conditions are equivalent*:

(i) *The faces are 4-colorable.*
(ii) *The edges are 3-colorable.*
(iii) *There exists a function $h: V \to \{1, -1\}$ such that $\sum_{v \in C} h(v) \equiv 0 \pmod 3$ for every circuit $C$ of the graph.*

*Proof.* The equivalence of (i) and (ii) is a consequence of **7.58** and **7.60**. To prove (ii) ⇔ (iii) we consider the edge-graph $\bar{G}(\bar{V}, \bar{S})$. It is easily seen that $\bar{G}$ is again a planar graph and that $\bar{G}(\bar{V}, \bar{S})$ can be drawn in such a way that $\bar{G}$ has $|F| + |V|$ faces. $|F|$ faces of $\bar{G}$ correspond to the faces of $G$, whereas the other $|V|$ faces correspond to the triangles generated by the vertices of $G$. We call these the "vertex-faces" (see Figure 7.25, where the vertex-faces are striped). Note that any edge of $\bar{G}$ is adjacent to precisely one face of each type. Hence, if we orient the edges of all vertex-faces cyclically clockwise, then all faces of the first type are oriented cyclically counterclockwise with the exception of the exterior face whose edges are again clockwise oriented. Applying **7.55** with $q = 3$ and $l = 1$ we have: The edges of $G(V, S)$ are 3-colorable ⇔ $\bar{G}(\bar{V}, \bar{S})$ is 3-colorable ⇔ $\mathbf{P}(\bar{G}(\bar{V}, \bar{S}))$ is 1-colorable over $GF(3)$ ⇔ There exists $\bar{h}: \bar{S} \to \{1, -1\}$ such that $\sum_{\bar{e} \in \partial \bar{C}} \bar{h}(\bar{e}) \equiv 0 \pmod 3$ for all faces $\bar{C}$ in $\bar{G}(\bar{V}, \bar{S})$. $\bar{h}$ is plainly constant on the edges of a vertex-face $\bar{v}$.

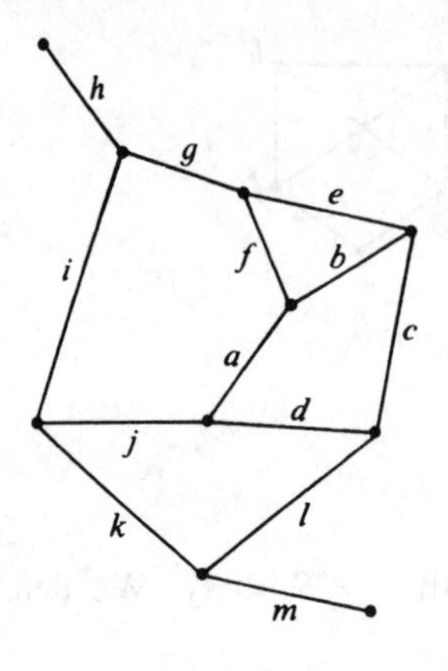


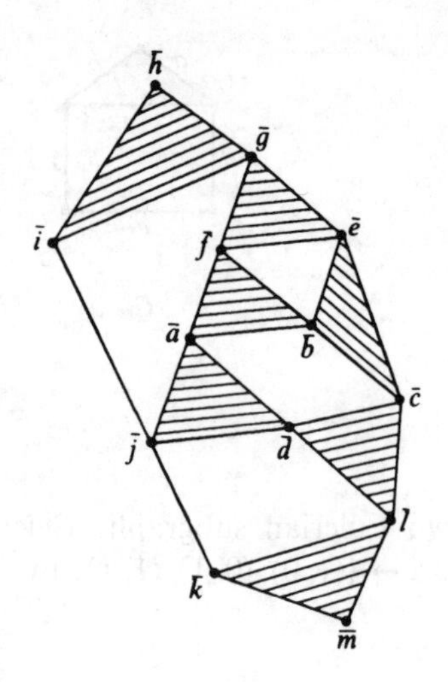


Figure 7.25

Hence we can define unambiguously $h: V \to \{1, -1\}$ by $h(v) := \bar{v}(\bar{e})$ for $\bar{e} \in \partial\bar{v}$. The function $h$ now has the property required in (iii). Since this last argument is clearly reversible, we are finished. □

By analogy to our previous characterizations, it is interesting to ask for obstructions relative to $k$-colorability. That is, we want to determine the minor-minimal loopless $q$-linear matroids which cannot be $k$-colored over $GF(q)$. Let us call these matroids *achromatic $k$-blocks over $GF(q)$*. By **7.55**, we have the following proposition.

**7.62 Proposition.** *Let $\mathbf{M}(S)$ be a $q$-linear matroid without loops. Then $\mathbf{M}(S)$ is not $k$-colorable over $GF(q)$ if and only if for any coordinatization $\phi: S \to V(n, q)$, $\phi(S)$ has a non-empty intersection with every $(n-k)$-dimensional subspace of $V(n, q)$.* □

An achromatic $k$-block $\mathbf{M}(S)$ over $GF(q)$ is clearly a geometry, whence we may regard $S$ as a set of points in the projective geometry $\mathbf{PG}(n-1, q)$, $n \geq r(S)$. The following theorem gives a characterization of those point sets in $\mathbf{PG}(n-1, q)$ which constitute achromatic $k$-blocks. Note that any matroid which is not $k$-colorable necessarily has rank at least $k+1$.

**7.63 Proposition** (Tutte). *Let $1 \leq k \leq n-1$. A set $S$ of points in $\mathbf{PG}(n-1, q)$ is an achromatic $k$-block over $GF(q)$ if and only if*

(i) *$S \cap W \neq \varnothing$ for every subspace $W$ of $\mathbf{PG}(n-1, q)$ with $r(W) = n-k$.*
(ii) *For every non-empty flat $U$ of the submatroid $\mathbf{M}(S)$ with $r(U) \leq n-k$, there exists a subspace $W$ of $\mathbf{PG}(n-1, q)$ with $r(W) = n-k$ and $W \cap S = U$.*

*Proof.* Let $S$ be an achromatic $k$-block. We have seen (i) in **7.62**. Let $U$ be a nonempty flat of $\mathbf{M}(S)$ and consider the contraction $\mathbf{M}(S)/U$. We know from **6.37** that $L(S/U) \subseteq [J(U), J(S)]$, where $J$ is the closure operator in $\mathbf{PG}(n-1, q)$. Hence, by the minimality of $\mathbf{M}(S)$, there must be an $(n-k)$-subspace $W$ of $\mathbf{PG}(n-1, q)$ with $W \cap J(S) = J(U)$; this implies $W \cap S = (W \cap J(S)) \cap S = J(U) \cap S = U$. Now assume (i) and (ii). By **7.62**, condition (i) ensures that $S$ is not $k$-colorable. Suppose $\mathbf{M}(S)$ possesses a proper minor $(\mathbf{M}(S)/U) \cdot A$ which is also not $k$-colorable. Then certainly $\mathbf{M}(S)/U$ is also not $k$-colorable. Hence it suffices to prove that all proper contractions $\mathbf{M}(S)/U$ and all proper restrictions $\mathbf{M}(S) \cdot A$ are $k$-colorable. The first is seen by reversing the argument above and by noticing that $\mathbf{M}(S)/U$ and $\mathbf{M}(S)/\bar{U}$ have the same lattice of flats. To prove the second claim, take any $p \in S - A$. By (ii), there is an $(n-k)$-subspace $W$ of $\mathbf{PG}(n-1, q)$ with $W \cap S = \{p\}$, and hence $W \cap A = \emptyset$. $A$ violates (i), whence $M(S) \cdot A$ can be $k$-colored. □

We specialize to the binary case. Clearly, $\mathbf{PG}(k, 2)$ is an achromatic $k$-block for every $k \geq 1$. (This is, of course, true for any $GF(q)$.) The following proposition shows that, for $k = 1$, there are no others.

**7.64 Proposition.** *The 3-point line* $\mathbf{PG}(1, 2)$ *is the only binary achromatic* 1*-block.*

*Proof.* By **7.26**, a binary matroid is 1-colorable over $GF(2)$ if and only if $S$ is a cocycle, which is the case if and only if $\mathbf{M}(S)$ has only even circuits (**7.27**(ii)). Hence if $\mathbf{M}(S)$ is not 1-colorable, then it contains some odd circuit and any such circuit is contractible to a circuit of length 3, i.e., to $\mathbf{PG}(1, 2)$. □

A complete list of achromatic 2-blocks would, of course, among other things, prove the 4-color theorem. Hence it is not surprising that the determination of the list of 2-blocks poses great difficulties. Let us quote two other conjectures in this connection. It is clear from **7.58** that $\mathbf{P}(K_5)$ is an achromatic 2-block over $GF(2)$.

**Hadwiger's conjecture** (for $n = 5$). $\mathbf{P}(K_5)$ *is the only graphic achromatic 2-block over* $GF(2)$.

In the language of graph theory the conjecture asserts that every loopless graph which is not 4-colorable possesses a minor isomorphic to $K_5$. By a result of Wagner [2] this conjecture is equivalent to the 4-color conjecture and can thus be considered true. The general conjecture reads: $G$ *not* $(n-1)$*-colorable* $\Rightarrow K_n$ *is a minor of* $G$.

There is an equally interesting cographic block. The *Petersen graph* $P$ is shown in Figure 7.26. It is easy to see that the edges of $P$ cannot be 3-colored; hence the bond matroid of $P$ is not 2-colorable over $GF(2)$ by **7.60**. $\mathbf{B}(P)$ can be shown to be an achromatic 2-block.

**Tutte's conjecture.** $\mathbf{B}(P)$ *is the only cographic achromatic 2-block over* $GF(2)$.

The Fano plane is, of course, another binary achromatic 2-block and it is known that there are no others beside $\mathbf{F}$, $\mathbf{P}(K_5)$, and $\mathbf{B}(P)$ up to rank 7.

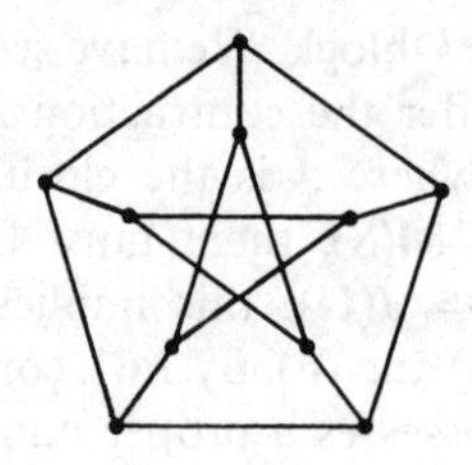

Figure 7.26. Petersen graph.

One last remark on 5-colorings. We observed earlier that every loopless planar graph can be 5-colored, or, what is the same, that every loopless planar matroid is 1-colorable over $GF(5)$. There are, of course, non-planar graphic matroids which are not 1-colorable over $GF(5)$ (for example, $\mathbf{P}(K_n)$ for $n \geq 6$), but another conjecture of Tutte's asserts that every loopless cographic matroid is 1-colorable over $GF(5)$. In terms of graphs: *For every bridgeless oriented graph $\vec{G}(V, S)$, there exists a function $f: S \to GF(5) - \{0\}$, called a* 5-flow, *such that, for all $v \in V$,*

$$\sum_{e^+ = v} f(e) - \sum_{e^- = v} f(e) \equiv 0 \pmod 5.$$

It is known that every loopless cographic matroid can be 1-colored over $GF(8)$. (Jaeger)

Exercises VII.3

1.* Let $G(V, S)$ be a finite undirected graph. We define the following two operations on $G$: (A) separate (or join) two subsets $V', V'' \subseteq V$ at a cut point; (B) Separate $V', V''$ at a vertex-cut set $\{u, v\}$ consisting of two vertices and rejoin $V', V''$ turning $V''$ upside down. (Cf. Figure 7.9 where the part to the right of the edge $e$ is turned upside down.) Any graph $G'$ that is obtained from $G$ by a sequence of operations (A) and (B) is called *2-isomorphic* to $G$. Prove: $\mathbf{P}(G) \cong \mathbf{P}(G') \Leftrightarrow G'$ is 2-isomorphic to $G$. (Whitney)

→ 2. Let $\mathbf{M}(S)$ be a finite simple connected matroid. Prove: $\mathbf{M}(S)$ is graphic if and only if there is a family $\mathfrak{H}_0$ of copoints such that

 (i) every $p \in S$ is not contained in precisely 2 members of $\mathfrak{H}_0$,
 (ii) $r(\inf \mathfrak{H}') \leq |\mathfrak{H}_0 - \mathfrak{H}'| - 1$ for all $\mathfrak{H}' \subsetneqq \mathfrak{H}_0$.

 (Sachs)

→ 3. Determine all maps with at most 5 edges and show that all of them have characteristic 2. (There are 12.)

4. Draw the dual to the map in Figure 7.13.

5. Prove **7.43**.

→ 6. Let $\vec{G}(V, S)$ be a finite directed graph. Prove:

(i) Every cycle $f \in \mathfrak{Z}(\vec{G}, W)$ is the sum of elementary cycles $f_i$ with $\| f_i \| \subseteq \| f \|$; every coboundary $h \in \mathfrak{C}(\vec{G}, W)$ is the sum of elementary coboundaries $h_i$ with $\|h_i\| \subseteq \|h\|$.

(ii) Every positive cycle (coboundary) $f$ of $\vec{G}(V, S)$ over $\mathbb{Z}$ (i.e., $f(e) \geq 0$ for all $e \in S$) is the sum of elementary positive cycles (coboundaries).

(iii) Decompose the cycle of the figure into elementary positive cycles according to (ii).

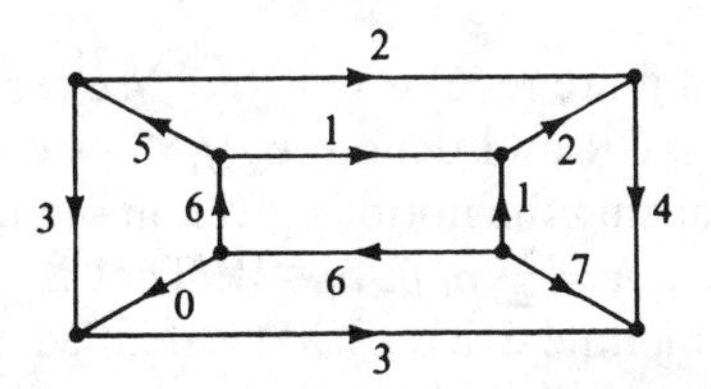


→ 7. A polygon $C$ in $\vec{G}(V, S)$ is said to be *consistently oriented* if $C = \| f \|$ for some positive cycle $f \in \mathfrak{Z}(\vec{G}, \mathbb{Z})$; a consistently oriented bond is defined similarly. Prove:

(i) For every edge $e \in S$, exactly one of two possibilities holds: Either $e$ is contained in a consistently oriented polygon or in a consistently oriented bond.

(ii) Let $u \neq v \in V$. Then there is a directed path from $u$ to $v$ or a consistently oriented bond $V = V_1 \cup V_2$ with $u \in V_1$, $v \in V_2$.

(iii) A connected graph $\vec{G}(V, S)$ is acyclic (i.e., $\vec{G}$ contains no consistently oriented polygon) if and only if every pair $u, v \in V$ is separated by a consistently oriented bond.

→ 8. Let $G(V, S)$ be a finite undirected graph and $S = S_R \cup S_B \cup \{e\}$ a partition of $S$ into red edges $S_R$, blue edges $S_B$, and a single green edge $e$. Prove: Exactly one of the following alternatives holds:

(i) There is a polygon $K$ with $e \in K \subseteq S_R \cup e$.

(ii) There is a bond C with $e \in C \subseteq S_B \cup e$.

9. Let $R$ be a rectangle with integral side lengths. We want to subdivide $R$ into squares with *distinct* integral side lengths. Example:

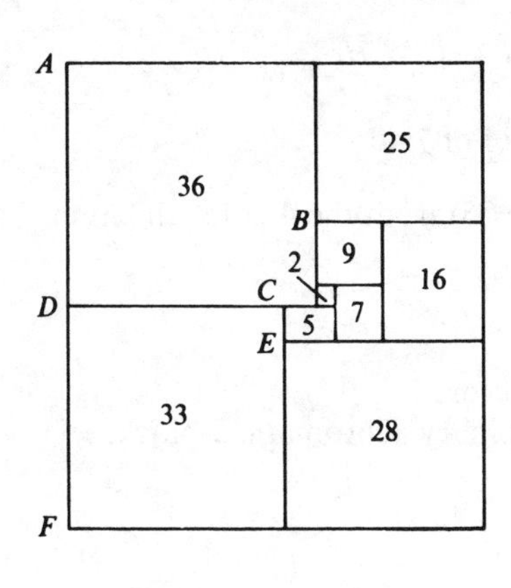


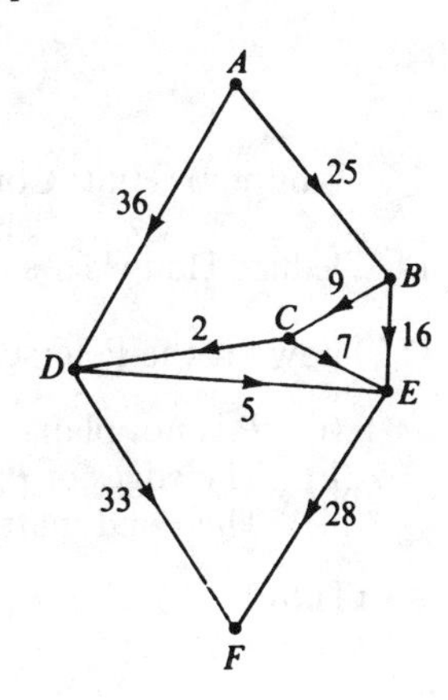

As in the figure, represent any horizontal line by a vertex, set $q$ = uppermost horizontal, $s$ = lowermost horizontal, and put an arrow $X \overset{w}{\rightarrow} Y$ if there is a square with $X$ as upper and $Y$ as lower horizontal and side length $w$. Prove:

(i) The function $w$ thus defined satisfies both Kirchhoff laws for $v \neq q, s$. Hence $\vec{G}(q, s)$ is, in particular, a network.
(ii) No rectangle can be divided into less than 9 squares and there is exactly one other rectangle apart from the one above which can be decomposed into 9 squares (minimal side lengths 32 and 33).

→10.* Let $\mathbf{M}(S)$ be a finite regular matroid. We represent $\mathbf{M}(S)$ as usual by a unimodular matrix $R$ and denote by $f_i: S \to \mathbb{Z}$ the function corresponding to row $i$. Generalizing the notions given in section B, call a mapping $h: S \to W$ a *coboundary* if $h = \sum a_i f_i$, $a_i \in W$. That is, $f_1, \ldots, f_r$ generate linearly the coboundary module $\mathfrak{C}(\mathbf{M})$. $\mathfrak{Z}(\mathbf{M})$ is defined analogously. Prove:

(i) The number of coboundaries $h \in \mathfrak{C}(\mathbf{M})$ over $\mathbb{Z}/k\mathbb{Z}$ with $\|h\| = S$ is $\chi(\mathbf{M}; k)$ where $\chi(\mathbf{M}; x)$ is the characteristic polynomial of ex. VI.3.1.
(ii) The number of cycles $g \in \mathfrak{Z}(\mathbf{M})$ over $\mathbb{Z}/k\mathbb{Z}$ with $\|g\| = S$ is $\chi(\mathbf{M}^{\perp}; k)$.

(Crapo) (Hint: Let $f(\mathbf{M})$ be the number of coboundaries $h$ with $\|h\| = S$. Show that $(-1)^{r(\mathbf{M})} f(\mathbf{M})$ is a chromatic invariant.)

11. Verify the equivalence of the following statements on a graph $G(V, S)$:

(i) $\mathbf{P}(G(V, S))$ is 1-colorable over $GF(3)$.
(ii) $G(V, S)$ is 3-colorable.
(iii) For any orientation $\vec{G}$ there is a function $h: S \to \{1, -1\}$ such that for every polygon $C$ we have $\sum_{e \in C_1} h(e) \equiv \sum_{e \in C_2} h(e) \pmod 3$ where $C = C_1 \dot\cup C_2$ is the partition of $C$ into its like-oriented parts.

→12. Let $\mathfrak{L} = (V, S, F)$ be a plane map with 3-regular skeleton $G(\mathfrak{L})$. Show that $\mathfrak{L}$ is 3-colorable if and only if every face is bounded by an even number of edges. (Kempe)

13. Let $G$ and $H$ be connected graphs and $\bar{G}$ and $\bar{H}$ their edge graphs. Prove that, with one exception,

$$G \cong H \Leftrightarrow \bar{G} \cong \bar{H}.$$

(Whitney) (Hint: Consider the proof of **7.40**.)

14.* Deduce Hadwiger's conjecture ($n = 5$) from the 4-color theorem.

15. Show for the Petersen graph $P$:

(i) $P$ is not planar.
(ii) The edges of $P$ cannot be 3-colored.
(iii)* The bond matroid $\mathbf{B}(P)$ is a binary achromatic 2-block.

(Tutte)

## 4. Transversal Matroids

The last class of matroids which we want to discuss in some detail is the class of transversal matroids. Since the family of transversal matroids is not closed with respect to minors or duals we cannot expect a simple excluded-minor characterization as for regular or graphic matroids. There is, however, an interesting geometric characterization using the copoints which we shall deduce first. After that, we study the smallest minor-closed class containing the transversal matroids, and finally, an interesting application to electrical networks.

### A. Characterization

We always interpret a transversal matroid as a matroid $\mathbf{T}(S; \mathfrak{A})$ induced by the partial transversals of the family $\mathfrak{A} = \{A_1, \ldots, A_n\}$ of subsets of the finite set $S$. Hence $B \subseteq S$ is independent if and only if $B$ is a system of distinct representatives of some subfamily of $\mathfrak{A}$. Whenever convenient, we shall write $\mathbf{T}(S; A_1, \ldots, A_n)$ for $\mathbf{T}(S; \mathfrak{A})$.

**7.65 Proposition.** *Let* $\mathbf{T}(S; A_1, \ldots, A_n)$ *be a transversal matroid and* $C \subseteq S$. *Then*

(i)
$$r(C) = \min_{B \subseteq C} (|\{j \in \mathbb{N}_n : |A_j \cap B| \neq \emptyset\}| + |C - B|)$$
$$= \min_{J \subseteq \mathbb{N}_n} \left( \left| C \cap \bigcup_{j \in J} A_j \right| + n - |J| \right).$$

(ii) *$C$ is a partial transversal* $\Leftrightarrow |B| \leq |\{j \in \mathbb{N}_n : A_j \cap B \neq \emptyset\}|$ *for all* $B \subseteq C$
$$\Leftrightarrow \left| C \cap \bigcup_{j \in J} A_j \right| \geq |C| - n + |J| \text{ for all } J \subseteq \mathbb{N}_n$$
$$\Leftrightarrow \left| C \cap \bigcap_{j \in J} A_j^c \right| \leq n - |J| \text{ for all } J \subseteq \mathbb{N}_n$$

*where* $A^c = S - A$ *for all* $A \subseteq S$.

*Proof.* For the binary relation $R \subseteq S \times \mathbb{N}_n$ defined by $(p, i) \in R :\Leftrightarrow p \in A_i$, we clearly have $R(B) = \{j \in \mathbb{N}_n : A_j \cap B \neq \emptyset\}$ and hence the first formula in (i) by **6.27**(iii). The rank of $C$ is also equal to the rank of the restriction

$$\mathbf{T}(C; A_1 \cap C, \ldots, A_n \cap C).$$

Now since for $J \subseteq \mathbb{N}_n$, $R(J) = \bigcup_{j \in J} (C \cap A_j)$ we obtain the second formula in (i) by another application of **6.27**(iii). The first two equivalences in (ii) are corollaries of (i); the third follows by taking complements. □

**Definition.** Let $\mathbf{M}(S)$ be a transversal matroid. Any family $\mathfrak{A}$ of subsets of $S$ with $\mathbf{M}(S) = \mathbf{T}(S; \mathfrak{A})$ is called a *representation* of $\mathbf{M}(S)$. Our first goal is to find a representation of a given transversal matroid which is as simple as possible.

**7.66 Proposition.** *Let* $\mathbf{M}(S) = \mathbf{T}(S; A_1, \ldots, A_n)$ *be a transversal matroid of rank* $r \leq n$ *and* $\mathfrak{B} = \{A_{i_1}, \ldots, A_{i_r}\}$ *a subfamily of* $\mathfrak{A}$ *which possesses a transversal. Then*

$$\mathbf{M}(S) = \mathbf{T}(S; A_{i_1}, \ldots, A_{i_r}).$$

*Proof.* Let $\mathbf{M}'(S) = \mathbf{T}(S; \mathfrak{B})$. Then $r(\mathbf{M}'(S)) = r$. Any basis of $\mathbf{M}'(S)$ is clearly independent in $\mathbf{M}(S)$. Now let $C$ be a circuit in $\mathbf{M}'(S)$. If $C$ is independent in $\mathbf{T}(S; A_1, \ldots, A_n)$ then there exist $p \in C$ and $A_j \notin \mathfrak{B}$ with $p \in A_j$. Hence if we enlarge $C - p$ to a basis $B$ of $\mathbf{M}'(S)$, i.e., to a transversal of $\mathfrak{B}$, then $B \cup p$ is a transversal of $\mathfrak{B} \cup A_j$ with $|\mathfrak{B} \cup A_j| = r + 1$, in contradiction to $r(\mathbf{M}) = r$. □

**7.66** says that we may confine ourselves to representations of a transversal matroid $\mathbf{M}$ which contain $n = r(\mathbf{M})$ sets. Next we want to find a minimal representation $\mathfrak{A} = \{A_1, \ldots, A_n\}$ of $\mathbf{M}(S)$, in the sense that $\mathbf{M}(S) \neq \mathbf{T}(S; \mathfrak{B})$ for all $\mathfrak{B} = \{B_1, \ldots, B_n\}$ with $B_i \subseteq A_i$ and $B_j \subsetneqq A_j$ for at least one $j$.

**7.67 Lemma.** *Let* $\mathbf{M}(S) = \mathbf{T}(S; A_1, \ldots, A_n)$. *Then* $A_i^c = S - A_i$ *is a flat of* $\mathbf{M}(S)$ *for all* $i$.

*Proof.* For $A_i = \varnothing$ there is nothing to prove. So assume $A_i \neq \varnothing$ and consider the restriction

$$\mathbf{M}(S) \cdot A_i^c = \mathbf{T}(A_i^c; A_1 - A_i, \ldots, A_n - A_i).$$

A basis $B$ of $\mathbf{M}(S) \cdot A_i^c$ is a partial transversal of $\{A_j - A_i : j \in \mathbb{N}_n, j \neq i\}$ and therefore $B \cup p$ a partial transversal of $\mathfrak{A}$ for any $p \in A_i$. This implies that $p \notin \overline{B} = \overline{A_i^c}$ for all $p \notin A_i^c$, and thus $\overline{A_i^c} = A_i^c$. □

**7.67** shows that the minimal possible sets $\neq \varnothing$ in any representation are cocircuits of the given transversal matroid. That, indeed, such a representation by cocircuits always exists is the content of the following theorem.

**7.68 Proposition** (Bondy–Welsh). *Let* $\mathbf{M}(S) = \mathbf{T}(S; A_1, \ldots, A_n)$ *be a transversal matroid with* $A_i \neq \varnothing$ *for all* $i$. *Then there are cocircuits* $C_1, \ldots, C_n$ *with* $C_i \subseteq A_i$ *for all* $i$ *such that* $\mathbf{M}(S) = \mathbf{T}(S; C_1, \ldots, C_n)$.

*Proof.* Let $Y$ be a basis of $A_1^c$ in $\mathbf{M}(S)$. We enlarge $Y$ to a basis $Z$ of $\mathbf{M}(S)$, where $Y \neq Z$ because of $A_1 \neq \varnothing$ and **7.67**. $Y$ is a maximal partial transversal of the family $\{A_2 - A_1, \ldots, A_n - A_1\}$ and we may choose $z \in Z \cap A_1$ such that $X = Z - z$ is a partial transversal of $\{A_2, \ldots, A_n\}$. We now set $C_1 = A_1 - X$ and claim that $C_1$ is a cocircuit of $\mathbf{M}(S)$ with

$$\mathbf{M}(S) = \mathbf{T}(S; C_1, A_2, \ldots, A_n).$$

We have $X \subseteq C_1^c$ and $B \nsubseteq C_1^c$ for any basis $B$ of $\mathbf{M}(S)$ with $B \supseteq X$; hence $r(C_1^c) = r(\mathbf{M}(S)) - 1$. Since the basis $X$ of $C_1^c$ is a partial transversal of $\{A_2, \ldots, A_n\}$, we infer from **7.66** that

$$\mathbf{M}(S) \cdot C_1^c = \mathbf{T}(C_1^c; C_1^c \cap A_1, \ldots, C_1^c \cap A_n) = \mathbf{T}(C_1^c; C_1^c \cap A_2, \ldots, C_1^c \cap A_n).$$

Let $\mathbf{M}'(S) = \mathbf{T}(S; C_1, A_2, \ldots, A_n)$. Any independent set in $\mathbf{M}'(S)$ is clearly independent in $\mathbf{M}(S)$. If, on the other hand, $D$ is dependent in $\mathbf{M}'(S)$, then, by **7.65**(ii), there exists $J \subseteq \mathbb{N}_n - \{1\}$ with

$$\left| D \cap C_1^c \cap \bigcap_{j \in J} A_j^c \right| > n - 1 - |J|.$$

Now

$$\begin{aligned} D \cap C_1^c \cap \bigcap_{j \in J} A_j^c &= (D \cap C_1^c) \cap \bigcap_{j \in J} (C_1 \cup A_j^c) \\ &= (D \cap C_1^c) \cap \bigcap_{j \in J} (C_1^c \cap A_j)^c. \end{aligned}$$

Applying **7.65**(ii) again we conclude that $D \cap C_1^c$ is dependent in $\mathbf{M}(S) \cdot C_1^c$ and hence that $D$ is dependent in $\mathbf{M}(S)$. $C_1$ is therefore a cocircuit by **7.67**, and we obtain the theorem by repeated application of this argument. □

The minimal representations yield an interesting characterization of transversal matroids by means of copoints, reminiscent of the characterization of graphic matroids in **7.42**.

**7.69 Theorem** (Ingleton). *Let $\mathbf{M}(S)$ be a finite matroid of rank $n$. Then $\mathbf{M}(S)$ is a transversal matroid if and only if there exist $n$ copoints $H_1, \ldots, H_n$ such that*

(i) $r(\bigcap_{j \in J} H_j) \leq n - |J|$ *for all* $J \subseteq \mathbb{N}_n$;
(ii) *For every circuit $C$ there exists $J \subseteq \mathbb{N}_n$ with $|J| = n - |C| + 1$ and $C \subseteq \bigcap_{j \in J} H_j$.*

*Proof.* If $\mathbf{M}(S)$ is a transversal matroid, then by **7.66** and **7.68** we can find cocircuits $C_1, \ldots, C_n$ with

$$\mathbf{M}(S) = \mathbf{T}(S; C_1, \ldots, C_n).$$

Set $H_i = C_i^c$, $i = 1, \ldots, n$, and suppose $B_J$ is a basis of $\bigcap_{j \in J} H_j$ for $J \subseteq \mathbb{N}_n$. Then, by **7.65**(ii),

$$r\left(\bigcap_{j \in J} H_j\right) = |B_J| = \left| B_J \cap \bigcap_{j \in J} C_j^c \right| \leq n - |J| \quad \text{for all } J \subseteq \mathbb{N}_n,$$

which is (i). If $C$ is a circuit of $\mathbf{M}(S)$ then, again by **7.65**(ii), there exists $J \subseteq \mathbb{N}_n$ with

$$\left|C \cap \bigcap_{j \in J} H_j\right| > n - |J|.$$

Suppose $C \nsubseteq \bigcap_{j \in J} H_j$. Then $C \cap \bigcap_{j \in J} H_j$ is independent, and thus

$$\left|C \cap \bigcap_{j \in J} H_j\right| = r\left(C \cap \bigcap_{j \in J} H_j\right) \le r(C) + r\left(\bigcap_{j \in J} H_j\right) - r\left(C \cup \bigcap_{j \in J} H_j\right)$$

$$\le |C| - 1 + n - |J| - |C| + 1 = n - |J|.$$

We conclude $C \subseteq \bigcap_{j \in J} H_j$ and, by **7.65**(ii),

$$|C - p| = \left|(C - p) \cap \bigcap_{j \in J} H_j\right| \le n - |J| \quad \text{for all } p \in S,$$

i.e.,

$$|C| = n - |J| + 1.$$

Assume, conversely, that the copoints satisfy the conditions of the theorem. We claim that

$$\mathbf{M}(S) = \mathbf{T}(S; C_1, \dots, C_n) \quad \text{with } C_i = H_i^c \text{ for } i = 1, \dots, n.$$

Let $B$ be independent in $\mathbf{M}(S)$. Then

$$\left|B \cap \bigcap_{j \in J} C_j^c\right| = \left|B \cap \bigcap_{j \in J} H_j\right| = r\left(B \cap \bigcap_{j \in J} H_j\right)$$

$$\le r\left(\bigcap_{j \in J} H_j\right) \le n - |J| \quad \text{for all } J \subseteq \mathbb{N}_n;$$

hence $B$ is a partial transversal of $\{C_1, \dots, C_n\}$ by **7.65**. If, on the other hand, $D$ is a circuit of $\mathbf{M}(S)$, then there exists $J \subseteq \mathbb{N}_n$ with $|J| = n - |D| + 1$ and $D \subseteq \bigcap_{j \in J} H_j$, whence

$$\left|D \cap \bigcap_{j \in J} C_j^c\right| = \left|D \cap \bigcap_{j \in J} H_j\right| = |D| = n - |J| + 1 > n - |J|.$$

It follows from **7.65** again that $D$ is not a partial transversal of $\{C_1, \dots, C_n\}$, and the proof is complete. □

**Example.** The polygon matroid $\mathbf{P}(K_4)$ is not transversal. If it were, then, by **7.69**(ii), any polygon $C$ of length 3 would have to be contained in some copoint $H_i$, i.e., $C = H_i$, since $C$ itself is a copoint. Since there are 4 polygons of length 3 but $r(\mathbf{P}(K_4)) = 3$, $\mathbf{P}(K_4)$ cannot be transversal.

It is clear that we can construct a whole class of non-transversal matroids by this method. On the other hand, we have the following result, whose proof is left to the reader.

**7.70 Proposition.**

(i) *Any finite matroid* $\mathbf{M}(S)$ *with* $|S| \leq 5$ *is transversal.*
(ii) *Any finite matroid* $\mathbf{M}(S)$ *with* $r(\mathbf{M}) \geq |S| - 2$ *is transversal.* □

$\mathbf{P}(K_4)$ shows that neither statement can be strengthened.

## B. Gammoids

Let us regard a transversal matroid as a matroid $\mathbf{T}(S, R, I)$ on the bipartite graph $G(S \cup I, R)$. A partial transversal $A \subseteq S$ can be defined as a set of vertices for which $|A|$ vertex-disjoint paths connecting $A$ with a subset of $I$ exist. Looked at in this way, the following generalization immediately comes to mind.

**Definition.** Let $\vec{G}$ be a finite directed graph with vertex-set $V$, and let $A, B \subseteq V$. We say $A$ is *linked* into $B$ if there exist $|A|$ vertex-disjoint directed paths in $\vec{G}$ leading from the vertices in $A$ into a subset of the vertices in $B$. (Note that $A$ and $B$ need not be disjoint.) If, in addition, $|A| = |B|$, then we say $A$ is *linked onto B.*

Our aim is to prove that, for fixed subsets $S, I \subseteq V$, the sets $A \subseteq S$ which are linked into $I$ induce, as family of independent sets, a matroid on $S$. The following idea is the key to the proof. Let $\vec{G}(V, E)$ be a finite directed graph. We associate with $\vec{G}$ a bipartite graph $G'(V \cup V', E')$ as follows. The defining vertex-sets of $G'$ are $V$ and $V'$, where $V'$ is a copy of $V$. We denote by $v' \in V'$ the vertex corresponding to $v \in V$ and, in general, by $B' \subseteq V'$ the set corresponding to $B \subseteq V$. The edge-set $E'$ of $G'$ is given by

$$E' = \{\{v, v'\} : v \in V\} \cup \{\{v, u'\} : (u, v) \in E\}.$$

See Figure 7.27 for an example, where the directed graph $\vec{G}$ appears in the left-hand side and the corresponding bipartite graph $\vec{G}'$ on the right hand side.

For the rest of this section, by a path in $\vec{G}$ we always mean a directed path.

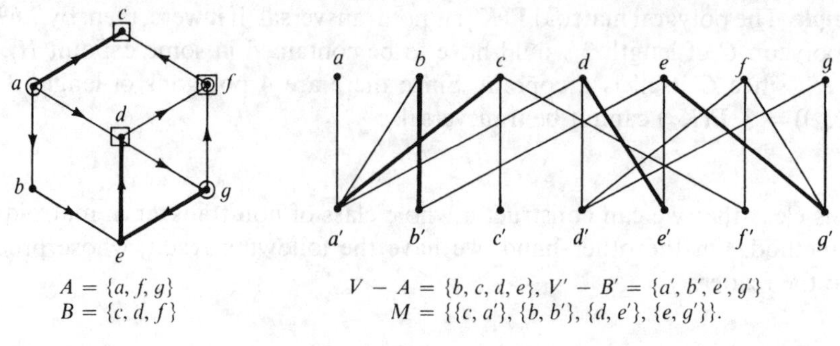


Figure 7.27

**7.71 Lemma** (Ingleton–Piff). *Let $\vec{G}(V, E)$ be a finite directed graph. Then $A$ is in $\vec{G}$ linked onto $B$ if and only if there exists a matching $M \subseteq E'$ in $G'(V \cup V', E')$ with* $\text{match}_V M = V - A$ *and* $\text{match}_{V'} M = V' - B'$.

*Proof.* Let $A$ be linked onto $B$ by the paths $P_i$. We define $\phi: V' - B' \to V - A$ by setting

$$\phi u' = \begin{cases} u & \text{if } u \text{ is not on any path } P_i \\ v & \text{if } u \text{ is on } P_i \text{ and } v \text{ is the successor of } u \text{ on } P_i. \end{cases}$$

$\Phi$ is clearly a bijection and we have

$$\{\phi u', u'\} \in E' \quad \text{for all } u' \in V' - B'.$$

$M = \{\{\phi u', u'\} : u' \in V' - B'\}$ is thus a matching with the required properties.

Conversely, let $\{\{\phi u', u'\} : u' \in V' - B'\}$ be a matching between $V - A$ and $V' - B'$. For $a \in A$ we define the following path $P_a$ in $\vec{G}$ from $a$ into $B$. If $a \in A \cap B$ then $P_a = \{a\}$. Otherwise we have $a' \in V' - B'$ with $\phi a' \in V - A$, and we take $a_1 = \phi a'$ to be the successor of $a$ in $P_a$. If $a_1 \in B$, we set $P_a = \{a, a_1\}$. If not, then $a_2 = \phi a'_1 \in V - A$ with $a_2 \neq a_1$ since $\phi$ is injective. We continue this process until we first reach a vertex $a_k \in B$ and then set $P_a = \{a, a_1, \ldots, a_k\}$. The injectivity of $\Phi$ now immediately implies that all paths $P_a$ are disjoint. □

In Figure 7.27, the three paths from $A$ into $B$, $P_1 = \{a, c\}$, $P_2 = \{f\}$, and $P_3 = \{g, e, d\}$, induce the matching as drawn in the graph on the right-hand side.

**7.72 Theorem** (Ingleton–Mason–Piff). *Let $\vec{G}(V, E)$ be a finite directed graph, $I \subseteq V$.*

(i) *$\mathfrak{B} = \{A \subseteq V: A$ linked onto $I\}$ is the family of bases of a matroid $\mathbf{L}(V, \vec{G}, I)$ on $V$, called the* strict gammoid on $\vec{G}$ with respect to $I$.

(ii) *A matroid is a strict gammoid if and only if its dual matroid is transversal.*

*Proof*. First we show that $\mathfrak{B}^{\perp} = \{V - A: A \text{ linked onto } I\}$ is the family of bases of a transversal matroid thereby proving (i) and half of (ii). By **7.71** we have:

> $A \in \mathfrak{B} \Leftrightarrow A$ is linked in $\vec{G}$ onto $I \Leftrightarrow$ there is a matching between $V - A$ and $V' - I'$ in $G' \Leftrightarrow V - A$ is a basis of the transversal matroid on $G'$ induced by the vertex-sets $V$ and $V' - I'$.

Conversely, let $\mathbf{M}(V) = \mathbf{T}(V, R, F)$ be the transversal matroid induced by the bipartite graph $G(V \cup F, R)$, where, by **7.66**, we may assume $|F| = r(\mathbf{M}(V))$. If $V - I$ is a basis of $\mathbf{M}(V)$, then there exists a bijection $\psi: V - I \to F$ with $(a, \psi(a)) \in R$ for all $a \in V - I$. We set $v' := \psi(v)$ for $v \in V - I$ and define the directed graph $\vec{G}(V, E)$ on $V$ by

$$E: = \{(u, v): u \neq v, (v, u') \in R\}.$$

$V - A$ is then a basis of $\mathbf{M}(V)$ if and only if there exists a bijection $\phi: F \to V - A$ with $(\phi v', v') \in R$ for all $v' \in F$. Using this fact, we construct, as in the proof of **7.71**, a linking of $A$ onto $I$ in $\vec{G}$; this proves that $A$ is a basis of the strict gammoid $\mathbf{L}(V, \vec{G}, I)$. The converse, that for any basis $A$ of $\mathbf{L}(V, \vec{G}, I)$, $V - A$ is a basis of $\mathbf{M}(V)$, follows as in **7.71**. Hence we obtain $\mathbf{M}^{\perp}(V) \cong \mathbf{L}(V, \vec{G}, I)$, which is what we wanted to prove. □

**Definition.** A matroid is called a *gammoid* if it is a restriction of a strict gammoid.

In summary, $\mathbf{M}(S)$ is a gammoid if and only if there exist a finite directed graph $\vec{G}(V, E)$ and sets $S, I \subseteq V$ such that $A \subseteq S$ is independent if and only if $A$ is linked into $I$. We denote this gammoid by $\mathbf{L}(S, \vec{G}, I)$.

**Examples.** Any transversal matroid is a gammoid, but not conversely. For example, the non-transversal matroid of Figure 6.12 is the strict gammoid induced by the graph in Figure 7.28, where $I$ are the two vertices on the right.

The smallest non-gammoid is $\mathbf{P}(K_4)$. To prove this, consider the basis $\{1, 2, 3\}$ in Figure 7.29. We know from the proof of **7.72** that any strict gammoid $\mathbf{M}(V)$

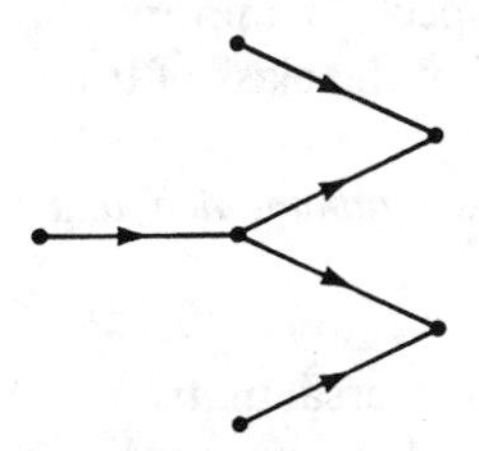

Figure 7.28

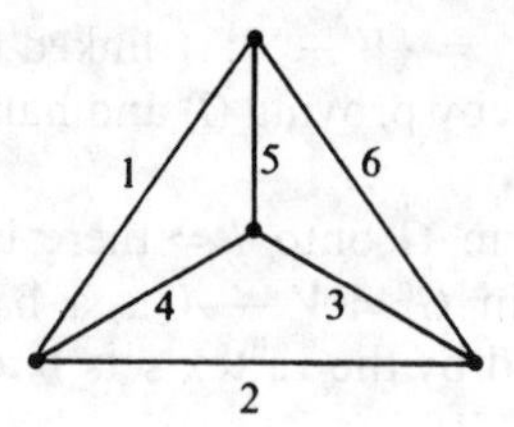


Figure 7.29

with basis $I$ may be realized as a matroid $\mathbf{L}(V, \vec{G}, I)$ for some directed graph $\vec{G}$ on $V$. Hence, if $\mathbf{P}(K_4)$ is a gammoid, then we may extend the basis $\{1, 2, 3\}$ of $\mathbf{P}(K_4)$ to the basis $I$ of $\mathbf{M}(V)$ and realize $\mathbf{P}(K_4)$ as the matroid $\mathbf{L}(S, \vec{G}, I)$, where $S = \{1, \ldots, 6\}, I \supseteq \{1, 2, 3\}$. Since $\{1, 2, 4\}, \{1, 3, 4\}$ are independent and $\{2, 3, 4\}$ is dependent, 4 is joined in $\vec{G}$ by a path $P$ to 2 and by a path $P'$ to 3, but to no other vertex of $I$. Similarly, 6 is joined by a path $Q$ to 1 and by a path $Q'$ to 2, but to no other vertex of $I$. $\{1, 2, 3, 5\}$ is a circuit; hence 5 is joined by paths $A$, $B$ and $C$ to 1, 2 and 3, respectively (see Figure 7.30). Since $\{1, 4, 5\}$ is a circuit we must have $P \cap C \neq \varnothing$ for every 4,2-path $P$ and 5,3-path $C$, and similarly $P' \cap B \neq \varnothing$ for every 4,3-path $P'$ and every 5,2-path $B$. By the same argument, we have $Q \cap B \neq \varnothing$ and $Q' \cap A \neq \varnothing$, since $\{3, 5, 6\}$ is a circuit. It follows that $\{4, 5, 6\}$ cannot be linked into $\{1, 2, 3\}$, contradicting the fact that $\{4, 5, 6\}$ is a basis of $\mathbf{P}(K_4)$.

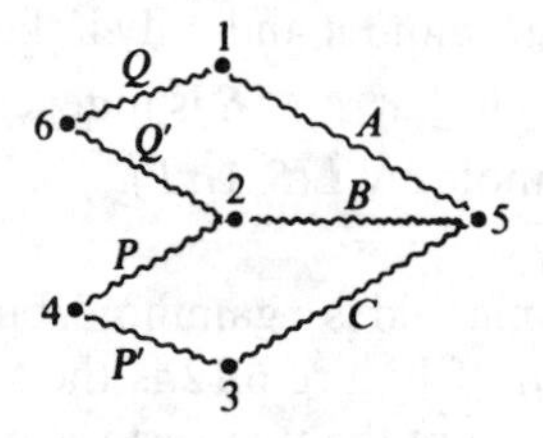


Figure 7.30

We already know that neither the contraction nor the dual of a transversal matroid need be transversal. The following theorems show that the class of gammoids is closed with respect to minors and duals and that it constitutes precisely the "minor-closure" of the class of transversal matroids.

**7.73 Proposition.** *A matroid is a gammoid if and only if it is the contraction of a transversal matroid.*

*Proof.* The contraction of a transversal matroid is, by **6.68**, the dual of a restriction of a cotransversal matroid, i.e., of a gammoid, by **7.72**(ii). The converse is similarly clear. □

**7.74 Corollary.** *The class of gammoids is closed with respect to minors and duals.* □

Applying **6.39**, **6.70** and **7.7** we have the following corollary of **7.73**.

**7.75 Corollary.** *A gammoid is coordinatizable over any sufficiently large finite field and, in particular, over any infinite field.* □

Just as we extended the construction of transversal matroids to bipartite graphs endowed with a matroid structure **(6.29)**, we may generalize **7.72** to the following theorem (Mason): Let $\vec{G}(V, E)$ be a finite directed graph and let $\mathbf{M}(V)$ be a matroid on $V$. The family $\mathfrak{I} = \{A \subseteq V: A$ is linked onto some independent set of $\mathbf{M}(V)\}$ is then the family of independent sets of a matroid.

## C. Series-parallel Networks

To conclude our matroid study of graphs and transversal systems, we briefly discuss an interesting class of graphic matroids arising from networks. Consider a finite undirected graph $G$ with vertex-set $V$ and edge-set $S$. The graph $sG(e)$, obtained from $G$ by subdividing the edge $e$ into two edges, is called the *series extension of G at e*. The graph $pG(e)$, obtained by adding an edge parallel to $e$, is called the *parallel extension of G at e*.

**Definition.** A *series-parallel network* is any graph which can be obtained from a bridge or a loop by a finite sequence of series and parallel extensions.

Figure 7.31 shows all series-parallel networks up to 3 edges. The concept of a series-parallel extension can be readily generalized to matroids.

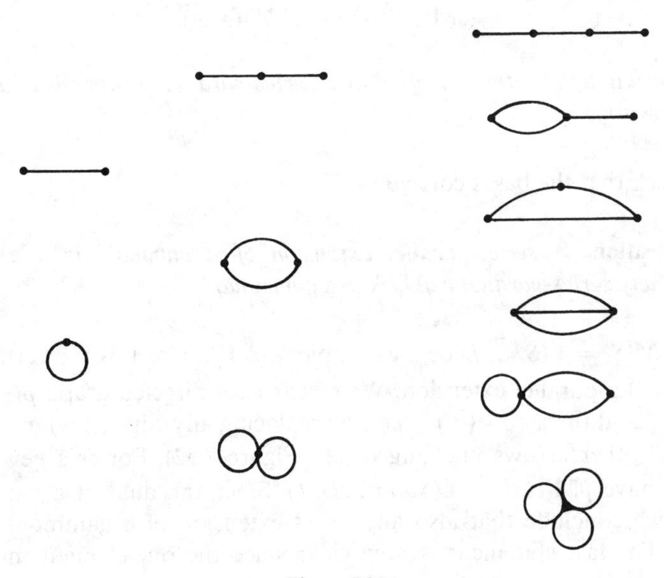

Figure 7.31

**Definition.** Let $\mathbf{M}(S)$ be a finite matroid, $e \in S$. The *series extension of* $\mathbf{M}$ *at e by* $e' \notin S$ is the matroid $s\mathbf{M}(e, e')$ on $S \cup e'$ which has as its bases all sets of the form

(i) $B \cup e'$, $B$ basis of $\mathbf{M}$,
(ii) $B \cup e$, $B$ basis of $\mathbf{M}$, $e \notin B$.

The *parallel extension of* $\mathbf{M}$ *at e by* $e' \notin S$ is the matroid $p\mathbf{M}(e, e')$ on $S \cup e'$ with bases

(i) $B$, $B$ basis of $\mathbf{M}$,
(ii) $(B - e) \cup e'$, $B$ basis of $\mathbf{M}$ with $e \in B$.

A *series-parallel extension* of a matroid $\mathbf{M}$ is any matroid which can be derived from $\mathbf{M}$ by a finite sequence of series and parallel extensions.

It is easy to see that $s\mathbf{M}$ and $p\mathbf{M}$ are indeed matroids and that for polygon matroids we obtain the former notions. As with graphs, we say that $\mathbf{M}$ is a *series-parallel matroid* if $\mathbf{M}$ is a series-parallel extension of a one-element matroid. Now since these one-element matroids are trivially graphic, and since any series or parallel extension of a graphic matroid is graphic, we have that $\mathbf{M}(S)$ is a series-parallel matroid if and only if $\mathbf{M}(S)$ is the polygon matroid of some series-parallel network. We shall henceforth, as is customary, use the term series-parallel network for both the graph and its polygon matroid.

It is natural to ask which properties of a matroid are preserved by series-parallel extensions. First, we have the following useful result.

**7.76 Proposition.** *Let* $\mathbf{M}(S)$ *be a finite matroid and let* $e \in S$, $e' \notin S$. *Then*

$$(s\mathbf{M}(e, e'))^{\perp} = (p\mathbf{M}^{\perp})(e, e').$$

*In particular, it follows that the dual of a series-parallel network is again a series-parallel network.*

*Proof.* Check that the bases coincide. □

**7.77 Proposition.** *A series-parallel extension of a gammoid is a gammoid. In particular, any series-parallel network is a gammoid.*

*Proof.* Let $\mathbf{M}(S) = \mathbf{L}(S, \vec{G}, I)$ be a gammoid, $v' \notin V$ where $V$ is the vertex-set of $\vec{G}$, and $p\mathbf{M}(v, v')$ a parallel extension. We construct a directed graph $p\vec{G}$ by adding the vertex $v'$ and the arrow $(v, v')$, and by replacing any edge $(v, w)$ in $\vec{G}$ by $(v', w)$, keeping the other arrows unchanged (see Figure 7.32). For this new graph $p\vec{G}$ we clearly have $p\mathbf{M}(v, v') = \mathbf{L}(S \cup v', p\vec{G}, I)$. Since the dual of a gammoid is a gammoid we conclude that also any series extension of a gammoid is again a gammoid. The last statement is now clear since the one-element matroids are gammoids. □

Figure 7.32

We come to the main theorem of this section, a "forbidden minor" characterization of series-parallel networks. First we need a graph-theoretic lemma.

**7.78 Lemma** (Dirac). *Suppose $G(V, S)$ is a finite simple 2-connected graph all of whose vertices have degree at least 3. Then $G(V, S)$ contains $K_4$ as a minor.*

*Proof.* Let $C = \{v_0, v_1, \ldots, v_t = v_0\}$ be a longest circuit in $G$. We call any path $\{v_i, w_1, w_2, \ldots, v_j\}$ which connects two distinct vertices $v_i$, $v_j$ of $C$ but otherwise contains only edges and vertices outside of $C$ a *chord* of $C$. Since the graph is 2-connected and $\gamma(v) \geq 3$ for all $v \in V$, there is a chord starting from any $v_i \in C$. Among all pairs of vertices in $C$ joined by a chord we pick one, say $v_i$, $v_j$, joined by the chord $A$, with minimal distance apart on $C$. There must be at least one vertex $v_k$ in the smaller part of $C$ between $v_i$ and $v_j$ since $C$ is a longest circuit in $G$ and $G$ has no parallel edges. Now $v_k$ is joined by a chord $B$ to some vertex $v_l$, where $v_l$ must lie on $C$ as shown in Figure 7.33, by the minimality condition on the pair $v_i$, $v_j$. If the chords $A$ and $B$ are vertex-disjoint then we obtain a $K_4$ on $\{v_i, v_j, v_k, v_l\}$ by contracting the paths between them. If, on the other hand, $A$ and $B$ have a common vertex and $x$ is the first such vertex on $B$ starting at $v_k$, then we clearly obtain a $K_4$ on $\{x, v_i, v_k, v_j\}$, again by suitably contracting paths. □

It is immediate from the definition that if a series-parallel network has a loop then it has no coloops, and dually, if it has a coloop then it has no loops. Let us call this fact the *loop-coloop condition*.

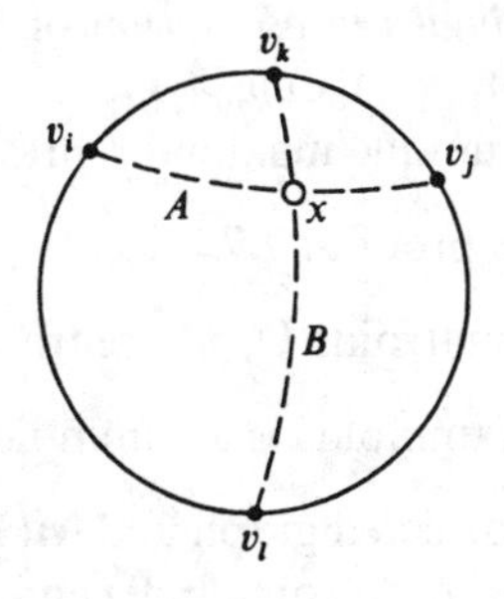


Figure 7.33

**7.79 Theorem** (Duffin). *Let* $\mathbf{M}(S)$ *be a finite matroid satisfying the loop-coloop condition. Then the following statements are equivalent.*

(i) $\mathbf{M}(S)$ *is a series-parallel network.*
(ii) $\mathbf{M}(S)$ *is a binary gammoid.*
(iii) $\mathbf{M}(S)$ *has no minor isomorphic to* $\mathbf{U}_2(4)$ *or* $\mathbf{P}(K_4)$.
(iv) $\mathbf{M}(S)$ *is the polygon matroid of a graph which contains no minor isomorphic to* $K_4$.

*Proof.* (i) $\Rightarrow$ (ii). Any series-parallel network is graphic and hence binary. By **7.77**, it is also a gammoid.

(ii) $\Rightarrow$ (iii). This follows from **7.22**, **7.74** and the fact that $\mathbf{P}(K_4)$ is not a gammoid.

(iii) $\Rightarrow$ (iv). Since $\mathbf{M}(S)$ contains no minor isomorphic to $\mathbf{U}_2(4)$, it is binary. By Tutte's theorem **7.49**, a binary but non-graphic matroid contains a minor isomorphic to $\mathbf{F}$, $\mathbf{F}^{\perp}$, $\mathbf{B}(K_5)$, or $\mathbf{B}(K_{3,3})$. Clearly, $\mathbf{F}$ contains $\mathbf{P}(K_4)$ as a minor as do $\mathbf{P}(K_5)$ and $\mathbf{P}(K_{3,3})$ by **6.40**. Using this, and the fact that $\mathbf{P}(K_4)$ is self-dual, we deduce that $\mathbf{F}^{\perp}$, $\mathbf{B}(K_5)$, and $\mathbf{B}(K_{3,3})$ contain $\mathbf{P}(K_4)$ as minor.

(iv) $\Rightarrow$ (i). We may assume that $\mathbf{M}(S) = \mathbf{P}(G(V, S))$, where $G$ is a connected graph satisfying the loop-coloop condition. Since such a graph is a series-parallel network if and only if all its 2-connected blocks are series-parallel networks (just arrange the initial edges of each block on a path by a sequence of series extensions or as a family of loops by a sequence of parallel extensions) we may suppose that $G$ is 2-connected. To prove (iv) $\Rightarrow$ (i), we use induction on the number $n$ of edges. For $n = 1$ there is nothing to prove. Assume $G$ has $n$ edges, $n > 1$. If $G$ contains a pair of parallel edges $e, e'$, then, by induction, $G' = G - e'$ is a series-parallel network and we have $G = pG'(e, e')$. If $G$ contains a vertex of degree 2 with incident edges $e, e'$, then $G = sG'(e, e')$ where $G' = G/e'$. By **7.76**, $G'$ contains no minor isomorphic to $K_4$ since $\mathbf{P}(K_4)$ is self-dual. Hence $G'$ and thus $G$ is a series-parallel network by induction. The final possibility is that $G$ is a graph without parallel edges (and loops since $n > 1$) and $\gamma(v) \geq 3$ for all $v \in V$. **7.78** shows, however, that in this case $G$ contains $K_4$ as a minor. □

EXERCISES VII.4

→ 1. Let $\mathbf{M}(S)$ be a transversal matroid of rank $r$. We say the family $\mathfrak{A} = \{A_1, \ldots, A_r\}$ is a *maximal representation* of $\mathbf{M}(S)$ if $\mathbf{M}(S) = \mathbf{T}(S; \mathfrak{A})$ and $\mathbf{M}(S) \not\cong \mathbf{T}(S; A_1, \ldots, A_{i-1}, A_i \cup p, A_{i+1}, \ldots, A_r)$ for any $i$ and $p \in S - A_i$. Prove that $\mathbf{M}(S)$ has a unique maximal representation. (Mason)

2. Fill in the details in the proof of **7.72**.

3. Show that all uniform matroids $\mathbf{U}_k(n)$ are transversal.

4. Show that the sum of gammoids is a gammoid.

→ 5. Let $\vec{G}(V, E)$ be a finite directed graph and $\mathbf{M}(V)$ a matroid on $V$. Show that $\mathfrak{I} = \{A \subseteq V: A$ is linked onto some independent set in $\mathbf{M}(V)\}$ is the family of independent sets of a matroid. (Mason)

6. A matroid $\mathbf{M}(S)$ is called *base orderable* if for any two bases $B$, $B'$ there exists a bijection $\sigma: B \to B'$ such that $(B - p) \cup \sigma(p)$ and $(B' - \sigma(p)) \cup p$ are both bases, for each $p \in S$. Prove:

   (i) Every minor of a base orderable matroid is base orderable.
   (ii) The dual of a base orderable matroid is base orderable.
   (iii) The sum of base orderable matroids is base orderable.

→ 7. Prove that gammoids are base orderable and hence, in particular, transversal matroids. (Brualdi–Scrimger–Mason)

8. Using ex. 6 show that the polygon matroids of the graphs depicted below are not transversal. Construct a whole class of non-transversal polygon matroids.

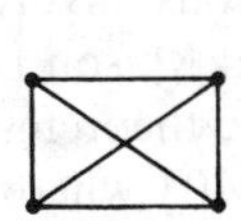
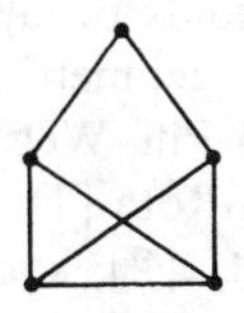

9. Prove that the Fano matroid is not base orderable and is not a proper series-parallel extension of any matroid.

10. Prove that for any matroid $\mathbf{M}$, $s\mathbf{M}(e, e')$ and $p\mathbf{M}(e, e')$ are matroids and verify **7.76**.

11. Prove that a series-parallel extension of a base-orderable matroid is base-orderable and that any series extension of a transversal matroid is transversal. What about parallel extensions of transversal matroids?

→12.* Let $G$ be a finite undirected graph. Prove that $\mathbf{P}(G)$ is transversal if and only if $G$ has no minor isomorphic to $K_4$ or $C_k^2 (k > 2)$, where $C_k^2$ is the circuit of length $k$ with all edges replaced by two parallel edges. For example, $C_3^2$ is the graph of fig. 6.12. (Bondy–Las Vergnas) (Hint: Use ex. 10 and the fact that if $\mathbf{M}' = s\mathbf{M}(e, e')$ is transversal, then so is $\mathbf{M}'/e'$.)

→13. Prove the equivalence of the following statements for a finite binary matroid $\mathbf{M}$:

   (i) $\mathbf{M}$ is a gammoid.
   (ii) $\mathbf{M}$ is base-orderable.
   (iii) $\mathbf{M}$ is a series-parallel network.
   (iv) $\mathbf{M}$ has no minor isomorphic to $\mathbf{P}(K_4)$.
   (v) $\mathbf{M}$ is the polygon matroid of a graph which has no minor isomorphic to $K_4$.

14. Let $G(V, S)$ be a finite undirected graph (not necessarily bipartite). A *matching* in $G$ is a set of edges in $G$ no two of which have a common endpoint. Prove that the family of subsets of $V$ which meet the edges of some matching in $G$ is the family of independent sets of a matroid on $V$, called the *matching matroid* of $G$. (Surprisingly, it can be shown that a matroid is a matching matroid if and only if it is transversal.) (Edmonds–Fulkerson)

→15. Let $\beta$ be the invariant defined in ex. VI.4.12 and let **M** be any finite matroid with at least two points. Prove that $\beta(\mathbf{N}) = \beta(\mathbf{M})$ for any series-parallel extension **N** of **M**, and deduce from this that any connected series-parallel network **M** on at least two points satisfies $\beta(\mathbf{M}) = 1$. (It turns out that connected series-parallel networks are characterized by this condition.) (Brylawski)

## Notes

Coordinatization theory, which began with Rado's paper [3] and Tutte's series of papers [5, 6, 10], has been one of the most active branches of matroid theory, with the binary matroids occupying the central position. For a good account of the algebraic and geometrical questions involved the reader may consult Ingleton [1]; see also Piff–Welsh [1] and Mason [2]. The critical problem was formulated by Crapo–Rota [1]. The basic theorems on binary matroids are due to Lehman [1], Minty [2], and Tutte [10] where also regular matroids were studied in depth. A very interesting general concept of orientability of arbitrary matroids has recently been proposed by Bland–Las Vergnas [1]. The applications of matroids to graphs presented in section 3 are undoubtedly among the most attractive features of the theory in so far as they permit a unified approach to a large part of graph theory from an algebraic point of view. For a good account of related topics the reader is advised to consult Biggs [1]. Other papers used in the preparation of this chapter are Whitney [1, 2, 6], Graver [1], Tutte [12], Ford–Fulkerson [1], and Duffin [3]. For transversal matroids see, e.g., Brualdi [1, 5, 6], Mason [3], Ingleton [2, 4], and Ingleton–Piff [1]. Welsh [1] gives a good general account.

Chapter VIII

# Combinatorial Order Theory

Under the general term "combinatorial order theory" we want to collect some results on posets by concentrating less on the structure of posets than on properties present in any poset, such as chains, antichains, matchings, etc. Typical problems to be considered are the determination of the minimal number of chains into which a finite poset can be decomposed or the existence of a matching between the points and copoints of a ranked poset. In fact, the importance of this branch of combinatorial mathematics derives to a large extent from the fact that most of the main results are *existence theorems* supplementing the many counting results established in chapters III to V. To testify to the broad range of applications, we have included a variety of examples from different sources (graphs, networks, 0,1-matrices, etc.). Each of the sections is headed by a basic theorem after which we study variations of the main theorem, applications, and the mutual dependence with other results.

## 1. Maximum-Minimum Theorems

We have already encountered several results asserting that the maximum of one quantity equals the minimum of another. Examples are **6.55**, where the maximal cardinality of a matching is expressed as a minimum, and **2.9**, where the minimal dimension of a coding is estimated in terms of antichains. Invariably, one half of the theorem, namely the proof of max $\leq$ min, is trivial. The most general result is the maximum flow—minimum cut theorem **7.53** which works for arbitrary ordered integral domains. The specialization of **7.53** to the integers is the starting point for our discussion.

### A. Graph Theorems

To apply the notions of flow and cut in a network, we have to give some equivalent definitions for directed graphs. Unless otherwise stated, all graphs are finite.

**Definition.** Let $\vec{G}(V, E)$ be a directed graph, $q \neq s \in V$. We say that $A \subseteq E$ *separates q from s* if, after deletion of $A$, there is no directed path $q \to v_1 \to v_2 \to \cdots \to s$ from $q$ to $s$. Similarly, we say that $U \subseteq V$ not containing $q$, $s$ *separates q from s* if the deletion of $U$ and its incident edges destroys every directed path from $q$ to $s$.

The following two theorems show that the minimal cardinality of separating sets is equal to the maximum of certain other graph invariants. The method of proof is typical and will be repeated several times. A word regarding the terminology: We say that two $u$, $v$-paths are *vertex-disjoint* if they have only the vertices $u$ and $v$ in common.

**8.1 Theorem.** *Let $\vec{G}(V, E)$ be a directed graph and $q \neq s \in V$. The minimal cardinality $\lambda_{\vec{G}}(q, s)$ of a $q, s$-separating edge-set is equal to the maximal number of edge-disjoint directed paths from $q$ to $s$.*

*Proof.* Since a $q,s$-separating edge-set must contain an edge from any directed $q, s$-path we must have min $\geq$ max. We regard $\vec{G}(V, E)$ as a network $\vec{G}(q, s)$ over $\mathbb{Z}$ with source $q$, sink $s$, and capacity $c$ identically $c \equiv 1$. For a cut $(X, Y)$, we set, as in the proof of **7.52**, $E(X, Y) := \{e \in E : e^- \in X, e^+ \in Y\}$. By the definition of $c$, this implies $c(X, Y) = |E(X, Y)|$. Since any such set $E(X, Y)$ separates $q$ from $s$, we conclude

$$\lambda_{\vec{G}}(q, s) \leq \min_{(X, Y)\text{cut}} c(X, Y).$$

By the max flow-min cut equality it remains to be shown that

$$\max_{f \text{ admissible}} w(f) \leq \text{maximal number of edge-disjoint } q, s\text{-paths.}$$

Let $f$ be an admissible flow with $w(f) = l$. Since $f$ takes only the values 0 or 1, it is clear that the elementary flows correspond bijectively to the directed paths from $q$ to $s$. Hence if $w(f) = l > 1$, we choose an elementary flow $f_1$ with $\|f_1\| \subseteq \|f\|$ (such a flow $f_1$ certainly exists). $f' = f - f_1$ is then again an admissible flow with $w(f') = l - 1$ and $\|f'\| \cap \|f_1\| = \varnothing$. By induction, we may assume that $\|f'\|$ contains $l - 1$ edge-disjoint $q, s$-paths; together with $\|f_1\|$ this gives $l$ such paths. □

**8.2 Theorem.** *Let $\vec{G}(V, E)$ be a directed graph, $q \neq s \in V$ with $(q, s) \notin E$. The minimal cardinality $\kappa_{\vec{G}}(q, s)$ of a $q, s$-separating vertex-set is equal to the maximal number of vertex-disjoint directed paths from $q$ to $s$.*

*Proof.* Again, we trivially have $\kappa_{\vec{G}}(q, s) \geq$ max. To see the converse inequality, we define the following directed graph $\vec{H}(V', E')$:

$$V' := \{q, s\} \cup \bigcup_{v \in V - \{q, s\}} \{v^{(1)}, v^{(2)}\},$$

$$E' := \begin{cases} (q, v^{(1)}) & \text{if } (q, v) \in E,\ v \neq q \\ (v^{(2)}, s) & \text{if } (v, s) \in E,\ v \neq s \\ (v^{(2)}, w^{(1)}) & \text{if } (v, w) \in E,\ v, w \neq q, s \\ (v^{(1)}, v^{(2)}) & \text{for all } v \in V - \{q, s\}. \end{cases}$$

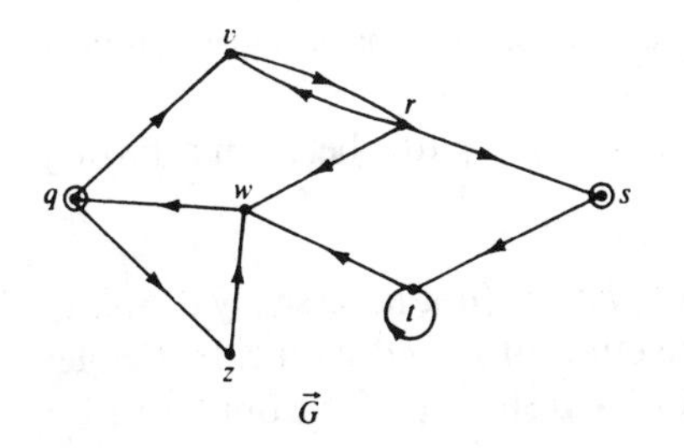


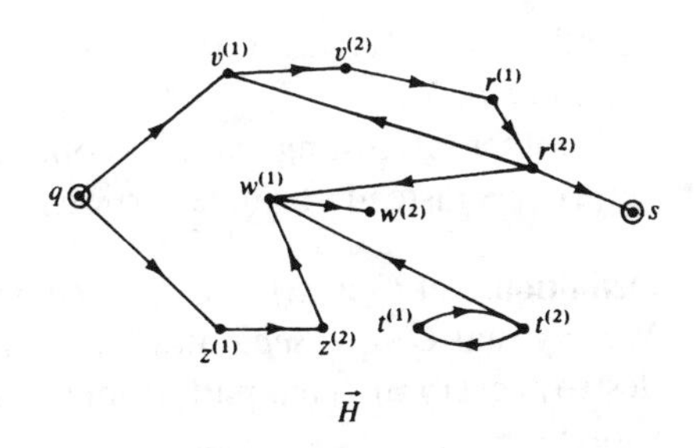


Figure 8.1

The capacity $c$ on $\vec{H}$ is again $c \equiv 1 \in \mathbb{Z}$ (see Figure 8.1 for an example). As in the previous proof, an admissible flow $f$ in $\vec{H}(q, s)$ with $w(f) = l$ gives rise to $l$ edge-disjoint directed paths from $q$ to $s$ which, by the construction of $\vec{H}$, correspond biuniquely to $l$ vertex-disjoint $q, s$-paths in $\vec{G}$. It remains to be shown that $\kappa_{\vec{G}}(q, s) \leq c(X, Y)$ for every cut $(X, Y)$ of the network $\vec{H}(q, s)$. Let $(X, Y)$ be a cut of minimal capacity and $E'(X, Y) = \{e \in E' : e^- \in X, e^+ \in Y\}$. Suppose there is an edge $e = (v^{(2)}, w^{(1)}) \in E'(X, Y)$. The cut $(X', Y')$ defined by

$$X' = (X - v^{(2)}) \cup v^{(1)}, \qquad Y' = (Y - v^{(1)}) \cup v^{(2)}$$

satisfies $c(X', Y') \leq c(X, Y)$ since $(v^{(1)}, v^{(2)})$ is the only edge $k \in E'$ with $k^+ = v^{(2)}$. For the same reason, we may put $x^{(1)}$ into $X$ if $(q, x^{(1)}) \in E'(X, Y)$ and $y^{(2)}$ into $Y$ if $(y^{(2)}, s) \in E'(X, Y)$. Repeatedly exchanging vertices in this way, we finally arrive at a minimal capacity cut $(\tilde{X}, \tilde{Y})$ with $E'(\tilde{X}, \tilde{Y}) \subseteq \bigcup_{v \in V - \{q, s\}} (v^{(1)}, v^{(2)})$. (Notice that at this point we make use of the assumption $(q, s) \notin E$.) The set $\{v \in V : v^{(1)} = e^- \text{ for some } e \in E'(\tilde{X}, \tilde{Y})\}$ is clearly a $q, s$-separating vertex-set in $\vec{G}(V, E)$ whence

$$\kappa_{\vec{G}}(q, s) \leq c(\tilde{X}, \tilde{Y}) = \min c(x, Y). \quad \square$$

The two graph theorems just proved imply conversely the max flow—min cut theorem **7.53** over $\mathbb{Z}$. All three theorems are thus equivalent in the sense that each directly implies the others. To prove **8.2** $\Rightarrow$ **8.1** we construct from $\vec{G}(V, E), q \neq s \in V$, a *directed edge graph* $\vec{\bar{G}}(\bar{V}, \bar{E})$. (Compare the corresponding definition for undirected graphs before **7.60**.) The vertices $\bar{e} \in \bar{V}$ are the edges $e \in E$ plus two new vertices $q^*, s^*$. The edges $u$ are

$$u \in \bar{E} :\Leftrightarrow \begin{cases} u = (\bar{e}, \bar{f}) & \text{if } e^+ = f^-, e, f \in E \\ u = (q^*, \bar{e}) & \text{if } e^- = q, e \in E \\ u = (\bar{e}, s^*) & \text{if } e^+ = s, e \in E. \end{cases}$$

Application of **8.2** to $\vec{\bar{G}}$ yields the edge-version **8.1** for the original graph $\vec{G}$.

To see that **8.1** implies the max flow—min cut theorem over $\mathbb{Z}$ we take an arbitrary network $\vec{G}(q, s)$ over $\mathbb{Z}$ with capacity function $c$ and replace each edge $e$ by $c(e)$

equally oriented edges and delete $e$ if $c(e) = 0$. Theorem **8.1** for this new graph is now precisely **7.53**.

Another interesting variant is obtained by considering paths between arbitrary vertex-sets instead of single vertices.

**Definition.** Let $\vec{G}(V, E)$ be a directed graph and $S, I \subseteq V$ (not necessarily disjoint). We say that $C \subseteq V$ *separates* $S$ *from* $I$ if the deletion of $C$ and its incident edges destroys every directed path from a vertex in $S$ to a vertex in $I$ (or from $S$ to $I$ for short).

Notice that this definition differs slightly from the previous ones since we do not require $S$ and $I$ to be disjoint. As a consequence we need make no hypothesis on $S$ and $I$ in the following statement.

**8.3 Proposition.** *Let $\vec{G}(V, E)$ be a directed graph and $S, I \subseteq V$. The minimal cardinality of an $S, I$-separating vertex-set equals the maximal number of vertex-disjoint directed paths from $S$ to $I$.*

*Proof.* We adjoin two new vertices $q^*$, $s^*$ and all edges

$$\{(q^*, u): u \in S\} \cup \{(w, s^*): w \in I\}$$

and apply **8.2** to this new graph. □

Of course, the maximum in **8.3** is just the rank of the gammoid $\mathbf{L}(S, \vec{G}, I)$. It is clear that **8.3** implies **8.2** (and hence **8.1** and **7.53** (over $\mathbb{Z}$)) by specializing $S$ and $I$ to single vertices.

The analogue of **8.2** for undirected graphs is called *Menger's theorem.* We prove it by again setting up a suitable network. Once the vertex version is established the edge-version follows by the same argument as in the proof of the implication **8.2** ⇒ **8.1**. The definition of a separating set carries over in the natural way.

**8.4 Theorem** (Menger). *Let $G(V, E)$ be an undirected graph and $q \neq s \in V$ with $\{q, s\} \notin E$. The minimal cardinality $\kappa_G(q, s)$ of a $q,s$-separating vertex-set equals the maximal number of vertex-disjoint paths from $q$ to $s$.*

*Proof.* We define $\vec{G}(V', E')$ as follows:

$$V' := \{q, s\} \cup \bigcup_{v \in V - \{q, s\}} \{v^{(1)}, v^{(2)}\},$$

$$E' := \begin{cases} (q, v^{(1)}) & \text{if } \{q, v\} \in E,\ v \neq q \\ (v^{(2)}, s) & \text{if } \{v, s\} \in E,\ v \neq s \\ (v^{(2)}, w^{(1)}), (w^{(2)}, v^{(1)}) & \text{if } \{v, w\} \in E,\ v, w \neq q, s \\ (v^{(1)}, v^{(2)}) & \text{for all } v \in V - \{q, s\}. \end{cases}$$

Now apply **8.2** to $\vec{G}$. □

**8.5 Proposition.** *Let $G(V, E)$ be an undirected graph, $q \neq s \in V$. The minimal cardinality $\lambda_G(q, s)$ of a $q, s$-separating edge-set equals the maximal number of edge-disjoint paths from $q$ to $s$.* □

Just as **8.2** implies **8.3** by adding $q^*$ and $s^*$, **8.4** can be expressed in the following way.

**8.6 Proposition.** *Let $G(V, S)$ be an undirected graph and $S, I \subseteq V$. The minimal cardinality of an $S, I$-separating vertex-set equals the maximal number of vertex-disjoint paths from $S$ to $I$.* □

As an application, we have the following theorem characterizing $k$-connected graphs (see **6.87**), whose proof is left to the reader.

**8.7 Proposition** (Whitney). *An undirected graph is $k$-connected if and only if any two distinct vertices are joined by at least $k$ vertex-disjoint paths.* □

## B. Matching Theorems

We now specialize the results of the previous section to bipartite graphs. Because of their 3-fold interpretation as statements about bipartite graphs, set systems, and 0,1-matrices, respectively, these theorems occupy a central position among all existence results in combinatorics. In fact, it was precisely this wide range of applications of matching theorems which generated a general interest in maximum-minimum problems in recent years.

Recall from section VI.2.C that a *matching* in a bipartite graph is a set of edges no two of which have a common endpoint, and that a *point cover* is a set of vertices which meets every edge. By definition, a matching in the bipartite graph $G(S \cup I, R)$ contains at most $\min(|S|, |I|)$ edges. We call the matching $M$ a *full matching* if $|M| = \min(|S|, |I|)$. Seen differently, a matching is just a collection of vertex-disjoint paths from $S$ to $I$, whereas the point covers are precisely the $S, I$-separating vertex-sets. Thus, by applying **8.6**, we obtain the following theorem, which is, of course, nothing but the rank formula **6.27**(iv).

**8.8 Theorem** (König). *Let $G(S \cup I, R)$ be a bipartite graph. Then*

$$\max_{M \text{ matching}} |M| = \min_{C \text{ point cover}} |C|. \quad \square$$

For completeness, let us restate the defect version **6.27**(iii) as well. Recall that $\delta(A) = |A| - |R(A)|$ for $A \subseteq S$.

**8.9 Theorem** (Ore). *Let $G(S \cup I, R)$ be a bipartite graph. Then*

$$\max_{M \text{ matching}} |M| = |S| - \max_{A \subseteq S} \delta(A). \quad \square$$

We now interpret **8.8** in terms of matrices. Let $M$ be a matrix over some set, where we are only interested in whether an entry is 0 or $\neq 0$. By a *line* we mean either a row or a column. A set of entries $\neq 0$ is called a *diagonal* if no two of them are on the same line. We say that a set $S$ of lines *covers* all entries $\neq 0$ if every non-zero entry is on at least one line of $S$.

**Example.**

$$\begin{bmatrix} 0 & 0 & 1 & 0 & 0 & 0 \\ 1 & \textcircled{-1} & 0 & 2 & 0 & -3 \\ 0 & 0 & 4 & 0 & 0 & \textcircled{-1} \\ 0 & 4 & 5 & 0 & \textcircled{-2} & 0 \\ 0 & 0 & \textcircled{-3} & 0 & 0 & -6 \end{bmatrix}$$

The encircled elements form a diagonal, the rows 2 and 4 together with the columns 3 and 6 cover all elements $\neq 0$. As usual, we associate with $M = [m_{ij}]$ a bipartite graph by taking as defining vertex-sets the rows and columns, respectively, setting

$$\{r_i, c_j\} \in R :\Leftrightarrow m_{ij} \neq 0.$$

**8.8** yields then the following result.

**8.10 Proposition.** *The maximal cardinality of a diagonal of elements $\neq 0$ in a matrix equals the minimal number of lines covering all elements $\neq 0$.* □

**8.11 Corollary.** *Let $M$ be an $n \times n$-matrix over a field. Then all terms in the determinant expansion* $\det M = \sum_{\sigma \in S_n} \operatorname{sign}\sigma\, m_{1\sigma(1)} \cdots m_{n\sigma(n)}$ *are equal to 0 if and only if there is an $r$, $1 \leq r \leq n$, such that $M$ has a 0-submatrix of type $r \times (n - r + 1)$.*

*Proof.* Let $m$ be the maximal cardinality of a diagonal of elements $\neq 0$. Clearly, all terms in $\det M$ vanish if and only if $m < n$, i.e., by **8.10**, if and only if there are $i$ rows and $j$ columns with $i + j < n$ which cover all non-zero entries. This is equivalent to the existence of a 0-submatrix of type $(n - i) \times (n - j)$ with $(n - i) + (n - j) = 2n - (i + j) > n$. □

Next, we ask when a bipartite graph $G(S \cup I, R)$ possesses a matching of maximal cardinality $|S|$. By **8.9**, this is the case when and only when $\delta(A) \leq 0$ for all $A \subseteq S$. Hence we again obtain Hall's theorem **6.25**.

**8.12 Theorem** (Hall). *Let $G(S \cup I, R)$ be a bipartite graph. Then*

$$\max_{M \text{ matching}} |M| = |S| \Leftrightarrow |A| \leq |R(A)| \quad \textit{for all } A \subseteq S. \quad \square$$

Hall's theorem is thus a trivial consequence of König's theorem **8.8**. On the other hand, it is also easy to derive **8.8** directly from **8.12** (see the exercises).

**8.12** is often called the "marriage theorem" for the following reason. Let $S$ and $I$ be sets of ladies and gentlemen, respectively. We set $\{u, v\} \in R$ if lady $u$ wishes to be wed to the gentleman $v$, assuming the ideal situation that the wish to get married exists either on both sides or on neither. Then it is possible for all ladies to find a suitable partner (without committing bigamy) if and only if for any $k$ ladies there are at least $k$ gentlemen who wish to marry one or more of them.

Another useful corollary is the following result on regular bipartite graphs.

**8.13 Corollary.** *Let $G(S \cup I, R)$ be a bipartite graph which is regular on $S$, say of degree $d \neq 0$, and regular on $I$, say of degree $e$. Then $G(S \cup I, R)$ possesses a full matching.*

*Proof.* Assume without loss of generality that $|S| \leq |I|$; thus $e \leq d$. By counting the edges between $A \subseteq S$ and $R(A)$ in two ways we obtain

$$d|A| \leq e|R(A)| \leq d|R(A)|,$$

hence $|A| \leq |R(A)|$. $\square$

The interpretation of matchings as partials transversals is not only useful in the study of matroids, as demonstrated in the previous chapters, but it also permits easy solutions to many other problems on set systems. We shall present a sample of these results in section 2.

C. Coding Theorems

Let us recall the coding problem stated in section II.1.B. Any embedding of a finite distributive lattice $L$ into a finite chain product is called a *coding* of $L$. The coding problem consists in determining the minimal number of chains (called the *dimension* of the coding) needed for a coding of $L$. **2.9** reduced this problem to the following question in arbitrary posets. What is the minimum number of disjoint chains into which a finite poset can be decomposed? The answer is provided by the following theorem.

**8.14 Theorem** (Dilworth). *Let $P$ be a finite poset. The minimal number of disjoint chains into which $P$ can be decomposed equals the maximal cardinality of an antichain in $P$.*

*Proof.* Again we trivially have max $\leq$ min. To show the other inequality let $\mathfrak{A}(P)$ be the family of antichains in $P$. We use induction on $|P|$. For $|P| = 1$ there is nothing to prove. Assume that the theorem is true for all posets $Q$ with $|Q| < |P|$ and set $n = \max_{A \in \mathfrak{A}(P)} |A|$.

*Case a.* There exists $U \in \mathfrak{A}(P)$, $|U| = n$, which contains neither all maximal elements of $P$ nor all minimal elements. We define $P^+, P^-$ by

$$P^+ := \{p \in P : p \geq u \text{ for some } u \in U\},$$
$$P^- := \{p \in P : p \leq u \text{ for some } u \in U\}.$$

The hypothesis on $U$ implies that

$$P^+ \neq P, \quad P^- \neq P \text{ and } P = P^+ \cup P^-, \quad U = P^+ \cap P^-.$$

By induction, $P^+$ and $P^-$ can each be decomposed into $n$ chains, whence by pasting these chains together at the points of $U$ we obtain a decomposition of $P$ into $n$ chains.

*Case b.* Each antichain of cardinality $n$ contains either all maximal elements or all minimal elements of $P$. Hence there are at most two such antichains, one consisting of all maximal elements the other of all minimal elements, respectively. Let $a$ be a maximal element and $b$ a minimal element with $b \leq a$. By induction, $P - \{a, b\}$ can be decomposed into $n - 1$ chains whence, by adding the chain $\{b \leq a\}$, $P$ can be decomposed into $n$ chains. □

**8.14** immediately implies the theorem of König by associating with a bipartite graph $G(S \cup I, R)$ the poset on $S \cup I$ whose order relation is given by

$$x < y :\Leftrightarrow \{x, y\} \in R.$$

Conversely, it is not hard to derive Dilworth's theorem from König's by defining a suitable bipartite graph for a given poset (see the exercises).

By interchanging the role of chains and antichains in **8.14**, we obtain another max-min theorem whose proof, surprisingly, is almost trivial.

**8.15 Proposition.** *The maximal cardinality of a chain in a finite poset $P$ (= length of the chain $+1$) equals the minimal number of disjoint antichains into which $P$ can be decomposed.*

*Proof.* Trivially, max $\leq$ min. Denote by $l(x)$ the length of the longest chain in $P$ with endpoint $x$. Since elements of the same length are clearly incomparable, $P = \dot{\bigcup}_i \{x : l(x) = i\}$ is a decomposition into antichains of the required sort. □

There is an interesting problem in graph theory connected with our last two results. To every finite poset $P_<$ we associate its *comparability graph* $G(P, R)$ with vertex-set $P$ and $\{u, v\} \in R$ if and only if $u < v$ or $v < u$. The complete subgraphs of $G(P, R)$ correspond bijectively to the chains in $P_<$, whereas the totally unconnected subgraphs (i.e., with no edges) correspond to the antichains in $P_<$. **8.14** can thus be stated as follows: Let $\mathscr{C}$ be the class of undirected graphs which are comparability graphs of some poset. If $G(V, E) \in \mathscr{C}$ then:

(+) *The minimal number of vertex-disjoint complete subgraphs into which $G$ can be decomposed equals the maximal cardinality of a totally unconnected subgraph.*

**8.15** states property (+) for the complementary graph $G^c(V', E')$ where $V' = V$ and $E' = V^{(2)} - E$. In terms of $G(V, E)$ this reads as follows:

(+ +) *The minimal number of vertex-disjoint totally unconnected subgraphs into which $G$ can be decomposed* (=chromatic number) *equals the maximal cardinality of a complete subgraph of $G$* (=clique number).

Notice that the inequality min $\geq$ max in (+) and (+ +) holds for any simple graph.

A simple graph is called *perfect* if $G$ as well as each induced subgraph of $G$ satisfies (+). A theorem of Lovász states that a graph is perfect if and only if its complement is perfect. Using this theorem and the trivial result **8.15**, we thus obtain a new proof of Dilworth's theorem which can be stated succinctly: *Comparability graphs are perfect.*

**Examples.** The graph of Figure 8.2 is easily seen to be perfect but it is not a comparability graph. For, if it were, then we may assume $a < b < c$. The element $b'$ is comparable with $b$, whence either $b' < c$ or $a < b'$, contradicting $\{b', c\}, \{b', a\} \notin E$. The smallest non-perfect graph is the circuit of length 5 which has chromatic number 3 and clique number 2. It is a still unproven conjecture of Berge's that a graph is perfect if and only if neither it nor its complement contains a circuit of odd length without chords.

We come to our final maximum-minimum theorem. Let $L$ be a finite distributive lattice and $P$ its subposet of irreducible elements $\neq 0$. It was shown in **2.9** that any partition of $P$ into $k$ disjoint chains yields a coding of $L$ of dimension $k$, and vice versa. Hence we have $\min_{\phi \text{ coding}} \dim \phi = d(P)$ where $d(P)$ is the minimum in **8.14**. To apply **8.14** we need to interpret $\max_{A \in \mathfrak{A}(P)} |A|$ in the lattice $L$.

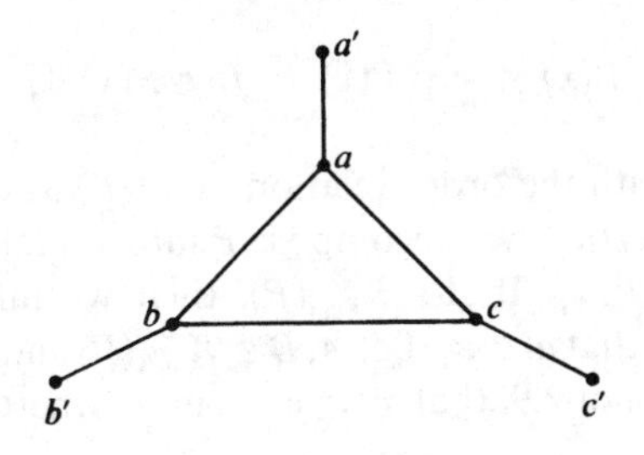


Figure 8.2

**8.16 Lemma.** *Let $L$ be a finite distributive lattice and $P$ its subposet of irreducible elements $\neq 0$. As in* **2.4**, *we set $P(x) = \{p \in P: p \le x, p \max\}$ for $x \in L$. Then:*

(i) *The antichains in $P$ are precisely the sets $P(x)$.*
(ii) *$A \to \sup A$, $x \to P(x)$ is a bijection between $\mathfrak{A}(P)$ and $L$.*
(iii) *$x$ covers in $L$ precisely $|P(x)|$ elements.*

*Proof.* (iii) was shown in **2.10**. Any set $P(x)$ is clearly in $\mathfrak{A}(P)$. Now assume $A \in \mathfrak{A}(P)$ and let $x = \sup A \in L$. If $p \in P$ with $p \le x$, then $p \le q$ for some $q \in A$, by **2.2**. Hence $P(x) \subseteq A$ and therefore $P(x) = A$ since $A$ is an antichain; this proves (i) and (ii). □

Since by **8.16**, $\max_{A \in \mathfrak{A}(P)} |A| = \max_{x \in L} |P(x)|$, we have thus proved the following theorem.

**8.17 Theorem** (Dilworth). *Let $L$ be a finite distributive lattice and $P$ its subposet of irreducible elements $\neq 0$. For $x \in L$, let $P(x) = \{p \in P: p \le x, p \max\}$ and $C(x) = \{q \in L: q \lessdot x\}$. Then*

$$\min_{\phi \text{ coding of } L} \dim \phi = \max_{x \in L} |P(x)| = \max_{x \in L} |C(x)|. \quad \square$$

The bijection in **8.16**(ii) suggests an order relation on the set $\mathfrak{A}(P)$ of antichains, by which $\mathfrak{A}(P)$ becomes a distributive lattice.

**8.18 Proposition.** *Let $P$ be a finite poset and $\mathfrak{A}(P)$ the family of antichains in $P$ (including $\varnothing$). The order relation*

$$A \le B :\Leftrightarrow \forall p \in A\ \exists q \in B \text{ with } p \le q$$

*defines a distributive lattice on $\mathfrak{A}(P)$; in fact, $\mathfrak{A}(P) \cong \mathfrak{J}(P)$ where $\mathfrak{J}(P)$ is the lattice of ideals of $P$. Furthermore, the set $\mathfrak{A}_{\max}(P) \subseteq \mathfrak{A}(P)$ of antichains of maximal size forms a sublattice of $\mathfrak{A}(P)$ and is thus also distributive.*

*Proof.* Setting $L = \mathfrak{J}(P)$ we may, by **2.5**, consider $P$ to be the subposet of irreducible elements $\neq 0$ in $L$. Using the bijection in **8.16**(ii) and **2.2** we have

$$x \le y \text{ in } L \Leftrightarrow \sup P(x) \le \sup P(y) \Leftrightarrow \forall p \in P(x)\ \exists q \in P(y) \text{ with } p \le q$$

and thus $L \cong \mathfrak{A}(P)_<$, with the order relation on $\mathfrak{A}(P)$ as defined in the theorem. To prove the second assertion, we decompose $P$ into a minimal number of disjoint chains $P = P_1 \,\dot\cup \cdots \dot\cup\, P_{d(P)}$. If $A \in \mathfrak{A}_{\max}(P)$, then we number the elements in $A = \{a_1, \ldots, a_{d(P)}\}$ such that $a_i \in P_i$. Let $A, B \in \mathfrak{A}_{\max}(P)$ and $x = \sup A$, $y = \sup B$. It follows, as in the proof of **2.9**, that $x \vee y = \sup C$ where

$$C = \{\max(a_1, b_1), \ldots, \max(a_{d(P)}, b_{d(P)})\}$$

and $x \wedge y = \sup D$ where $D = \{\min(a_1, b_1), \ldots, \min(a_{d(P)}, b_{d(P)})\}$. Since $C$ and $D$ are clearly in $\mathfrak{A}_{\max}(P)$ again, the assertion follows. □

Let us note another interesting consequence of **8.16**. Let $L$ and $P \subseteq L$ be as before and denote by $Q$ the poset of meet-irreducible elements $\neq 1$ in $L$. Since, by **2.7**, $P$ and $Q$ are isomorphic there are as many antichains in $P$ of cardinality $k$ as there are antichains in $Q$ of cardinality $k$, for all $k$. Dualizing **8.16**(i) and **8.16**(iii), we have the following result.

**8.19 Corollary.** *In a finite distributive lattice the number of elements which cover precisely k elements is the same as the number of elements which are covered by precisely k elements, for all k.* □

It is a remarkable fact that the statement of **8.19** holds in arbitrary finite *modular* lattices; hence (taking $k = 1$) any finite modular lattice contains, in particular, the same number of join-irreducible elements as meet-irreducible elements (although their orders need not be isomorphic any more). See Dilworth [6] and Ganter–Rival [1].

Let us summarize the theorems of this section. We have seen that the max flow—min cut theorem **7.53** over $\mathbb{Z}$ and the theorems **8.1**, **8.2** and **8.3** on directed graphs are all equivalent and that they imply Menger's theorem **8.4** on undirected graphs. **8.4**, in turn, trivially implies **8.8** and hence the equivalent theorems **8.12**, **8.14** and **8.17**. By the argument used in the proof of **7.72**, one can show that **8.8** also directly implies **8.3**, whence all our maximum-minimum results are equivalent.

## Exercises VIII.1

1. Verify in detail the equivalence of **8.1**, **8.2** and **7.53** (over $\mathbb{Z}$).

2. Try to deduce **8.4** from **8.5**.

→ 3. We defined in **6.87** an undirected graph $G(V, E)$ to be $k$-connected if $\kappa(G) := \min_{T \subseteq V} (|T|: T \text{ separating}) \geq k$, with the convention $\kappa(K_n) := n - 1$. Show that $\kappa(G) = \min_{\{u,v\} \notin E} \kappa_G(u, v)$ and that $G(V, E)$ is $k$-connected if and only if any two distinct vertices are joined by at least $k$ vertex-disjoint paths. (Whitney)

→ 4. Prove the following dual form of **8.5**. In an undirected graph the maximal number of edge-disjoint bonds separating $q$ and $s$ is equal to the minimal number of edges in a path joining $q$ to $s$. (Fulkerson)

5. Verify for an undirected graph $G(V, E)$, $q \neq s \in V$:
   (i) $\kappa_G(q, s) \leq \lambda_G(q, s) \leq \min(\gamma(q), \gamma(s))$,
   (ii) $\kappa(G) \leq \lambda(G) \leq \min_{v \in v} \gamma(v)$.
   (iii) Construct a graph $G$ with $\kappa(G) < \lambda(G) < \min_{v \in V} \gamma(v)$.

6. Determine $\kappa(K_{m,n})$ and $\lambda(K_{m,n})$.

7.* Generalize **6.87**: If $G(V, E)$ is $k$-connected, then any $k$ vertices lie on a circuit. (Dirac) What about the converse?

→ 8. Deduce Königs' theorem **8.8** from Hall's theorem **8.12**. (Hint: Let $D = X \cup Y$, $X \subseteq S$, $Y \subseteq I$ be a point cover of minimal size; now consider the subgraphs $G(V, R')$, $G(V, R'')$ where

$$R' = R \cap (X \times (I - Y)),\ R'' = R \cap ((S - X) \times Y).)$$

→ 9. Generalize **8.11**: Let $M$ be an $m \times n$-matrix over a field and let $1 \leq r \leq \min(m, n)$. Then every term in the determinant expansion of every $r \times r$-submatrix of $M$ vanishes if and only if $M$ possesses a 0-submatrix of type $(c, d)$ with $c + d = m + n - r + 1$.

10. Let $M$ be an $m \times n$-matrix without 0-lines. A set $L$ of lines is called *non-intersecting* if any two lines in $L$ are either parallel or else intersect in a 0-element. Prove that the maximal number of non-intersecting lines of $M$ equals the minimal number of non-zero elements incident with all lines.

→ 11. Deduce Dilworth's theorem **8.14** from König's theorem **8.8**. (Hint: For a poset $P$ define a bipartite graph $G(P \cup P', R)$ where $P'$ is a copy of $P$ and $\{x, y'\} \in R :\Leftrightarrow x < y$ in $P$.)

12. Let $G(V, E)$ be a perfect graph. For $v \in V$ denote by $H$ the graph obtained from $G$ by adding a new vertex $v'$ and joining $v'$ to all neighbors of $v$. Show that $H$ is also perfect.

→ 13.* Assume the following generalization of ex. 12: Let $G(V, E)$ be a graph and denote by $G_A$ the subgraph induced by $A \subseteq V$. The clique number of $G_A$ is denoted by $\omega(G_A)$ and the maximal number of vertices in $G_A$ no two of which are joined by an edge by $\alpha(G_A)$. Then, if all proper subgraphs of $G$ are perfect and if $\omega(G_A)\alpha(G_A) \geq |A|$ for all $A \subseteq V$, any graph $H$ obtained from $G$ by replacing each $x \in V$ by a set $X = \{x_1, x_2, \ldots\}$, and by joining $x_i$ to $y_j$ if and only if $x$ and $y$ are neighbors in $G$, also satisfies $\omega(H_A)\alpha(H_A) \geq |A|$ for all vertex-sets $A$. Deduce from this Lovász' Theorem: The following conditions are equivalent for a simple graph $G$:
(i) $\omega(G_A)\alpha(G_A) \geq |A| \quad (A \subseteq V)$.
(ii) $G$ is perfect.
(iii) The complement of $G$ is perfect.
(Hint: Only (i) ⇒ (ii) needs some care.)

14.* Prove Dilworth's theorem: In a finite modular lattice the number of elements which cover precisely $k$ elements equals the number of elements which are covered by precisely $k$ elements, for every $k$. (Hint: Use ex. II.3.9 to prove that the number of $(n + 1)$-tuples $(p; q_1, \ldots, q_n)$ with $p \prec q_i$ for all $i$, equals the number of $(n + 1)$-tuples $(p; q_1, \ldots, q_n)$ with $q_i \prec p$ for all $i$, for every $n$. From this, deduce the theorem by induction on $k$.)

→15. Generalize the concept of an antichain as follows. Let $P$ be a poset. $A \subseteq P$ is called an *h-family* if $A$ contains no chains of cardinality $h + 1$ (=length $h$). Thus, the antichains are just the 1-families. Denote by $\mathfrak{A}_h(P)$ the set of all $h$-families. Prove:

(i) For $A \in \mathfrak{A}_h(P)$ and $x \in A$, let $d_A(x)$ be the cardinality of the largest chain in $A$ with $x$ as first element. The sets $A_i := \{x \in A : d_A(x) = i\}$, $i = 1, \ldots, h$, form a partition of $A$ into $h$ antichains.

(ii) Among all partitions of $A$ into $h$ antichains, the one in (i) is characterized by $A_h \leq A_{h-1} \leq \cdots \leq A_1$, where $<$ is the order relation in $\mathfrak{A}(P)$ as defined in **8.18**.

(iii) By defining $A \leq B :\Leftrightarrow A_i \leq B_i$ for $i = 1, \ldots, h$, $\mathfrak{A}_h(P)$ becomes a semimodular (but, in general, not distributive) lattice. (Greene–Kleitman)

## 2. Transversal Theorems

As mentioned in several places above, it was Hall's theorem **8.12** in its interpretation as a theorem on set systems and their transversals which stimulated intensive research in the field that we today call transversal theory. This section presents some results on families of sets and their transversals with various applications in section C.

### A. Set Systems and Transversals

Let us restate Hall's theorem **6.25** and Rado's theorem **6.28** for set systems. As in section VII.4.A, we consider a set $S$ and a system of subsets $\{A_1, \ldots, A_n\}$, denoted by $T(S; A_1, \ldots, A_n)$ or $T(S; \mathfrak{A})$, where the $A_i$'s need not be different. Notice that since we are considering only finitely many subsets we shall need no further assumptions on $S$. Defining the binary relation $R \subseteq \mathbb{N}_n \times S$ as usual by

$$(i, p) \in R :\Leftrightarrow p \in A_i,$$

we obtain $R(J) = \bigcup_{j \in J} A_j$ for all $J \subseteq \mathbb{N}_n$, and thus the following theorems.

**8.20 Theorem** (Hall). *Let $T(S; A_1, \ldots, A_n)$ be a set system. Then $\{A_1, \ldots, A_n\}$ possesses a transversal (or system of distinct representatives) if and only if*

$$\left| \bigcup_{j \in J} A_j \right| \geq |J| \quad \textit{for all } J \subseteq \mathbb{N}_n. \quad \square$$

**8.21 Theorem** (Rado). *Let $T(S; A_1, \ldots, A_n)$ be a set system and $\mathbf{M}(S)$ a matroid on $S$ with rank function $r$. Then $\{A_1, \ldots, A_n\}$ possesses an independent transversal if and only if*

$$r\left( \bigcup_{j \in J} A_j \right) \geq |J| \quad \textit{for all } J \subseteq \mathbb{N}_n. \quad \square$$

We call the conditions in **8.20** and **8.21** *Hall's condition* and *Rado's condition*, respectively.

See Mirsky [1] for an account of the myriad of theorems refining or generalizing **8.20** and **8.21**. We illustrate three lines of investigation. First we study the existence of partial transversals of prescribed length, then of those with prescribed subsets, and finally the conditions under which two families of sets posses a common transversal.

A simple but useful result is the reformulation of **8.13** for set systems.

**8.22 Proposition.** *Let $T(S; A_1, \ldots, A_n)$ be a set system. Suppose that $|A_j| = k$ for all $j$ and that every $p \in S$ is contained in precisely $l$ of the sets $A_j$, $l \leq k$. Then $\mathfrak{A}$ possesses a transversal.* $\square$

**Definition.** Let $T(S; A_1, \ldots, A_n)$ be a set system. We call $B \subseteq S$ a *transversal with defect $d$* if $B$ is a partial transversal of $\mathfrak{A}$ and $|B| = n - d$. Transversals of defect 0 are thus transversals in the usual sense.

We ask under what conditions pairwise disjoint transversals with prescribed defects exist. This is plainly a packing problem as discussed in section VI.3.B and could be answered using the theorems there. We give a direct proof using **8.21**. For the connection of all these questions with the sum of matroids see the exercises and Brualdi [5].

**8.23 Proposition.** *Let $T(S; A_1, \ldots, A_n)$ be a set system and $\mathbf{M}(S)$ a matroid on $S$ with rank function $r$. There exist $m$ pairwise disjoint transversals $B_1, \ldots, B_m$ with defects $d_1, \ldots, d_m$ whose union $B_1 \cup \cdots \cup B_m$ is independent in $\mathbf{M}(S)$ if and only if*

$$r\left(\bigcup_{j \in J} A_j\right) \geq (|J| - d_1)^+ + \cdots + (|J| - d_m)^+ \quad \textit{for all } J \subseteq \mathbb{N}_n,$$

*where $x^+ = \max(x, 0)$.*

*Proof.* It is easy to see that the condition is necessary. To prove sufficiency, we define sets $D_i$ with $|D_i| = d_i$ for $i = 1, \ldots, m$, such that $S, D_1, \ldots, D_m$ are pairwise disjoint. Consider the free matroid $\mathbf{FM}(D_i)$ on each $D_i$ and let $\mathbf{H}(U) = \mathbf{M}(S) \times \prod_{i=1}^m \mathbf{FM}(D_i)$ be the product of matroids on $U = S \cup \bigcup_{i=1}^m D_i$. We want to verify Rado's condition for the set system

$$T(U; A_1 \cup D_1, A_1 \cup D_2, \ldots, A_1 \cup D_m, \ldots, A_n \cup D_m)$$

together with the matroid $\mathbf{H}(U)$. Let $A_{j_1} \cup D_{i_1}, \ldots, A_{j_k} \cup D_{i_k}$ be $k$ of these sets, $J$ the set of distinct indices $j_l$ and similarly $I$ the set of distinct indices $i_l$. Then $|J||I| \geq k$ (why?) and thus, by the hypothesis

$$r_{\mathbf{H}}\left(\bigcup_{l=1}^{k}(A_{j_l} \cup D_{i_l})\right) = r\left(\bigcup_{j \in J} A_j\right) + \sum_{i \in I} d_i \geq \sum_{j \notin I} (|J| - d_j)^+ + \sum_{i \in I} \max(|J|, d_i) \geq k.$$

By **8.21**, $\mathbf{H}(U)$ contains an independent set $\{t_{ij}: i = 1, \ldots, m, j = 1, \ldots, n\}$ with $t_{ij} \in A_j \cup D_i$. Thus the set $\{t_{ij}: j = 1, \ldots, n\} \cap S$ is, for each fixed $i$, a transversal of $\mathfrak{A}$ with defect at most $d_i$, and the theorem follows. □

We study next the existence of transversals containing a prescribed subset.

**8.24 Proposition.** *Let $T(S; A_1, \ldots, A_n)$ be a set system on the finite set $S$, and $X \subseteq S$. Then $\mathfrak{A}$ possesses a transversal $B$ with $B \supseteq X$ if and only if*

(i) $|\bigcup_{i \in I} A_i| \geq |I|$ *for all* $I \subseteq \mathbb{N}_n$, *and*
(ii) $|X \cap \bigcup_{i \in I} A_i| \geq |X| - n + |I|$ *for all* $I \subseteq \mathbb{N}_n$.

*Proof.* Condition (i) says that $\mathfrak{A}$ possesses a transversal or, equivalently, that $r(\mathbf{T}(S; \mathfrak{A})) = n$. By **7.65**(ii), (ii) states that $X$ is independent in the matroid $\mathbf{T}(S; \mathfrak{A})$ and hence can be enlarged to a basis, i.e., to a transversal of $\mathfrak{A}$. □

The last result suggests the following "symmetrical" version.

**8.25 Proposition** (Mendelsohn–Dulmage). *Let $T(S; A_1, \ldots, A_n)$ be a set system on the finite set $S$. Then the following statements are equivalent:*

(i) *$X \subseteq S$ is a partial transversal and $\mathfrak{B} \subseteq \mathfrak{A}$ possesses a transversal.*
(ii) *There exist $X_0$, $\mathfrak{B}_0$ with $X \subseteq X_0 \subseteq S$, $\mathfrak{B} \subseteq \mathfrak{B}_0 \subseteq \mathfrak{A}$ so that $X_0$ is a transversal of $\mathfrak{B}_0$.*

*Proof.* Suppose $r(\mathbf{T}(S; \mathfrak{A})) = r \leq n$. Since $\mathfrak{B}$ is independent in the transversal matroid $\mathbf{T}(\mathfrak{A}; S)$ on $\mathfrak{A}$, it can be extended to a subfamily $\mathfrak{B}_0 \supseteq \mathfrak{B}$, $|\mathfrak{B}_0| = r$, which possesses a transversal, and we have $\mathbf{T}(S; \mathfrak{A}) = \mathbf{T}(S; \mathfrak{B}_0)$ by **7.66**. Since $X$ is a partial transversal of $\mathfrak{A}$, it is independent in $\mathbf{T}(S; \mathfrak{A})$, hence also in $\mathbf{T}(S; \mathfrak{B}_0)$, and can thus be enlarged to a transversal $X_0$ of $\mathfrak{B}_0$. □

In our final variant of Hall's theorem we study two families and their possible transversals.

**Definition.** Let $T(S; A_1, \ldots, A_n)$ and $T(S; B_1, \ldots, B_n)$ be two set systems. We call $\{x_i: i = 1, \ldots, n\}$, $x_i \in S$, a *system of common representatives* of $\mathfrak{A}$ and $\mathfrak{B}$ if

$$x_i \in A_i \cap B_{\Theta(i)} \qquad (i = 1, \ldots, n)$$

for some permutation $\Theta$ of $\{1, \ldots, n\}$. If all elements $x_i$ are distinct then $\{x_1, \ldots, x_n\}$ is called a *common transversal* of $\mathfrak{A}$ and $\mathfrak{B}$.

**8.26 Proposition.** *The set systems $T(S; A_1, \ldots, A_n)$ and $T(S; B_1, \ldots, B_n)$ have a system of common representatives if and only if the union of any $k$ sets of $\mathfrak{A}$ intersects at least $k$ sets of $\mathfrak{B}$, for $k = 1, \ldots, n$.*

*Proof.* We regard $\mathfrak{A}$ and $\mathfrak{B}$ as the defining vertex-sets of a bipartite graph $G(\mathfrak{A} \cup \mathfrak{B}, R)$ with

$$\{A_i, B_j\} \in R :\Leftrightarrow A_i \cap B_j \neq \emptyset.$$

Then the matchings in $G(\mathfrak{A} \cup \mathfrak{B}, R)$ correspond precisely to the partial systems of common representatives of $\mathfrak{A}$ and $\mathfrak{B}$. For $\mathfrak{A}_0 \subseteq \mathfrak{A}$ we have

$$R(\mathfrak{A}_0) = \{B_l : B_l \text{ intersects some set of } \mathfrak{A}_0\};$$

hence the theorem follows from **8.12**. □

**8.26** is especially interesting when $\mathfrak{A}$ and $\mathfrak{B}$ are both partitions of $S$. In this case, a system of common representatives is, of course, a common transversal.

**8.27 Corollary.** *Let $S = A_1 \dot\cup \cdots \dot\cup A_n = B_1 \dot\cup \cdots \dot\cup B_n$ be two partitions of $S$ into non-empty sets. $\mathfrak{A}$ and $\mathfrak{B}$ possess a common transversal if and only if the union of any $k$ of the sets in $\mathfrak{A}$ contains at most $k$ sets of $\mathfrak{B}$, for $k = 1, \ldots, n$. In particular, if $|A_i| = |B_i| = t$ for all $i$, then $\mathfrak{A}$ and $\mathfrak{B}$ possess a common transversal.* □

The last result has an interesting corollary for finite groups.

**8.28 Corollary.** *Let $G$ be a finite group and $H$ a subgroup. Then the family of left cosets of $H$ and the family of right cosets of $H$ possess a common transversal.* □

Finally, we come to the existence of common transversals for arbitrary families.

**8.29 Theorem.** *Let $T(S; A_1, \ldots, A_m)$ and $T(S; B_1, \ldots, B_n)$ be set systems and let $k \in \mathbb{N}$. Then $\mathfrak{A}$ and $\mathfrak{B}$ possess a common partial transversal of length $k$ if and only if*

$$\left| \bigcup_{i \in I} A_i \cap \bigcup_{j \in J} B_j \right| \geq |I| + |J| - (m + n - k) \quad \text{for all } I \subseteq \mathbb{N}_m, J \subseteq \mathbb{N}_n.$$

*In particular, if $|\mathfrak{A}| = |\mathfrak{B}| = n$ then $\mathfrak{A}$ and $\mathfrak{B}$ possess a common transversal if and only if $|\bigcup_{i \in I} A_i \cap \bigcup_{j \in J} B_j| \geq |I| + |J| - n$ for all $I, J \subseteq \mathbb{N}_n$.*

*Proof.* Consider the transversal matroid $\mathbf{M}(S) = \mathbf{T}(S; \mathfrak{B})$. Let $r$ be its rank function. $\mathfrak{A}$ and $\mathfrak{B}$ possess a common partial transversal of length $k$ if and only if $\mathfrak{A}$ possesses an independent partial transversal in $\mathbf{M}(S)$ of length $k$, i.e., of defect $m - k$. By **8.23**, this is the case if and only if

$$r\left(\bigcup_{i \in I} A_i\right) \geq |I| - (m - k) \quad \text{for all } I \subseteq \mathbb{N}_m.$$

Applying **7.65**(i) to $\mathbf{T}(S; \mathfrak{B})$ and $C = \bigcup_{i \in I} A_i$, we infer that this last condition for $I$ is equivalent to

$$\left| \bigcup_{i \in I} A_i \cap \bigcup_{j \in J} B_j \right| \geq |I| - (m - k) + |J| - n \quad \text{for all } J \subseteq \mathbb{N}_n. \quad \square$$

One may now pose questions about common transversals with prescribed subsets or prescribed defects; see the exercises for further details.

## B. Rado's Selection Principle

Up to now we have considered families containing only finitely many sets. In this section we make a brief detour into the infinite case, trying to determine whether there are infinite analogues of some of our basic theorems, such as Hall's and Dilworth's. In the course of these investigations, the general result below, discovered by Rado, has proved to be of fundamental importance.

**Definition.** Let $\mathfrak{A} = \{A_i : i \in I\}$ be an arbitrary family of subsets of a set $S$. We denote by $\mathscr{J} := \{J \subseteq I : |J| < \infty\}$ the family of all finite subsets of the index set $I$. Let $J \in \mathscr{J}$. Any function $\Theta_J : J \to S$ with $\Theta_J(j) \in A_j$ for all $j \in J$ is called a *local choice function.* Similarly, $\Theta : I \to S$ with $\Theta(i) \in A_i$ for all $i \in I$ is called a *global choice* function.

Our aim is to prove that for every system $\{\Theta_J : J \in \mathscr{J}\}$ of local choice functions there exists a global choice function $\Theta$ which, when restricted to $J$, simulates, in a certain sense, the behaviour of $\Theta_J$, for all $J$. In other words, the selection principle permits us to deduce from the existence of local configurations (transversals, partitions, etc.) the existence of a global configuration.

**8.30 Theorem** (Rado). *Let $\mathfrak{A} = \{A_i : i \in I\}$ be a family of finite subsets of a set $S$ and suppose there is a local choice function $\Theta_J$ for each $J \in \mathscr{J}$. Then there exists a global choice function $\Theta$ with the following property: For every $J \in \mathscr{J}$ there is some $K \in \mathscr{J}$ with $K \supseteq J$ such that $\Theta(j) = \Theta_K(j)$ for all $j \in J$.*

*Proof.* Let $\Omega$ be the totality of families $\mathfrak{B} = \{B_i : i \in I\}$ with $B_i \subseteq A_i$ for all $i$ such that, for every $J \in \mathscr{J}$, there is some $K$ with $J \subseteq K \in \mathscr{J}$ and $\Theta_K(j) \in B_j$ for all $j \in J$. $\Omega$ is non-empty since $\mathfrak{A} \in \Omega$. We partially order $\Omega$ by

$$\mathfrak{B} \leq \mathfrak{C} :\Leftrightarrow B_i \subseteq C_i \quad \text{for all } i \in I.$$

The proof consists in showing that $\Omega$ contains a minimal family $\mathfrak{M}$ from which $\Theta$ can be constructed. Let $\{\mathfrak{B}(\lambda) : \lambda \in \Lambda\}$ be a chain in $\Omega$. For $\lambda \in \Lambda$ set

$$\mathfrak{B}(\lambda) = \{B_i(\lambda) : i \in I\}, \qquad B_i(\lambda) \subseteq A_i,$$

and

$$\mathfrak{B}^* = \{B_i^* : i \in I\}, \quad \text{where } B_i^* = \bigcap_{\lambda \in \Lambda} B_i(\lambda) \subseteq A_i.$$

We claim that $\mathfrak{B}^* \in \Omega$. Since the $A_i$'s are all finite we infer that for every $i \in I$

$$B_i^* = B_i(\lambda_i) \quad \text{for some } \lambda_i \in \Lambda.$$

This implies that for every $J \in \mathscr{J}$ and $\lambda_0 = \min(\lambda_j : j \in J)$

$$B_j(\lambda_0) \subseteq B_j(\lambda_j) = B_j^* \subseteq B_j(\lambda_0),$$

and thus

$$B_j^* = B_j(\lambda_0) \quad \text{for all } j \in J.$$

Since $\mathfrak{B}(\lambda_0) \in \Omega$ there exists for every $J \in \mathscr{J}$ some $K \in \mathscr{J}$ with $J \subseteq K$ and

$$\Theta_K(j) \in B_j(\lambda_0) = B_j^* \quad \text{for all } j \in J.$$

This is precisely the condition for $\mathfrak{B}^* \in \Omega$, whence $\Omega$ contains a minimal family $\mathfrak{M} = \{M_i : i \in I\}$ by Zorn's lemma.

We now prove $|M_i| = 1$ for every $i$. Suppose otherwise. Then there exists $i_0 \in I$ with $x \neq y \in M_{i_0}$. Since $\mathfrak{M} \in \Omega$ we know that for every $J \in \mathscr{J}$ there is some $K \in \mathscr{J}$ with $J \subseteq K$ and $\Theta_K(j) \in M_j$ for all $j \in J$. Let us call any such $K$ an *associate* of $J$. Now consider $\mathfrak{M}' = \{M_i' : i \in I\}$ where

$$M_i' := \begin{cases} M_i & \text{for } i \neq i_0 \\ M_{i_0} - \{x\} & \text{for } i = i_0. \end{cases}$$

The minimality of $\mathfrak{M}$ implies the existence of $J_x \in \mathscr{J}$ with $i_0 \in J_x$ such that for all associates $K$ of $J_x$ there is some $j_x \in J_x$ with $\Theta_K(j_x) \notin M'_{j_x}$. It follows that $j_x = i_0$ and hence that

$$\Theta_K(i_0) = x \quad \text{for all associates } K \text{ of } J_x.$$

Similarly, there exists some $J_y \in \mathscr{J}$ with $i_0 \in J_y$ such that

$$\Theta_L(i_0) = y \quad \text{for all associates } L \text{ of } J_y.$$

Finally, let $N$ be an associate of $J_x \cup J_y$. Then $N$ is also an associate of both $J_x$ and $J_y$, and we obtain

$$\Theta_N(i_0) = x, \qquad \Theta_N(i_0) = y,$$

which is impossible.

To conclude the proof, let $M_i = \{z_i\}$ for all $i$ and define $\Theta: I \to S$ by

$$\Theta(i) = z_i.$$

Since $\mathfrak{M} \in \Omega$, every $J \in \mathscr{J}$ has an associate $K \in \mathscr{J}$ and for this $K$ we have

$$\Theta_K(j) = z_j = \Theta(j) \quad \text{for all } j \in J. \quad \square$$

**8.31 Corollary.** *Let $\mathfrak{A} = \{A_i : i \in I\}$ be a family of finite subsets of a set S and assume the same situation as in* **8.30**. *Then if all local choice functions are injective then so is the global choice function.*

*Proof.* Let $i \neq j \in I$. For $J = \{i, j\}$ there exists an associate $K$, whence $\Theta(i) = \Theta_K(i)$, $\Theta(j) = \Theta_K(j)$. Since $\Theta_K$ is injective we conclude $\Theta(i) \neq \Theta(j)$. $\square$

As applications of the selection principle we can now extend our fundamental theorems of sections 1 and 2.

**8.32 Theorem** (M. Hall). *Let $\mathfrak{A} = \{A_i : i \in I\}$ be an arbitrary family of finite subsets of a set S. Then the following conditions are equivalent:*

(i) *$\mathfrak{A}$ possesses a transversal.*
(ii) *Every finite subfamily $\mathfrak{A}_J = \{A_j : j \in J\}$ possesses a transversal.*
(iii) *Every finite subfamily $\mathfrak{A}_J$ satisfies Hall's condition.*

*Proof.* We know from **8.20** that (ii) $\Leftrightarrow$ (iii), whereas the implication (i) $\Rightarrow$ (ii) is trivially true. (ii) $\Rightarrow$ (i) is now precisely the content of **8.31**. $\square$

The correct extension of Hall's theorem to families of arbitrary subsets is still unknown. The obvious conjecture that **8.32**(iii) is also sufficient in the general case is false. For example, take $\mathfrak{A} = \{\mathbb{N}, \{1\}, \{2\}, \ldots\}$. Every finite subfamily of $\mathfrak{A}$, in fact every proper subfamily of $\mathfrak{A}$, possesses a transversal, but the whole family $\mathfrak{A}$ does not. For sufficient conditions see, e.g., Milner–Shelah [1] or Folkman [2].

To extend Rado's theorem **8.21** to arbitrary families we have to first weaken the definition of a matroid suitably, in order to allow infinite independent sets. (See ex. II.3.3.)

**Definition.** A set $S$ together with a family $\mathfrak{I} \subseteq 2^S$ (whose members are called independent sets) is called a *finitary matroid* if

(i) $\varnothing \in \mathfrak{I}$; $I \in \mathfrak{I}, J \subseteq I \Rightarrow J \in \mathfrak{I}$.
(ii) $I, J \in \mathfrak{I}, |I| < |J| < \infty \Rightarrow \exists p \in J - I$ such that $I \cup p \in \mathfrak{I}$.
(iii) $I \in \mathfrak{I} \Leftrightarrow J \in \mathfrak{I}$ for all $J \subseteq I, |J| < \infty$.

The proof of the following theorem follows the same lines as that of **8.30**.

**8.33 Theorem** (Rado). *Let* **M**$(S)$ *be a finitary matroid and $\mathfrak{A} = \{A_i : i \in I\}$ a family of subsets of S of finite rank. Then the following conditions are equivalent:*

(i) *$\mathfrak{A}$ possesses an independent transversal.*
(ii) *Every finite subfamily possesses an independent transversal.*
(iii) *Every finite subfamily satisfies Rado's condition.* $\square$

A natural question is whether, by analogy with **6.26**, the partial transversals of an arbitrary family $\mathfrak{A}$ of finite subsets of $S$ induce a finitary matroid on $S$. Properties (i) and (ii) above are certainly satisfied; (iii), however, is not true in general. Take $\mathfrak{A} = \{\{1, 2\}, \{1, 3\}, \{1, 4\}, \ldots\} \subseteq 2^{\mathbb{N}}$. Every finite subset of $\mathbb{N}$, indeed every proper subset of $\mathbb{N}$, is a partial transversal, but not $\mathbb{N}$.

Next, we demonstrate a transfinite version of Dilworth's theorem **8.14**.

**8.34 Theorem** (Dilworth). *Let $P_<$ be an arbitrary poset in which the cardinalities of antichains are bounded above by some number $m \in \mathbb{N}$. Then $P$ can be decomposed into $m$ disjoint chains.*

*Proof.* We define the set system $\mathfrak{A} = \{A_p : p \in P\}$ by $A_p := \{1, \ldots, m\}$ for all $p \in P$. Let $J \subseteq P$ be finite. By **8.14**, we may decompose $J = J_1 \cup \cdots \cup J_m$ into disjoint chains (some of which may be empty). Now we define the choice function $\Theta_J : J \to \{1, \ldots, m\}$ by

$$\Theta_J(p) := k,$$

where $k$ is the unique integer with $p \in J_k$. If $p, q \in J$ with $\Theta_J(p) = \Theta_J(q)$, then $p$ and $q$ are in the same chain and hence comparable. We choose such a function $\Theta_J$ for every finite subposet $J \subseteq P$, and then $\Theta : P \to \{1, \ldots, m\}$ according to **8.30**. To conclude the proof we claim that $P = P_1 \cup \cdots \cup P_m$ with

$$P_i = \{p \in P : \Theta(p) = i\}$$

is a required chain decomposition of $P$. Let $p, q \in P_i$. Then $\Theta(p) = \Theta(q)$. By **8.30**, there exists a finite subposet $K \subseteq P$ with $\{p, q\} \subseteq K$ and $\Theta(p) = \Theta_K(p)$, $\Theta(q) = \Theta_K(q)$. Hence $\Theta_K(p) = \Theta_K(q)$, and $p$ and $q$ are comparable. □

Our final application of the selection principle concerns the coloring of graphs; other applications are contained in the exercises.

**8.35 Proposition** (deBruijn–Erdös). *An arbitrary graph $G(V, E)$ is $k$-colorable if and only if every finite subgraph is $k$-colorable.*

*Proof.* Suppose chrom $H \leq k$ for every finite subgraph $H$. We define the family $\mathfrak{A} = \{A_v : v \in V\}$ by $A_v := \{1, \ldots, k\}$ for all $v \in V$. Let $J \subseteq V$, $|J| < \infty$. By the hypothesis, there exists a function $\Theta_J : J \to \{1, \ldots, k\}$ with $\{u, v\} \in E \Rightarrow \Theta_J(u) \neq \Theta_J(v)$, for all $u, v \in J$. The global function $\Theta$ constructed as in **8.30** is then the required coloring. □

## C. Applications

The importance of Hall's theorem and its variants lies in the simple formulation of the statement and the manifold interpretability of the conclusion. In many situations, it is useful to set up a suitable family of sets and ask for the transversals. Hall's theorem produces either a bijection between sets or a certain configuration. The following examples will illustrate both aspects.

*Latin Rectangles.* Consider the set $\{1, \ldots, n\}$ (or any other $n$-set). A *Latin square* of order $n$ is an $n \times n$-matrix with entries from $\{1, \ldots, n\}$ in which every row and column contains each of the numbers $1, \ldots, n$ precisely once. More generally, a *Latin rectangle* of type $r \times n$, $r \leq n$, is an $r \times n$-matrix in which the elements appear in every row and column at most once.

**8.36 Proposition.** *Let $L$ be a Latin rectangle of type $r \times n, r \leq n$. Then $L$ can always be extended to a Latin square by adding $n - r$ suitable rows.*

*Proof.* Let $A_j$ be the set of numbers which do not appear in column $j$, for $j = 1, \ldots, n$. Clearly $|A_j| = n - r$ for all $j$; furthermore every number $i$ is contained in precisely $n - r$ of the sets $A_j$. Thus $\{A_1, \ldots, A_n\}$ possesses a transversal $B$, by **8.22**, which can be added as the $r + 1$-th row. After $n - r$ steps the required square is obtained. □

0, 1-*matrices.* We have often had occasion to use incidence matrices $M = [m_{ij}]$ where $m_{ij} = 1$ or 0 depending on whether the element $i$ is in the set $j$ or not. The row sums $r_i$ and column sums $s_j$ therefore count the number of appearances of the element $i$ and the cardinality of the set $j$, respectively. Now let $R = (r_1, \ldots, r_m)$ and $S = (s_1, \ldots, s_n)$ be two vectors of non-negative integers. Under what conditions is there a 0, 1-matrix of type $m \times n$ with row sums $r_1, \ldots, r_m$ and column sums $s_1, \ldots, s_n$? Quite obviously, $\sum_{i=1}^m r_i = \sum_{j=1}^n s_j$. That this condition is not enough is shown by the following example: Let $R = (3, 3, 2, 1)$ and $S = (4, 4, 1)$. Any 0, 1-matrix with row sum vector $R$ and column sum vector $S$ would have to contain only ones in rows 1 and 2 and in columns 1 and 2 and hence would have to contain at least 10 ones.

We need some notation. Let $r_1, \ldots, r_m \in \mathbb{Z}$ with $0 \leq r_i \leq n$. For $j = 1, \ldots, n$ we set

$$r_j^* := |\{i : r_i \geq j\}|.$$

Clearly $\sum_{i=1}^m r_i = \sum_{j=1}^n r_j^*$ (we have already used this notation in **3.17**).

**8.37 Lemma.** *Let $r_1, \ldots, r_m \in \mathbb{Z}$ with $0 \leq r_i \leq n$ for all $i$ and let $A_1, \ldots, A_n$ be pairwise disjoint finite sets with $|A_j| = s_j$ where we assume $s_1 \geq s_2 \geq \cdots \geq s_n$. Then $\{A_1, \ldots, A_n\}$ possesses $m$ pairwise disjoint transversals of length $r_1, \ldots, r_m$, respectively, if and only if*

$$\sum_{j=n-k+1}^{n} s_j \geq \sum_{j=n-k+1}^{n} r_j^* \quad \textit{for } k = 1, \ldots, n.$$

*Proof.* The existence of partial transversals as required in the theorem is, by **8.23**, equivalent to

$$\left| \bigcup_{j \in J} A_j \right| \geq \sum_{i=1}^{m} (|J| - (n - r_i))^+ \quad \text{for all } J \subseteq \mathbb{N}_n$$

and thus, because $|A_1| \geq \cdots \geq |A_n|$, to

$$|A_{n-k+1} \dot\cup \cdots \dot\cup A_n| = \sum_{j=n-k+1}^{n} s_j \geq \sum_{i=1}^{m} (k-n+r_i)^+ \quad \text{for all } k = 1, \ldots, n.$$

The conclusion now follows from

$$\sum_{i=1}^{m} (k-n+r_i)^+ = \sum_{i=1}^{m} \sum_{j=n-k+1}^{r_i} 1 = \sum_{j=n-k+1}^{n} \sum_{\substack{1 \leq i \leq m \\ r_i \geq j}} 1 = \sum_{j=n-k+1}^{n} r_j^*. \quad \square$$

Let $(x_1, \ldots, x_n), (y_1, \ldots, y_n) \in \mathbb{R}^n$, and $\bar{x}_1 \geq \bar{x}_2 \geq \cdots \geq \bar{x}_n$, $\bar{y}_1 \geq \cdots \geq \bar{y}_n$ be rearrangements of their coordinates according to magnitude. We obtain a partial order $\prec$ on $\mathbb{R}^n$, setting

$$(x_1, \ldots, y_n) \prec (y_1, \ldots, y_n) :\Leftrightarrow \sum_{i=1}^{k} \bar{x}_i \leq \sum_{i=1}^{k} \bar{y}_i$$

for $k = 1, \ldots, n$ with equality for $k = n$.

**8.38 Theorem** (Gale–Ryser). *Let $r_1, \ldots, r_m \in \mathbb{Z}$ with $0 \leq r_i \leq n$ for all $i$ and $s_1, \ldots, s_n \in \mathbb{N}$. Then there exists an $m \times n$-matrix with entries 0 and 1, and row sums $r_1, \ldots, r_m$ and column sums $s_1, \ldots, s_n$, respectively, if and only if*

$$(s_1, \ldots, s_n) \prec (r_1^*, \ldots, r_n^*).$$

*Proof.* Suppose such a matrix $M = [m_{ij}]$ exists. Then clearly $\sum_{j=1}^{n} r_j^* = \sum_{i=1}^{m} r_i = \sum_{j=1}^{n} s_j$. Notice that we trivially have $r_1^* \geq r_2^* \geq \cdots \geq r_n^*$. Let $\mathfrak{A} = \{A_1, \ldots, A_n\}$ where

$$A_j = \{(i,j) : 1 \leq i \leq m, m_{ij} = 1\}, \qquad j = 1, \ldots, n.$$

The sets $A_1, \ldots, A_n$ are pairwise disjoint with $|A_j| = s_j$ for $j = 1, \ldots, n$. Furthermore, $\mathfrak{A}$ possesses $m$ disjoint partial transversals of length $r_1, \ldots, r_m$, respectively. Hence by **8.37**,

$$\sum_{j=n-k+1}^{n} \bar{s}_j \geq \sum_{j=n-k+1}^{n} r_j^* \quad \text{for } k = 1, \ldots, n,$$

which is equivalent to

$$\sum_{j=1}^{k} \bar{s}_j \leq \sum_{j=1}^{k} r_j^* \quad \text{for } k = 1, \ldots, n,$$

whence $(s_1, \ldots, s_n) \prec (r_1^*, \ldots, r_n^*)$.

Now suppose $(s_1, \ldots, s_n) \prec (r_1^*, \ldots, r_n^*)$. By reversing the argument just made, the condition in **8.37** holds. Take any family $\mathfrak{A} = \{A_1, \ldots, A_n\}$ of disjoint sets with $|A_j| = s_j$ and partial transversals $T_1, \ldots, T_m$ as specified there, and define the $m \times n$-matrix $M = [m_{ij}]$ by

$$m_{ij} = \begin{cases} 1 & \text{if } T_i \cap A_j \neq \varnothing \\ 0 & \text{if } T_i \cap A_j = \varnothing. \end{cases}$$

The row sums of $M$ are then $r_1, \ldots, r_m$, whereas the column sums $s_1', \ldots, s_n'$ satisfy $s_j' \leq s_j$ for $j = 1, \ldots, n$. Now clearly $\sum_{j=1}^n s_j' = \sum_{i=1}^m r_i = \sum_{j=1}^n r_j^* = \sum_{j=1}^n s_j$ whence $s_j' = s_j$ for all $j$, and the proof is complete. □

**8.38** concerns the existence of a 0, 1-matrix with given row and column sums. A much more difficult problem is the determination of their *number*. For further details see Ryser [1] or Snapper [1].

*Stochastic Matrices.* In statistics, matrices whose row sums or column sums (or both) are constant are of interest. A trivial but important example is that of *permutation matrices*, which are $n \times n$-matrices containing precisely one 1 in each row and each column and 0's elsewhere.

**8.39 Proposition.** *Let $M = [m_{ij}]$ be an $n \times n$-matrix whose entries are non-negative real numbers and suppose that all row sums and all column sums are equal to s. Then*

$$M = c_1 P_1 + \cdots + c_t P_t$$

*where each $P_i$ is a permutation matrix and the $c_i$'s are non-negative reals with $\sum_{i=1}^t c_i = s$. In particular, if $M \neq 0$ then $M$ contains a diagonal of length $n$ consisting of non-zero elements.*

*Proof.* If $M = 0$ then there is nothing to prove, so assume $M \neq 0$. We claim that $M$ contains a diagonal of length $n$ of non-zero entries. If not, then, by **8.10**, $M$ could be covered by $c$ rows and $d$ columns with $c + d < n$. We infer that

$$ns = \sum_{i,j=1}^{n} m_{ij} \leq (c + d)s < ns,$$

which is a contradiction. This proves the last assertion of the theorem. Now let $P_1$ be the permutation matrix with the 1's in exactly the positions of the elements of the diagonal, and let $c_1$ be the smallest of these $n$ non-zero entries. The matrix $M - c_1 P_1$ satisfies the conditions of the theorem with constant row and column sums $s - c_1$ and contains at least one more 0-entry than $M$. By induction, we finally arrive at the equation $M - c_1 P_1 - c_2 P_2 - \cdots - c_t P_t = 0$. □

**8.39** has an important corollary.

**Definition.** An $n \times n$-matrix $M$ is called *doubly stochastic* if the entries of $M$ are non-negative reals and if all row and column sums are equal to 1.

**8.40 Corollary** (Birkhoff). *A matrix $M$ is doubly stochastic if and only if $M = \sum_{i=1}^{t} c_i P_i$ with $\sum_{i=1}^{t} c_i = 1$ for certain permutation matrices $P_i$ and non-negative reals $c_i$. In particular, the product of two doubly stochastic matrices is again doubly stochastic.* $\square$

Recall the definition of the permanent of an $n \times n$-matrix $M = [m_{ij}]$ from section IV.2.B:

$$\operatorname{per}(M) = \sum_{\sigma \in S_n} m_{1\sigma(1)} \dots m_{n\sigma(n)}.$$

Since any non-zero doubly stochastic matrix possesses a diagonal of length $n$ by **8.39**, we have

$$0 < \operatorname{per}(M) \leq 1,$$

where the inequality on the right is a consequence of $\operatorname{per}(M) \leq$ product of the row sums (proof?). The $n \times n$-matrix $N = [n_{ij}]$ with $n_{ij} = 1/n$ for all $i$ and $j$ is doubly stochastic with

$$\operatorname{per}(N) = \frac{n!}{n^n},$$

and a still unproven conjecture of Van der Waerden asserts that this is the minimal possible value, i.e., that

$$\operatorname{per}(M) \geq \frac{n!}{n^n},$$

for every doubly stochastic matrix $M$.

*Inequalities Between Symmetric Functions.* Let $x_1, \dots, x_n$ be non-negative real numbers. The well-known inequality between the *geometric* and *arithmetic mean* states that

$$\sqrt[n]{x_1 x_2 \dots x_n} \leq \frac{x_1 + x_2 + \cdots + x_n}{n}.$$

If $b = (b_1, \dots, b_n) \in \mathbb{R}^n$ is any non-negative vector, then we denote by $[b]$ by the expression

$$[b] := \frac{1}{n!} \sum_{\sigma \in S_n} x_1^{b_{\sigma(1)}} \dots x_n^{b_{\sigma(n)}}.$$

For example, the geometric mean $[g]$ and the arithmetic mean $[a]$ result from the vectors $g = (1/n, 1/n, \ldots, 1/n)$ and $a = (1, 0, \ldots, 0)$. In general, we call any such expression $[b]$ a *symmetric mean* in the $x_i$'s. The vector $c$ is called an *average* of the vector $d$ (both with non-negative coordinates) if there exists a doubly stochastic matrix $M$ with $c = Md$. For instance,

$$g = Ma$$

where $M = [m_{ij}]$ with $m_{ij} = 1/n$ for all $i, j$. Generalizing the classical inequality stated above, we have the following theorem.

**8.41 Theorem** (Muirhead). *Let* $[c]$ *and* $[d]$ *be symmetric means. Then*

$$[c] \leq [d] \quad \text{for all non-negative values of } x_1, \ldots, x_n$$

*if and only if* $c$ *is an average of* $d$.

*Proof*. Suppose $c$ is an average of $d$. We rewrite $[c]$ in the form

$$[c] = \frac{1}{n!} \sum_Q \exp\left( \sum_{i=1}^{n} (Qc)_i \log x_i \right),$$

where $Q$ runs through all $n \times n$-permutation matrices. We assume $x_i > 0$ for all $i$; the general result follows from standard continuity arguments. The main step of the proof uses the well-known convexity property of exp

$$\exp\left( \sum_{i=1}^{n} \lambda_i x_i \right) \leq \sum_{i=1}^{n} \lambda_i \exp(x_i) \quad \text{for all } x_1 \geq 0, \ldots, x_n \geq 0, \lambda_i \geq 0, \text{ with } \sum_{i=1}^{n} \lambda_i = 1.$$

By hypothesis, there exists a doubly stochastic matrix $M$ with $c = Md$. **8.40** now implies

$$c = \sum_P \lambda_P Pd, \quad \lambda_P \geq 0 \text{ with } \sum_P \lambda_P = 1,$$

with $P$ running through all permutation matrices. Thus

$$\begin{aligned} n![c] &= \sum_Q \exp\left( \sum_{i=1}^{n} (Qc)_i \log x_i \right) = \sum_Q \exp\left( \sum_{i=1}^{n} \sum_P \lambda_P (QPd)_i \log x_i \right) \\ &\leq \sum_Q \sum_P \lambda_P \exp\left( \sum_{i=1}^{n} (QPd)_i \log x_i \right) = \sum_P \lambda_P \sum_Q \exp\left( \sum_{i=1}^{n} (Qd)_i \log x_i \right) \\ &= \sum_P \lambda_P n![d] = n![d], \end{aligned}$$

since $\sum_P \lambda_P = 1$ and the matrices $QP$ run through all permutation matrices for a fixed $P$.

Now suppose $[c] \leq [d]$ for all non-negative $x_1, \ldots, x_n \in \mathbb{R}$. Since, in the definition of a symmetric mean, all possible permutations occur, we may assume $c_1 \geq c_2 \geq \cdots \geq c_n \geq 0$ and similarly $d_1 \geq d_2 \geq \cdots \geq d_n \geq 0$. Set $x_1 = \cdots = x_k = x \geq 0$ and $x_{k+1} = \cdots = x_n = 0$. $[c] \leq [d]$ then reduces to the polynomial inequality

$$\sum_{\sigma \in S_n} x^{c_{\sigma(1)} + \cdots + c_{\sigma(k)}} \leq \sum_{\sigma \in S_n} x^{d_{\sigma(1)} + \cdots + d_{\sigma(k)}}.$$

For large $x$ this implies that

$$c_1 + \cdots + c_k \leq d_1 + \cdots + d_k \quad \text{for } k = 1, \ldots, n.$$

Furthermore, for small $x$, say $x = 1/n$, the polynomial inequality implies, for $k = n$, that $d_1 + \cdots + d_n \leq c_1 + \cdots + c_n$, and thus altogether

$$(c_1, \ldots, c_n) \prec (d_1, \ldots, d_n),$$

in the sense of **8.38**. A doubly stochastic matrix $M$ with $c = Md$ is now easily constructed by induction on $n$. $\square$

EXERCISES VIII.2

1. Let $T(S; A_1, \ldots, A_n)$ be a finite set system. We call two transversals $\{x_1, \ldots, x_n\}$ and $\{y_1, \ldots, y_n\}$ with $x_i, y_i \in A_i$ *distinct* if $x_i \neq y_i$ for at least one $i$. Show: If $m \leq |A_1| \leq |A_2| \leq \cdots \leq |A_n|$, then there are at least $\prod_{k=1}^{\min(m,n)} (|A_k| - k + 1)$ distinct transversals. Extend to independent transversals. (Rado)

2. Demonstrate with an example that the obvious generalization of **8.29** to three families $\mathfrak{A}, \mathfrak{B}, \mathfrak{C} \subseteq 2^S$ with $|\mathfrak{A}| = |\mathfrak{B}| = |\mathfrak{C}| = n$ (i.e., $\mathfrak{A}, \mathfrak{B}, \mathfrak{C}$ possess a common partial transversal if and only if $|\bigcup_{i \in I} A_i \cap \bigcup_{j \in J} B_j \cap \bigcup_{k \in K} C_k| \geq |I| + |J| + |K| - 2n$ for all $I, J, K \subseteq \mathbb{N}_n$) is not valid.

→ 3. Let $T(S; \mathfrak{A})$ and $T(S; \mathfrak{B})$ be two finite set systems and $\mathfrak{A}' \subseteq \mathfrak{A}$, $\mathfrak{B}' \subseteq \mathfrak{B}$. Prove that the following statements are equivalent:

   (i) $\mathfrak{A}'$ and a subfamily of $\mathfrak{B}$ possess a common transversal, and $\mathfrak{B}'$ and a subfamily of $\mathfrak{A}$ possess a common transversal.
   (ii) There exist families $\mathfrak{A}_0, \mathfrak{B}_0$ with $\mathfrak{A}' \subseteq \mathfrak{A}_0 \subseteq \mathfrak{A}$, $\mathfrak{B}' \subseteq \mathfrak{B}_0 \subseteq \mathfrak{B}$ which possess a common transversal.
   (Hint: Consider the proof of **8.29**.)

4. Let $T(S; A_1, \ldots, A_n)$ be a finite set system and $S = \dot{\bigcup}_{i=1}^{p} X_i$ a partition of $S$. Further, let $c_j, d_j \in \mathbb{Z}$ with $0 \leq c_j \leq d_j \leq |X_j|, j = 1, \ldots, p$. Show that $\mathfrak{A}$ possesses a transversal $E$ with

$$c_j \leq |E \cap x_j| \leq d_j \qquad (j = 1, \ldots, p)$$

if and only if

$$\left|\bigcup_{i \in I} A_i \cap \bigcup_{j \in J} X_j\right| \geq |I| - \min\left(n - \sum_{j \in J} c_j, \sum_{j \notin J} d_j\right)$$

for all pairs $I \subseteq \mathbb{N}_n, J \subseteq \mathbb{N}_p$.

(Hoffman–Kuhn) (Hint: Consider the families $X(c)$ anb $X(d)$ consisting of $c_j$ and $d_j$ copies of $X_j$, respectively, $j = 1, \ldots, p$. Now apply ex.3.)

5. Consider the following generalization of set systems and their transversals. Consider the pair $(S; \mathfrak{M})$ where $\mathfrak{M} = \{\mathbf{M}_1, \ldots, \mathbf{M}_n\}$ is a family of matroids on the finite set $S$ with rank functions $r_i$. A subset $T \subseteq S$ is called a *transversal* of $\mathfrak{M}$ if there is a mapping $\phi: T \to \mathbb{N}_n$ such that $\phi^{-1}(i)$ is a basis of $\mathbf{M}_i$ for all $i$. $T \subseteq S$ is called a *partial transversal* if $\phi: T \to \mathbb{N}_n$ exists such that $\phi^{-1}(i)$ is independent in $\mathbf{M}_i$, for all $i$. Hence if $T(S; A, \ldots, A_n)$ is a set system and $\mathbf{M}_i$ has as bases all one-element subsets of $A_i$, then we obtain the usual definitions. For a given system $(S; \mathfrak{M})$ prove:
   (i) The partial transversals of $(S; \mathfrak{M})$ are the family of independent sets of a matroid on $S$, namely the matroid $\sum_{i=1}^{n} \mathbf{M}_i$.
   (ii) $\mathfrak{M}$ possesses a transversal if and only if

$$|B| \geq \sum_{i=1}^{n} (r_i(S) - r_i(S - B)) \quad \text{for all } B \subseteq S.$$

   (iii) $\max_{A \text{ part. tr. of } \mathfrak{M}} |A| = \min_{B \subseteq S} (\sum_{i=1}^{n} r_i(B) + |S - B|)$.
   (iv) $S$ is transversal of $\mathfrak{M}$ if and only if

$$\sum_{i=1}^{n} (r_i(S) - r_i(S - A)) \leq |A| \leq \sum_{i=1}^{n} r_i(A) \quad \text{for all } A \subseteq S.$$

   (Brualdi)

→ 6. Generalize Rado's theorem **8.21**: Let $\mathfrak{M} = \{\mathbf{M}_1, \ldots, \mathbf{M}_n\}$ be a family of matroids on the finite set $S$ and $\mathbf{M}(S)$ another matroid on $S$ with rank function $r$. Then $\mathfrak{M}$ possesses a transversal which is independent in $\mathbf{M}(S)$ if and only if

$$r(A) \geq \sum_{i=1}^{n} (r_i(S) - r_i(S - A)) \quad \text{for all } A \subseteq S.$$

→ 7. Let $T(S; A_1, \ldots, A_n)$ be a finite set system. Show that there exists a disjoint union $X = \bigcup_{i=1}^{n} B_i$, $B_i \subseteq A_i$, $|B_i| = d_i$, $i = 1, \ldots, n$, if and only if $|A_i| \geq d_i$ for all $i$ and

$$|B| \geq \sum_{i=1}^{n} \max(0, d_i - |A_i - B|) \quad \text{for all } B \subseteq S.$$

(Hint: Use ex.5(ii).)

8. Prove Szpilrajn's theorem **1.4** by means of the selection principle.

→ 9. Prove with the aid of the selection principle: An infinite graph is a comparability graph if and only if every finite induced subgraph is a comparability graph. (Wolk)

→10. Prove:
(i) Finitary matroids possess bases.
(ii) Any two bases have the same cardinality.
(Hint: Use the selection principle.)

11. Prove **8.33**.

12. Show that there exist at least $n!(n-1)! \cdots (n-r+1)!$ Latin rectangles over $\mathbb{N}_n$ of type $r \times n$.

→13.* Let $1 \leq r, s \leq n$ and let $R$ be a Latin rectangle over $\mathbb{N}_n$ of type $r \times s$. Show that $R$ can be extended to a Latin square of order $n$ if and only if every $i \in \mathbb{N}_n$ occurs at least $r + s - n$ times in $R$. (Ryser)

14. Complete the proof of **8.41**.

→15. Let $c = (c_1, \ldots, c_n) \in \mathbb{R}^n$ with non-negative coordinates and $\sum_{i=1}^{n} c_i = 1$. Show that then $[g] \leq [c] \leq [a]$, where $[g]$ and $[a]$ are the geometric and arithmetic means, respectively.

## 3. Sperner Theorems

We continue our discussion of combinatorial problems on finite posets begun in section 1.C. The central result on arbitrary posets was Dilworth's theorem: The minimal number $d(P)$ of disjoint chains into which a finite poset $P$ can be decomposed equals the maximal size of an antichain in $P$. Whereas in section 1 we studied primarily chain decompositions, we now concentrate on the second aspect, namely maximal antichains. As before we denote by $\mathfrak{A}(P)$ the family of antichains in $P$ and set

$$s(P) := \max_{A \in \mathfrak{A}(P)} |A|.$$

$s(P)$ is called the *Sperner number*; **8.14** says simply that $d(P) = s(P)$.

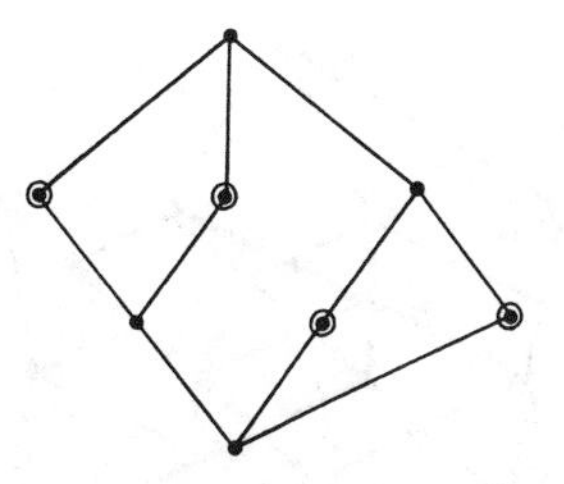

Figure 8.3

**Definition.** Let $P$ be a finite poset with rank function $r$ and let $P^{(k)} := \{x \in P : r(x) = k\}$ be the $k$-th level of $P$ with $W_k(P) := |P^{(k)}|$, $k = 0, \ldots, r(P)$. $P$ is said to possess the *Sperner property* (S) if

$$s(P) = \max_k W_k(P).$$

Since every level is an antichain, we always have $s(P) \geq \max_k W_k(P)$. By **8.14**, (S) is equivalent to (D): $P$ can be decomposed into $\max_k W_k(P)$ disjoint chains.

**Example.** The poset of Figure 8.3 has a rank function but does not possess (S) since the circled elements form an antichain of size 4 whereas $\max_k W_k = 3$.

Property (D) suggests the method by which we seek to prove the Sperner property for a given ranked poset $P$. We study certain canonical decompositions of $P$, trying to find one among them which contains only $\max W_k(P)$ members. Hence we shall prove, in general, not only the existence of some chain decomposition having the required number of chains, but also the existence of a very special minimal decomposition.

To get a clear picture of how the theorems are interrelated, we postulate a few other properties.

**Definition.** Let $P$ be a finite poset with rank function $r$, $r(P) = n$, and $W_0, \ldots, W_n$ the level numbers. We denote by $G^{(k)} = G(P^{(k)} \cup P^{(k+1)}, R^{(k)})$ the bipartite graph induced by the covering relation on $P^{(k)} \cup P^{(k+1)}$.

(i) $P$ possesses the *unimodular property* (U) if $\{W_k : k = 0, \ldots, n\}$ is a unimodal sequence, i.e., if

$$W_0 \leq W_1 \leq \cdots \leq W_m \geq \cdots \geq W_n.$$

(ii) $P$ possesses the *matching property* (M) if the bipartite graph

$$G(P^{(k)} \cup P^{(k+1)}, R^{(k)})$$

contains a full matching for all $k = 0, \ldots, n - 1$.

(iii) $P$ is said to be *symmetric* or to possess property (SYM) if there exists a partition of $P$ into disjoint chains $C_i = \{a_i <\cdot \cdots <\cdot b_i\}$ which are not refinable and where $r(a_i) + r(b_i) = n$ for all $i$.

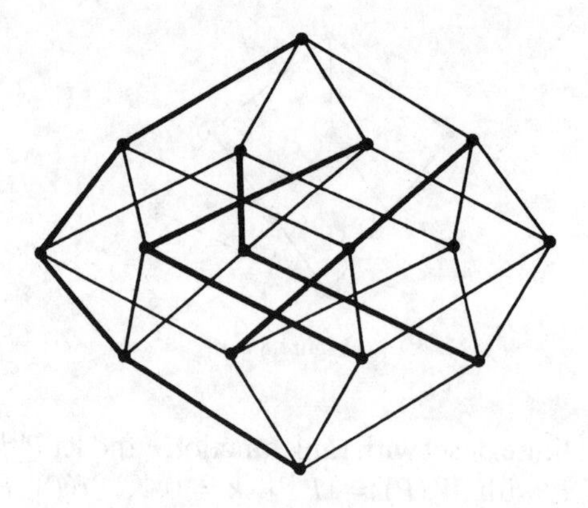

Figure 8.4

**Example.** We already know that $\mathscr{B}(n)$, $\mathscr{L}(n, q)$ and $\mathscr{P}(n)$ possess property (U); see **3.7**, **3.13**, and **3.30**. Figure 8.4 gives a symmetric decomposition of $\mathscr{B}(4)$ by using the heavily marked covering relations. Notice that these relations also induce full matchings for all bipartite graphs $G^{(k)}$, a fact which we now prove in general.

**8.42 Proposition.** *For a finite poset $P$ with rank function the following implications hold*:

$$(\mathrm{SYM}) \Rightarrow (\mathrm{U}) \wedge (\mathrm{M}) \Rightarrow (\mathrm{S}) \Leftrightarrow (\mathrm{D}).$$

*Proof*. We already know the equivalence of (S) and (D). Let $P$ satisfy (U) and (M). By adjoining the edges of the full matchings in $G(P^{(k)} \cup P^{(k+1)}, R^{(k)})$, we obtain (because of (U)) a chain decomposition of $P$ into $\max_k W_k$ chains. Now suppose $P = C_1 \dot{\cup} \cdots \dot{\cup} C_t$ is a symmetric chain decomposition. Every element $x \in P$ with $r(x) = k < r(P)/2$ is then contained in some chain $C_i$ whose last element has rank at least $r(P) - k > r(P)/2$. This implies $W_k \leq W_{k+1}$ for $0 \leq k < r(P)/2$ and, by symmetry, $W_k \geq W_{k+1}$ for $r(P) > k \geq r(P)/2$. Hence $P$ satisfies (U) with the largest levels lying in the middle and $W_k = W_{r(P)-k}$ for all $k$. By the same argument we conclude that the chains $C_i$ induce a full matching in each graph $G(P^{(k)} \cup P^{(k+1)}, R^{(k)})$. □

Notice that neither (U) nor (M) alone implies the Sperner property. The poset in Figure 8.3 is an example for (U) $\not\Rightarrow$ (S) whereas the poset $P$ in Figure 8.5 shows that (M) $\not\Rightarrow$ (S), since $P$ satisfies (M) but $s(P) = 4 > 3 = \max_k W_k$. The uniform lattice $U_3(\mathscr{B}(4))$ shown in Figure 2.16 illustrates the fact that (U) $\wedge$ (M) $\not\Rightarrow$ (SYM).

We divide this section into 3 parts, first considering property (S), then (U) and (M), and finally (SYM).

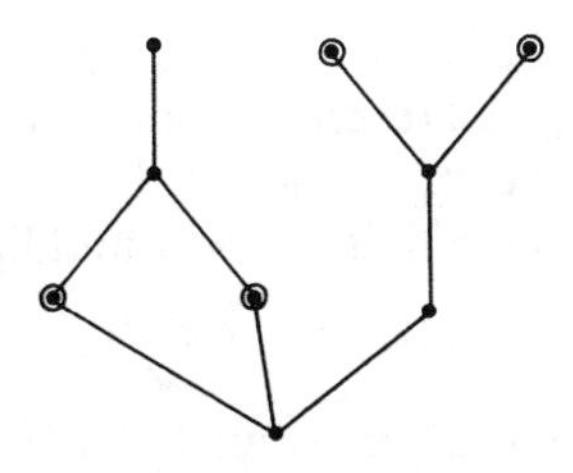

Figure 8.5

## A. The Sperner Property

The original theorem of Sperner states that all Boolean algebras possess property (S).

**8.43 Theorem** (Sperner). *The Boolean algebras $\mathscr{B}(n)$ possess property* (S) *for all n.*

*Proof* (Lubell). The level numbers of $\mathscr{B}(n)$ are the binomial coefficients $\binom{n}{k}$, $k = 0, 1, \ldots, n$. Let $S$ be an $n$-set. We count the maximal chains in $\mathscr{B}(S)$. Clearly there are $n!$ such chains. If $\{A_1, \ldots, A_m\}$ is an arbitrary antichain in $\mathscr{B}(S)$, then a maximal chain contains at most one of the sets $A_i$. Since the number of maximal chains passing through a fixed $A_i$ is precisely $|A_i|!(n - |A_i|)!$, we obtain

$$\sum_{i=1}^{m} |A_i|!(n - |A_i|)! \leq n!,$$

and thus

$$\sum_{i=1}^{m} \frac{1}{\binom{n}{|A_i|}} \leq 1.$$

If $d_k$ denotes the number of $A_i$'s with $|A_i| = k$, then we can write this last inequality as

$$(+) \qquad \sum_{k=0}^{n} \frac{d_k}{\binom{n}{k}} \leq 1,$$

whence

$$\frac{m}{\max\binom{n}{k}} \leq 1,$$

which is precisely property (S). □

Notice that we have not used either of the properties (U) or (M) in deriving the *stronger* inequality (+). We shall return to condition (+) in section B. Notice also that (+) implies that $m = \max_k \binom{n}{k}$ can hold only when $d_k = 0$ for $k \neq \lfloor n/2 \rfloor, \lceil n/2 \rceil$. From this it is easy to conclude that the two middle levels are, in fact, the only maximal sized antichains.

**8.44 Corollary.** *The only maximal sized antichains in $\mathscr{B}(n)$ are the two middle levels of rank $\lfloor n/2 \rfloor$ and $\lceil n/2 \rceil$ (which coincide for even n).* □

Let us note two interesting applications in number theory and analysis.

**8.45 Proposition.** *Let $m = p_1 p_2 \dots p_n$ be a square-free integer. The maximal number of divisors of m which do not divide one another is $\binom{n}{\lfloor n/2 \rfloor}$.*

*Proof.* We know that $[1, m] \subseteq \mathscr{T}$ is isomorphic to $\mathscr{B}(n)$. Now apply **8.43**. □

**8.46 Proposition.** *Let $a_1, \dots, a_n \in \mathbb{R}$ with $a_i \geq 1$ for all i and let $I = (b, b+2)$ be an open interval in $\mathbb{R}$. Then of the $2^n$ numbers $\sum_{i=1}^n \varepsilon_i a_i$, $\varepsilon_i = \pm 1$ for all i, at most $\binom{n}{\lfloor n/2 \rfloor}$ lie in I.*

*Proof.* Suppose $a = \sum \varepsilon_i a_i$ and $a' = \sum \varepsilon_i' a_i$ lie in $I$, $a \neq a'$. We set $A = \{i : \varepsilon_i = 1\}$ and $A' = \{i : \varepsilon_i' = 1\}$. If $A \subseteq A'$ then

$$a' - a = \sum \varepsilon_i' a_i - \sum \varepsilon_i a_i = 2 \sum_{i \in A' - A} a_i \geq 2.$$

Hence the index sets $A$ form an antichain in $\mathscr{B}(n)$ and the result follows from **8.43**. □

The method of proof used in **8.43** can be carried over verbatim to vector space lattices $\mathscr{L}(n, q)$.

**8.47 Proposition.** *The vector space lattices $\mathscr{L}(n, q)$ possess property* (S) *for all n and q and the only maximal sized antichains are the middle levels (which coincide for even n).* □

By the same argument one can verify property (S) for arbitrary chain products. As we shall deduce the stronger condition (SYM) for chain products in section C (which curiously enough is easier to prove), we defer a discussion until then. It was recently proved by Canfield [2] that the partition lattices $\mathscr{P}(n)$ do not possess property (S), and hence not (M), for large $n$.

We remarked at the beginning of this section that we often seek special minimal chain decompositions, or, what is the same, certain maximal sized antichains, satisfying some additional requirement. The following theorem is a typical example.

**8.48 Proposition.** *Let $P$ be a finite poset and $G$ a permutation group on $P$ which preserves the order relation on $P$. Then $P$ contains a maximal sized antichain which is a union of $G$-orbits.*

*Proof*. We know from **8.18** that the set $\mathfrak{A}_{\max}(P)$ of maximal sized antichains forms a distributive lattice. $G$ induces on $\mathfrak{A}_{\max}(P)$ an order preserving permutation group $G'$. It follows that every $g' \in G'$ must fix the maximal element $A \in \mathfrak{A}_{\max}(P)$ (and similarly the minimal element), whence $A$ must be a union of $G$-orbits. □

Since evidently all elements of a $G$-orbit must be of the same rank, every level decomposes into $G$-orbits. Therefore, if $G$ acts transitively on the levels of $P$, in which case the levels themselves are orbits, then $P$ must satisfy the Sperner property. This remark furnishes new proofs of **8.43** and **8.47**. The order preserving permutations of a partition lattice $\mathscr{P}(N)$ are induced by the permutations on $N$; this implies that two partitions are in the same orbit if and only if they have the same *type*. By the remark at the end of **2.67** the types correspond bijectively to the *number partitions* of $n = |N|$ which, in turn, are ordered by the dominance relation (see Figure 1.9). Hence we have the following corollary.

**8.49 Corollary.** *There exists a maximal sized antichain in $\mathscr{P}(n)$ consisting of all partitions of certain types. Furthermore, these types form an antichain in the dominance order $\mathscr{D}(n)$.* □

It was thought for a while that any finite geometric lattice possesses the Sperner property. The following was the first counterexample to this conjecture.

**Example** (Dilworth–Greene). Let $G_n$ be the graph of Figure 8.6, $n \geq 3$, and let $P_n$ be the lattice of flats of its polygon matroid $\mathbf{P}(G_n)$. Consider the flats $U$ of rank $k$. If $U$ does not contain $p$ then $U$ contains at most one of $u_i$ or $v_i$ for every $i$. This gives $2^k\binom{n-1}{k}$ flats. Since, by **6.40**, $[p, 1] \cong \mathscr{B}(n-1)$, there are $\binom{n-1}{k-1}$ flats $U$ containing $p$. Hence, altogether we have

$$W_k(P_n) = 2^k\binom{n-1}{k} + \binom{n-1}{k-1} \qquad (k = 1, \ldots, n).$$

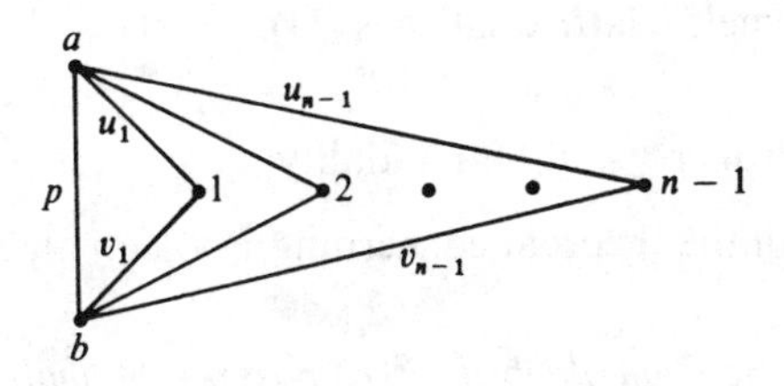


Figure 8.6

The vertex-automorphism group of the graph $G_n$ is plainly generated by all permutations of $\{1, \ldots, n-1\}$ together with the reflection $(a, b)$. It follows that the $k$-level $P_n^{(k)}$ splits into two sets $A_k$ and $B_k$ depending on whether a flat contains $p$ or not. Thus

$$|A_k| = 2^k \binom{n-1}{k}, \qquad |B_k| = \binom{n-1}{k-1}.$$

By **8.48**, there exists a maximal sized antichain $A$ of $P_n$ of the form

$$A = A_{i_1} \cup \cdots \cup A_{i_s} \cup B_{j_1} \cup \cdots \cup B_{j_t}.$$

Now for $k < l$ there obviously exist $x \in A_k$, $y \in A_l$ with $x \leq y$, similarly for $B_k$, $B_l$ and $A_k$, $B_l$. Hence $A$ must be of the form

$$A = A_k \cup B_l \quad \text{with } k \geq l$$

and, conversely, any such set is an antichain.

The maximal value of $|A_k|$ occurs approximately at $k = \lfloor 2n/3 \rfloor$, whereas that of $|B_l|$ at approximately $l = \lceil n/2 \rceil$. Hence we obtain the Sperner number $s(P_n)$ by separately maximizing $|A_k|$ and $|B_l|$, i.e.,

$$s(P_n) = \max_k 2^k \binom{n-1}{k} + \max_k \binom{n-1}{k-1}.$$

For $n = 11$ this gives

$$\max_k W_k(P_{11}) = W_7(P_{11}) = 15570,$$
$$s(P_{11}) = 15612,$$

and thus $\max_k W_k(P_{11}) < s(P_{11})$.

It is easily verified that $P_n$ possesses (U). Applying **8.42**, we conclude that $P_n$ is a geometric lattice which satisfies neither (M) nor (S). It is conjectured, however, that every finite geometric lattice satisfies (U).

### B. The Matching Property and Unimodality

Again, we begin with the theorem concerning Boolean algebras.

**8.50 Theorem.** *The Boolean algebras $\mathscr{B}(n)$ possess the matching property* (M) *and the unimodular property* (U) *for all n.*

*Proof.* (U) was shown in **3.7**. The bipartite graph $G(\mathscr{B}^{(k)}(n) \cup \mathscr{B}^{(k+1)}(n), R^{(k)})$ is regular on $\mathscr{B}^{(k)}(n)$ of degree $n - k$, and on $\mathscr{B}^{(k+1)}(n)$ of degree $k + 1$. Hence there exists a full matching by **8.13**. □

For vector space lattices, (M) is verified by an identical argument, whereas (U) was shown in **3.13**.

**8.51 Proposition.** *The vector space lattices* $\mathscr{L}(n, q)$ *possess properties* (M) *and* (U) *for all n and q.* □

Notice that in the proof of **8.50** only the regularity of the bipartite graph $G^{(k)}$ was used, which, when applied to all levels, says that the number of maximal chains extending from 0 to an element $a$ (and dually) depends only on the rank $r(a)$. This, on the other hand, was also the argument used to establish (+) in the proof of **8.43**. We are thus led to the following definitions.

**Definition.** Let $P$ be a finite poset with rank function $r$ and level numbers $W_0, W_1, \ldots, W_n$.

(i) $P$ is called *regular* or said to possess property (R) if any two elements of the same rank cover the same number of elements and are covered by the same number of elements.

(ii) $P$ is called *normal* or said to possess property (N) if, for every antichain $A$ in $P$,

$$\sum_{a \in A} \frac{1}{W_{r(a)}} \leq 1.$$

This last property is also called the LYM-condition, named after Lubell–Yamamoto–Meshalkin.

Boolean algebras and finite vector space lattices are thus regular (and normal) whereas, in general, finite chain products or partition lattices are not.

**8.52 Proposition.** *Let P be a finite poset with rank function. Then*

(i) (R) ⇒ (N).
(ii) (N) ⇒ (M) ∧ (S).

*Proof.* Let $P$ be regular and suppose that every element of rank $k$ covers $a_k$ elements and is covered by $b_k$ elements, $k = 0, 1, \ldots, n = r(P)$. To verify (i), we proceed as in the proof of **8.43**. Every element of rank $k$ is contained in $a_1 a_2 \ldots a_k b_k b_{k+1} \ldots b_{n-1}$ maximal chains. Hence this number is the same for all $x \in P^{(k)}$; denote it by $f(k)$. On the other hand, since every maximal chain contains exactly one element of rank $k$ we must have

$$f(k) = \frac{t}{W_k},$$

where $t$ is the total number of maximal chains. Now let $A$ be an antichain in $P$ and let $d_k = |\{x \in A: r(x) = k\}|$. Since every maximal chain contains at most one element of $A$, we conclude that

$$\sum_{k=0}^{n} d_k f(k) \leq t,$$

and thus

$$\sum_{a \in A} \frac{1}{W_{r(a)}} = \sum_{k=0}^{n} \frac{d_k}{W_k} \leq 1.$$

The proof of (N) $\Rightarrow$ (S) can be carried over verbatim from **8.43**. It remains to be shown that (N) $\Rightarrow$ (M). Suppose $W_k = |P^{(k)}| \leq W_{k+1} = |P^{(k+1)}|$. We have to show that all of $P^{(k)}$ can be matched into $P^{(k+1)}$ or, in other words, that Hall's condition holds in the graph $G(P^{(k)} \cup P^{(k+1)}, R^{(k)})$. For $A \subseteq P^{(k)}$ let $R^{(k)}(A) \subseteq P^{(k+1)}$ be the set of elements in $P$ which cover at least one element of $A$. Since $A \cup (P^{(k+1)} - R^{(k)}(A))$ is an antichain, we have, by hypothesis,

$$\frac{|A|}{W_k} + \frac{|P^{(k+1)} - R^{(k)}(A)|}{W_{k+1}} = \frac{|A|}{W_k} + 1 - \frac{|R^{(k)}(A)|}{W_{k+1}} \leq 1,$$

and thus

$$\frac{|A|}{W_k} \leq \frac{|R^{(k)}(A)|}{W_{k+1}} \leq \frac{|R^{(k)}(A)|}{W_k},$$

i.e.

$$|A| \leq |R^{(k)}(A)|.$$

The case $W_k \geq W_{k+1}$ is dealt with similarly by considering $B \subseteq P^{(k+1)}$ and $R^{(k)}(B) \subseteq P^{(k)}$. □

**8.53 Corollary.** *The following lattices are regular and thus satisfy* (N), (M), *and* (S):

(i) *Boolean algebras* $\mathcal{B}(n)$.
(ii) *Vector space lattices* $\mathcal{L}(n, q)$.
(iii) *Affine lattices* $\mathcal{A}(n, q)$.
(iv) *Uniform lattices* $U_k(\mathcal{B}(n))$. □

We remark that neither implication in **8.52** can be reversed. The chain product $\mathscr{C}(2) \times \mathscr{C}(1)$ is normal but not regular. The poset in Figure 8.7 satisfies (M) and (S) (as a matter of fact it is symmetric) but is not normal, as is seen from the antichain consisting of the circled elements.

The method used in the proof of **8.52** can easily be extended from antichains to arbitrary subsets in $P$.

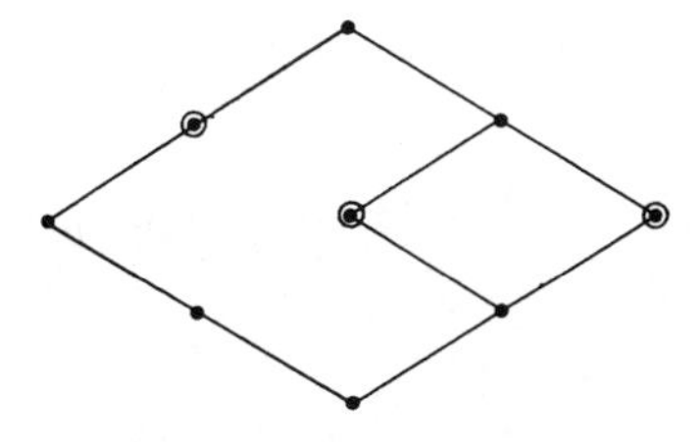

Figure 8.7

**8.54 Proposition.** *Let $P$ be a finite poset which is regular and let $W_0, W_1, \ldots, W_n$ be the level numbers. If $F$ is any subset of $P$, then*

$$|F| \leq \max_{C \in \mathscr{C}} \left( \sum_{a \in C \cap F} W_{r(a)} \right),$$

*where $\mathscr{C}$ is the set of maximal chains in $P$.*

*Proof*. Define $\lambda(a, C)$ for all $a \in F$ and $C \in \mathscr{C}$ by

$$\lambda(a, C) := \begin{cases} W_{r(a)} & \text{if } a \in C \\ 0 & \text{if } a \notin C. \end{cases}$$

By counting $\sum_{a \in F, C \in \mathscr{C}} \lambda(a, C)$ in two ways we obtain, with the notation of the preceding proof,

$$\sum_{k=0}^{n} d_k f(k) W_k = \sum_{C \in \mathscr{C}} \left( \sum_{a \in C \cap F} W_{r(a)} \right).$$

Now, by the argument in **8.52**, the left-hand side is just $|F|t$. Hence

$$|F| = \frac{1}{t} \sum_{C \in \mathscr{C}} \left( \sum_{a \in C \cap F} W_{r(a)} \right) \leq \max_{C \in \mathscr{C}} \left( \sum_{a \in C \cap F} W_{r(a)} \right). \quad \square$$

Call $F$ an *h-family* of the ranked poset $P$ if $F$ contains no chain of cardinality $h + 1$ (= length $h$). Any union of $h$ levels is an example of an $h$-family. The 1-families are just the antichains.

**8.55 Corollary.** *Let $P$ be a finite regular poset. Then the maximal cardinality of an $h$-family equals the sum of the $h$ largest level numbers.* $\square$

For Boolean algebras this yields some interesting results. **8.56** is an example; others are contained in the exercises.

**8.56 Proposition.** *Let $\mathfrak{F} = \{A_1, \ldots, A_m\}$ be a family of subsets of an n-set S and let $h \in \mathbb{N}$. If*

$$A_i \subseteq A_j \Rightarrow |A_j - A_i| \leq h - 1 \qquad \textit{for all } 1 \leq i, j \leq n,$$

*then*

$$m \leq \sum_{l=1}^{h} \binom{n}{\lfloor (n+l)/2 \rfloor}. \quad \square$$

**8.52** shows that condition $(N)$ is stronger than the matching property (M). The following theorem, suggested by the proof of **8.52**, shows how to strengthen (M) in order to characterize normal posets.

**8.57 Theorem** (Harper–Kleitman). *Let P be a finite poset with rank function r and level numbers $W_0, W_1, \ldots, W_n$. Then P is normal if and only if, for every $k = 0, 1, \ldots, n-1$,*

$$\frac{|A|}{W_k} \leq \frac{|R^{(k)}(A)|}{W_{k+1}} \qquad \textit{for all } A \subseteq P^{(k)}.$$

*This condition is called the* normalized matching condition. *Notice that by duality, P is also normal if and only if, for every $k = 0, \ldots, n-1$,*

$$\frac{|B|}{W_{k+1}} \leq \frac{|R^{(k)}(B)|}{W_k} \qquad \textit{for all } B \subseteq P^{(k+1)}.$$

*Proof.* That a normal poset satisfies the condition of the theorem was shown in the proof of **8.52**. For $k$ between 0 and $n-1$ let $P^{(k)} = \{a_1, \ldots, a_{W_k}\}$, $P^{(k+1)} = \{b_1, \ldots, b_{W_{k+1}}\}$ and set $W = \prod_{i=0}^{n} W_i$. We define the bipartite graph $\tilde{G}(\tilde{P}^{(k)} \cup \tilde{P}^{(k+1)}, \tilde{R}^{(k)})$ as follows:

$$\tilde{P}^{(k)} := \left\{a_{ij} : i = 1, \ldots, W_k,\ j = 1, \ldots, \frac{W}{W_k}\right\}$$

$$\tilde{P}^{(k+1)} := \left\{b_{uv} : u = 1, \ldots, W_{k+1},\ v = 1, \ldots, \frac{W}{W_{k+1}}\right\}$$

with

$$\{a_{ij}, b_{uv}\} \in \tilde{R}^{(k)} :\Leftrightarrow a_i <\!\cdot\, b_u \quad \text{in } R^{(k)}.$$

In other words, $\tilde{P}^{(k)}$ consists of $W/W_k$ copies of $P^{(k)}$, and $\tilde{P}^{(k+1)}$ of $W/W_{k+1}$ copies of $P^{(k+1)}$, with the inherited covering relations. In particular, we have $|\tilde{P}^{(k)}| = |\tilde{P}^{(k+1)}| = W$. We claim that $\tilde{G}$ contains a full matching. Let $\tilde{A} \subseteq \tilde{P}^{(k)}$ with

$\tilde{R}^{(k)}(\tilde{A}) \subseteq \tilde{P}^{(k+1)}$ and let $A \subseteq P^{(k)}$ be the set of elements $a_i$ for which there is at least one $j$ with $a_{ij} \in \tilde{A}$. Then clearly

$$|\tilde{A}| \leq \frac{W}{W_k}|A| \quad \text{and} \quad |\tilde{R}^{(k)}(\tilde{A})| = \frac{W}{W_{k+1}}|R^{(k)}(A)|,$$

whence, by the normalized matching property, Hall's condition

$$|\tilde{A}| \leq |\tilde{R}^{(k)}(\tilde{A})| \quad \text{for all } \tilde{A} \subseteq \tilde{P}^{(k)}$$

results. Putting these matchings together for all consecutive ranks we obtain a collection of $W$ maximal chains which between them contain every element in $P$ of rank $k$ precisely $W/W_k$ times. Hence, if $A$ is an arbitrary antichain, then, with the notation $d_k$ as in **8.52**,

$$\sum_{k=0}^{n} d_k \frac{W}{W_k} \leq W,$$

i.e.,

$$\sum_{a \in A} \frac{1}{W_{r(a)}} = \sum_{k=0}^{n} \frac{d_k}{W_k} \leq 1. \quad \square$$

While it is somewhat harder to show that the partition lattices $\mathscr{P}(n)$ do not, in general, possess (M), it is easy to prove that they do not satisfy the normalized matching property as seen in the following example.

**Example** (Spencer). Let $n$ be even and let $X$ be the set of all partitions of an $n$-set $S$ into 2 blocks each of cardinality $n/2$. Then $|X| = \frac{1}{2}\binom{n}{n/2}$. Since any partition in $X$ covers precisely $2S_{n/2,2}$ partitions, and any partition into 3 blocks is covered by at most one member of $X$, we have $|R(X)| = \frac{1}{2}\binom{n}{n/2}2S_{n/2,2} = \binom{n}{n/2}S_{n/2,2}$. If (N) holds then

$$\frac{|X|}{S_{n,2}} \leq \frac{|R(X)|}{S_{n,3}},$$

which reduces to

$$(+) \qquad S_{n,3} \leq 2S_{n/2,2} \cdot S_{n,2}.$$

Using the recursion **3.29**(ii) one easily computes $S_{n,2} = 2^{n-1} - 1$ and $S_{n,3} = (3^{n-1} - 2^n + 1)/2$. Substitution into (+) now yields

$$3^{n-1} - 1 \leq (2^{n/2} - 1)(2^n - 2),$$

which is false for $n \geq 20$. A similar argument shows that $\mathscr{P}(n)$ is not normal for odd $n \geq 20$. It was, however, verified by computer that $\mathscr{P}(n)$ is indeed normal for $n \leq 19$ (Graham–Harper).

To conclude our discussion of the matching property we briefly consider arbitrary finite geometric lattices. We have shown that every such lattice possesses at least as many copoints as points. The following result strengthens this fact by exhibiting special matchings.

**8.58 Proposition** (Greene). *Let L be a finite geometric lattice with point set S and copoint set C. Then there exist injections f and g such that*

(i) $f: S \to C$ *with* $p \leq f(p)$ *for all* $p \in S$,
(ii) $g: S \to C$ *with* $p \nleq g(p)$ *for all* $p \in S$.

*Proof*. Consider the bipartite graph $G(S \cup C, R)$ with $\{p, h\} \in R :\Leftrightarrow p \leq h$. Let $A = \{p_1, \ldots, p_k\} \subseteq S$. To prove (i), we have to show that $|\{h \in C : h \geq p_i$ for some $i\}| \geq k$. If $p_1 \vee \cdots \vee p_k = 1$, then we apply **4.54** to the restriction $L(A)$. Now suppose $p_1 \vee \cdots \vee p_k = x < 1$ and let $y$ be a minimal complement of $x$ in $L$. Again by **4.54** applied to $L(A)$, there are at least $k$ copoints $c_1, \ldots, c_k$ in $[0, x]$ such that for every $i$ we have $c_i \geq p_j$ for at least one $j$. By **2.44**, $c_1 \vee y, \ldots, c_k \vee y$ are $k$ distinct copoints in $L$ as required. (ii) is verified by an analogous argument. □

**8.58** can be strengthened to the statement that in any finite geometric lattice $L(S)$ there exist $|S|$ disjoint maximal chains (Mason).

We turn to property (U). One of the difficulties in proving (U) for a particular poset lies in finding a suitable algebraic formulation of the rising-falling property of the level numbers. For this reason, (U) is usually replaced by the following stronger, but handier, condition.

**Definition.** A sequence $\{W_0, W_1, \ldots, W_n\}$ of non-negative real numbers is called *logarithmically concave* if

$$W_k^2 \geq W_{k-1} W_{k+1} \quad \text{for } k = 1, \ldots, n-1.$$

Clearly, any log-concave sequence is unimodal. Furthermore, the definition implies that

$$\frac{W_k}{W_{k+1}} \leq \frac{W_{k+1}}{W_{k+2}} \leq \frac{W_{k+2}}{W_{k+3}} \leq \cdots,$$

and hence, in general, that

$$\frac{W_k}{W_{k+1}} \leq \frac{W_{k+j}}{W_{k+1+j}}.$$

Now

$$\frac{W_k}{W_{k+2}} = \frac{W_k}{W_{k+1}} \frac{W_{k+1}}{W_{k+2}} \leq \frac{W_{k+j}}{W_{k+1+j}} \frac{W_{k+1+j}}{W_{k+2+j}} = \frac{W_{k+j}}{W_{k+2+j}}$$

and thus

$$(+) \qquad \frac{W_k}{W_{k+l}} \leq \frac{W_{k+j}}{W_{k+l+j}} \quad \text{for all } k, l, j \geq 0.$$

**8.59 Proposition.** *The sequences* $\{\binom{n}{k}: k = 0, \ldots, n\}$, $\{\binom{n}{k}_q: k = 0, \ldots, n\}$ *and* $\{S_{n,k}: k = 1, \ldots, n\}$ *are logarithmically concave.*

*Proof.* For the binomial and Gaussian numbers, this follows immediately from **3.4** and **3.11**, respectively. For the Stirling numbers we use induction on $n$ and the recursion **3.29**(ii). We have

$$S_{n+1,k}^2 = S_{n,k-1}^2 + k^2 S_{n,k}^2 + 2k S_{n,k-1} S_{n,k}$$

and

$$S_{n+1,k-1} S_{n+1,k+1} = S_{n,k} S_{n,k-2} + (k^2 - 1) S_{n,k+1} S_{n,k-1} + ((k-1) S_{n,k} S_{n,k-1} + (k+1) S_{n,k+1} S_{n,k-2}).$$

The inequality $\geq$ follows in the first two summands by induction and in the third summand from (+) above and induction. □

**8.60 Proposition.** *Let $P$ and $Q$ be finite posets with rank function whose sequences of level numbers are log-concave. Then the sequence of level numbers in $P \times Q$ is also log-concave.*

*Proof.* The proof of this result is not difficult. One just has to notice that

$$W_k(P \times Q) = \sum_{i=0}^{k} W_i(P) W_{k-i}(Q)$$

and to group the coefficients together carefully. □

As a corollary we have the result **8.59** for chain products.

**8.61 Proposition.** *The sequence of level numbers of a finite chain product is logarithmically concave.* □

For finite geometric lattices it is conjectured that the level number sequence is always log-concave; in fact, inspection of small lattices makes it seem plausible that the Boolean algebras minimize the quotient $W_k^2/W_{k-1}W_{k+1}$, i.e., it might be true that for every finite geometric lattice on $n$ points

$$W_k^2 \geq \frac{k+1}{k}\frac{n-k+1}{n-k} W_{k-1}W_{k+1}.$$

For our fundamental lattices this is easily verified; in general, only partial results are known.

## C. Symmetric Posets

We begin by relating property (SYM) to the properties dealt with in the previous sections. We already know that any symmetric poset satisfies (U) and (M). As a matter of fact, the sequence of level numbers of a symmetric poset must obviously be of the form

$$(+) \qquad W_0 = W_n \leq W_1 = W_{n-1} \leq \cdots \leq W_{\lfloor n/2 \rfloor} = W_{\lceil n/2 \rceil}.$$

The following theorem shows that (+) together with property (N) is enough to assure symmetric decomposability.

**8.62 Proposition** (Griggs). *Let $P$ be a finite poset with rank function and level numbers $W_0, W_1, \ldots, W_n$. Suppose that*

(i) $W_0 = W_n \leq W_1 = W_{n-1} \leq \cdots \leq W_{\lfloor n/2 \rfloor} = W_{\lceil n/2 \rceil}$
(ii) *$P$ is normal.*

*Then $P$ is symmetric.*

*Proof.* We use induction on $n$. For $n = 0$ there is nothing to prove. Suppose $n \geq 1$ is odd, i.e., $\lceil n/2 \rceil = \lfloor n/2 \rfloor + 1$. By the normality of $P$, there exists a matching between the levels $P^{(\lfloor n/2 \rfloor)}$ and $P^{(\lceil n/2 \rceil)}$. The new poset $P'$ obtained from $P$ by identifying the matched pairs in the middle levels and keeping the rest unchanged again satisfies (i) and (ii). By induction, $P'$ is symmetric and hence so is $P$ by extending the symmetric chains in $P'$ in the obvious way. Now suppose $n \geq 2$ is even and consider the three middle levels $P^{((n/2)-1)}$, $P^{(n/2)}$ and $P^{((n/2)+1)}$. For every $p \in P^{((n/2)-1)}$ we define the subset $A_p \subseteq P^{(n/2)}$ by

$$A_p := \{x \in P^{(n/2)}: p <\cdot\, x\}.$$

Similarly, for every $q \in P^{((n/2)+1)}$ we define $B_q \subseteq P^{(n/2)}$ by

$$B_q := \{x \in P^{(n/2)}: x <\cdot\, q\}.$$

Now we claim the families $\mathfrak{A}$ and $\mathfrak{B}$ possess a common transversal. Once we have proved this we may partition the three middle levels into symmetric chains of cardinality 1 and 3 and then proceed as before by induction on $n$. We have to verify condition **8.29** for $\mathfrak{A}$ and $\mathfrak{B}$. Let $I \subseteq P^{((n/2)-1)}$ and $J \subseteq P^{((n/2)+1)}$. By the normalized matching property we have

$$\frac{|I|}{W_{(n/2)-1}} \leq \frac{|\bigcup_{p \in I} A_p|}{W_{n/2}}, \qquad \frac{|J|}{W_{(n/2)+1}} \leq \frac{|\bigcup_{q \in J} B_q|}{W_{n/2}},$$

and thus, because $W_{(n/2)-1} = W_{(n/2)+1} \leq W_{n/2}$,

$$\begin{aligned} \left|\bigcup_{p \in I} A_p \cap \bigcup_{q \in J} B_q\right| &= \left|\bigcup_{p \in I} A_p\right| + \left|\bigcup_{q \in J} B_q\right| - \left|\bigcup_{p \in I} A_p \cup \bigcup_{q \in J} B_q\right| \\ &\geq \frac{W_{n/2}}{W_{(n/2)-1}}(|I| + |J|) - W_{n/2} \geq |I| + |J| - W_{(n/2)-1}. \quad \square \end{aligned}$$

We remark that the converse of **8.62** is false, as shown by the poset of Figure 8.7 which is symmetric but not normal. Also, (N) cannot be weakened to (M). The poset of Figure 8.8 satisfies (M) and condition (i) in **8.62** but is plainly not symmetric. Since regular posets are normal, any regular poset with symmetric level number sequence is symmetric. By **8.53** this gives the following corollary.

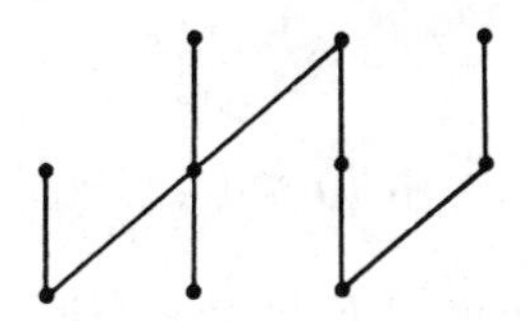

Figure 8.8

**8.63 Corollary.** *Boolean algebras $\mathscr{B}(n)$ and vector space lattices $\mathscr{L}(n, q)$ are symmetric.* $\square$

As simple and elegant as the proof of **8.62** is, it does not provide an explicit partitioning of a given symmetric poset into chains. The following theorem shows that the symmetry of posets is inherited by products, a fact which will provide just such an explicit chain decomposition for Boolean algebras and, more generally, for finite chain products. For vector space lattices a comparable result has not yet been found.

To facilitate the proof we use the following notation. Let $P$ be a poset with rank function $r$ and $C = \{c_0 <\cdot\, c_1 \quad <\cdot \cdots <\cdot\, c_l\}$ an unrefinable chain in $P$. We set

$$C_{\leq j} := \{c_0 <\cdot\, c_1 <\cdot \cdots <\cdot\, c_j\} \quad \text{and} \quad C_{\geq j} := \{c_j <\cdot\, c_{j+1} <\cdot \cdots <\cdot\, c_l\}.$$

If $C = \{c_0 \lessdot c_1 \lessdot \cdots \lessdot c_l\}$ and $D = \{d_0 \lessdot d_1 \lessdot \cdots \lessdot d_m\}$ are two unrefinable chains with $c_l \lessdot d_0$, then we set

$$C \cup D := \{c_0 \lessdot c_1 \lessdot \cdots \lessdot c_l \lessdot d_0 \lessdot \cdots \lessdot d_m\}.$$

**8.64 Proposition.** *Let P and Q be finite posets. If P and Q are both symmetric then so is $P \times Q$.*

*Proof.* Let $P = C_1 \dot\cup \cdots \dot\cup C_m$ and $Q = D_1 \dot\cup \cdots \dot\cup D_n$ be symmetric chain decompositions. We pick a pair of these chains, say $C = \{c_0 \lessdot \cdots \lessdot c_k\}$ and $D = \{d_0 \lessdot \cdots \lessdot d_l\}$. Suppose $r(c_0) = r$, $r(c_k) = R$ and $r(d_0) = s$, $r(d_l) = S$; then by property (SYM)

$$r + R = r(P) \quad \text{and} \quad s + S = r(Q).$$

We decompose the product $C \times D$ into chains $E_j$ by the following procedure. Set

$$(C', d) := \{(c, d) : c \in C'\} \quad \text{for all } C' \subseteq C, d \in D,$$

and similarly $(c, D')$ for all $c \in C$ and $D' \subseteq D$. Now

$$E_j := (C_{\leq k-j}, d_j) \cup (c_{k-j}, D_{\geq j+1}) \qquad (j = 0, \ldots, l).$$

In other words,

$$E_j = \{(c_0, d_j) \lessdot \cdots \lessdot (c_{k-j}, d_j) \lessdot (c_{k-j}, d_{j+1}) \lessdot \cdots \lessdot (c_{k-j}, d_l)\}.$$

The rank of the smallest element of $E_j$ in $P \times Q$ is $r + s + j$, and the rank of the largest element is $R - j + S$. Thus

$$(r + s + j) + (R - j + S) = (r + R) + (s + S) = r(P) + r(Q) = r(P \times Q).$$

Hence $E_j$ is a symmetric chain in $P \times Q$ for every $j$. Furthermore, the $E_j$'s are clearly pairwise disjoint and exhaust $C \times D$. By constructing this decomposition for every pair of chains $C_i$, $D_j$ we obtain the required symmetric decomposition of $P \times Q$. □

**8.65 Corollary** (deBruijn et al.). *Any finite chain product is symmetric.* □

**Example.** Let $C = \{1 \lessdot 2 \lessdot 3 \lessdot 4 \lessdot 5\}$ and $D = \{1 \lessdot 2 \lessdot 3 \lessdot 4\}$. The symmetric chains $E_j$ of $C \times D$ as constructed in the preceding proof are marked heavily in Figure 8.9.

Applying **2.56** and ex. II.3.8 (for the case of planes) together with **8.64**, we obtain the corresponding result for modular geometric lattices.

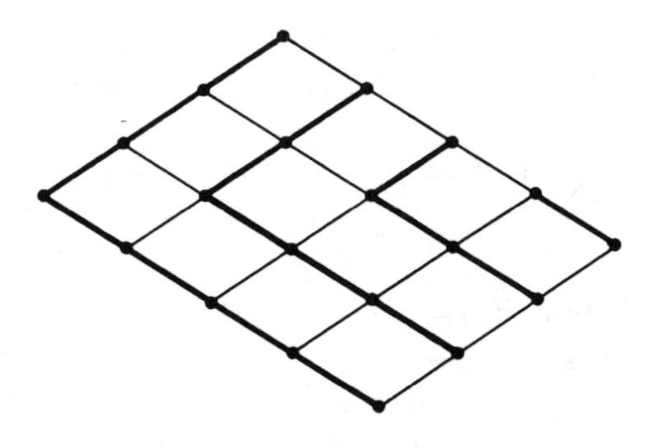

Figure 8.9

**8.66 Corollary.** *Any finite modular geometric lattice is symmetric.* □

We close our discussion by presenting an explicit symmetric partitioning of a chain product. Let $P = \prod_{i=1}^{n} \mathscr{C}(d_i)$ with $\mathscr{C}(m) = \{0 <\cdot 1 <\cdot \cdots <\cdot m\}$. We write the elements $x \in P$ in vector form:

$$x = (l_1, \ldots, l_n) \qquad (0 \le l_i \le d_i).$$

Then $r(x) = \sum_{i=1}^{n} l_i$ and thus

$$P^{(k)} = \left\{(l_1, \ldots, l_n): 0 \le l_i \le d_i, \sum_{i=1}^{n} l_i = k\right\}.$$

An element $y = (l_1, \ldots, l_n)$ covers $x = (k_1, \ldots, k_n)$ if and only if their coordinates are the same except for one for which $l_i = k_i + 1$. We define functions $f_i: P \to \mathbb{Z}, i = 0, \ldots, n+1$, as follows. Set $l_0 = d_0 = l_{n+1} = 0$. If $a = (l_1, \ldots, l_n) \in P$, then

$$(+) \qquad \begin{aligned} f_0(a) &:= 0 \\ f_i(a) &:= \sum_{j=0}^{i-1} (d_j - l_j) - \sum_{j=0}^{i} l_j \qquad (i = 1, \ldots, n+1). \end{aligned}$$

Let

$$(++) \qquad \begin{aligned} M(a) &:= \{i \in \{0, 1, \ldots, n+1\}: f_i(a) = \min\} \\ \underline{m}(a) &= \min_{i \in M(a)} i, \qquad \overline{m}(a) = \max_{i \in M(a)} i. \end{aligned}$$

Notice that we always have $\overline{m}(a) \ge 1$ and $\underline{m}(a) \le n$.

We now construct a chain decomposition of $P$ by giving the upper and lower neighbor of each $a \in P$. Let $a = (l_1, \ldots, l_n)$. Then the upper neighbor $\bar{a}$ is

$$\bar{a} = (\bar{l}_1, \ldots, \bar{l}_n) \quad \text{with} \quad \begin{cases} \bar{l}_j = l_j & \text{for} \quad j \neq \overline{m}(a) \\ \bar{l}_{\overline{m}(a)} = l_{\overline{m}(a)} + 1 \end{cases} \quad \text{if} \quad 1 \leq \overline{m}(a) \leq n;$$

$a$ has no upper neighbor if $\overline{m}(a) = n + 1$.

The lower neighbor $\underline{a}$ is

$$\underline{a} = (\underline{l}_1, \ldots, \underline{l}_n) \quad \text{with} \quad \begin{cases} \underline{l}_j = l_j & \text{for} \quad j \neq \underline{m}(a) \\ \underline{l}_{\underline{m}(a)} = l_{\underline{m}(a)} - 1 \end{cases} \quad \text{if} \quad 1 \leq \underline{m}(a) \leq n;$$

$a$ has no lower neighbor if $\underline{m}(a) = 0$.

To justify these definitions we have to show that $a$ is the lower neighbor of $b$ if $b$ is the upper neighbor of $a$, and conversely. Now suppose $a <\cdot\, b$ and assume $a$ and $b$ differ in coordinate $t$. Then

$$\begin{aligned} f_j(b) &= f_j(a) & (j &= 0, \ldots, t-1) \\ f_t(b) &= f_t(a) - 1 & & \\ f_i(b) &= f_i(a) - 2 & (i &= t+1, \ldots, n+1). \end{aligned}$$

If $b$ is the upper neighbor of $a$ then $t = \overline{m}(a)$. This implies $f_i(a) > f_t(a)$ for all $i > t$ and thus $f_i(b) \geq f_t(b)$ for all $i \geq t$; hence $\underline{m}(b) \leq t$. On the other hand, $f_j(a) \geq f_t(a)$ for all $j \leq t$, and thus $f_j(b) \geq f_t(b) + 1$ for all $j < t$. Altogether we have $\underline{m}(b) = t$ whence $a$ is the lower neighbor of $b$. The reverse implication is shown similarly.

**8.67 Proposition.** *Let $P = \prod_{i=1}^{n} \mathscr{C}(d_i)$ be a finite chain product. The upper-lower neighbor construction partitions $P$ into symmetric chains and this decomposition is precisely the inductive decomposition defined by consecutively decomposing $\mathscr{C}(d_1), \mathscr{C}(d_1) \times \mathscr{C}(d_2), \ldots, \mathscr{C}(d_1) \times \cdots \times \mathscr{C}(d_n)$ as in the proof of* **8.64**.

*Proof.* Denote by $[a]$ the chain containing $a \in P$ obtained by the upper-lower neighbor construction. Let the smallest element in $[a]$ have rank $r$ and the largest have rank $R$. For the upper neighbor $b$ of $a$ we have shown that

$$\min_i f_i(b) = \min_i f_i(a) - 1,$$

$$f_{n+1}(b) = f_{n+1}(a) - 2;$$

hence

$$R - r(a) = f_{n+1}(a) - \min_i f_i(a) = r(P) - 2r(a) - \min_i f_i(a).$$

Similarly, if $c$ is the lower neighbor of $a$ then

$$\min_i f_i(c) = \min_i f_i(a) + 1,$$

$$f_0(c) = f_0(a) = 0,$$
$$r(a) - r = -\min_i f_i(a),$$

and thus

$$r + R = r(P).$$

That the two decompositions are identical now follows easily by induction on $n$. □

**Example.** Let $P = \mathscr{C}(2) \times \mathscr{C}(3) \times \mathscr{C}(2)$. The symmetric decomposition constructed as in **8.66** is shown below.

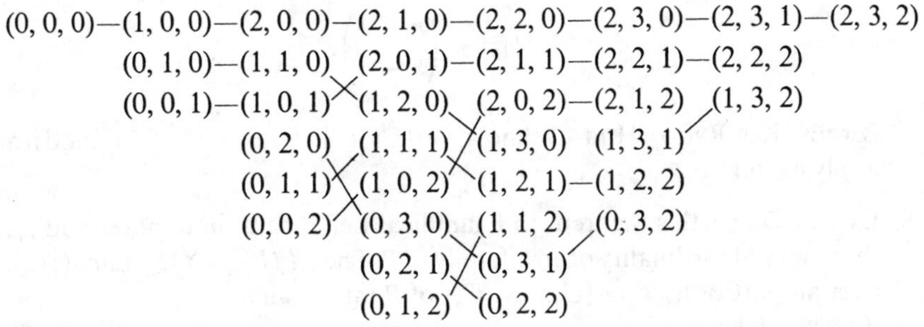

(0, 0, 0)—(1, 0, 0)—(2, 0, 0)—(2, 1, 0)—(2, 2, 0)—(2, 3, 0)—(2, 3, 1)—(2, 3, 2)
(0, 1, 0)—(1, 1, 0) (2, 0, 1)—(2, 1, 1)—(2, 2, 1)—(2, 2, 2)
(0, 0, 1)—(1, 0, 1) (1, 2, 0) (2, 0, 2)—(2, 1, 2) (1, 3, 2)
(0, 2, 0) (1, 1, 1) (1, 3, 0) (1, 3, 1)
(0, 1, 1) (1, 0, 2) (1, 2, 1)—(1, 2, 2)
(0, 0, 2) (0, 3, 0) (1, 1, 2) (0, 3, 2)
(0, 2, 1) (0, 3, 1)
(0, 1, 2) (0, 2, 2)

EXERCISES VIII.3

1. Prove that in $\mathscr{B}(n)$ and $\mathscr{L}(n, q)$ the only maximal sized antichains are the middle levels.

2.* Let $S_1$, $S_2$ be disjoint sets with $|S_i| = n_i$, $n_1 \leq n_2$. Suppose $\mathfrak{A} = \{A_1, \ldots, A_m\} \subseteq 2^{S_1 \cup S_2}$ is a family in which no relation

$$A_i \cap S_1 = A_j \cap S_1, A_i \cap S_2 \supseteq A_j \cap S_2 \qquad (i \neq j)$$

$$A_i \cap S_1 \supseteq A_j \cap S_1, A_i \cap S_2 = A_j \cap S_2 \qquad (i \neq j)$$

holds. Show that $m \leq \binom{n}{\lfloor n/2 \rfloor}$ with $n = n_1 + n_2$.

3. Prove by means of the previous exercise the following generalization of **8.46**: Let $a_1, \ldots, a_n \in \mathbb{C}$ with $|a_i| > 1$. Then at most $\binom{n}{\lfloor n/2 \rfloor}$ of the $2^n$ numbers $\sum_{i=1}^{n} \varepsilon_i a_i$, $\varepsilon_i = \pm 1$, lie in the unit disk. (Katona–Kleitman)

→ 4.* Let $P$ be a finite poset and $\mathfrak{A}_h(P)$ the lattice of $h$-families as defined in ex. VIII.1.15. Show that the set $\mathfrak{S}_h(P)$ of maximal sized $h$-families forms a distributive sublattice of $\mathfrak{A}_h(P)$. (Greene–Kleitman)

→ 5. Prove that a ranked poset is normal if and only if there exists a collection $\mathscr{C}$ of (not necessarily) distinct maximal chains such that for every $k$, every element of rank $k$ occurs in the same number of chains in $\mathscr{C}$.

→ 6. An *ordered set system* on $\mathbb{N}_n$ is a sequence $\vec{\mathfrak{A}} = \{A_1, \ldots, A_t\}$ of subsets of $\mathbb{N}_n$. $\mathrm{Sym}(\vec{\mathfrak{A}})$ is the collection of ordered set systems which are obtained from $\mathfrak{A}$ by permuting the elements $1, \ldots, n$. Let $\mathfrak{F}$ be a family of subsets of $\mathbb{N}_n$ which has at most $p$ members in common with each $\vec{\mathfrak{B}} \in \mathrm{Sym}(\vec{\mathfrak{A}})$, and let $d_k = \#\{A \in \mathfrak{F} : |A| = k\}$. Prove that

$$\sum_{i=1}^{t} \frac{d_{|A_i|}}{\binom{n}{|A_i|}} \le p.$$

(Hint: Consider the proof of **8.54**.)

→ 7. Using the previous exercise derive the following result: Let $k \le n/2$ and $\mathfrak{F}$ be a family of $k$-subsets of $\mathbb{N}_n$ no two members of which are disjoint. Then

$$|\mathfrak{F}| \le \binom{n-1}{k-1}.$$

(Erdös–Ko–Rado)(Hint: Define $A_i = \{i, i+1, \ldots, i+(k-1)\} \bmod n$ and apply ex. 6.)

8.* Extend Dilworth's theorem to $h$-families: Let $P$ be a finite poset and $d_h(P)$ the maximal cardinality of an $h$-family in $P$. Then $d_h(P) = \sum_{i=1}^{n} \min_{\mathscr{C}}(|C_i|, h)$ over all partitions $\mathscr{C} = \{C_1, \ldots, C_n\}$ of $P$ into chains.
(Greene–Kleitman)

→ 9. Let $G(S \cup I, R)$ be a bipartite graph with $|S| = n$, $|I| = t$. Let $\bar{p} := |R(p)|$ for $p \in S$, $\bar{h} = |R(h)|$ for $h \in I$. Show that (i) and (ii) below imply $t \ge n$.
(i) $(p, h) \notin R \Rightarrow \bar{p} \ge \bar{h}$.
(ii) $\bar{h} < n$ for all $h$.
(Motzkin)

10. Use ex. 9 to give a new proof of $W_1(P) \le W_{n-1}(P)$ for any finite geometric lattice $P$ of rank $n$.

11.* Let $L(S)$ be a finite geometric lattice with $|S| = n$. Prove:
(i) There exist $n$ pairwise disjoint maximal chains in $L$.
(ii) The same holds for finite semimodular lattices whenever the 1-element is the supremum of points.
(iii) Give an example to show the necessity of the assumption in (ii).
(Mason) (Hint: Use **8.3** and consider the proof of **8.58**.)

→12.* Strengthen **8.60**: Let $P$ and $Q$ be normal posets with log-concave level number sequences. Then $P \times Q$ is normal and has a log-concave level number sequence.

13. Let $\{v_0, \ldots, v_n\}$ be a sequence of non-negative real numbers and suppose the polynomial $\sum_{i=0}^{n} v_i x^i$ has only real roots. Prove that $\{v_0, \ldots, v_n\}$ is then unimodal with a unique maximum or two maxima, and that, in fact,

$$v_k^2 \geq v_{k-1}v_{k+1}\frac{k}{k-1}\frac{n-k+1}{n-k} \qquad (k = 2, \ldots, n-1).$$

Apply the result to $\{|s_{n,k}|\}$, $\{S_{n,k}\}$. (Hint: Apply Rolle's theorem to the polynomial $\sum v_i x^i y^{n-i}$.)

14. Let $G(S \cup I, R)$ be a finite bipartite graph and $\lambda$ and $\mu$ total orderings of $S$ and $I$, respectively. Define a function $\phi_L$ from $S$ to $I$ (i.e., $\phi_L$ may be defined only on a subset of $S$) by the following procedure:
(a) $\phi_L$ is defined on $S$ according to the order $\lambda$ starting with the $\lambda$-minimal element.
(b) Let $a \in S$. Then $\phi_L(a) = b$ where $b$ is the $\mu$-minimal element in $R(a)$ which is not already the image of some $a' <_\lambda a$. If such $b$'s do not exist, then $\phi_L$ is not defined on $a$.
$L = \{(a, \phi_L(a)) : a \in S\}$ is called the *lexicographic matching* induced by $\lambda$ and $\mu$. Let $P = \prod_{i=1}^{n} \mathscr{C}(d_i)$ be a chain product as in **8.67**. We order each level $P^{(k)}$ by

$$(l_1, \ldots, l_n) \underset{\lambda_k}{<} (l'_1, \ldots, l'_n) :\Leftrightarrow l_i > l'_i$$

where $i$ is the smallest index with $l_i \neq l'_i$.

Show that the lexicographic matchings induced by the $\lambda_k$'s yield precisely the symmetric chain decomposition in **8.67**. (Aigner)

→15.* Let $P = \prod_{i=1}^{n} \mathscr{C}(d_i)$ as in ex. 14. We use the symbols "(" for "no" and ")" for "yes" and associate with $a = l_1 l_2 \ldots l_n \in P$ the word

$$a = l_1 l_2 \ldots l_n \leftrightarrow \underbrace{))\ldots)}_{l_1}\underbrace{((\ldots(}_{d_1-l_1}\underbrace{)\ldots)}_{l_2}\ldots\underbrace{)\ldots)}_{l_n}\underbrace{(\ldots(}_{d_n-l_n},$$

written for short as

$$a = )^{l_1}(^{d_1-l_1})^{l_2}(^{d_2-l_2}\ldots)^{l_n}(^{d_n-l_n}.$$

In any such expression we remove iteratively every pair ( ) retaining the original positions. The remaining "reduced" expression $\bar{a}$ is of the form $\bar{a} = )^u(^v$ with empty places.
Example:

$$)(()())( \rightarrow )(..())( \rightarrow )(....)( \rightarrow )......(.$$

The chain [a] is now obtained by successively turning around the $u + v$ "free" parentheses:

$$[a] = \{(^{u+v} <\cdot)^1(^{u+v-1} <\cdot)^2(^{u+v-2} <\cdots <\cdot)^{u+v}\}.$$

Prove:

(i) This gives a symmetric decomposition of $P$. (Leeb–Schönheim)

(ii) The decomposition is the same as the one obtained in **8.67** and in the previous exercise.

## 4. Ramsey Theorems

To conclude this book, we report on a theorem which, next to Hall's theorem, is perhaps the most important existence theorem in combinatorics yet discovered. Just as Sperner's theorem **8.43** on Boolean algebras initiated a general study of what we today call Sperner-type results, so Ramsey's theorem on Boolean algebras gave rise to a whole class of similar theorems. We again consider posets with a rank function. Suppose $P$ has rank $n$, and let $P^{(k)}$ denote the $k$-th level of $P$ as before. For $a \in P$ with $r(a) \geq k$ we set $P^{(k)}(a) := \{x \in \mathrm{P}: x \leq a, r(x) = k\}$. We are interested in partitioning $P^{(k)}$ into $r$ classes (some of which may be empty), $P^{(k)} = A_1 \cup \cdots \cup A_r$. To make the language more expressive, we sometimes say that $P^{(k)}$ has been *colored* with $r$ colors and that $A_i$ is the $i$-th color class. An element $a \in P$ of rank $r(a) \geq k$ is called *monochromatic* if all elements of rank $k$ below $a$ are colored with the same color, i.e., if $P^{(k)}(a) \subseteq A_i$ for some $i$.

Let $\mathscr{P} = \{P_0, P_1, P_2, \ldots\}$ be a sequence of ranked posets and consider the following statement $R(k; l; r)$, where $k$, $l$ and $r$ are positive integers.

$R(k; l; r)$: *There exists a number $n_0(k; l; r)$ depending only on $k$, $l$, $r$ such that whenever $n \geq n_0(k; l; r)$ and $P_n^{(k)}$ is colored with $r$ colors, then $P_n$ contains a monochromatic element of rank $l$.*

**Definition.** The sequence $\mathscr{P} = \{P_0, P_1, P_2, \ldots\}$ of ranked posets is said to possess the *Ramsey property* if $R(k; l; r)$ holds for all positive integers $k$, $l$, and $r$. The smallest numbers $N(k; l; r)$ among the numbers $n_0(k; l; r)$ are called the *Ramsey numbers*.

Notice that the truth of $R(k; l; r)$ implies the truth of $R(k; l'; r)$ for all $l' \leq l$, and that $R(k; l; r)$ holds vacuously for $l < k$.

A *Ramsey theorem*, roughly speaking, is one asserting the Ramsey property for a given sequence of posets. In this section we prove Ramsey's original theorem for Boolean algebras and then consider some variants and applications.

A. Ramsey's Theorem for Boolean Algebras

To get an idea of the statement $R(k; l; r)$, consider the sequence

$$\mathcal{B} = \{\mathcal{B}(0), \mathcal{B}(1), \mathcal{B}(2), \ldots\}$$

of Boolean algebras and take $k = 1$. $R(1; l; r)$ then says the following: Given $l$ and $r$, there exists an integer $N(1; l; r)$ such that if the elements of an $n$-set, $n \geq N(1; l, r)$, are partitioned into $r$ classes then at least one of the classes contains $l$ (or more) elements. Hence $R(1; l; r)$ is just what is usually called the *pigeon-hole principle*. In this case the Ramsey numbers are trivially given by

$$N(1; l; r) = r(l - 1) + 1.$$

In this spirit, Ramsey theorems are asymptotic existence theorems stating that if we distribute a large enough set over not too many classes then at least one of the classes contains many elements.

To expedite the ensuing induction arguments, it is convenient to formulate the following stronger statement $R(k; l_1, \ldots, l_r)$ where $k$, $r$ and $l_1, \ldots, l_r$ are positive integers. As before, let $\mathcal{P}$ be a sequence of ranked posets.

$R(k; l_1, \ldots, l_r)$: *There exists a number $n_0(k; l_1, \ldots, l_r)$ depending only on $k, l_1, \ldots, l_r$ such that whenever $n \geq n_0(k; l_1, \ldots, l_r)$ and $P_n^{(k)} = A_1 \cup \cdots \cup A_r$ is an $r$-coloring, then $P_n$ contains a monochromatic element $a$ with $r(a) = l_i$ and $P^{(k)}(a) \subseteq A_i$ for some $i$.*

The smallest integers $N(k; l_1, \ldots, l_r)$ among the $n_0$'s are again called the *Ramsey numbers*. Clearly, if $\mathcal{P}$ satisfies $R(k; l_1, \ldots, l_r)$ for all $k$ and $l_1, \ldots, l_r$ then it possesses the Ramsey property. The converse is also true by simply taking $l = \max(l_1, \ldots, l_r)$ and recalling that $R(k; l; r) \Rightarrow R(k; l'; r)$ for $l' \leq l$.

To prove Ramsey's theorem for Boolean algebras it is convenient to consider the case $r = 2$ separately ($r = 1$ is trivial).

**8.68 Proposition.** *The sequence $\mathcal{B} = \{\mathcal{B}(0), \mathcal{B}(1), \ldots\}$ of Boolean algebras satisfies $R(k; l_1, l_2)$ for all positive integers $k, l_1, l_2$.*

*Proof.* We use induction on $k$. For $k = 1$ the assertion is clear, the Ramsey number being $l_1 + l_2 - 1$. Suppose $\mathcal{B}$ satisfies $R(k - 1; l'_1, l'_2)$ for all $l'_1, l'_2$. To prove $R(k; l_1, l_2)$ we use induction on $l_1 + l_2$. If $l_1 < k$ or $l_2 < k$ then $R(k; l_1, l_2)$ is vacuously true. If $l_1 = k$ and $l_2 \geq k$, then $N(k; k, l_2) = l_2$ since, in any partition $\mathcal{B}^{(k)}(S) = A_1 \cup A_2$ with $|S| \geq l_2$, either $A_1 \neq \emptyset$, in which case any member of $A_1$ is a required $k$-subset, or $A_1 = \emptyset$, i.e., $A_2 = \mathcal{B}^{(k)}(S)$, in which case any $l_2$-subset of $S$ will do. Similarly, we have $N(k; l_1, k) = l_1$ for all $l_1 \geq k$. Now suppose $R(k; l'_1, l'_2)$ holds for all pairs $(l'_1, l'_2)$ with $k \leq l'_1 \leq l_1 - 1$, $k \leq l'_2 \leq l_2$ and for all pairs $(l'_1, l'_2)$ with $k \le l'_1 \le l_1$, $k \le l'_2 \le l_2 - 1$. Set $p_1 := N(k; l_1 - 1, l_2)$, $p_2 := N(k; l_1, l_2 - 1)$. We then claim that $R(k; l_1, l_2)$ holds with

$$(+) \qquad N(k; l_1, l_2) \le N(k-1; p_1, p_2) + 1.$$

Let $S$ be an $n$-set with $n \ge N(k-1; p_1, p_2) + 1$ and $\mathscr{B}^{(k)}(S) = A_1 \cup A_2$ any partition of its $k$-subsets. Let $a \in S$ and set $T = S - a$. We define the partition $\mathscr{B}^{(k-1)}(T) = B_1 \dot\cup B_2$ by

$$B_1 := \{R \subseteq T: R \cup a \in A_1\},$$
$$B_2 := \{R \subseteq T: R \cup a \in A_2\}.$$

Since $|T| \ge N(k-1; p_1, p_2)$ there exists in $T$ either a $p_1$-subset $X$ with $\mathscr{B}^{(k-1)}(X) \subseteq B_1$, or a $p_2$-subset $Y$ with $\mathscr{B}^{(k-1)}(Y) \subseteq B_2$. Assume the former. From $|X| = p_1 = N(k; l_1 - 1, l_2)$ and the induction hypothesis, we infer that $X$ contains either an $(l_1 - 1)$-subset $U$ with $\mathscr{B}^{(k)}(U) \subseteq A_1$, or an $l_2$-subset $V$ with $\mathscr{B}^{(k)}(V) \subseteq A_2$. If the second alternative holds, then $V$ is a required $l_2$-subset of $S$. If the first alternative holds then $U \cup a$ has all its $k$-subsets in $A_1$ and hence is a required $l_1$-subset of $S$. The case where $Y$ exists above is dealt with similarly. □

**8.69 Theorem** (Ramsey). *The sequence $\mathscr{B} = \{\mathscr{B}(0), \mathscr{B}(1), \mathscr{B}(2), \ldots\}$ of Boolean algebras possesses the Ramsey property, that is, for any positive integers $k$, $r$ and $l_1, \ldots, l_r$ there exists a smallest integer $N(k; l_1, \ldots, l_r)$ depending only on $k, r, l_1, \ldots, l_r$ such that whenever the $k$-subsets of an $n$-set $S$, $n \ge N(k; l_1, \ldots, l_r)$, are partitioned into $r$ classes $A_1, \ldots, A_r$, then there exists an $l_i$-subset $T$ of $S$ with $\mathscr{B}^{(k)}(T) \subseteq A_i$ for some $i$.*

*Proof.* We use induction on $r$. We have proved the theorem for $r = 1$ and 2. Suppose it is true for all integers up to $r - 1$ and let $r \ge 3$. We claim that $R(k; l_1, \ldots, l_r)$ holds with

$$(++) \qquad N(k; l_1, \ldots, l_r) \le N(k; l_1, \ldots, l_{r-2}, p_{r-1}),$$

where $p_{r-1} := N(k; l_{r-1}, l_r)$.

Let $S$ be an $n$-set with $n \ge N(k; l_1, \ldots, l_{r-2}, p_{r-1})$ and $\mathscr{B}^{(k)}(S) = A_1 \dot\cup \cdots \dot\cup A_r$ an arbitrary $r$-coloring. Consider the partition

$$\mathscr{B}^{(k)}(S) = A_1 \dot\cup \cdots \dot\cup A_{r-2} \dot\cup (A_{r-1} \dot\cup A_r).$$

By the induction hypothesis, there exists either an $l_i$-subset $T_i$ with $\mathscr{B}^{(k)}(T_i) \subseteq A_i$ for some $i$ between 1 and $r - 2$, in which case we are through, or a $p_{r-1}$-subset $T$ with $\mathscr{B}^{(k)}(T) \subseteq A_{r-1} \cup A_r$. Since $|T| = N(k; l_{r-1}, l_r)$ we can now apply **8.68** to finish the proof. □

Notice that the formulas (+) and (++) not only prove the existence of the Ramsey numbers $N(k; l_1, \ldots, l_r)$ but also furnish a recursion for them. The exact values for the Ramsey numbers are, however, known only for very small values of $k$ and $l_1, \ldots, l_r$.

After $k = 1$ (the pigeon-hole principle), $k = 2$ is the simplest case. When $k = 2$ one may conveniently interpret the $n$-set $S$ as the vertices of the complete graph $K_n$ and $\mathscr{B}^{(2)}(S) = A_1 \cup \cdots \cup A_r$ as an $r$-coloring of the edges. **8.69** then reads as follows: If $n \geq N(k; l_1, \ldots, l_r)$ and if the edges of $K_n$ are colored in any way with $r$ colors, then, for some $i$, there is a complete monochromatic subgraph on $l_i$ vertices.

Consider the case of two colors, say red and blue. From our previous discussion we have

$$(+)\qquad \begin{aligned} &N(2; 2, l_2) = l_2, N(2; l_1, 2) = l_1 \quad \text{for } l_1 \geq 2, l_2 \geq 2, \\ &N(2; l_1, l_2) \leq N(2; l_1 - 1, l_2) + N(2; l_1, l_2 - 1). \end{aligned}$$

Also clearly,

$$N(2; l_1, l_2) = N(2; l_2, l_1).$$

For the first non-trivial case $l_1 = l_2 = 3$, the recursion (+) yields

$$N(2; 3, 3) \leq 3 + 3 = 6.$$

On the other hand, $N(2; 3, 3) > 5$ since $K_5$ possesses a 2-coloring of its edges without monochromatic triangles as shown in Figure 8.10 where the bold lines are the red edges and the dotted lines the blue edges. Hence $N(2; 3, 3) = 6$. A popular interpretation of this fact is that at any party with six people there are always 3 people who mutually know one another, or 3 of whom no one knows another.

It can be easily shown that in the recursion (+) above we have strict inequality if both numbers on the right-hand side are even. The following table summarizes the numbers $N(2; l_1, l_2)$ for small $l_1$ and $l_2$.

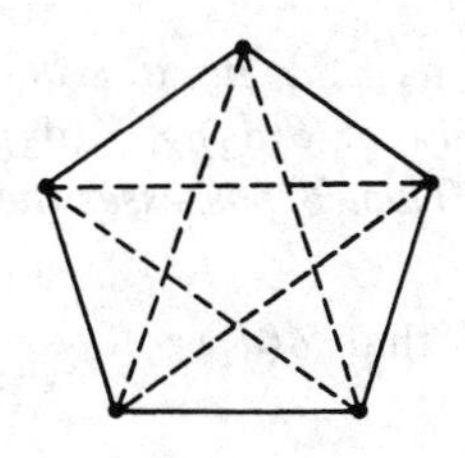

Figure 8.10

**8.70** Table of Ramsey numbers $N(2; l_1, l_2)$

| $l_1 \backslash l_2$ | 2 | 3 | 4 | 5 | 6 |
|---|---|---|---|---|---|
| 2 | 2 | 3 | 4 | 5 | 6 |
| 3 | 3 | 6 | 9 | 14 | 18 |
| 4 | 4 | 9 | 18 | $\frac{25}{28}$ | $\frac{34}{45}$ |
| 5 | 5 | 14 | $\frac{25}{28}$ | $\frac{38}{55}$ | $\frac{38}{94}$ |
| 6 | 6 | 18 | $\frac{34}{45}$ | $\frac{38}{94}$ | $\frac{102}{178}$ |

$a/b$ means that the number lies between the bounds $a$ and $b$.

A natural generalization of Ramsey's problem for $k = 2$ is the question whether, for every finite graph $H$, there exists a finite graph $G$ such that, if the edges of $G$ are arbitrarily 2-colored, then $G$ contains an induced monochromatic subgraph isomorphic to $H$. This has been answered affirmatively by Deuber [1]; see Graham–Rothschild [1, 3] and Van Lint [1, ch.VI] for further results in this direction.

B. Some Applications

**8.71 Proposition.** *The sequence* $\mathscr{P} = \{\mathscr{P}(1), \mathscr{P}(2), \mathscr{P}(3), \ldots\}$ *of finite partition lattices possesses the Ramsey property.*

*Proof.* Let $k, l, r \in \mathbb{N}$ and $n \geq 2N_{\mathscr{B}}(k; l; r) - 1$, where $N_{\mathscr{B}}$ is the Ramsey number for Boolean algebras. Suppose the partitions of an $(n + 1)$-set $S$ of rank $k$ (i.e., with $n + 1 - k$ blocks) are $r$-colored in any way. Let $\pi \in \mathscr{P}(S)$ consist of $N_{\mathscr{B}}(k; l; r)$ pairs and $n - 2N_{\mathscr{B}}(k; l; r) + 1$ singleton blocks. Then $[0, \pi] \cong \mathscr{B}(N_{\mathscr{B}}(k; l; r))$ and the result follows from **8.69**. $\square$

Using the same argument we see that the sequence $\mathscr{U} = \{\mathscr{U}_0(1), \mathscr{U}_1(2), \mathscr{U}_2(3), \ldots\}$ of uniform lattices also possesses the Ramsey property. The inverted partition lattices $\mathscr{P}^*(n)$ present more difficulties. For a proof that $\mathscr{P}^* = \{\mathscr{P}^*(1), \mathscr{P}^*(2), \mathscr{P}^*(3), \ldots\}$ possesses the Ramsey property, see Graham–Leeb–Rothschild [1].

**8.72 Proposition.** *Let* $\{d_1, d_2, d_3, \ldots\}$ *be a sequence of positive integers and* $\mathscr{C} = \{\mathscr{C}(d_1), \mathscr{C}(d_1) \times \mathscr{C}(d_2), \mathscr{C}(d_1) \times \mathscr{C}(d_2) \times \mathscr{C}(d_3), \ldots\}$ *the corresponding sequence of finite chain products. Then* $\mathscr{C}$ *possesses the Ramsey property.*

*Proof.* We just have to notice that $\mathscr{C}(d_1) \times \cdots \times \mathscr{C}(d_n)$ contains an element $a$ with $[0, a] \cong \mathscr{B}(n)$. $\square$

We next consider two applications, one in arithmetic and one in geometry.

**8.73 Proposition** (Schur). *For every $r \in \mathbb{N}$ there exists a smallest number $S_r$ such that if $n \geq S_r$ and $\{1, \ldots, n\}$ is partitioned in any way into r classes, then there exists a class which contains l numbers $x_1, \ldots, x_l$ with $x_l = \sum_{i=1}^{l-1} x_i$.*

*Proof.* Let $n \geq N_{\mathscr{B}}(2; l, \ldots, l)$ be the Boolean Ramsey number with $r$ $l$'s on the right-hand side, and let $\mathbb{N}_n = A_1 \cup \cdots \cup A_r$ be an $r$-coloring. We define the partition $\mathscr{B}^{(2)}(\mathbb{N}_n) = B_1 \cup \cdots \cup B_r$ by

$$\{a, b\} \in B_i :\Leftrightarrow |a - b| \in A_i \qquad (i = 1, \ldots, r).$$

By **8.69**, there exists a subset $T = \{a_1, \ldots, a_l\} \subseteq \mathbb{N}_n$ and some index $h$ such that $\{a_i, a_j\} \subseteq B_h$ for all $1 \leq i < j \leq l$. Without loss of generality, assume that $a_1 < a_2 < \cdots < a_l$. The numbers $x_i = a_{i+1} - a_i$ for $i = 1, \ldots, l-1$ and $x_l = a_l - a_1$ are then all in $A_h$ and satisfy $\sum_{i=1}^{l-1} x_i = x_l$. $\square$

Notice that we have proved $S_r \leq N_{\mathscr{B}}(2; l, \ldots, l)$.

Now the geometrical problem. Consider $s$ points $P_1, \ldots, P_s$ in the Euclidean plane $\mathbb{R}^2$. $P_1, \ldots, P_s$ are the vertices of a *convex s-gon* if the region bounded by the edge sequence $\overline{P_1P_2}, \overline{P_2P_3}, \ldots, \overline{P_sP_1}$ is convex (after suitably numbering the points $P_i$). (For the definitions, see Grünbaum [1].)

**8.74 Proposition** (Erdös–Szekeres). *For every $s \geq 3$ there exists a smallest number $C_s$ such that, for $n \geq C_s$, if n points in the Euclidean plane have no three points collinear, then s of them are the vertices of a convex s-gon.*

*Proof.* Obviously $C_3 = 3$ since every triangle is convex. For $s = 4$, there are two possibilities depending on whether the convex hull is a quadrilateral or a triangle; hence $C_4 > 4$. If we place 5 points in the plane, no three of which are collinear, then the convex hull contains either 3 or 4 or 5 vertices. In the latter two cases any 4 of those form a convex 4-gon. In the case of a triangle two of the exterior points must lie in the same half plane generated by the line containing the two interior points. (Cf. Figure 8.11.) These four points then span a convex 4-gon; hence $C_4 = 5$.

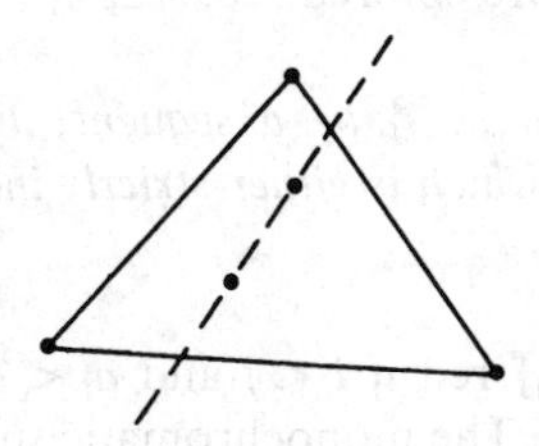

Figure 8.11

*Claim*: Given $s$ points in $\mathbb{R}^2$ no three of which are collinear and any 4 of which form a convex 4-gon, then they form a convex $s$-gon. To see this let $P_1, \ldots, P_m$ be the vertices of the convex hull; we have to show $m = s$. Now if $m < s$ then $P_{m+1}$ is in the interior of one of the triangles $P_1P_2P_3$, $P_1P_3P_4, \ldots, P_1P_{m-1}P_m$, say $P_1P_{i-1}P_i$. Then $P_1$, $P_{i-1}$, $P_i$, $P_{m+1}$ do not form a convex 4-gon.

To finish the proof suppose $s \geq 4$ and consider an $n$-set $S \subseteq \mathbb{R}^2$, $n \geq N_{\mathscr{B}}(4; s, 5)$, with no three points collinear. We decompose $\mathscr{B}^{(4)}(S) = A_1 \,\dot{\cup}\, A_2$, putting all convex 4-gons into $A_1$ and the rest into $A_2$. Ramsey's theorem **8.69** implies that there exists either an $s$-subset all of whose 4-gons are convex or a 5-subset all of whose 4-gons are not convex. As we have seen, the second alternative is impossible; hence the first alternative holds whence the proof is complete by our proven claim. □

Notice that we have proved $C_s \leq N_{\mathscr{B}}(4; s, 5)$. It is known that

$$C_s = 2^{s-2} + 1 \quad \text{for } s = 3, 4, 5, 6,$$

and it may be true that this formula holds in general.

As a final variant we consider an infinite version of **8.69** for $k = 2$.

**8.75 Proposition.** *Let $K_\infty$ be the complete graph with countably many vertices and let the edges of $K_\infty$ be colored with $r$ colors. Then there exists an infinite monochromatic complete subgraph.*

*Proof*. We assume $r = 2$; the general statement follows easily by induction. Label the vertices $v_1, v_2, \ldots$ and denote by $r(v_i)$ the number of red edges incident with $v_i$.

*Case a*. There exist infinitely many vertices $v_{i_1}, v_{i_2}, \ldots$ with $r(v_{i_j}) < \infty$. We define the vertices $w_1, w_2, \ldots$ inductively. Set $w_1 = v_{i_1}$; $w_2$ is the first vertex among the $v_{i_j}$'s which is not joined to $w_1$ by a red edge; $w_3$ is the first vertex among the $v_{i_j}$'s after $w_2$ which is not joined to $w_1$ or $w_2$ by a red edge, and so on. The $w_i$'s clearly generate a complete graph all of whose edges are colored blue.

*Case b*. Suppose (*a*) fails and let $V_1 := \{v : r(v) = \infty\}$; thus $|V_1| = \infty$. Choose $w_1 \in V_1$ and denote by $V_1' \subseteq V_1$ the (infinite) set of vertices in $V_1$ which are joined to $w_1$ by a red edge. Let $V_2 \subseteq V_1'$ be the set of vertices in $V_1'$ with $r(v) = \infty$ in $V_2$; again $|V_2| = \infty$ since otherwise we may construct infinite blue complete subgraphs on $V_2$ by (*a*). Choose $w_2 \in V_2$, and so on. This gives an infinite complete subgraph all of whose edges are colored red. □

**8.76 Corollary**. *Let $\{a_1, a_2, a_3, \ldots\}$ be a sequence of real numbers. Then there exists an infinite subsequence which is either strictly increasing, strictly decreasing, or constant.*

*Proof*. Color the edge $\{a_i, a_j\}$ red if $i < j$ and $a_i < a_j$, blue if $i < j$ and $a_i > a_j$ and green if $i < j$ and $a_i = a_j$. The monochromatic subgraphs correspond then to the sequences as required in the proposition. □

## C. Further Results

When we examine the proof of **8.68**, we see that the crucial step consists in looking at the partition $\mathscr{B}^{(k-1)}(T) = B_1 \dot{\cup} B_2$ induced by the given partition $\mathscr{B}^{(k)}(S) = A_1 \dot{\cup} A_2$. The induction works because, by the binomial recursion

$$\binom{l}{k} = \binom{l-1}{k-1} + \binom{l-1}{k},$$

we may reduce the assertion for $(k, l)$ to $(k-1, l-1)$ and $(k, l-1)$. In general, to prove a Ramsey theorem for a given sequence $\mathscr{C}$ of ranked posets, one has to construct a whole class $\mathfrak{C}$ of sequences (containing $\mathscr{C}$) such that for every $\mathscr{B} \in \mathfrak{C}$ there exists $\mathscr{A} \in \mathfrak{C}$ which is connected to $\mathscr{B}$ in a certain way so as to allow an induction argument. The conclusion is then that all sequences in $\mathfrak{C}$ possess the Ramsey property, and hence in particular $\mathscr{C}$. Using this approach Graham, Leeb and Rothschild were able to prove a very general theorem which among other things implies the Ramsey property for vector space lattices and inverted partition lattices.

**8.77 Theorem** (Graham–Leeb–Rothschild). *The following sequences possess the Ramsey property*:

(i) $\mathscr{L} = \{\mathscr{L}(0, q), \mathscr{L}(1, q), \mathscr{L}(2, q), \ldots\}$ *of vector space lattices for any* $q$.
(ii) $\mathscr{A} = \{\mathscr{A}(0, q), \mathscr{A}(1, q), \mathscr{A}(2, q), \ldots\}$ *of affine lattices for any* $q$.
(iii) $\mathscr{P}^* = \{\mathscr{P}^*(1), \mathscr{P}^*(2), \mathscr{P}^*(3), \ldots\}$ *of inverted partition lattices.* □

To give an idea of this general induction argument we present a short proof of Van der Waerden's theorem that any $r$-coloring of the integers contains a monochromatic arithmetic progression of arbitrary length. First we need some terminology. Let $[a, b]$ denote the set of integers $x$ with $a \le x \le b$. We call two vectors $(x_1, \ldots, x_m), (y_1, \ldots, y_m) \in [0, l]^m$ *l-equivalent* if they agree up to and including the last occurrence of $l$. Clearly, this is an equivalence relation on $[0, l]^m$ and we notice that all vectors $(x_1, \ldots, x_m) \in [0, l-1]^m$ are $l$-equivalent. Let $l, m \ge 1$ and denote by $A(l; m)$ the following statement:

$A(l; m)$: *For any positive integer* $r$, *there exists a smallest number* $N(l, m, r)$ *such that, for any* $n \ge N(l, m, r)$ *and for any* $r$*-coloring of* $[1, n]$, *there are positive integers* $a, d_1, \ldots, d_m$ *with* $a + l\sum_{i=1}^m d_i \le n$ *such that, for each* $l$*-equivalence class* $E$ *of* $[0, l]^m$, *the set* $\{a + \sum_{i=1}^m x_i d_i : (x_1, \ldots, x_m) \in E\}$ *is monochromatic.*

Consider $A(l; m)$ for $m = 1$. In this case there are only two equivalence classes, one consisting of all numbers $0, 1, \ldots, l-1$ and the other consisting of $l$. Hence $A(l; 1)$ says that for $n \ge N(l, 1, r)$ and any $r$-coloring of $[1, n]$ there exists a monochromatic arithmetic progression $a, a + d, a + 2d, \ldots, a + (l-1)d$ of length $l$. This is precisely Van der Waerden's theorem.

**8.78 Theorem** (Graham–Rothschild). *$A(l; m)$ holds for all $l, m \ge 1$.*

*Proof*. We use induction on $l$ and $m$. $A(1; 1)$ is trivially true.

*Claim*: $A(l; m)$ for some $m \geq 1 \Rightarrow A(l; m + 1)$. For a fixed $r$ let $M := N(l, m, r)$, $M' := N(l, 1, r^M)$ (which exist by induction) and let $C: [1, MM'] \to [1, r]$ be an $r$-coloring. We define an $r^M$-coloring $C'$ of $[1, M']$ by using $[1, r]^M$ as color set and setting

$$C'(k) := (C((k - 1)M + 1), C((k - 1)M + 2), \ldots, C(kM)) \quad \text{for } k \in [1, M'].$$

Hence $C'(k) = C'(k')$ if and only if $C((k - 1)M + j) = C((k' - 1)M + j)$ for $j = 1, \ldots, M$. By induction there exist $a'$ and $d'$ with $a' + ld' \leq M'$ such that $C'(a' + xd')$ is constant for $x = 0, \ldots, l - 1$. Applying $A(l; m)$ to the interval $[(a' - 1)M + 1, a'M]$, we obtain integers $a, d_1, \ldots, d_m$ with $(a' - 1)M + 1 \leq a$, $a + \sum_{i=1}^m ld_i \leq a'M$, such that $C(a + \sum_{i=1}^m x_i d_i)$ is constant on all $l$-equivalence classes. We set $d_i' = d_i$ for $i = 1, \ldots, m$ and $d_{m+1}' = d'M$, and claim that $\{a, d_1', \ldots, d_{m+1}'\}$ is a set required by $A(l; m + 1)$. First, we have $a + \sum_{i=1}^{m+1} ld_1' = a + \sum_{i=1}^m ld_i + ld'M \leq a'M + ld'M \leq MM'$. Now let $X = a + \sum_{i=1}^{m+1} x_i d_i'$ and $Y = a + \sum_{i=1}^{m+1} y_i d_i'$, where $(x_1, \ldots, x_{m+1}), (y_1, \ldots, y_{m+1}) \in [0, l]^{m+1}$ are $l$-equivalent, and set $X_0 = a + \sum_{i=1}^m x_i d_i'$, $Y_0 = a + \sum_{i=1}^m y_i d_i'$. We have to show $C(X) = C(Y)$. If $x_{m+1} = y_{m+1} = l$ then $x_i = y_i$ for all $i$, and there is nothing to prove. So assume $x_{m+1}, y_{m+1} \in [0, l - 1]$ and hence that $(x_1, \ldots, x_m)$ and $(y_1, \ldots, y_m)$ are $l$-equivalent. Since $C'(a') = C'(a' + x_{m+1}d')$ it follows that $C((a' - 1)M + j) = C((a' + x_{m+1}d' - 1)M + j)$ for $j = 1, \ldots, M$. Suppose $X_0 = (a' - 1)M + h$ for some $h \in [1, M]$. Then

$$X = X_0 + x_{m+1}d'M = (a' + x_{m+1}d' - 1)M + h$$

and hence $C(X) = C(X_0)$. By the same argument, $C(Y) = C(Y_0)$ and thus $C(X) = C(Y)$ since $C(X_0) = C(Y_0)$.

*Claim*. $A(l; m)$ for all $m \geq 1 \Rightarrow A(l + 1; 1)$. For a fixed $r$, consider an $r$-coloring $C$ of $[1, N(l, r, r)]$. Then there exist positive integers $a, d_1, \ldots, d_r$ with $a + \sum_{i=1}^r ld_i \leq N(l, r, r)$ such that $C(a + \sum_{i=1}^r x_i d_i)$ is constant on the $l$-equivalence classes. Among the $r + 1$ vectors $(0, 0, \ldots, 0), (l, 0, \ldots, 0), (l, l, 0, \ldots, 0), \ldots, (l, l, \ldots, l)$ in $[0, l]^r$ there must be two, say with $u$ and $v$ $l$'s, respectively, for which

$$C\left(a + \sum_{i=1}^{u} ld_i\right) = C\left(a + \sum_{i=1}^{v} ld_i\right).$$

Assume $u < v$. Since $(\underbrace{l, \ldots, l}_{u}, \underbrace{x, \ldots, x}_{v-u}, 0, \ldots, 0)$ is $l$-equivalent to $(\underbrace{l, \ldots, l}_{u}, 0, \ldots, 0)$ for any $x \in [0, l - 1]$ we conclude that all $l + 1$ numbers

$$\left(a + \sum_{i=1}^{u} ld_i\right) + x\left(\sum_{i=u+1}^{v} d_i\right) \qquad (x = 0, \ldots, l)$$

are colored alike. This is precisely the requirement for $A(l + 1; 1)$. □

**8.79 Corollary** (Van der Waerden). *For every $r$ and $l$ there exists a smallest integer $N(l, r)$ such that in any $r$-coloring of $[1, n]$, $n \geq N(l, r)$, there exists a monochromatic arithmetic progression of length $l$.* $\square$

As with all Ramsey numbers, the exact value of $N(l, r)$ is known only for the very smallest values of $l$ and $r$. One way to estimate $N(l, r)$ is to find out how large a set can be without containing an arithmetic progression of length $l$. Define $r_l(n)$ to be the maximal size of a subset of $[1, n]$ which contains no arithmetic progression of length $l$. Using the recursion

$$r_l(m + n) \leq r_l(m) + r_l(n)$$

it is easy to see that $\lim_{n \to \infty}(r_l(n)/n)$ exists for every $l$. A celebrated combinatorial theorem proved recently asserts that all these limits are in fact 0, as conjectured by Erdös and Turan.

**8.80 Theorem** (Szemerédi). *We have $\lim_{n \to \infty}(r_l(n)/n) = 0$ for every $l$. Hence any set $S$ of integers with $\limsup_{n \to \infty}(|S \cap [1, n]|/n) > 0$ contains arbitrarily long arithmetic progressions.* $\square$

We close with what is perhaps the most interesting open problem in this area bordering on combinatorics and number theory. Let $\pi(n)$ be the number of primes $\leq n$. Is it true that $r_l(n) < \pi(n)$ for sufficiently large $l$ and $n$? If so, the prime numbers must contain arithmetic progressions of arbitrary length.

## Exercises VIII.4

→ 1. Prove: Given $k$ and $r$ and a countably infinite set $S$, for any $r$-coloring of $\mathscr{B}^{(k)}(S)$ there exists an infinite subset $T \subseteq S$ all of whose $k$-subsets are colored alike. Deduce **8.69** from this.

2. Show that

$$N(2; l_1, l_2) \leq \binom{l_1 + l_2 - 2}{l_1 - 1}.$$

→ 3. Verify that $N(2; l_1, l_2) < N(2; l_1 - 1, l_2) + N(2; l_1, l_2 - 1)$ if both numbers on the right-hand side are even.

4. By ex. 3, $N(2; 3, 4) \leq 9$. Show $N(2; 3, 4) > 8$, and hence $N(2; 3, 4) = 9$, by taking as vertices of $K_8$ the residues modulo 8 and coloring the edge $\{i, j\}$ depending on the difference $i - j \pmod 8$.

5. By ex. 3, $N(2; 4, 4) \leq 18$. Show $N(2; 4, 4) = 18$ by taking as vertices of $K_{17}$ the residues modulo 17 and proceeding as in the previous exercise.

→ 6. Prove : $2^{l/2} \leq N(2; l, l) \leq 2^{2l}$ for $l \geq 3$.

7. Let the edges of $K_n$ be colored red or blue and let $r_i$ be the number of red edges emanating from $v_i$. Show that for the number $m$ of monochromatic triangles in $K_n$,

$$m = \binom{n}{3} - \frac{1}{2}\sum_{i=1}^{n} r_i(n - 1 - r_i).$$

→ 8. Suppose $G(V, E)$ is a finite simple graph on $n$ vertices which contains no complete subgraph on $k$ vertices. Prove

$$|E| \leq \frac{k-2}{2(k-1)}(n^2 - r^2) + \binom{r}{2},$$

where $r$ is defined by $n = q(k-1) + r$, $1 \leq r \leq k-1$, with equality if and only if $G(V, E)$ is of the form $V = V_1 \dot\cup \cdots \dot\cup V_{k-1}$ with $|V_i| = q + 1$ for $i = 1, \ldots, r$, $|V_i| = q$ for $i = r + 1, \ldots, k - 1$, and $\{u, v\} \in E$ if and only if $u$ and $v$ are in different $V_i$'s. (Turan) (Hint: Consider a graph $G$ with this property and the maximal possible number of edges. Take a $K_{k-1}$ in $G$ and consider the possible edges outside $K_{k-1}$.)

9. For $k = 3$, ex. 8 says that a graph $G(V, E)$ without triangles can have at most $\lfloor n^2/4 \rfloor$ edges, with equality if and only if $G$ is bipartite. Hence if $G$ has $\lfloor n^2/4 \rfloor + 1$ edges, then $G$ contains at least one triangle. Show that, in fact, $G$ must contain at least $\lfloor n/2 \rfloor$ triangles.

10. Prove the Ramsey property for the sequence $\mathscr{D} = \{\mathscr{D}_1, \mathscr{D}_2, \ldots\}$ of dominance orders.

→ 11. Using **8.73**, prove the following statement: For every $n \in \mathbb{N}$ there is a number $B(n)$ such that, for every prime number $p > B(n)$, the congruence $x^n + y^n \equiv z^n \pmod{p}$ has a non-trivial solution in $x, y, z$.

12. Verify the values $C_5$ and $C_6$ in **8.74**.

13. Using **8.75**, prove the following statement: Let $\{f_1, f_2, \ldots\}$ be a uniformly bounded sequence of continuous real functions on $[0, 1] \subseteq \mathbb{R}$. Suppose there is an integer $k \geq 0$ such that for all $i \neq j$, the equation $f_i(x) = f_j(x)$ has at most $k$ solutions. Then there exists a subsequence which is convergent on $[0, 1]$. (Hint: Take the $f_i$'s as vertices and color $\{f_i, f_j\}$ depending on whether $f_i(x) = f_j(x)$ has exactly $k$ solutions in $[0, \frac{1}{2}]$. Now use induction.)

14. Consider 0, 1-matrices. The notation

$$\begin{bmatrix} * & j \\ i & * \end{bmatrix}$$

indicates that the entries below the main diagonal are all equal to $i$ and above the main diagonal are all equal to $j$. The entries on the main diagonal may be 0 or 1. Show that given $m \in \mathbb{N}$ there exists a number $M_m$ such that any $n \times n$-matrix with $n \geq M_m$ contains a principal $m \times m$-submatrix of the form

$$\begin{bmatrix} * & 0 \\ 0 & * \end{bmatrix}, \quad \begin{bmatrix} * & 1 \\ 0 & * \end{bmatrix}, \quad \begin{bmatrix} * & 0 \\ 1 & * \end{bmatrix}, \quad \text{or} \quad \begin{bmatrix} * & 1 \\ 1 & * \end{bmatrix}.$$

→15.* Prove: Given integers $l$ and $r$, there exists an integer $N(l, r)$ such that if $\mathbb{N}_n$, $n \geq N(l, r)$, is partitioned into $r$ classes, then there exist $l$ integers $a_1, \ldots, a_l$ such that all non-zero sums $\sum_{i=1}^{l} \varepsilon_i a_i$, $\varepsilon_i = 0$ or 1, are in the same class. (Sanders) (Hint: Consider the proof of **8.78**.)

## Notes

Transversal theory, which saw its birth with the papers by P. Hall [1] and Rado [1, 2], witnessed rapidly accelerating growth both in depth and diversity after the appearance of the survey articles by Mirsky–Perfect [1, 2]. For a very detailed exposition the reader is referred to Mirsky [1]; an interesting collection of applications is contained in Harper–Rota [1]. Harary [1] and Berge [1] are good sources for the graph theorems involved. Sperner-type problems have, especially in recent years, been at the center of combinatorial research. For good reviews of these and related questions see the survey papers by Greene–Kleitman [3] and Katona [2, 5]. After Ramsey's fundamental theorem [1], Ramsey-type theorems have been more or less confined to graphs (with the notable exception of the theorems by Erdös–Szekeres and Schur), and to infinite sets as expounded in Erdös–Rado–Hajnal [1]. Only in very recent years has the theory been extended to other poset sequences; see the survey articles by Graham–Rothschild [1, 3]. The reader should also attempt to read Szemerédi's great paper [1].

# Bibliography


Abel, N. H.

1. Beweis eines Ausdruckes, von welchem die Binomial-Formel ein einzelner Fall ist. *Crelle* **1**, 159–160 (1826).

Aigner, M.

1*. *Kombinatorik*. Berlin-Heidelberg-New York: Springer-Verlag, vol. I (1975), vol. II (1976).
2*. *Higher Combinatorics* (ed.). Dordrecht: Reidel (1977).
3. Lexicographic matching in Boolean algebras. *J. Comb. Theory* (B) **14**, 187–194 (1973).
4. Uniformität des Verbandes der Partitionen. *Math. Ann.* **207**, 1–22 (1974).
5. Symmetrische Zerlegung von Kettenprodukten. *Monatsh. Math.* **79**, 177–189 (1975).

Aigner, M.–Dowling, T. A.

1. Matching theory for combinatorial geometries. *Trans. Amer. Math. Soc.* **158**, 231–245 (1971).

Aigner, M.–Prins, G.

1. Segment preserving maps of partial orders. *Trans. Amer. Math. Soc.* **166**, 351–360 (1972).

Alder, H. L.

1. Partition identities—from Euler to the present. *Amer. Math. Monthly* **76**, 733–746 (1969).

Al-Salaam, W. A.

1. Operational representations for the Laguerre and other polynomials. *Duke Math. J.* **31**, 127–142 (1964).

André, D.

1. Sur les permutations alternées. *J. Math. Pures Appl.* **7**, 167–184 (1881).

Andrews, G. E.

1. On the foundations of combinatorial theory V: Eulerian differential operators. *Studies in Appl. Math.* **50**, 345–375 (1971).
2. Partition identities, *Advances Math.* **9**, 10–51 (1972).

Appel, K.–Haken, W.–Koch, J.

1. Every planar map is four colorable. *Ill. Math. J.* **21**, 429–567 (1977).

Artin, E.

1. *Coordinates in Affine Geometry*. Reports Math. Coll. Notre Dame (1940).

Baclawski, K.

1. *Homology and Combinatorics of Ordered Sets*. Ph.D. thesis Harvard Univ. (1976).

Baer, R.

1*. *Linear Algebra and Projective Geometry*. New York: Academic Press (1952).

Baker, K.

1. A generalization of Sperner's lemma. *J. Comb. Theory* **6**, 224–225 (1969).

Basterfield, J.G.–Kelly, L. M.

1. A characterisation of sets of $n$ points which determine $n$ hyperplanes. *Camb. Phil. Soc.* **64**, 585–588 (1968).

Baxter, G.

1. An analytic problem whose solution follows from a simple algebraic identity. *Pac. J. Math.* **10**, 731–742 (1960).

Beckenbach, E. F.

1*. *Applied Combinatorial Mathematics* (ed.). New York: Wiley (1964).

Becker, H. W.–Riordan, J.
1. The arithmetic of Bell and Stirling numbers. *Amer. J. Math.* **70**, 385–394 (1934).

Bell, E. T.
1. Exponential polynomials. *Annals Math.* **35**, 258–277 (1934).
2. Postulational bases for the umbral calculus. *Amer. J. Math.* **62**, 717–724 (1940).

Bender, E. A.
1. The asymptotic number of non-negative integer matrices with given row and column sums. *Discrete Math.* **10**, 217–223 (1974).

Bender, E. A.–Goldman, J. R.
1. Enumerative uses of generating functions. *Indiana Univ. Math. J.* **20**, 753–765 (1971).

Berge, C.
1.* *The Theory of Graphs.* London: Methuen and New York: Wiley (1962).
2.* *Principles of Combinatorics.* New York, London: Academic Press (1971).
3. Some classes of perfect graphs. Chap. 5, *Graph Theory and Theoretical Physics* (Harary, ed.). New York: Academic Press (1967).
4. Perfect graphs. *Studies in Math.* vol. 11 (Fulkerson, ed.), 1–22. Math. Assoc. Amer. (1975).

Berlekamp, E. R.
1.* *Algebraic Coding Theory.* New York: McGraw-Hill (1968).

Biggs, N.
1.* *Algebraic Graph Theory.* London: Cambridge University Press (1974).

Birkhoff, G.
1.* *Lattice Theory.* 3rd edition. Providence: Amer. Math. Soc. Coll. Publ. vol. **25** (1967).
2. On the combination of subalgebras. *Proc. Camb. Phil. Soc.* **29**, 441–464 (1933).
3. Combinatorial relations in projective geometries. *Annals Math.* **36**, 743–748 (1935).
4. On the structure of abstract algebras. *Proc. Camb. Phil. Soc.* **31**, 433–454 (1935).
5. Abstract linear dependence in lattices. *Amer. J. Math.* **57**, 800–804 (1935).
6. Rings of sets. *Duke Math. J.* **3**, 442–454 (1937).
7. Tres observaciones sobre el algebra lineal. *Univ. Nac. Tucumán Rev. Ser.* **A5**, 147–151 (1946).

Birkhoff, G. D.
1. A determinental formula for the number of ways of colouring a map. *Annals Math.* **14**, 42–46 (1912).

Bixby, R. E.
1. Kuratowski's and Wagner's theorems for matroids. *J. Comb. Theory* (B) **22**, 31–53 (1977).
2. On Reid's characterisation of the matroids representable over $GF(3)$. To appear.

Blackburn J. A.-Crapo, H. H.–Higgs, D. A.
1. A catalogue of combinatorial geometries. *Math. Comput.* **27**, 155–166 (1973).

Blakely, N.
1. Algebra of formal power series. *Duke Math. J.* **31**, 341–345 (1964).

Bland, R. G.–Las Vergnas, M.
1. Orientability of matroids. *J. Comb. Theory* (B) **24**, 94–123 (1978).

Bondy, J. A.
1. Transversal matroids, base orderable matroids, and graphs. *Quart. J. Math.* (Oxford) **23**, 81–89 (1972).
2. Presentations of transversal matroids. *J. London Math. Soc.* (2) **5**, 289–292 (1972).

Bondy, J. A.–Welsh, D. J. A.
1. Some results on transversal matroids and constructions for identically self-dual matroids. *Quart. J. Math.* (Oxford) **22**, 435–451 (1972).

Bott, R.–Duffin, R. J.
1. On the algebra of networks. *Trans. Amer. Math. Soc.* **74**, 99–109 (1953).

Brooks, R. L.–Smith, C. A. B.–Stone, A. H.–Tutte, W. T.
1. The dissection of rectangles into squares. *Duke Math. J.* **7**, 312–340 (1940).

Brown, T. J.
1. Transversal theory and F-products. *J. Comb. Theory* (A) **17**, 290–299 (1974).

Brualdi, R. A.
1. A very general theorem on systems of distinct representatives. *Trans. Amer. Math. Soc.* **140**, 149–160 (1969).
2. Comments on bases in dependence structures. *Bull. Austr. Math. Soc.* **1**, 161–167 (1969).

3. Admissible mappings between dependence spaces. *Proc. London Math. Soc.* (3) **21**, 307–329 (1970).
4. Induced matroids. *Proc. Amer. Math. Soc.* **29**, 213–221 (1971).
5. Generalized transversal theory. *Théorie des Matroides* (Bruter, ed.), 5–31. Lecture Notes Math. **211**, Springer-Verlag (1971).
6. On families of finite independence structures. *Proc. London Math. Soc.* **22**, 265–293 (1971).

Brualdi, R. A.–Dinolt, G. W.
1. Characterizations of transversal matroids and their presentations. *J. Comb. Theory* **12**, 268–286 (1972).

Brualdi, R. A.–Scrimger, E. B.
1. Exchange systems, matchings, and transversals. *J. Comb. Theory* **5**. 244–257 (1968).

deBruijn, N. G.
1. Generalization of Polya's fundamental theorem in enumeration combinatorial analysis. *Indag. Math.* **21**, 59–69 (1959).
2. Enumerative combinatorial problems concerning structures. *Nieuw. Arch. Wisk.* **11**, 142–161 (1963).
3. Polya's theory of counting. *Applied Combinatorial Mathematics* (Beckenbach, ed.), 144–184. New York: Wiley (1964).
4. Colour patterns which are invariant under a given permutation of the colours. *J. Comb. Theory* **2**, 418–421 (1967).
5. Enumeration of tree-shaped molecules. *Recent Progress in Combinatorics* (Tutte, ed.), 59–68. New York, London: Academic Press (1969).

deBruijn, N. G.–Erdös, P.
1. A colour problem for infinite graphs and a problem in the theory of relations. *Kon. Ned. Akad. Wetensch. Proc.* (A) **54**, 371–373 (1951).

deBruijn, N. G.–van E. Tengbergen, C. A.–Kruyswijk, D. R.
1. On the set of divisors of a number. *Nieuw. Arch. Wisk.* (2) **23**, 191–193 (1952).

Bruter, C. P.
1.* *Théorie des Matroides* (ed.). Lecture Notes Math. **211**, Springer-Verlag (1971).

Brylawski, T. H.
1. A combinatorial model for series-parallel networks. *Trans. Amer. Math. Soc.* **154**, 1–22 (1971).
2. A decomposition for combinatorial geometries. *Trans. Amer. Math. Soc.* **171**, 235–282 (1972).
3. The lattice of integer partitions. *Discrete Math.* **6**, 201–219 (1973).
4. A note on Tutte's unimodular representation theorem. *Proc. Amer. Math. Soc.* **52**, 499–502 (1975).

Burnside, W.
1.* *Theory of Groups of Finite Order.* New York: Dover Publications (1955).

Busacker, R. G.–Saaty, T. L.
1.* *Finite Graphs and Applications.* New York: McGraw-Hill (1965).

Camion, P.
1. Modules unimodulaires. *J. Comb. Theory* **4**, 301–362 (1968).

Canfield, R.
1. On the location of the maximum Stirling numbers of the second kind. To appear.
2. On a problem of Rota. *Bull. Amer. Math. Soc.* **84**, 164 (1978).

Carlitz, L.
1. Eulerian numbers and polynomials. *Math. Mag.* **30**, 203–214 (1958).
2. A note on Laguerre polynomials. *Michigan Math. J.* **7**, 219–223 (1960).
3. Some inversion formulas. *Rand. Circ. Mat. Palermo* (2) **12**, 183–199 (1963).
4. Rings of arithmetic functions. *Pac. J. Math.* **14**, 1165–1171 (1964).
5. Permutations and sequences. *Advances Math.* **14**, 92–120 (1974).

Carlitz, L.–Riordan, J.
1. The divided central differences of zero. *Can. J. Math.* **15**, 94–100 (1964).

Carlitz, L.–Roselle, D. P.–Scoville, R. A.
1. Permutations and sequences with repetitions by number of increases. *J. Comb. Theory* **1**, 350–374 (1966).

Cartier, P.–Foata, D.
1. *Problèmes Combinatoires de Commutation et Réarrangement.* Lecture Notes Math. **85**, Springer-Verlag (1969).

Cayley, A.
1. On the theory of the analytical forms called trees. *Philos. Mag.* **13**, 19–30 (1857).
2. On the analytical form called trees. *Amer. Math. J.* **4**, 266–268 (1881).
3. A theorem on trees. *Quart. J. Math.* **23**, 376–378 (1889).

Chowla, S.–Herstein, I. N.–Scott, W. R.
1. The solution of $x^d = 1$ in symmetric groups. *Norske Vid. Selsk. Fach.* (Trondheim) **25**, 29–31 (1952).

Comtet, L.
1.* *Advanced Combinatorics.* Dordrecht and Boston: Reidel (1974).

Crapo, H. H.
1. Single element extensions of matroids. *J. Res. Nat. Bur. Stand.* **69B**, 57–65 (1965).
2. The Möbius function of a lattice. *J. Comb. Theory* **1**, 126–131 (1966).
3. Structure theory for geometric lattices. *Rend. Sem. Math. Univ. Padova* **38**, 14–22 (1967).
4. A higher invariant for matroids. *J. Comb. Theory* **2**, 406–417 (1967).
5. Möbius inversion in lattices. *Archiv Math.* **19**, 595–607 (1968).
6. The Tutte polynomial. *Aequationes Math.* **3**, 211–229 (1969).
7. Erecting geometries. Proc. 2nd Chapel Hill Conf. on Comb. Math., 74–99. Univ. North Carolina (1970).
8. Constructions in combinatorial geometries. NSF Advanced Science Seminar in Comb. Theory. Bowdoin College (1971).

Crapo, H. H.–Rota, G.-C.
1.* *On the Foundations of Combinatorial Theory II: Combinatorial Geometries.* Cambridge Mass.: MIT Press (1970).
See also Blackburn–Crapo–Higgs.

Crawley, P.–Dilworth, R. P.
1.* *Algebraic Theory of Lattices.* Englewood Cliffs: Prentice-Hall Inc. (1973).

Davis, R. L.
1. The number of structures of finite relations. *Proc. Amer. Math. Soc.* **4**, 486–495 (1953).
2. Order algebras. *Bull. Amer. Math. Soc.* **76**, 83–87 (1970).

Dembowski, P.
1.* *Finite Geometries.* Berlin-Heidelberg-New York: Springer-Verlag (1968).
2.* *Kombinatorik.* BI Hochschultext 741a. Mannheim-Wien-Zürich (1970).

Dénes, J.
1. The representation of a permutation as the product of a minimal number of transpositions. *Publ. M.I. Hung. Acad. Sci.* **4**, 63–70 (1959).

Désarméniens, J.–Kung, J. P. S.–Rota, G.-C.
1. Invariant theory, Young bitableaux, and Combinatorics. *Advances Math.* **27**, 63–92 (1978).

De Sousa, J.–Welsh, D. J. A.
1. A characterisation of binary transversal matroids. *J. Math. Anal. Appl.* **40** (1), 55–59 (1972).

Deuber, W.
1. Generalizations of Ramsey's theorem. *Infinite and Finite Sets.* Colloqu. honoring P. Erdös, 323–332. Keszthely: Colloq. Math. Soc. J. Bolyai (1975).

Dilworth, R. P.
1. The arithmetical theory of Birkhoff lattices. *Duke Math. J.* **8**, 286–299 (1941).
2. Dependence relations in a semimodular lattice. *Duke Math. J.* **11**, 575–587 (1944).
3. Note on the Kurosh–Ore theorem. *Bull. Amer. Math. Soc.* **52**, 659–663 (1946).
4. A decomposition theorem for partially ordered sets. *Annals Math.* **51**, 161–166 (1950).
5. The structure of relatively complemented lattices. *Annals Math.* **51**, 348–359 (1950).
6. Proof of a conjecture on finite modular lattices. *Annals Math.* **60**, 359–364 (1954).
7. Some combinatorial problems on partially ordered sets. *Combinatorial Analysis* (Bellman–Hall, eds.), 85–90. Providence: Amer. Math. Soc. (1960).

Dilworth, R. P.–Greene, C.
1. A counterexample to the generalization of Sperner's theorem. *J. Comb. Theory* **10**, 18–20 (1971).
See also Crawley–Dilworth, Hall–Dilworth.

Dinolt, G. W. See Brualdi–Dinolt.

Dirac, G. A.
1. A property of 4-chromatic graphs and some remarks on critical graphs. *J. London Math. Soc.* **27**, 85–92 (1952).

2. In abstrakten Graphen vorhandene vollständige 4-Graphen und ihre Unterteilungen. *Math. Nachr.* **22**, 61–85 (1960).
3. Généralisations du théorème de Menger. *C. R. Acad. Sci.* (Paris) **250**, 4252–4253 (1960).

Dobinski, G.
1. *Grunert's Archiv* **61**, 333–336.

Doubilet, P.–Rota, G.-C.–Stanley, R. P.
1. On the foundations of combinatorial theory VI: The idea of generating function. Proc. 6th Berkeley Symp. on Math. Stat. and Prob. vol. II: Probability Theory, 267–318. Univ. Calif. (1972).

Dowling, T. A.
1. Codes, packings and the critical problem. Atti del Covegno di Geometria Comb. e sue Appl., 210–224. Univ. Perugia (1971).
2. A $q$-analog of the partition lattice. *A Survey of Combinatorial Theory* (Srivastava, ed.), 101–115. Amsterdam: North-Holland Publ. Co. 1973.
3. A class of geometric lattices based on finite groups. *J. Comb. Theory* (B) **14**, 61–86 (1973).
4. Complementing permutations in finite lattices. *J. Comb. Theory* (B) **23**, 223–226 (1977).

Dowling, T. A.–Kelly, D. G.
1. Elementary strong maps and transversal geometries. *Discrete Math.* **7**, 209–225 (1974).

Dowling, T. A.–Wilson, R. M.
1. Whitney number inequalities for geometric lattices. *Proc. Amer. Math. Soc.* **47**, 504–512 (1975).
See also Aigner–Dowling.

Duffin, R.
1. Non-linear networks IIa. *Bull. Amer. Math. Soc.* **53**, 963–971 (1947).
2. An analysis of the Wang algebra of networks. *Trans. Amer. Math. Soc.* **93**, 114–131 (1959).
3. Topology of series-parallel networks. *J. Math. Anal. Appl.* **10**, 303–318 (1965).
4. Electrical network models. *Studies in Math.* vol. **11** (Fulkerson ed.), 94–138 (1975).
See also Bott–Duffin.

Dulmage, A. L. See Mendelsohn–Dulmage.

Edelberg M. See Kleitman–Edelberg–Lubell.

Edmonds, J.
1. Minimum partition of a matroid into independent subsets. *J. Res. Nat. Bur. Stand.* **69B**, 65–72 (1965).
2. Lehman's switching game and a theorem of Tutte and Nash–Williams. *J. Res. Nat. Bur. Stand.* **69B**, 73–77 (1965).
3. Submodular functions, matroids and certain polyhedra. *Combinatorial Structures and their Applications* (Guy *et al.*, eds.), 69–87. Paris: Gordon & Breach (1970).

Edmonds, J.–Fulkerson, D. R.
1. Transversals and matroid partition. *J. Res. Nat. Bur. Stand.* **69B**, 147–153 (1965).
2. Bottleneck extrema. *J. Comb. Theory* **8**, 299–306 (1970).
See also Young–Murty–Edmonds.

Erdös, P.
1. On a lemma of Littlewood and Offord. *Bull. Amer. Math. Soc.* **51**, 898–502 (1945).
2. Some remarks on the theory of graphs. *Bull. Amer. Math. Soc.* **53**, 292–294 (1947).
3. Some remarks on Ramsey's theorem. *Can. Math. Bull.* **7**, 619–630 (1964).

Erdös, P.–Ko, C.–Rado, R.
1. Intersection theorems for systems of finite sets. *Quart. J. Math.* (Oxford) (2) **12**, 313–318 (1961).

Erdös, P.–Rado, R.–Hajnal, A.
1. Partition relations for cardinal numbers. *Trans. Amer. Math. Soc.* **16**, 93–196 (1965).

Erdös, P.–Szekeres, G.
1. A combinatorial problem in geometry. *Compositio Math.* **2**, 463–470 (1939).

Erdös, P.–Turan, P.
1. On some sequences of integers. *J. London Math. Soc.* **11**, 261–264 (1936).
See also deBruijn–Erdös.

Fan, K.
1. On Dilworth's coding theorem. *Math. Z.* **127**, 92–94 (1972).

Foata, D.
1. On the Netto inversion number of a sequence. *Proc. Amer. Math. Soc.* **19**, 236–240 (1968).

2. *La Série Génératrice Exponentielle dans les Problèmes d'énumération.* Montréal: Les Presses de l'Université de Montréal (1974).
3. *Studies in enumeration.* Dept. of Stat. Univ. of North Carolina Mimeo Series No. 974 (1975).

Foata, D.–Schützenberger, M. P.
1. *Théorie Géométrique des Polynômes Eulériens.* Lectures Notes Math. **138**, Springer-Verlag (1970). See also Cartier–Foata.

Folkman, J.
1. The homology groups of a lattice. *J. Math. Mech.* **15**, 631–636 (1966).
2. Transversals of infinite families with finitely many infinite members. RC Memo RM-5676-PR (1968).

Ford, L. R.–Fulkerson, D. R.
1.* *Flows in Networks.* Princeton: Princeton Univ. Press (1962).
2. Maximal flow through a network. *Can. J. Math.* **8**, 399–404 (1956).
3. Network flows and systems of representatives. *Can. J. Math.* **10**, 78–84 (1958).

Fournier, J. C.
1. Sur la représentation sur un corps des matroides à sept et huit éléments. *C. R. Acad. Sci.* (Paris) Ser. **270**, 810–813 (1970).
2. Représentation sur un corps des matroides d'ordre $\leq 8$. *Théorie des Matroides* (Bruter, ed.), 50–61. Lecture Notes Math. **211**, Springer-Verlag (1971).

Frame, J. S.–Robinson, G. de B.–Thrall, R. M.
1. The hook lengths of $S_n$. *Can. J. Math.* **6**, 316–325 (1954).

Françon, J.
1. Preuves combinatoires des identités d'Abel. *Discrete Math.* **8**, 331–343 (1974).

Freese, R.
1. An application of Dilworth's lattice of maximal antichains. *Discrete Math.* **7**, 107–109 (1974).

Frobenius, G.
1. Über Matrizen aus nicht-negativen Elementen. *Sitzungsber. Preuss. Akad. Wiss.* 456–477 (1912).

Fujiwara, S. See Sasaki–Fujiwara.

Fulkerson, D. R.
1.* *Studies in Graph Theory* (ed.). Studies in Math. vol. 11/12. Math. Ass. of America (1975).
2. Notes on combinatorial mathematics: Anti-blocking polyhedra (mimeo). RM 620/1-PR (1970).
3. Blocking and anti-blocking pairs of polyhedra. *Math. Programming* **1**, 168–194 (1971). See also Edmonds–Fulkerson, Ford–Fulkerson.

Gale, D.
1. A theorem on flows in networks. *Pac. J. Math.* **7**, 1073–1082 (1957).

Ganter, B.–Rival, I.
1. Dilworth's covering theorem for modular lattices. A simple proof. *Algebra Universalis* **3**, 348–350 (1973).

Garsia, A. M.
1. An exposé of the Mullin–Rota theory of polynomials of binomial type. *J. Linear and Multilinear Algebra* **1**, 47–66 (1973).

Garsia, A. M.–Joni, S. A.
1. A new expression for umbral operators and power series inversion. *Proc. Amer. Math. Soc.* **64**, 179–185 (1977).

Geissinger, L.
1. Valuations on distributive lattices I, II, III. *Archiv Math.* **24**, 230–239; 337–345; 475–481 (1973).

Gleason, A. M. See Greenwood–Gleason.

Goldman, J. R.–Rota, G. C.
1. On the foundations of combinatorial theory IV: Finite vector spaces and Eulerian generating functions. *Studies in Applied Math.* **49**, 239–258 (1970). See also Bender–Goldman.

Gomory, R. E.–Hu, T. C.
1. Multi-terminal network flows. *J. of SIAM* **9**, 551–570 (1961).

Goodman, A. J.
1. On sets of aquaintances and strangers at any party. *Amer. Math. Monthly* **66**, 778–783 (1959).

Gordon, B.–Houton, L.
1. Note on plane partitions. I, II, *J. Comb. Theory* **4**, 72–80; 81–99 (1968).

Gould, H. W.
1. Some generalizations of Vandermonde's convolution. *Amer. Math. Monthly* **63**, 84–91 (1956).
2. Stirling number representation problems. *Proc. Amer. Math. Soc.* **11**, 443–451 (1960).
3. A series transformation for finding convolution identities. *Duke Math. J.* **28**, 193–202 (1961).
4. A new convolution formula and some new orthogonal relations for inversion of series. *Duke Math. J.* **29**, 393–404 (1962).
5. Theory of binomial sums. *Proc. West Virginia Ac. Sci.* **34**, 158–161 (1963).
6. An identity involving Stirling numbers. *Ann. I. Statist. M.* **17**, 265–269 (1965).

Graham, R. L.–Harper, L. H.
1. Some results on matchings in bipartite graphs. *SIAM J.* **17**, 1017–1022 (1969).

Graham, R. L.–Leeb, K.–Rothschild, B. L.
1. Ramsey's theorem for a class of categories. *Advances Math.* **8**, 417–433 (1972).

Graham, R. L.–Rothschild, B. L.
1. A survey of finite Ramsey theorems. Proc. 2nd Louisiana Conf. Comb. Graph Th. and Comp., 21–41. Louisiana State Univ. (1971).
2. Ramsey's theorem for $n$-parameter sets. *Trans. Amer. Math. Soc.* **159**, 257–291 (1971).
3. Some recent developments in Ramsey theory. *Combinatorics* (Hall–VanLint, eds.). Math. Centre Tracts **56**, 61–76. Amsterdam (1974).
4. A short proof of Van der Waerden's theorem on arithmetic progressions. *Proc. Amer. Math. Soc.* **42**, 385–386 (1974).

Graver, J. E.
1.* *Lectures on the Theory of Matroids.* Univ. Alberta (1966).

Graver, J. E.–Yackel, J.
1. Some graph theoretic results associated with Ramsey's theorem. *J. Comb. Theory* **4**, 125–175 (1968).

Greene, C.
1. A rank inequality for finite geometric lattices. *J. Comb. Theory* **9**, 357–364 (1970).
2. An inequality for the Möbius function of a geometric lattice. Proc. Conf. on Möbius algebra. Univ. Waterloo (1971).
3. A multiple exchange property for bases. *Proc. Amer. Math. Soc.* **39**, 45–50 (1973).
4. On the Möbius algebra of a partially ordered set. *Advances Math.* **10**, 177–187 (1973).

Greene, C.–Kleitman, D. J.
1. The structure of Sperner $k$-families. *J. Comb. Theory* (A) **20**, 41–68 (1976).
2. Strong versions of Sperner's theorem. *J. Comb. Theory* (A) **20**, 80–88 (1976).
3. Proof techniques in the theory of finite sets. *MAA Survey of Combinatorics.* To appear.
See also Dilworth–Greene.

Greenwood, R. E.–Gleason, A. M.
1. Combinatorial relations and chromatic graphs. *Can. J. Math.* **7**, 1–7 (1955).

Griggs, J. R.
1. Sufficient conditions for a symmetric chain order. *SIAM J. Appl. Math.* **32**, 807–809 (1977).

Gross, O. A.
1. Preferential arrangements. *Amer. Math. Monthly* **69**, 4–8 (1962).

Grünbaum, B.
1.* *Convex Polytopes.* London: Wiley (1967).

Guilbaud, G. Th.–Rosenstiehl, P.
1. Analyse algebrique d'un scrutin. *Math. et Sci. Humaines* **4**, 9–33 (1960).

Hadwiger, H.
1. Gruppierung mit Nebenbedingungen. *Mitt. Verein Schweizer Vers. Math.* **43**, 113–222 (1943).
2. Über eine Klassifikation der Streckenkomplexe. *Viertelj. Schr. Naturforsch. Ges. Zürich* **88**, 133–142 (1943).

Hajnal, A. See Erdös–Rado–Hajnal.

Haken, W. See Appel–Haken–Koch.

Hales, A.–Jewett, R. I.
1. Regularity and positional games. *Trans. Amer. Math. Soc.* **106**, 222–229 (1963).

Hall, M., Jr.
1.* *The Theory of Groups.* New York: MacMillan (1959).
2.* *Combinatorial Theory.* Waltham, Toronto, London: Blaisdell Publ. Co. (1967).

3. An existence theorem for Latin squares. *Bull. Amer. Math. Soc.* **51**, 387–388 (1945).
4. Distinct representatives of subsets. *Bull. Amer. Math. Soc.* **54**, 922–926 (1948).

Hall, M., Jr.–Dilworth, R. P.
1. The imbedding problem for modular lattices. *Annals Math.* **45**, 450–456 (1944).

Hall, M., Jr.–Van Lint, J. H.
1.* *Combinatorics* (eds.). Math. Centre Tracts. Amsterdam (1974).

Hall, P.
1. On representatives of subsets. *J. London Math. Soc.* **10**, 26–30 (1935).
2. The Eulerian functions of a group. *Quart. J. Math.* (Oxford) **7**, 134–151 (1936).

Hansel, G.
1. Problèmes de dénombrement et d'évaluation de bornes concernant les éléments du treillis distributif libre. *Publ. Inst. Stat. Paris* **16**, 163–294 (1967).

Harary, F.
1.* *Graph Theory.* Reading: Addison-Wesley (1969).
2. The number of linear, directed, rooted, and connected graphs. *Trans. Amer. Math. Soc.* **78**, 445–463 (1955).
3. On the number of bicolored graphs. *Pac. J. Math.* **3**, 743–755 (1958).
4. Exponentiation of permutation groups. *Amer. Math. Monthly* **66**, 572–575 (1959).

Harary, F.–Palmer, E. M.
1.* *Graphical Enumeration.* Reading: Addison-Wesley (1973).
2. The power group enumeration theorem. *J. Comb. Theory* **1**, 157–173 (1966).

Harary, F.–Prins, G.
1. The number of homeomorphically irreducible trees, and other species. *Acta Math.* **101**, 141–162 (1959).

Harborth, H.
1. Über das Maximum bei Stirlingschen Zahlen 2. Art. *J. Reine Angew. Math.* **230**, 213–214 (1968).

Hardy, G. H.–Littlewood, J. E.–Polya, G.
1.* *Inequalities.* Cambridge Univ. Press (1934).

Hardy, G. H.–Wright, E. M.
1.* *An Introduction to the Theory of Numbers.* Oxford: Clarendon Press (1965).

Harper, L. H.
1. Stirling behaviour is asymptotically normal. *Ann. Math. Stat.* **38**, 410–414 (1967).
2. The morphology of partially ordered sets. *J. Comb. Theory* (A) **17**, 44–59 (1974).

Harper, L. H.–Rota, G. C.
1. Matching theory: an introduction. *Advances in Probability* **1**, 169–213. New York: Decker (1971).
See also Graham–Harper.

Hartmanis, J.
1. Lattice theory of generalized partitions. *Can. J. Math.* **11**, 97–106 (1959).

Heawood, P. J.
1. Map-colour theorems. *Quart. J. Math.* Oxford Ser. **24**, 322–338 (1890).

Heron, A. P.
1. A property of the hyperplanes of a matroid and an extension of Dilworth's theorem. *J. Math. Anal. Appl.* **42**, 119–132 (1973).

Herstein, I. See Chowla–Herstein–Scott.

Higgins, P. J.
1. Disjoint transversals of subsets. *Can. J. Math.* **11**, 280–285 (1959).

Higgs, D. A.
1. Maps of geometries. *J. London Math. Soc.* **41**, 612–618 (1966).
2. Strong maps of geometries. *J. Comb. Theory* **5**, 185–191 (1968).
3. Matroids and duality. *Colloq.* XX, 215–220 (1969).
See also Blackburn–Crapo–Higgs.

Hoffman, A. J.–Kuhn, H. W.
1. Systems of distinct representatives and linear programming. *Amer. Math. Monthly* **63**, 455–460 (1956).
2. Systems of distinct representatives. *Linear Inequalities and Related Systems.* Annals of Math. Studies No. **38**, 199–206. Princeton Univ. (1956).

Houten, L. See Gordon–Houten.

Hsieh, W. N.–Kleitman, D. J.

1. Normalized matching in direct products of partial orders. *Studies Appl. Math.* **52**, 285–289 (1973).

Hu, T. C. See Gomory–Hu.

Ingleton, A. W.

1. Representation of matroids. *Combinatorial Math. and its Applications* (Welsh, ed.), 149–169. London, New York: Acad. Press (1971).
2. A geometrical characterization of transversal independence structures. *Bull. London Math. Soc.* **3**, 47–51 (1971).
3. Conditions for representability and transversality of matroids. *Théorie des Matroides* (Bruter, ed.), 62–67. Lecture Notes Math. **211**, Springer-Verlag (1971).
4. Transversal matroids and related structures. *Higher Combinatorics* (Aigner, ed.), 117–131. Dordrecht: Reidel (1977).

Ingleton, A. W.–Piff, M. J.

1. Gammoids and transversal matroids. *J. Comb. Theory* **15**, 51–68 (1973).

Iri, M.

1. Comparison of matroid theory with algebraic topology with special reference to applications to network theory. *RAAG Res. Notes, Univ. Tokyo* **83** (1964).

Jacobson, N.

1*. *Structure of Rings.* Providence: Amer. Math. Soc. Coll. Publ. vol. **37** (1956).

Jaeger, F.

1. On nowhere-zero flows in mutigraphs. Proc. 5th British Comb. Conf. (Nash–Williams–Sheehan, eds.), 373–378. Univ. Aberdeen (1976).

Jewett, R. I. See Hales–Jewett.

Joni, S. A. See Garsia–Joni.

Jónsson, B.

1. On the representation of lattices. *Math. Scand.* **1**, 193–206 (1953).
2. Modular lattices and Desargues' theorem. *Math. Scand.* **2**, 295–314 (1954).
3. Lattice-theoretic approach to projective and affine geometry. *The axiomatic method* (Henkin–Suppes–Tarski, eds.). Amsterdam: Studies in logic, 188–203 (1959).
4. Representation of modular lattices and relation algebras. *Trans. Amer. Math. Soc.* **92**, 449–464 (1959).
5. Representations of complemented modular lattices. *Trans. Amer. Math. Soc.* **97**, 64–94 (1960).
6. Sublattices of a free lattice. *Can. J. Math.* **13**, 256–264 (1961).

Jung. H. A.

1. Zu einem Isomorphiesatz von Whitney für Graphen. *Math. Ann.* **164**, 270–271 (1966).

Kahaner, D. See Rota–Kahaner–Odlyzko.

Kanold, H.-J.

1. Einige neuere Abschätzungen bei Stirlingschen Zahlen zweiter Art. *J. Reine Angew. Math.* **238**, 148–160 (1968).

Kantor, W. M.

1. Dimension and embedding theorems for geometric lattices. *J. Comb. Theory* (A) **17**, 173–196 (1974).

Kaplansky, I.

1. Solution of the "problème des ménages." *Bull. Amer. Math. Soc.* **49**, 784–785 (1943).

Katona, G.

1. On a conjecture of Erdös and a stronger form of Sperner's theorem. *Studia Sci. Math. Hung.* **1**, 59–63 (1966).
2. *Sperner Type Theorems.* Dept. Stat. Univ. North Carolina Mimeo Series 600.17 (1969).
3. A simple proof of the Erdös–Ko–Rado theorem. *J. Comb. Theory* (B) **13**, 183–184 (1972).
4. Families of subsets having no subset containing another with small difference. *Nieuw. Arch. Wisk.* (3) **20**, 54–67 (1972).
5. Extremal problems for hypergraphs. *Combinatorics* (Hall–Van Lint, eds.). Math. Centre Tracts **56**, 13–42. Amsterdam (1974).

Kempe, A. D.

1. On the geographical problem for four colors. *Amer. J. Math.* **2**, 193–204 (1879).

Kerber, A.

1*. *Representations of Permutation Groups* II. Lecture Notes Math. **495**, Springer-Verlag (1975).

Kelly, D. G. See Dowling–Kelly.

Kelly, L. M. See Basterfield–Kelly.

Kirchhoff, G.

1. Über die Auflösung der Gleichungen, auf welche man bei der Untersuchung der linearen Verteilung galvanischer Ströme geführt wird. *Ann. Phys. Chem.* **72**, 497–508 (1847).

Klee, V.

1. The Euler characteristic in combinatorial geometry. *Amer. Math. Monthly* **70**, 119–127 (1963).

Kleitman, D.

1. On a lemma of Littlewood and Offord on the distribution of certain sums. *Math. Z.* **90**, 251–259 (1965).
2. On Dedekind's problem: The number of monotone Boolean functions. *Proc. Amer. Math. Soc.* **21**, 677–682 (1969).
3. On a lemma of Littlewood and Offord on the distributions of linear combination of vectors. *Advances Math.* **5**, 1–3 (1970).
4. On an extremal property of antichains in partial orders. The LYM property and some of its implications and applications. *Combinatorics* (Hall–van Lint, eds.). Math. Centre Tracts **56**, 77–90. Amsterdam (1974).

Kleitman, D. J.–Edelberg, M.–Lubell, D.

1. Maximal sized antichains in partial orders. *Discrete Math.* **1**, 47–53 (1971).

Kleitman, D. J.–Markowsky, G.

1. On Dedekind's problem: The number of monotone Boolean functions II. *Trans. Amer. Math. Soc.* **213**, 373–390 (1975).

See also Greene–Kleitman, Hsieh–Kleitman.

Knuth, D. E.

1.* *The Art of Computer Programming*, vol. 1 (1968), vol. 2 (1969), vol. 3 (1973). Reading: Addison-Wesley.
2. Permutations, matrices and generalized Young tableaux. *Pac. J. Math.* **34**, 709–727 (1970).

Ko, C. See Erdös–Ko–Rado.

Koch, J. See Appel–Haken–Koch.

König, D.

1.* *Theorie der Endlichen und Unendlichen Graphen.* Leipzig (1936). Reprinted New York: Chelsea (1950).
2. Über Graphen und ihre Anwendungen auf Determinantentheorie und Mengenlehre. *Math. Ann.* **77**, 453–465 (1916).
3. Graphen und Matrizen. *Mat. Fiz. Lapok* **38**, 116–119 (1931).
4. Über trennende Knotenpunkte in Graphen (nebst Anwendungen auf Determinanten und Matrizen). *Acta Lit. Sci. Sect. Math.* (Szeged) **6**, 155–179 (1932–1934).

Kreweras, G.

1. Sur une classe de problèmes de dénombrement liés au treillis des partitions d'entier. *Cahiers Buro* **6**, 1–107 (1965).
2. Sur les partitions non croissées d'un cycle. *Discrete Math.* **1**, 333–350 (1972).

Kruyswijk, D. R. See deBruijn–Van E. Tengbergen–Kruyswijk.

Kuhn, H. W. See Hoffman–Kuhn.

Kung, J. P. S. See Désarméniens–Kung–Rota.

Kuratowski, K.

1. Sur le problème des courbes gauches en topologié. *Fund. Math.* **15**, 271–283 (1930).

Lah, I.

1. Eine neue Art von Zahlen, ihre Eigenschaften und Anwendung in der mathematischen Statistik. *Mitteilungsblatt Math. Stat.* **7**, 203–212 (1955).

Las Vergnas, M.

1. Sur les systèmes des représents distincts d'une famille d'ensembles. *C. R. Acad. Sci.* (Paris) **270**, 501–503 (1970).
2. Sur la dualité en théorie des matroides. *Théorie des Matroides* (Bruter, ed.), 67–86. Lecture Notes Math. **211**, Springer-Verlag (1971).
3. Matroides orientables. *C. R. Acad. Sci.* (Paris) **280 A**, 61–64 (1975).

See also Bland–Las Vergnas.

Lazarson, T.

1. The representation problem for independence functions. *J. London Math. Soc.* **33**, 21–25 (1958).

Lawler, E. F.
1*. *Combinatorial Optimization. Networks and Matroids.* New York: Holt, Rinehart, Winston (1976).
Leeb, K.
1. Sperner theorems with choice functions. Preprint.
See also Graham–Leeb–Rothschild.
Lehman, A.
1. A solution of the Shannon switching game. *SIAM J.* **12**, 687–725 (1964).
Lieb, E. H.
1. Concavity properties and a generating function for Stirling numbers. *J. Comb. Theory* **5**, 203–206 (1968).
Lindström, B.
1. On the realization of convex polytopes, Euler's formula and Möbius functions. *Aequationes Math.* **6**, 235–240 (1971).
Van Lint, J.
1*. *Combinatorial Theory Seminar* (ed.). Lecture Notes Math. **382**, Springer-Verlag (1974).
See also Hall–Van Lint.
Littlewood, J. E. See Hardy–Littlewood–Polya.
Lloyd, E. K.
1. Polya's theorem in combinatorial analysis applied to enumerate multiplicative partitions. *J. London Math. Soc.* **43**, 224–230 (1968).
Lovász, L.
1. Normal hypergraphs and the perfect graph conjecture. *Discrete Math.* **2**, 253–267 (1972).
2. A characterization of perfect graphs. *J. Comb. Theory* **13**, 95–98 (1972).
Lubell, D.
1. A short proof of Sperner's theorem. *J. Comb. Theory* **1**, 299 (1966).
See also Kleitman–Edelberg–Lubell.
Mac Lane, S.
1. Some interpretations of abstract linear dependence in terms of projective geometry. *Amer. J. Math.* **58**, 236–240 (1936).
2. A combinatorial condition for planar graphs. *Fund. Math.* **28**, 22–32 (1937).
3. A structural characterization of planar combinatorial graphs. *Duke Math. J.* **3**, 340–372 (1937).
MacMahon, P. A.
1*. *Combinatory Analysis.* London: Cambridge Univ. Press 1915 (1916), reprinted by Chelsea, New York (1960).
2. The indices of permutations and the derivation therefrom of functions. *Amer. J. Math.* **35**, 281–322 (1913).
3. Two applications of general theorems in combinatory analysis. *Proc. London Math. Soc.* **15**, 314–321 (1916).
Mac Williams, F. J.–Sloane, N. J. A.
1. *The Theory of Error-Correcting Codes.* Amsterdam: North Holland (1977).
Maeda, F.
1*. *Kontinuierliche Geometrien.* Berlin, Göttingen, Heidelberg: Springer-Verlag (1958).
2. Lattice theoretic characterization of abstract geometries. *J. Sci. Hiroshima Univ.* **15** A, 87–96 (1951).
3. Perspectivity of points in matroid lattices. *J. Sci. Hiroshima Univ.* **28** A, 101–112 (1964).
Markowsky, G.
1. Some combinatorial aspects of lattice theory. Proc. Univ. Houston Lattice Theory Conf., 36–68. Univ. Houston (1973).
See also Kleitman–Markowsky.
Mason, J. H.
1. *Representations of Independence Spaces.* Ph.D. Diss. Univ. Wisconsin (1969).
2. Geometrical realization of combinatorial geometries. *Proc. Amer. Math. Soc.* **30** (1), 15–21 (1971).
3. On a class of matroids arising from paths in graphs. *Proc. London Math. Soc.* (3) **25**, 55–74 (1972).
4. Matroids: Unimodal conjectures and Motzkin's theorem. *Combinatorics Inst. of Math. and Appl.* (Welsh–Woodall, eds.), 207–220 (1972).
5. Maximal families of pairwise disjoint proper chains in a geometric lattice. *J. London Math. Soc.* **6**, 539–542 (1973).
Mendelsohn, N. S.
1. Permutations with confined displacements. *Can. Math. Bull.* **4**, 29–38 (1961).

Mendelsohn, N. S.–Dulmage, A. L.
1. Some generalizations of the problem of distinct representatives. *Can. J. Math.* **10**, 230–241 (1958).

Menger, K.
1. Zur allgemeinen Kurventheorie. *Fund. Math.* **10**, 96–115 (1927).

Meshalkin, L. D.
1. A generalization of Sperner's theorem on the number of subsets of a finite set. *Theory Probability Appl.* **8**, 203–204 (1963).

Milner, E. C.–Shelah, S.
1. Sufficiency conditions for the existence of transversals. *Can. J. Math.* **26**, 948–961 (1974).

Minty, G. J.
1. Monotone networks. *Proc. Roy. Soc.*, Ser. A **257**, 194–212 (1960).
2. On the axiomatic foundations of the theories of directed linear graphs, electrical networks and network programming. *Journ. Math. Mech.* **15**, 485–520 (1960).

Mirsky, L.
1.* *Transversal Theory.* New York, London: Academic Press (1971).
2. A theorem on common transversals. *Math. Ann.* **177**, 49–53 (1968).

Mirsky, L.–Perfect, H.
1. Systems of representatives. *J. Math. Anal. Appl.* **15**, 520–568 (1966).
2. Applications of the notion of independence to combinatorial analysis. *J. Comb. Theory* **2**, 327–357 (1967).

Möbius, A. F.
1. Über eine besondere Art der Umkehrung der Reihen. *J. Reine Angew. Math.* **9**, 105–129 (1832).

Moon, J. W.
1.* *Counting Labeled Trees, a Survey of Methods and Results.* Univ. Alberta (1969).

Motzkin, T.
1. The lines and planes connecting the points of a finite set. *Trans. Amer. Math. Soc.* **70**, 451–469 (1951).

Mullin, R.
1. On Rota's problem concerning partitions. *Aequationes Math.* **2**, 98–104 (1969).

Mullin, R.–Rota, G. C.
1. On the foundations of combinatorial theory III: theory of binomial enumeration. *Graph theory and its Applications* (Harris, ed.), 167–213. New York: Academic Press (1970).

Muirhead, R.
1. Some methods applicable to identities and inequalities of symmetric algebraic functions on $n$ letters. *Proc. Edin. Math. Soc.* **21**, 144–157 (1903).

Murty, U. S. R.
1. Equicardinal matroids and finite geometries. *Combinatorial Structures and their Applications* (Guy *et al.*, eds.), 289–293. New York, London, Paris: Gordon & Breach (1970).
2. Equicardinal matroids. *J. Comb. Theory* **11**, 120–126 (1971).
See also Young–Murty–Edmonds.

Nash Williams, C. St. J. A.
1. Edge-disjoint spanning trees of finite graphs. *J. London Math. Soc.* **36**, 445–450 (1961).
2. An application of matroids to graph theory. *Theory of Graphs.* Internat. Symp. (Rome), 263–265. Paris: Dunond (1966).

Něsetřil, J.–Rödl, V.
1. The Ramsey property for graphs with forbidden complete subgraphs. *J. Comb. Theory* (B) **20**, 243–249 (1976).

Nijenhuis, A.–Wilf, H. S.
1.* *Combinatorial Algorithms.* New York: Academic Press (1975).

Niven, I.
1. Formal power series. *Amer. Math. Monthly* **76**, 871–889 (1969).

Nörlund, N. E.
1.* *Vorlesungen über Differenzenrechnung.* New York: Chelsea (1954).

Oberschelp, W.
1. Kombinatorische Anzahlbestimmung in Relationen. *Math. Ann.* **174**, 53–78 (1967).

Odlyzko, A. See Rota–Kahaner–Odlyzko.

Ore, O.
1. On the foundations of abstract algebra II. *Annals Math.* **37**, 265–292 (1936).
2. On the theorem of Jordan–Hölder. *Trans. Amer. Math. Soc.* **41**, 266–275 (1937).
3. Theory of equivalence relations. *Duke Math. J.* **9**, 573–627 (1942).
4. Graphs and matching theorems. *Duke Math. J.* **22**, 625–639 (1955).

Otter, R.
1. The number of trees. *Annals. Math.* **49**, 583–599 (1948).

Palmer, E. M. See Harary–Palmer.

Perfect, H.
1. Applications of Menger's graph theorem. *J. Math. Analysis Appl.* **22**, 96–111 (1968).
See also Mirsky–Perfect, Pym–Perfect.

Petersen, J.
1. Die Theorie der regulären Graphen. *Acta Math.* **15**, 193–220 (1891).

Piff, M. J.
1. *Some Problems in Combinatorial Theory.* D. Phil. thesis. Oxford (1972).

Piff, M. J.–Welsh, D. J. A.
1. On the vector representation of matroids. *J. London Math. Soc.* **2**, 284–288 (1970).
2. The number of combinatorial geometries. *Bull. London Math. Soc.* **3**, 55–56 (1971).
See also Ingleton–Piff.

Polya, G.
1. Kombinatorische Anzahlbestimmung für Gruppen, Graphen und chemische Verbindungen. *Acta Math.* **68**, 145–254 (1937).
2. Sur les types des propositions composées. *J. Symb. Logic* **5**, 98–103 (1940).
See also Hardy–Littlewood–Polya.

Prins, G. See Aigner–Prins, Harary–Prins.

Pudlák, P.–Tůma, J.
1. Every finite lattice can be embedded in the lattice of all equivalences over a finite set. *Comm. Math. Univ. Carol.* **18** (2), 409–414 (1977).

Pym, J. S.–Perfect, H.
1. Submodular functions and independence structures. *J. Math. Anal. Appl.* **30**, 1–31 (1970).

Rado, R.
1. A theorem on independence relations. *Quart. J. Math.* (Oxford) **13**, 83–89 (1942).
2. Axiomatic treatment of rank in infinite sets. *Can. J. Math.* **1**, 337–343 (1949).
3. Note on independence functions. *Proc. London Math. Soc.* **7**, 300–320 (1957).
4. Note on the transfinite case of Hall's theorem on representatives. *J. London Math. Soc.* **42**, 321–324 (1967).
5. On the number of systems of distinct representatives of sets. *J. London Math. Soc.* **42**, 107–109 (1967).
See also Erdös–Ko–Rado, Erdös–Rado–Hajnal.

Ramsey, F. P.
1. On a problem of formal logic. *Proc. London Math. Soc.* (2) **30**, 264–286 (1930).

Raney, G. N.
1. Functional composition patterns and power series reversion. *Trans. Amer. Math. Soc.* **94**, 441–450 (1960).

Read, R.
1. The number of $k$-colored graphs on labeled nodes. *Can. J. Math.* **12**, 410–414 (1960).
2. On the number of self-complementary graphs and digraphs. *J. London Math. Soc.* **38**, 99–104 (1963).
3. An introduction to chromatic polynomials. *J. Comb. Theory* **4**, 52–71 (1968).

Redfield, J. H.
1. The theory of group-reduced distributions. *Amer. J. Math.* **49**, 433–455 (1927).

Riordan, J.
1.* *An Introduction to Combinatorial Mathematics.* New York, London, Sydney: John Wiley & Sons (1958).
2.* *Combinatorial Identities.* New York, London, Sidney: John Wiley & Sons (1968).
3. The number of two-terminal series-parallel networks. *J. Math. Phys.* **21**, 83–93 (1942).
4. Inverse relations and combinatorial identities. *Amer. Math. Monthly* **73**, 91–95 (1966).
See also Becker–Riordan, Carlitz–Riordan.

Rival, I. See Ganter–Rival.

Robinson, G. de B.

1. On the representation of the symmetric group I–III. *Amer. J. Math.* **60**, 745–760 (1938); **69**, 286–298 (1947); **70**, 277–294 (1948).
   See also Frame–Robinson–Thrall.

Rödl, V. See Nešetřil–Rödl.

Roman, S. M.–Rota, G.-C.

1. The umbral calculus. *Advances Math.* **27**, 95–188 (1978).

Roselle, D. P. See Carlitz–Roselle–Scoville.

Rosenstiehl, P. See Guilbaud–Rosenstiehl.

Rota, G.-C.

1. On the foundations of combinatorial theory I: theory of Möbius functions. *Z. Wahrscheinlichkeitsrechnung u. verw. Geb.* **2**, 340–368 (1964).
2. The number of partitions of a set. *Amer. Math. Monthly* **71**, 499–504 (1964).
3. Baxter algebras and combinatorial identities I, II. *Bull. Amer. Math. Soc.* 325–329 (1969); 330–334 (1969).
4. On the combinatorics of the Euler characteristic. *Studies in Pure Math.* (Mirsky, ed)., 221–233. London: Academic Press (1971).

Rota, G.-C.–Kahaner, D.–Odlyzko, A.

1. On the foundations of combinatorial theory VIII: Finite operator calculus. *J. Math. Anal. Appl.* **42**, 684–760 (1973).

Rota, G.-C.–Smith, D. A.

1. Fluctuation theory and Baxter algebras. *Symp. Math. Ist. Naz. Alta Mat.* **9**, 179–201 (1972).
   See also Crapo–Rota, Désarméniens–Kung–Rota, Doubilet–Rota–Stanley, Goldman–Rota, Harper–Rota, Mullin–Rota, Roman–Rota.

Rothschild, B. L. See Graham–Rothschild, Graham–Leeb–Rothschild.

Rutherford, D. E.

1.* *Substitutional Analysis.* Edinburgh: Oliver and Body (1948).

Ryser, H. J.

1.* *Combinatorial Mathematics.* Carus Math. Monographs Nr. 14, New York: Math. Ass. Amer. (1963).
2. A combinatorial theorem with applications to Latin rectangles. *Proc. Amer. Math. Soc.* **2**, 550–552 (1951).
3. Combinatorial properties of matrices of zeros and ones. *Can. J. Math.* **9**, 371–377 (1957).
4. Matrices of zeros and ones. *Bull. Amer. Math. Soc.* **66**, 442–464 (1960).

Saaty, T. L. See Busacker–Saaty.

Sachs, D.

1. Graphs, matroids and geometric lattices. *J. Comb. Theory* **9**, 192–199 (1970).

Sanders, J.

1. *A Generalization of a Theorem of Schur.* Ph. D. Diss. Yale Univ. (1968).

Sasaki, U.–Fujiwara, S.

1. The decomposition of matroid lattices. *J. Sci. Hiroshima Univ.* **15**, 183–188 (1952).
2. The characterization of partition lattices. *J. Sci. Hiroshima Univ.* **15**, 189–201 (1952).

Scheid, H.

1. Einige Ringe zahlentheoretischer Funktionen. *J. Reine Angew. Math.* **237**, 1–11 (1969).
2. Über ordnungstheoretische Funktionen. *J. Reine Angew. Math.* **238**, 1–13 (1969).
3. Über die Möbiusfunktion einer lokal endlichen Halbordnung. *J. Comb. Theory* **13**, 315–331 (1972).

Schensted, C.

1. Longest increasing and decreasing subsequences. *Can. J. Math.* **13**, 179–191 (1961).

Schur, I.

1. Über die Kongruenz $x^m + y^m = z^m \pmod p$. *Jahresbericht DMV* **25**, 114 (1916).

Schützenberger, M. P.

1. Contributions aux applications statistiques de la théorie de l'information. *Publ. Inst. Stat. Univ. Paris* **3**, 5–117 (1954).
2. Quelques remarques sur une construction de Schensted. *Math. Scand.* **12**, 117–128 (1963).
   See also Foata–Schützenberger.

Scott, W. R. See Chowla–Herstein–Scott.

Scoville, R. A. See Carlitz–Roselle–Scoville.

Scrimger, E. B. See Brualdi–Scrimger.

Seymour, P. D.

1. The max-flow min-cut property in matroids. Proc. Fifth British Comb. Conf. (Nash-Williams–Sheehan, eds.), 545–550. Winnipeg: Utilitas (1975).
2. The forbidden minors of binary clutters. *J. London Math. Soc.* (2) **12**, 356–360 (1976).
3. Matroid representation over *GF* (3). To appear.

Sheffer, I. M.

1. Some properties of polynomials of type zero. *Duke Math. J.* **5**, 590–622 (1939).

Shelah, S. See Milner–Shelah.

Slepian, D.

1. On the number of symmetry types of Boolean functions of $n$ variables. *Can. J. Math.* **5**, 185–193 (1953).

Smith, D. A.

1. Incidence functions as generalized arithmetic functions I, II, III. *Duke Math. J.* **34**, 617–634 (1967); **36**, 15–30 (1969); **36**, 343–368 (1969).
2. Multiplication operators on incidence algebras. *Indiana Univ. Math. J.* **20**, 369–383 (1970/71).
   See also Rota–Smith.

Smith, C. A. B.

1. Electrical currents in regular matroids. *Combinatorics Inst. of Math. and Appl.* (Welsh–Woodall, eds.), 262–284 (1972).
   See also Brooks–Smith–Stone–Tutte.

Snapper, E.

1. Group characters and non-negative integral matrices. *J. Algebra* **19**, 520–535 (1971).

Solomon, L.

1. The Burnside algebra of a finite group, *J. Comb. Theory* **2**, 603–615 (1967).

Spencer, J.

1. A generalized Rota conjecture for partitions. To appear.

Sperner, E.

1. Ein Satz über Untermengen einer endlichen Menge. *Math. Z.* **27**, 544–548 (1928).

Spitzer, F.

1. A combinatorial lemma and its applications to probability theory. *Trans. Amer. Math. Soc.* **82**, 323–339 (1965).

Stanley, R. P.

1. Structure of incidence algebras and their automorphism groups. *Bull. Amer. Math. Soc.* **76**, 1236–1239 (1970).
2. Modular elements in geometric lattices. *Algebra Universalis* **1**, 214–217 (1971).
3. Theory and application of plane partitions. *Studies in Appl. Math. I*, **50**, 167–188 (1971), II, **50**, 259–279 (1971).
4. *Ordered structures and partitions. Memoirs Amer. Math. Soc.* **119** (1972).
5. Supersolvable lattices. *Algebra Universalis* **2**, 197–217 (1972).
6. Acyclic orientations of graphs. *Discrete Math.* **5**, 171–178 (1973).
7. Combinatorial reciprocity theorems. *Advances Math.* **14**, 194–253 (1974).
8. The Fibonacci lattice. *Fibonacci Quarterly* **13**, 215–232 (1975).
9. Binomial posets, Möbius inversion, and permutation enumeration. *J. Comb. Theory* (A) **20**, 336–356 (1976).
10. Generating Functions. *MAA Survey of Combinatorics*. To appear.
    See also Doubilet–Rota–Stanley.

Steffensen, J. F.

1.* *Interpolation*. New York: Chelsea (1950).
2. The poweroid, an extension of the mathematical notion of power. *Acta Math.* **73**, 333–366 (1941).

Stein, S. K.

1.* *Mathematics, the Man-made Universe*. San Francisco: Freeman (1963).

Stone, A. H. See Brooks–Smith–Stone–Tutte.

Stonesifer, J. R.
1. Logarithmic concavity for edge lattices of graphs. *J. Comb. Theory* (A) **18**, 36–46 (1975).

Szekeres, G. See Erdös–Szekeres.

Szemerédi, E.
1. On sets of integers containing $k$ elements in arithmetic progression. *Acta Arith.* **27**, 199–245 (1975).

Szpilrajn, E.
1. Sur l'extension de l'ordre partiel. *Fund. Math.* **16**, 386–389 (1930).

Tait, P. G.
1. Remarks on the colouring of maps. *Proc. Royal Soc. Edinburgh* **10**, 729 (1880).

Tainiter, M.
1. A characterization of idempotents in semigroups. *J. Comb. Theory* **5**, 370–373 (1968).

Van E. Tengbergen, C. A. See deBruijn–Van E. Tengbergen–Kruyswijk.

Thrall, R. M. See Frame–Robinson–Thrall.

Touchard, J.
1. Sur la théorie des différences. Proc. Int. Congr. Math. Toronto, 623–629 (1928).
2. Nombres exponentiels et nombres de Bernoulli. *Can. J. Math.* **8**, 305–320 (1950).

Tůma, J. See Pudlák–Tůma.

Turan, P.
1. Eine Extremalaufgabe aus der Graphentheorie. *Math. Fiz. Lapok* **48**, 436–452 (1941).
See also Erdös–Turan.

Tutte, W. T.
1.* *Recent Progress in Combinatorics* (ed.). New York, London: Academic Press (1969).
2. A ring in graph theory. *Proc. Camb. Phil. Soc.* **43**, 26–40 (1947).
3. A contribution to the theory of chromatic polynomials. *Can. J. Math.* **6**, 80–91 (1954).
4. A class of Abelian groups. *Can. J. Math.* **8**, 13–28 (1956).
5. A homotopy theorem for matroids, I and II. *Trans. Amer. Math. Soc.* **88**, 144–174 (1958).
6. Matroids and graphs. *Trans. Amer. Math. Soc.* **90**, 527–552 (1959).
7. An algorithm for deciding whether a given binary matroid is graphic *Proc. Amer. Math. Soc.* **11**, 905–917 (1960).
8. On the problem of decomposing a graph into $n$ connected factors. *J. London Math. Soc.* **36**, 221–230 (1961).
9. A theory of 3-connected graphs. *Indag. Math.* **23**, 441–455 (1961).
10. Lectures on matroids. *J. Res. Nat. Bur. Stand.* **69B**, 1–48 (1965).
11. Connectivity in matroids. *Can. J. Math.* **18**, 1301–1324 (1966).
12. On the algebraic theory of graph colorings. *J. Comb. Theory* **1**, 15–50 (1966).
See also Brooks–Smith–Stone–Tutte.

Tverberg, H.
1. On Dilworth's decomposition theorem for partially ordered sets. *J. Comb. Theory* **3**, 305–306 (1967).

Vamos, P.
1. A necessary and sufficient condition for a matroid to be linear. Proc. Conf. on Möbius algebra. Univ. Waterloo (1971).
2. Linearity of matroids over division rings. Proc. Conf. on Möbius algebra. Univ. Waterloo (1971).

Veblen, O.–Young, J. W.
1.* *Projective Geometry*, vol. 1. Boston: Ginn and Co. (1946).

Van der Waerden, B. L.
1.* *Algebra*. 6th edition. Berlin, Göttingen, Heidelberg: Springer-Verlag (1964).
2. Aufgabe 45. Jahresber. DMV **35**, 117 (1926).
3. Ein Satz über Klasseneinteilungen von endlichen Mengen. *Abh. Math. Sem. Hamburg, Univ.* **5**, 185–188 (1927).
4. Beweis einer Baudet'schen Vermutung. *Nieuw. Arch. Wisk.* **15**, 212–216 (1927).

Wagner, K.
1. Über eine Eigenschaft der ebenen Komplexe. *Math. Ann.* **114**, 570–590 (1937).
2. Bemerkungen zu Hadwigers Vermutung. *Math. Ann.* **141**, 433–451 (1960).

Ward, M.
1. Arithmetic functions on rings. *Ann. Math.* **38**, 725–732 (1937).
2. The algebra of lattice functions. *Duke Math. J.* **5**, 357–371 (1939).

Weisner, L.
1. Abstract theory of inversion of finite series. *Trans. Amer. Math. Soc.* **38**, 474–484 (1935).

Welsh, D. J. A.
1.* *Matroid Theory*. London, New York, San Francisco: Academic Press (1976).
2. A bound for the number of matroids. *J. Comb. Theory* **6**, 313–316 (1969).
3. Euler and bipartite matroids. *J. Comb. Theory* **6**, 375–377 (1969).
4. On matroid theorems of Edmonds and Rado. *J. London Math. Soc.* **2**, 251–256 (1970).

Welsh, D. J. A.–Woodall, D. R.
1.* *Combinatorics* (eds.). Combinatorics Inst. of Math. and Appl. Oxford (1972).
See also Bondy–Welsh, de Sousa–Welsh, Piff–Welsh.

White, N.
1. Coordinatization of combinatorial geometries. Proc. 2nd Chapel Hill Conference on Comb. Math. and its Appl., 484–486. Univ. North Carolina (1970).

Whitman, P.
1. Lattices, equivalence relations, and subgroups. *Bull. Amer. Math. Soc.* **52**, 507–522 (1946).

Whitney, H.
1. Non-separable and planar graphs. *Trans. Amer. Math. Soc.* **34**, 339–362 (1932).
2. Congruent graphs and the connectivity of graphs. *Amer. J. Math.* **54**, 150–168 (1932).
3. A logical expansion in mathematics. *Bull. Amer. Math. Soc.* **38**, 572–579 (1932).
4. The coloring of graphs. *Annals Math.* **33**, 688–718 (1932).
5. 2-isomorphic graphs. *Amer. J. Math.* **55**, 245–254 (1933).
6. Planar graphs. *Fund. Math.* **21**, 73–84 (1933).
7. On the abstract properties of linear dependence. *Amer. J. Math.* **57**, 509–533 (1935).

Wielandt, H.
1.* *Finite Permutation Groups*. New York, London: Academic Press (1964).

Wilf, H. S.
1. Hadamard determinants, Möbius functions and the chromatic number of a graph. *Bull. Amer. Math. Soc.* **74**, 960–964 (1968).
2. A mechanical counting method and combinatorial applications. *J. Comb. Theory* **4**, 246–258 (1968).
See also Nijenhuis–Wilf.

Wille, R.
1.* *Kongruenzklassengeometrien*. Lecture Notes Math. **113**, Springer-Verlag (1970).
2. Verbandstheoretische Charakterisierung *n*-stufiger Geometrien. *Arch. Math.* **18**, 465–468 (1967).
3. On incidence geometries of grade n. Atti del Convegno di Geom. Comb. e sue Appl., 421–426. Univ. Perugia (1971).
4. Aspects of finite lattices. *Higher Combinatorics* (Aigner, ed.), 79–100. Dordrecht: Reidel (1977).

Wilson, R. See Dowling–Wilson.

Wolk, E. S.
1. The comparability graph of a tree. *Proc. Amer. Math. Soc.* **13**, 789–795 (1962).
2. A note on "The comparability graph of a tree." *Proc. Amer. Math. Soc.* **16**, 17–20 (1965).

Woodall, D. R. See Welsh–Woodall.

Worpitzky, N.
1. Studien über die Bernoullischen und Eulerschen Zahlen. *Crelle J.* **94**, 203–232 (1883).

Wright, E. M. See Hardy–Wright.

Yackel, J. See Graver–Yackel.

Yamamoto, K.
1. Logarithmic order of free distributive lattices. *J. Math. Soc. Japan* **6**, 343–353 (1954).

Young, A.
1. Quantitative Substitutional Analysis I–IX. *Proc. London Math. Soc.* (1) **33**, 97–146 (1901); **34**, 361–397 (1902); (2) **28**, 255–292 (1928); **31**, 253–272 (1930); **31**, 273–288 (1930); **34**, 196–230 (1932); **36**, 304–368 (1933); **37**, 441–495 (1934); **54**, 219–253 (1952).

Young, J. W. See Veblen–Young.

Young, P.–Murty, U. S. R.–Edmonds, J.
1. Equicardinal matroids and matroid designs. Proc. 2nd Chapel Hill Conference on Comb. Math. and its Appl., 498–542. Univ. North Carolina (1970).

Zaslavsky, T.
1. *Facing up to arrangements: face count formulas for partitions of space by hyperplanes. Memoirs Amer. Math. Soc.* **154** (1975).

# List of Symbols

## 1. Sets

| | |
|---|---|
| $\mathbb{N}$ | natural numbers |
| $\mathbb{N}_0$ | natural numbers including 0 |
| $\mathbb{Z}$ | integers |
| $\mathbb{Q}$ | rational numbers |
| $\mathbb{R}$ | real numbers |
| $\mathbb{C}$ | complex numbers |
| $\varnothing$ | empty set |

## 2. Mappings

| | |
|---|---|
| $\mathrm{Map}(N, R)$ | mappings from $N$ to $R$ |
| $\mathrm{Inj}(N, R)$ | injective mappings from $N$ to $R$ |
| $\mathrm{Sur}(N, R)$ | surjective mappings from $N$ to $R$ |
| $\mathrm{Bij}(N, R)$ | bijective mappings from $N$ to $R$ |
| $\mathrm{Mon}(N, R)$ | monotone mappings from $N$ to $R$ |
| $\mathrm{Hom}(N, R)$ | homomorphisms from $N$ to $R$ |

## 3. Lattices

| | |
|---|---|
| $\mathscr{C}(n)$ | chain of length $n$ |
| $\mathscr{C}(\infty)$ | chain $\cong \mathbb{N}_0$ |
| $\mathscr{B}(S)$ | Boolean algebra on $S$ |
| $\mathscr{B}(n)$ | Boolean algebra of rank $n$ |
| $\mathscr{B}(\infty)$ | Boolean algebra of all finite subsets of a countable set |
| $2^S$ | lattice of all subsets of $S$ |
| $\mathscr{M}(R)$ | lattice of finite multisets of $R$ |
| $\mathscr{T}$ | divisor lattice |
| $\mathscr{L}(V)$ | subspace lattice of a vector space $V$ |
| $\mathscr{L}(n, K)$ | subspace lattice of $K^n$ |

| | |
|---|---|
| $\mathscr{L}(n, q)$ | subspace lattice of $GF(q)^n$ |
| $\mathscr{L}(\infty, q)$ | lattice of all finite-dimensional subspaces of a countable-dimensional vector space over $GF(q)$ |
| $\mathscr{A}(V)$ | lattice of affine subspaces of a vector space $V$ |
| $\mathscr{A}(n, K)$ | affine subspace lattice of $K^n$ |
| $\mathscr{A}(n, q)$ | affine subspace lattice of $GF(q)^n$ |
| $\mathscr{P}(S)$ | lattice of partitions of $S$ |
| $\mathscr{P}(n)$ | partition lattice of an $n$-set |
| $\mathscr{P}(\infty)$ | lattice of finite partitions of a countable set |
| $\mathscr{U}(G)$ | lattice of subgroups of a group $G$ |
| $\mathscr{N}(G)$ | lattice of normal subgroups of a group $G$ |
| $\mathscr{C}\mathscr{R}(A)$ | lattice of congruence relations of $A$ |
| $\mathscr{C}\mathscr{C}(A)$ | lattice of congruence classes of $A$ |

## 4. Families of Sets

| | |
|---|---|
| $\mathfrak{J}(P)$ | lattice of ideals of a poset $P$ |
| $\mathfrak{F}(P)$ | lattice of filters of a poset $P$ |
| $\mathfrak{A}(P)$ | family of antichains of a poset $P$ |
| $\mathfrak{M}$ | patterns |
| $\mathfrak{M}_{\mathscr{F}}$ | patterns in $\mathscr{F}$ |
| $\mathfrak{B}$ | family of bases of a matroid |
| $\mathfrak{I}$ | family of independent sets |
| $\mathfrak{K}$ | family of circuits |
| $\mathfrak{H}$ | family of copoints |
| $\mathfrak{C}$ | family of cocircuits |
| $\mathfrak{S}$ | group of sets |
| $T(S; \mathfrak{A})$ | set system $\mathfrak{A}$ on $S$ |
| $(\mathfrak{P}, \mathfrak{G})$ | projective incidence system |

## 5. Special Permutation Groups and Lattices

| | |
|---|---|
| $S(N)$ | symmetric group on $N$ |
| $S_n$ | symmetric group of degree $n$ |
| $E(N)$ | identity group on $N$ |
| $E_n$ | identity group of degree $n$ |
| $A_n$ | alternating group |
| $C_n$ | cyclic group |
| $D_n$ | dihedral group |
| $\mathscr{V}(G)$ | vertex group of a graph $G$ |
| $\mathscr{E}(G)$ | edge group of a graph $G$ |
| $\mathscr{Y}$ | Young lattice |
| $\mathscr{D}_n$ | dominance order |

## 6. Counting Functions and Numbers

| | |
|---|---|
| $[x]_n$ | falling factorials |
| $[x]^n$ | rising factorials |
| $g_n(x)$ | Gaussian polynomials |
| $e_n(x)$ | exponential polynomials |
| $l_n(x)$ | Laguerre polynomials |
| $\binom{n}{k}$ | binomial coefficients |
| $\binom{n}{k_1 \ldots k_r}$ | multinomial coefficient |
| $\binom{n}{k}_q$ | Gaussian coefficients |
| $G_{n,q}$ | Galois numbers |
| $s_{n,k}$ | Stirling numbers of the first kind |
| $S_{n,k}$ | Stirling numbers of the second kind |
| $B_n$ | Bell numbers |
| $L_{n,k}$ | Lah numbers |
| $L'_{n,k}$ | signless Lah numbers |
| $P_{n,r}$ | partition numbers |
| $P_n$ | number of all partitions of $n$ |
| $\omega(P; x)$ | order polynomial |
| $\overline{\omega}(P; x)$ | strict order polynomial |
| $c(G; x)$ | chromatic polynomial |
| $\chi(P; x)$ | characteristic polynomial |
| $Z(P; x)$ | zeta polynomial |
| $A_{n,k}$ | Eulerian numbers |
| $\mathrm{Per}(b_1, \ldots, b_n)$ | number of permutations of type $1^{b_1} \ldots n^{b_n}$ |
| $P(b_1, \ldots, b_n)$ | number of partitions of type $1^{b_1} \ldots n^{b_n}$ |
| $\mathrm{chrom}(G)$ | chromatic number of a graph $G$ |
| $\mathrm{crit}(\mathbf{M})$ | critical exponent of $\mathbf{M}$ |
| $\mathrm{St}(\alpha)$ | standard tableaux with frame $\alpha$ |

## 7. Operators

| | |
|---|---|
| $\Delta$ | forward difference operator |
| $\nabla$ | backward difference operator |
| $\delta$ | central difference operator |
| $E^a$ | translation |
| $I$ | identity |
| $D$ | differential operator |
| $\underline{x}$ | multiplication by $x$ |
| $L_a$ | evaluation at $a$ |
| $D_{\leq}, D_{\geq}$ | lower (upper) difference operator |
| $S_{\leq}, S_{\geq}$ | lower (upper) sum operator |

## 8. Incidence Algebras and Functions

| | |
|---|---|
| $\mathbb{A}_K(P)$ | incidence algebra of $P$ over $K$ |
| $\mathbb{S}(P)$ | standard algebra |
| $\mathbb{F}(P, \approx)$ | reduced algebra modulo $\approx$ |
| $\mathbb{M}(P)$ | semigroup of multiplicative functions |
| $\zeta$ | zeta function |
| $\mu$ | Möbius function |
| $\delta$ | delta function |
| $\lambda$ | lambda function |
| $\kappa$ | cover function |
| $\eta$ | chain function |
| $\rho$ | length function |
| $\chi$ | characteristic |
| $\text{Möb}(P)$ | Möbius algebra |

## 9. Generating Functions

| | |
|---|---|
| $w(f)$ | weight of $f$ |
| $\gamma(S)$ | enumerator of $S$ |
| $\gamma(\mathscr{F}; G)$ | enumerator of $\mathscr{F}$ under the group $G$ |
| $\gamma(\mathscr{F}; G, H)$ | enumerator of $\mathscr{F}$ under $G$, $H$ |
| $Z(G)$ | cycle indicator of $G$ |
| $\varphi$ | Euler function |
| $\mathscr{P}(N, G)$ | lattice of coclosed partitions |

## 10. Matroids

| | |
|---|---|
| $\mathbf{FM}(n)$ | free matroid |
| $\mathbf{M}(V(n, K))$ | vector space matroid |
| $\mathbf{PG}(n, K)$ | projective geometry of $g$-dimension $n$ over $K$ |
| $\mathbf{PG}(n, q)$ | projective geometry of $g$-dimension $n$ over $GF(q)$ |
| $\mathbf{F}$ | Fano plane |
| $\mathbf{AG}(n, K)$ | affine geometry of $g$-dimension $n$ over $K$ |
| $\mathbf{AG}(n, q)$ | affine geometry of $g$-dimension $n$ over $GF(q)$ |
| $\mathbf{G}(\mathfrak{P}, \mathfrak{R}, \mathfrak{F})$ | incidence geometry |
| $\mathbf{P}(G(V, S))$ | polygon matroid of a graph $G$ |
| $\mathbf{B}(G(V, S))$ | bond matroid of a graph $G$ |
| $\mathbf{M}(F(S, K))$ | function space matroid |
| $\mathbf{T}(S, R, I)$ | transversal matroid induced by $R \subseteq S \times I$ |
| $\mathbf{T}(S, R, \mathbf{M}(I))$ | transversal matroid induced by $R \subseteq S \times I$, $\mathbf{M}(I)$ |
| $\mathbf{T}(S; \mathfrak{A})$ | transversal matroid on $S$ induced by $\mathfrak{A}$ |
| $\mathbf{T}(\mathfrak{A}; S)$ | transversal matroid on $\mathfrak{A}$ induced by $S$ |
| $\mathbf{L}(S, \vec{G}, I)$ | gammoid |
| $\mathbf{U}_k(n)$ | uniform matroid of rank $k$ on an $n$-set |

| | |
|---|---|
| $\mathbf{M}(S) \cdot A$ | restriction to $A$ |
| $\mathbf{M}(S)/A$ | contraction through $A$ |
| $U_k(\mathbf{M})$ | upper cut of the matroid $\mathbf{M}$ |
| $L_k(\mathbf{M})$ | lower cut of the matroid $\mathbf{M}$ |
| $D_k(\mathbf{M})$ | Dilworth completion |
| $\mathbf{M}(R)$ | matrix matroid |
| $\mathbf{M}^{\perp}$ | dual matroid |

## 11. Graphs

| | |
|---|---|
| $G(V, E)$ | graph with vertex-set $V$ and edge-set $E$ |
| $\vec{G}(V, E)$ | directed graph |
| $\vec{G}(q, s)$ | network with source $q$ and sink $s$ |
| $G^*$ | dual graph |
| $K_n$ | complete graph |
| $K_{m,n}$ | complete bipartite graph |
| $P$ | Petersen graph |
| $\mathfrak{L}(V, S, F)$ | map |
| $\mathfrak{L}^*$ | dual map |
| $\mathfrak{C}(\vec{G}, K)$ | coboundary module |
| $\mathfrak{Z}(\vec{G}, K)$ | cycle module |
| $\mathrm{st}(v)$ | star centered at $v$ |
| $\gamma(v)$ | degree of $v$ |
| $\kappa_{\vec{G}}$ | vertex connectivity number |
| $\lambda_{\vec{G}}$ | edge connectivity number |
| $w(f)$ | value of the flow $f$ |
| $c(X, Y)$ | capacity of the cut $(X, Y)$ |
| $M$ | matching |

## 12. Posets

| | |
|---|---|
| $d(P)$ | Dilworth number |
| $s(P)$ | Sperner number |
| $\mathfrak{A}(P)$ | family of antichains |
| $P^{(k)}$ | $k$-level of the poset $P$ |
| $W_k$ | $k$-th level number |
| $G^{(k)}$ | bipartite graph on $P^{(k)} \cup P^{(k+1)}$ |
| $\mathrm{Int}(P)$ | set of intervals of $P$ |

## 13. Conventions

| | |
|---|---|
| $i, j, k, l, m, n, \ldots$ | natural numbers |
| $A, B, S, T, \ldots$ | subsets |
| $U, V, W, Z, \ldots$ | subspaces |

| | |
|---|---|
| $\pi, \rho, \sigma, \tau, \ldots$ | partitions |
| $\phi, \psi, \Phi, \Psi, \ldots$ | mappings |
| $f, g, h, k, \ldots$ | functions |
| $\mathbf{M}, \mathbf{N}, \ldots$ | matroids |
| $\mathfrak{A}, \mathfrak{B}, \mathfrak{C}, \mathfrak{D}, \ldots$ | families of sets |
| $P, Q, R, \ldots$ | posets |
| $L, M, N, \ldots$ | lattices |
| $G, H, \ldots$ | groups |

# Subject Index

Druck: STRAUSS OFFSETDRUCK, MÖRLENBACH
Verarbeitung: SCHÄFFER, GRÜNSTADT